Sediment Flux to Basins; Causes, Controls and Consequences

Special Publication reviewing procedures

The Society makes every effort to ensure that the scientific and production quality of its books matches that of its journals. Since 1997, all book proposals have been refereed by specialist reviewers as well as by the Society's Books Editorial Committee. If the referees identify weaknesses in the proposal, these must be addressed before the proposal is accepted.

Once the book is accepted, the Society has a team of Book Editors (listed above) who ensure that the volume editors follow strict guidelines on refereeing and quality control. We insist that individual papers can only be accepted after satisfactory review by two independent referees. The questions on the review forms are similar to those for *Journal of the Geological Society*. The referees' forms and comments must be available to the Society's Book Editors on request.

Although many of the books result from meetings, the editors are expected to commission papers that were not presented at the meeting to ensure that the book provides a balanced coverage of the subject. Being accepted for presentation at the meeting does not guarantee inclusion in the book.

Geological Society Special Publications are included in the ISI Index of Scientific Book Contents, but they do not have an impact factor, the latter being applicable only to journals.

More information about submitting a proposal and producing a Special Publication can be found on the Society's web site: www.geolsoc.org.uk.

It is recommended that reference to all or part of this book should be made in one of the following ways:

JONES, S. J. & FROSTICK, L. E. (eds) 2002. *Sediment Flux to Basins: Causes, Controls and Consequences.* Geological Society, London, Special Publications, **191.**

BREWER, P. A. & PASSMORE, D. G. 2002. Sediment budgeting techniques in gravel-bed rivers *In*: JONES, S. J. & FROSTICK, L. E. (eds) *Sediment Flux to Basins: Causes, Controls and Consequences.* Geological Society, London, Special Publications, **191,** 97–113.

GEOLOGICAL SOCIETY SPECIAL PUBLICATION NO. 191

Sediment Flux to Basins: Causes, Controls and Consequences

EDITED BY

S. J. JONES
University of Durham, UK

&

L. E. FROSTICK
University of Hull, UK

2002
Published by
The Geological Society
London

THE GEOLOGICAL SOCIETY

The Geological Society of London (GSL) was founded in 1807. It is the oldest national geological society in the world and the largest in Europe. It was incorporated under Royal Charter in 1825 and is Registered Charity 210161.

The Society is the UK national learned and professional society for geology with a worldwide Fellowship (FGS) of 9000. The Society has the power to confer Chartered status on suitably qualified Fellows, and about 2000 of the Fellowship carry the title (CGeol). Chartered Geologists may also obtain the equivalent European title, European Geologist (EurGeol). One fifth of the Society's fellowship resides outside the UK. To find out more about the Society, log on to www.geolsoc.org.uk.

The Geological Society Publishing House (Bath, UK) produces the Society's international journals and books, and acts as European distributor for selected publications of the American Association of Petroleum Geologists (AAPG), the American Geological Institute (AGI), the Indonesian Petroleum Association (IPA), the Geological Society of America (GSA), the Society for Sedimentary Geology (SEPM) and the Geologists' Association (GA). Joint marketing agreements ensure that GSL Fellows may purchase these societies' publications at a discount. The Society's online bookshop (accessible from www.geolsoc.org.uk) offers secure book purchasing with your credit or debit card.

To find out about joining the Society and benefiting from substantial discounts on publications of GSL and other societies worldwide, consult www.geolsoc.org.uk, or contact the Fellowship Department at: The Geological Society, Burlington House, Piccadilly, London W1J 0BG: Tel. +44 (0)20 7434 9944; Fax +44 (0)20 7439 8975; Email: enquiries@geolsoc.org.uk.

For information about the Society's meetings, consult *Events* on www.geolsoc.org.uk. To find out more about the Society's Corporate Affiliates Scheme, write to enquiries@geolsoc.org.uk.

Published by The Geological Society from:
The Geological Society Publishing House
Unit 7, Brassmill Enterprise Centre
Brassmill Lane
Bath BA1 3JN, UK

(*Orders*: Tel. +44 (0)1225 445046
Fax +44 (0)1225 442836)
Online bookshop: http://bookshop.geolsoc.org.uk

British Library Cataloguing in Publication Data
A catalogue record for this book is available from the British Library.

ISBN 1-86239-095-9

Typeset by E & M Graphics, Midsomer Norton, Bath
Printed by Antony Rowe, Chippenham, UK.

Distributors

USA
AAPG Bookstore
PO Box 979
Tulsa
OK 74101-0979
USA
Orders: Tel. + 1 918 584-2555
Fax +1 918 560-2652
E-mail *bookstore@aapg.org*

India
Affiliated East-West Press PVT Ltd
G-1/16 Ansari Road, Daryaganj,
New Delhi 110 002
India
Orders: Tel. +91 11 327-9113
Fax +91 11 326-0538
E-mail *affiliat@nda.vsnl.net.in*

Japan
Kanda Book Trading Co.
Cityhouse Tama 204
Tsurumaki 1-3-10
Tama-shi
Tokyo 206-0034
Japan
Orders: Tel. +81 (0)423 57-7650
Fax +81 (0)423 57-7651

Contents

Acknowledgements

We would like to thank all participants and contributors to the volume for their assistance in providing papers for this Special Publication. This title evolved from a two day British Sedimentological Research Group meeting on the 22–23 June 2000 held at the Southampton Oceanography Centre in the UK. The meeting was organized in order to stimulate and bring together Earth scientists who are working on the many aspects of sediment flux to basins and provide a statement of current research into sediment supply to basins at the beginning of this new millennium.

The meeting was sponsored by the British Sedimentological Research Group and the Southampton Oceanography Centre. We would like to thank these organizations, together with Rian Torres, Emma Bennett and Kate Davis and several daily helpers for their generous assistance in making this meeting such a great success. Finally we would like to thank the many reviewers who so effectively reviewed the papers to maintain a high standard, and Helen Knapp, Angharad Hills and Adrian Hartley for their continuous support in steering the book through the final stages to eventual publication.

Stuart J. Jones, University of Durham
Lynne E. Frostick, University of Hull

Introduction

STUART J. JONES[1] & LYNNE E. FROSTICK[2]

[1]*Department of Geological Sciences, University of Durham, South Road, Durham DH1 3LE, UK*

[2]*Centre for Waste and Pollution Research, Department of Geography, University of Hull, Hull HU6 7RX, UK*

Abstract: The stratigraphic record stored in sedimentary basins had traditionally been interpreted in terms of tectonic subsidence, climate change and sea-level rise or fall. During the last decade, however, the fundamental control of sediment flux in creating sedimentary sequences, and thus the need to understand variations in sediment flux in order to interpret the sedimentary record, has been recognized. Sediment flux is a first-order control on the pattern and distribution of sedimentary facies in depositional basins. Consequently, basin fills reflect the sediment flux across their margins and the hinterland conditions driving the sediment flux provide evidence of dynamic geomorphic processes. In context, the drainage systems and sedimentary basin can be regarded as a 'production line' with the sedimentary record giving valuable insight into long term landscape evolution and geomorphological processes illuminating the evolution of sedimentary basins. This special publication contains a set of papers on sediment supply to basins, with a focus on clastic sediments. It presents a mix of hinterland and sedimentary basin studies with a gradation from orogenic belts to the deep marine. The papers present a current perspective on controls and constraints on sediment supply, a model and empirically based driven understanding of sediment flux and the interaction of geomorphology, landscape evolution and sedimentary geology to provide a more complete picture of the Earth system.

An holistic approach to sediment supply

The rate of sediment supply to any depositional basin is governed by a complex interaction of several parameters most notably bedrock uplift, weathering, erosion and the transportation of the sediment through drainage pathways. In most situations the sediment flux to a basin is driven by the rate of bedrock uplift and accompanying river incision which removes the sediment. However, from such a basic approach it becomes readily apparent that there are many other controls on the supply of sediment to basins. These include hinterland structure and lithology, precipitation, vegetation, temperature, local and regional gradients and hydraulic conditions in a channel. Further complications may arise through spatial and temporal heterogeneities in processes which carry sediment within the drainage basin and from the drainage to the depositional basin. However, several attempts have been made, through empirical studies of modern drainage basins, to produce universal relationships amongst sediment yield and the general catchment characteristics (e.g. Schumm & Hadley 1961; Milliman & Meade 1983; Jansson 1988; Milliman & Syvitski 1992). The filtering effect of the depositional basin through which a river flows should be kept in mind when considering sediment yield data, and it is clear that any data for erosion of sediment from a catchment and subsequent sediment supply calculated from these data is in fact a minimum estimate. In addition any data from such studies can not be universally applied as the sediment efflux from a catchment is controlled by a series of tectonic, climatic, geomorphic and hydraulic processes operating on different time scales and tractable in sediment yield analyses.

Nevertheless, taking the many limitations into account, studies of modern drainage basins do provide a valuable understanding of sediment flux to basins by contributing detailed observations of processes and sediment loads. These observations are short-term and must also be considered in context with longer time scale controls on sediment supply to basins (see **Frostick & Jones**). Tectonic and climatic changes result in new sediment budgets with a shift in the quantity and distribution of sediment movement, with important consequences for landscape development, denudation,

From: JONES, S. J. & FROSTICK, L. E. (eds) 2002. *Sediment Flux to Basins: Causes, Controls and Consequences*. Geological Society, London, Special Publications, **191**, 1–4. 0305-8719/02/$15.00

basin filling architectural patterns and sediment yields. As a result it is expedient to consider sediment flux to basins on all time scales, with the long-term averaged controls being integrated with the short-term geomorphic processes for a better holistic understanding of sediment supply particularly important for numerical modelling.

In this special publication three principal themes are followed in appreciating sediment supply to basins from orogenic belts to the deep marine. These provide a contemporary statement of research into sediment flux at the beginning of this new millennium.

(i) the importance of allocyclic controls affecting the rate and quantity of sediment supply to a depositional basin;
(ii) the role of drainage basins and networks in providing valuable data on erosional landscape development and appreciating the 'upstream' perspective of sediment flux and;
(iii) quantitative and qualitative modelling of sediment supply to continental and marine basins validated from short time scale data in order to facilitate the modeling of sediment flux on geological and geomorphological time-scales.

Hall & Nichols consider the Cenozoic tectonic development of Borneo and the relationship to the sedimentary record. They suggest that Borneo is an important example of mountain belts formed in regions of high erosion and sediment flux as it differs from orogenic belts in other parts of the world. They explore the effects of high erosion rates in the tropical climate of Borneo and how palaeoclimatic information is required to understand the erosion of ancient orogens. **Stokes *et al.*** used the Aguas and Feos rivers of the Sorbas Basin SE Spain to examine the impact of river capture in terms of spatial and temporal variations in rates of incision, sediment flux and link it with surface lowering through base level change. They demonstrate that the lowering of base level resulted in a dramatic increase in incision upstream of the capture sites and in excess of 60% of the landscape change can be accounted for by valley-constrained erosion and renewed sediment flux as opposed to overall surface lowering. **Thamo-Bazo *et al.*** continue the theme of allocyclic controls on sediment flux through the study of long-term dynamics of Quaternary fluvial systems under the influence of thrust tectonics in the southeastern portion of the inter-mountain Pannonian Basin of Hungary. They invoke tectonic activity as a strong control on the drainage pattern evolution and hence an important influence on sediment supply to a depositional basin. **Tropeano *et al.*** presents a study of the Quaternary infilling and cannibalization of the Bradanic Trough of southern Italy. They illustrate how the stratigraphy is dominated by a basin-wide erosional regression that caused cannibalisation of the sediment rather than a filling or overfilling phase often seen in other foredeep basins. This has important implications for sediment flux to adjacent basins and the modelling of depositional basin infill.

Once sediment is delivered to a sedimentary basin, it can be distributed, deposited or redirected out of the basin according to the prevailing transport and allocyclic conditions. The sediment supply rate, distribution, and facies are controlled firstly through tectonics and/or sea-level change, creating the accommodation space (see **Tropeano *et al.***). In many cases fluvial systems are responsible for the sediment flux in and out of depositional basins. The flux is governed by the volumes and grain sizes of sediment available for transportation, and the hydraulic conditions. **Frostick & Jones** explore some of the causes of temporal changes in sediment fluxes from the drainage to depositional basin. Using many examples they identify the need for shorter time scale processes to be incorporated into basin models. **Brewer & Passmore** and **Fuller *et al.*** continue the appreciation of geomorphic processes over shorter time scales in governing sediment flux to basins. They discuss the use of sediment budgeting methods based on the analysis of simple two-dimensional planform and more complex three-dimensional cross profile morphological changes in gravel-bed rivers. These studies represent a means of estimating the long-term residence time of stored sediment and the rates and patterns of reach scale sediment transfer. **Fuller *et al.*** identify substantial within-reach sediment transfer, with minimal net export down stream, crucial for a larger scale appreciation of basin sediment budgets. In contrast **Milan *et al.*** use tracer pebbles to consider the influence of flow character upon loci of scour and deposition in a gravel bed river. They conclude that any significant changes in the flood hydrograph result in alterations in the dominant spatial zonation of scour and deposition, upsetting the quasi-equilibrium in the channel and the sediment flux within a basin. **Richards** provides a detailed perspective on the role drainage basins and networks play in controlling sediment storage and flux. It is recognized that there is a need to return to the catchment scale in fluvial geomorphology in order to understand how drainage basin structure modulates the effects of environmental change when generating a record of sediment yield response.

Tipper has developed a numerical fractionation model for sediment delivery to basins. The model is described in terms of the variants and contexts in which the variants might be applicable, in particular the analysis of bed thickness throughout a basin, with obvious implications for sediment flux estimates. **Jones** presents a detailed study of modern transverse rivers draining the Spanish Pyrenees and identifies the important factors that control large-scale basin-wide patterns of erosion and deposition. The results highlight the importance of incorporating hydraulic data into geomorphological and geological models to assist interpretations of the mass distribution of sediment. **Bogaart *et al.*** review a number of simple process-based models to illucidate how climate impacts on the sediment supply to basins. They limited the examples to alluvial rivers within the NW-European lowlands during the last few thousand years of the Quaternary, as their evolution is well understood.

Evans & Archie document the Late Neogene development of the Ebro delta and the related drainage basin. The combination of geological and modern records are used in a reconstruction of the Ebro system that accounts for several episodes of sediment flux.

McManus takes an historical approach to the study of sediment flux delivered to the Atachafalaya Bay, Louisiana. Recent sedimentation in this bay is closely linked to the history of changing flows within the Atchafalaya River system and he shows how the diversion of up to 30% of the Mississippi waters led to an enormous rise in sediment flux to the Atchafalaya system. **Poulos & Collins** use a quantitative approach to fluviatile sediment fluxes to the Mediterranean Sea. Based upon a large number of data sets they reassess existing sedimentological information for the coastal zone, offshore zone and the floor of the Mediterranean Sea. Through information from dam sediments they infer that more than 40% of the total area of Mediterranean catchment has been removed causing a dramatic decrease in sediment supply and coastal retreat. **Poulos & Collins** suggest that the reduction in sediment supply is likely to be augmented though the effects of climatic changes and associated sea-level rise. In a linked paper on the Mediterranean Sea by **Poulos *et al.*** sediment fluxes and the evolution of a fluvially-supplied, tectonically active, coastal margin is investigated. Detailed sediment budgets for a portion of coastal margin of the Eastern Mediterranean Sea are presented and it is demonstrated that the majority of sediment flux is supplied by one river. They also infer that some 50% of the fluvial sediment flux accumulates on the shelf, whilst another 25% is transported over the slope to deeper ocean waters. In the final paper of this special publication **Reeder *et al.*** use radiocarbon dates made on planktonic foraminifers and pteropod shells providing dates for major turbidity events in the east Mediterranean Basin. They infer two partly independent cycles that can be recognized in the Late Quaternary turbidites. Firstly, climate controlled the sediment flux from the Nile delta with changes in flux correlating with pluvial and interpluvial climatic periods. Secondly sea-level changes focused the turbidite location. However, the desposition of megaturbidites occur independent of any particular sea-level stand or climatic condition.

The future

The growth in the number and range of topographical, stratigraphical and geochronological data sets that has occurred over the last two decades, coupled with the increase in ability to numerically model and simulate complex processes of sediment flux, sediment routing and landscape evolution have provided new opportunities for addressing questions of long-term trends in sediment supply to basins. It is therefore of critical importance to continue the drive to develop time-integrated data sets spread over geomorphological and geological time scales in order to appreciate the impacts of changes in processes in drainage and depositional basins.

Data sets of sediment supply to basins covering time periods over 10^3–10^7 years are rare (see **Frostick & Jones**, fig. 1). This may in part reflect the intermediate nature and overlap of several processes that interrelate through complex interactions. It is necessary to further develop techniques and detailed studies of sedimentary and drainage basins to allow a quantitative record of sediment supply rates to be established. This may involve the application of geochemical techniques such as the use of cosmogenic isotopes (e.g. Burbank *et al.* 1996), apatite fission track thermochronology (e.g. Kohn & Bishop 1999), or as yet undeveloped techniques.

Modelling of long-term landscape evolution and sediment flux not only requires the incorporation of the impacts of tectonic mechanisms and the response of the lithosphere to changes in load, but raises the question of how surface processes can be dealt with over the extended spatial and temporal time scales meaningful for sedimentary basins. As highlighted in this publication the best way forward is most likely through concentrating our attention on key end-member scenarios controlling sediment flux to basins. These key elements include the effects of tectonic uplift and renewed sediment supply, climate change and the impact on weathering, and the hydraulic conditions which transport the sediment through an evolving system.

The challenge ahead can only be met through an

integrated approach to understanding sediment supply to basins. This will involve long-term monitoring of the controlling processes at a range of spatial scales. These data can then be incorporated into models. Control over the progress towards understanding sediment flux to basins lies not only with the research funders, who must think beyond a three year time scale, but also with the researchers who must cross the discipline boundaries and work together to solve the problems.

References

BURBANK, D. W., LELAND, J., FIELDING, E., ANDERSON, R. S., BROZOVIC, N., REID, M. R. & DUNCAN, C. 1996. Bedrock incision, rock uplift and threshold hillslopes in the northwestern Himalayas. *Nature*, **379**, 505–510.

JANSSON, M. B. 1988. A global survey of sediment yields. *Geographical Annaler*, **70**, 81–98.

KOHN, B. & BISHOP, P. (EDS) 1999. Apatite fission track thermochronology and geomorphology. *Australian Journal of Earth Sciences*, **46**, 155–233.

MILLIMAN, J. D. & MEADE, R. H. 1983. World wide delivery of river sediment to the oceans. *Journal of Geology*, **91**, 1–21.

MILLIMAN, J. D. & SYVITSKI, J. P. M. 1992. Geomorphic/tectonic control of sediment discharge to the ocean: the importance of small mountainous rivers. *Journal of Geology*, **100**, 525–544.

SCHUMM, S. A. & HADLEY, R. F. 1961. Progress in the application of landform analysis in studies of semiarid erosion. *United States Geological Survey Circular*, **437**, 14.

Cenozoic sedimentation and tectonics in Borneo: climatic influences on orogenesis

ROBERT HALL & GARY NICHOLS

SE Asia Research Group, Department of Geology, Royal Holloway University of London, Egham, Surrey TW20 0EX, UK

(e-mail: robert.hall@gl.rhul.ac.uk, g.nichols@gl.rhul.ac.uk)

Abstract: The volume of sediment deposited in the basins around Borneo indicates that at least 6 km of crust has been removed by erosion during the Neogene. The amount of tectonic uplift implied by this is not reflected in a large area of high mountains on the island, which has an average elevation much lower than that of the Alps or Himalayas. High weathering and erosion rates in the tropical climate of SE Asia are likely to have been an important factor governing the formation of relief in Borneo, and consequently, controlled the structural development of the orogenic belt. Very rapid removal of material by erosion prevented tectonic denudation by faulting: around Borneo there was no lithospheric flexure due to thrust loading and no true foreland basins were developed. The sediment was deposited adjacent to the orogenic belt in older, deep oceanic basins. In terms of sediment yield, the Borneo mountains are comparable in importance to mountain ranges such as the Alps or Himalayas. However, the differences in elevation and structural style suggest that mountain belts formed in regions of high erosion rates may be different from those formed in other settings and the effects of climate need to be considered to understand orogenic evolution.

Borneo (Fig. 1) is one of the largest islands in the world but the geology of the area is poorly known because of difficulties of access to the tropical rainforest of much of the interior, a lack of reliable dating of igneous rocks, poorly fossiliferous sedimentary rocks and an absence of a coherent stratigraphic scheme for many parts of the island. Hydrocarbon exploration offshore has provided a limited database of information for the basins surrounding Borneo, although the links between the evolution of the sedimentary basins on land and offshore are poorly understood. In recent years the extrusion of continental fragments from Asia driven by India's collision, large-scale rotation of Borneo and effects of collision of Australia with SE Asia have all been implicated in the development of Borneo. Models for the plate tectonic evolution of SE Asia that have been developed in recent years have provided a regional context for the Cenozoic history of Borneo (e.g. Rangin *et al.* 1990; Daly *et al.* 1991; Briais *et al.* 1993; Lee & Lawver 1994, 1995; Hall 1996, 1997, 1998). Some elements of the tectonic evolution of Borneo and its surrounding basins, such as timing of deformation events, uplift and erosion, and ages of sediments, are now better understood. The Neogene development of the Borneo region has some implications for tectonic models of orogenic development in other parts of the world, but particularly in tropical regions.

In this paper we attempt to summarize some key features of the Cenozoic tectonic development of Borneo and relate them to the sedimentary record. We suggest that Borneo may be an important model for mountain belts formed in regions of high erosion rates because it differs from orogenic belts in other parts of the world, and its structural style is dependent on the climatic influences of its tropical setting. The implication of this analysis of Borneo is that we need to take climatic setting into account when considering modern and ancient examples of orogenic belts.

The geographic and geological setting of Borneo

The island of Borneo (Fig. 2) is a region of rugged terrain, but most of the central mountains of Borneo are below 2500 m and there are about a dozen peaks of about 2500–3000 m, mostly in the half of the island north of the equator. Mount Kinabalu is the only exception at just over 4000 m. Thus, on a

From: JONES, S. J. & FROSTICK, L. E. (eds) 2002. *Sediment Flux to Basins: Causes, Controls and Consequences.* Geological Society, London, Special Publications, **191**, 5–22. 0305-8719/02/$15.00

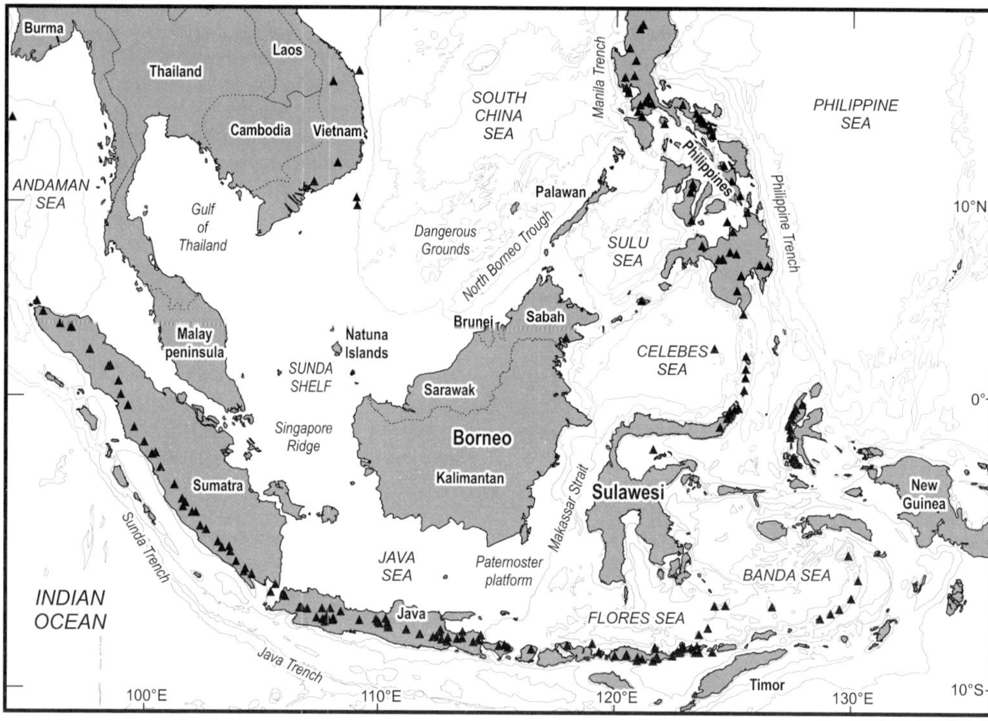

Fig. 1. Location of Borneo. Black filled triangles are volcanoes from the Smithsonian database (http://www.nmnh.si.edu/gvp/) and bathymetry is from the GEBCO digital atlas (IOC, IHO, BODC 1997). Bathymetric contours at 200, 2000, 4000 and 6000 m.

global scale, Borneo is not an island of major mountains and although the present area of highlands and mountains is large, at about 450 000 km^2, most of this area (about 95%) is well below 1000 m in height (Fig. 2). To the north and NE of Borneo are three deep basins (the South China, Sulu and Celebes Seas) and to the east is the narrow Makassar Strait (Fig. 1). The continental margins of all of these areas have relatively narrow shelves before descending to depths of several kilometres. In contrast, to the west and south is the Sunda Shelf, an area of very shallow depths connecting Borneo to Indochina, peninsula Malaysia and Thailand, Sumatra and Java. Much of the Sunda Shelf was emergent at intervals during the Pleistocene (Umbgrove 1938; van Bemmelen 1949) and probably during significant periods of the Cenozoic (Hall 1998; Moss & Wilson 1998).

Borneo is the result of Mesozoic accretion of ophiolitic, island arc crust and microcontinental fragments of south China and Gondwana origin, with their sedimentary cover, onto the Palaeozoic continental core of the Schwaner Mountains in the SW of the island (Hamilton 1979; Hutchison 1989, 1996*a*; Metcalfe 1996) (Fig. 3). Similarities between the pre-Cenozoic rocks of western Sulawesi and east Kalimantan suggest that they were accreted onto the eastern margin of Sundaland before the Cenozoic (Audley-Charles 1978; Sukamto 1978; Hamilton 1979; van Leeuwen 1981; Hasan 1991; Parkinson 1991; Wakita *et al.* 1996). At the beginning of the Cenozoic, Borneo formed a promontory of Sundaland at the eastern margin of Eurasia (Hall 1996), partly separated from Asia by oceanic crust of a proto-South China Sea. Palaeomagnetic (Fuller *et al.* 1999) and other geological evidence (discussed in Hall 1996) indicates that Borneo has rotated counterclockwise by approximately 45° since the Early Miocene (Fig. 4). Thus, for most of the Oligocene and Early Miocene the northwest Borneo margin would have been oriented NW–SE. This configuration suggests that much of the Palaeogene sediment in north Borneo was fed southwards across the Sunda Shelf from Indochina (Hall 1996; Hutchison *et al.* 2000, 2001). At about 25 Ma there was a second important period of change in plate boundaries and motions, probably a result of major collision events far from Borneo (Hall 1996), but with important consequences for Borneo and its margins. During the latest Oligocene and Early Miocene there was mountain building, erosion and change in character of sedimentation throughout much of Borneo, although the exact timing varies from place to place in the island as does the inferred cause (e.g. Rangin *et al.* 1990; Tan & Lamy 1990; van de Weerd & Armin 1992; Hazebroek & Tan 1993; Hutchison 1996*b*; Moss & Chambers 1999; Hutchison *et al.* 2000).

The Borneo margins

At present Borneo is an island, but the history and character of the continental margins on different sides of the island are not the same.

Sunda Shelf

The Sunda Shelf is a wide shallow region between Borneo and the land of Indochina and Malaya to the

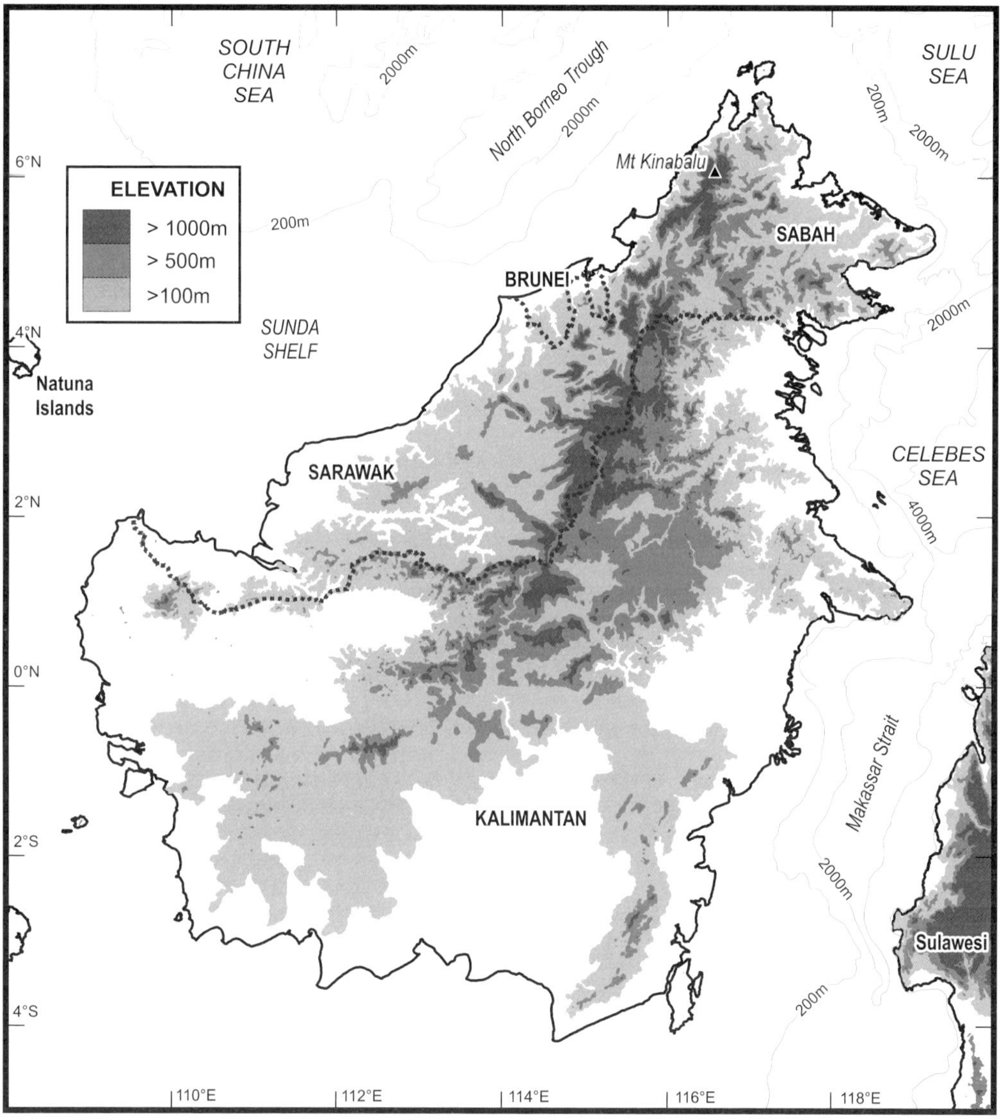

Fig. 2. The topography of Borneo based on the GTOPO30 global digital elevation model. Borneo has a total area of 734 900 km^2; 447 649 km^2 (61% of total area) is above 100 m, 126 159 km^2 (17% of total area) is above 500 m and 34 220 km^2 (5% of total area) is above 1000 m.

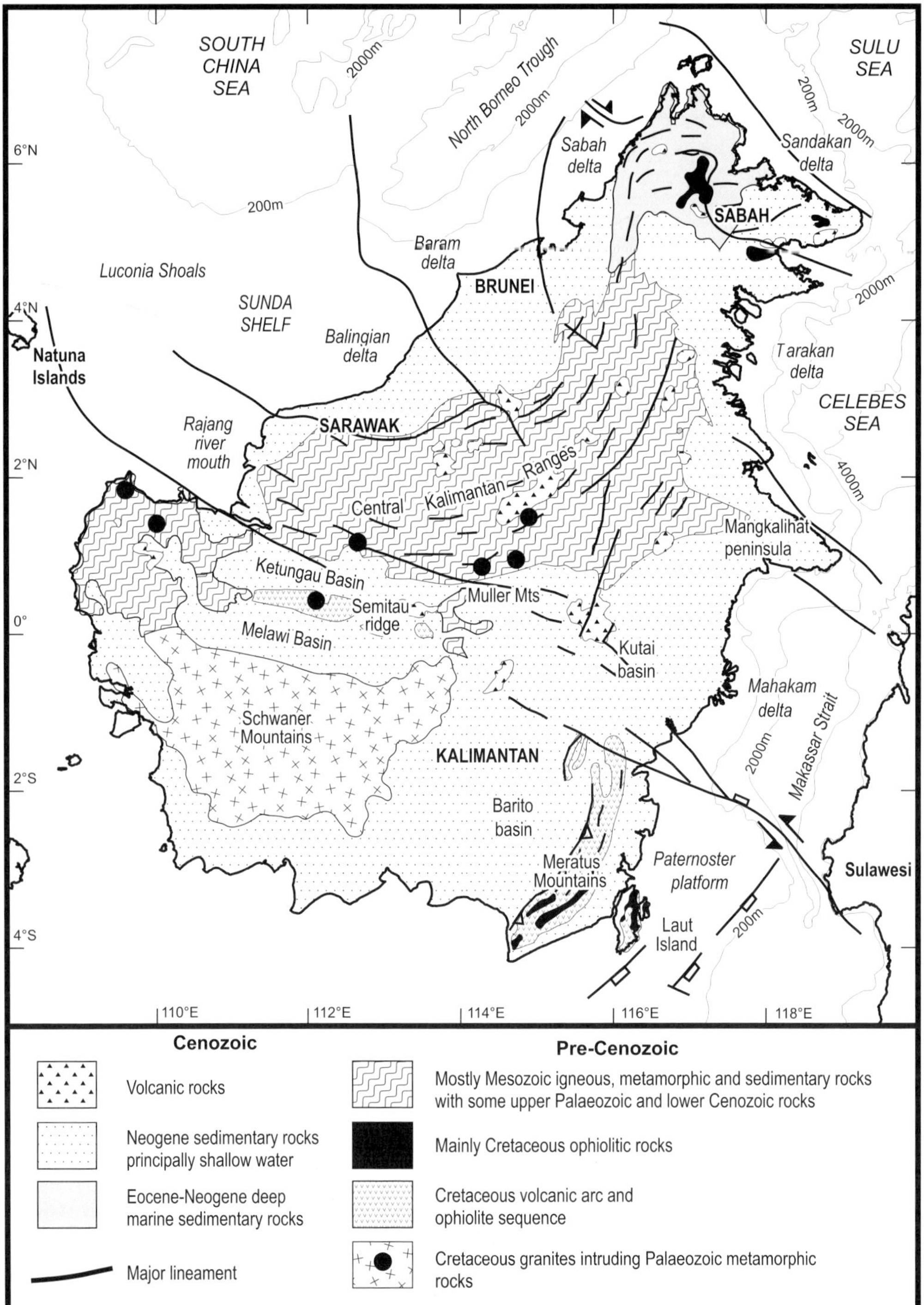

Fig. 3. Simplified geology of Borneo (modified from Hamilton 1979, Moss and Wilson 1998 and other sources).

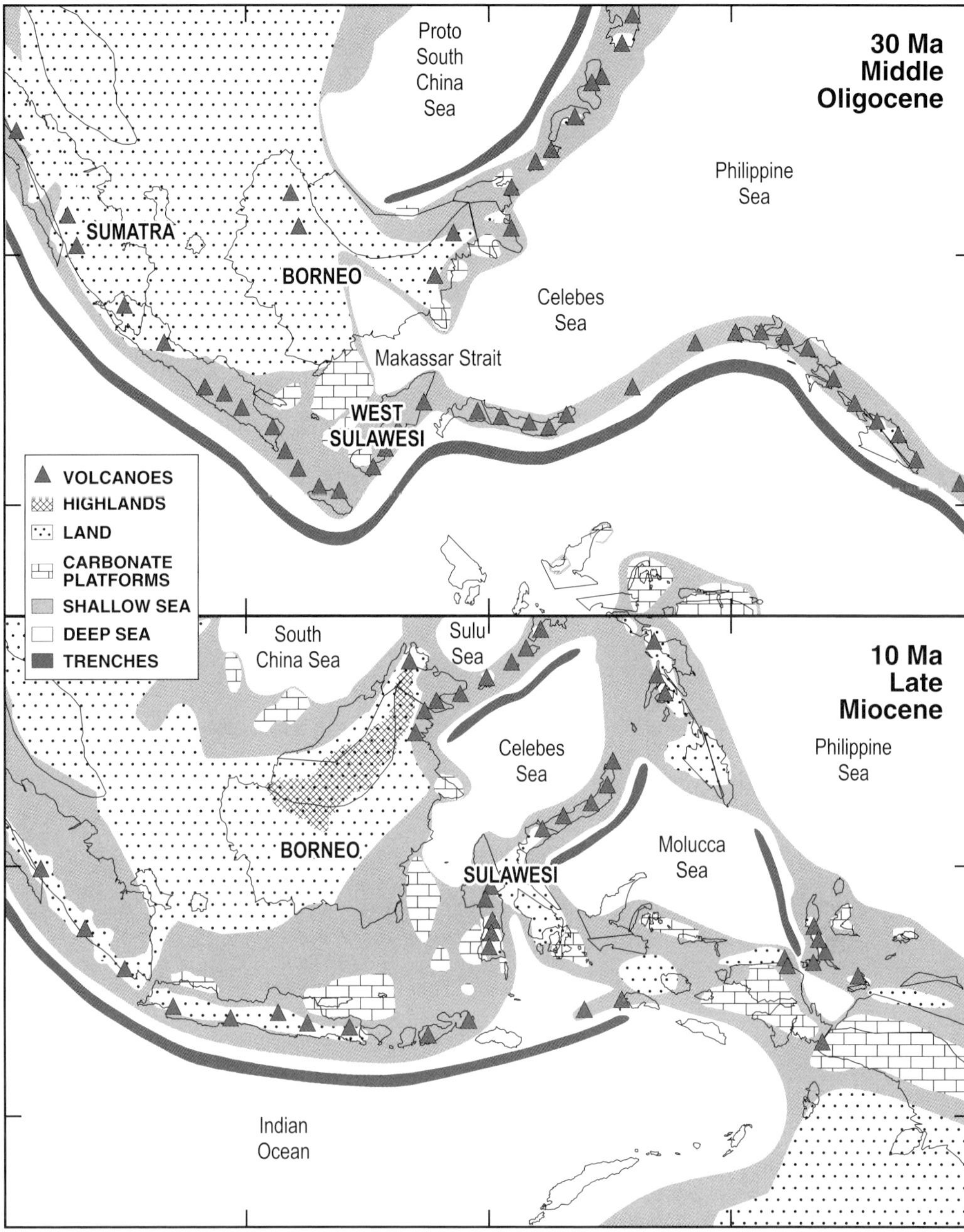

Fig. 4. Plate tectonic reconstructions of SE Asia at 30 Ma and 10 Ma (from Hall 1996). Distribution of land and sea, topography and bathymetry is from Hall (2001).

north and west (Figs 1 and 2). Much of the Sunda Shelf was emergent at intervals during the Pleistocene and there is a widespread assumption that this region was a stable area during the Cenozoic. It is probable that there was a land connection between Borneo and Sundaland throughout much of the Cenozoic (Moss & Wilson 1998), which may have been present since the Jurassic (Lloyd 1978). A thin cover of Quaternary sediments is reported to overlie pre-Cenozoic rocks

(Ben-Avraham & Emery 1973). Although western Borneo and Sundaland were probably part of a single plate during the Cenozoic, this need not mean that there has been no relative motion between Borneo and Sundaland, as argued for example by Lee & Lawver (1995). Numerous onland and offshore faults could have permitted distributed deformation and allowed rotation of all or parts of Borneo and other parts of Sundaland (Hall 1996), and this is widely accepted for other parts of the region (e.g. Tapponnier *et al.* 1982; England & Houseman 1986; Peltzer & Tapponnier 1988; Hutchison 1996*a*).

Makassar Strait–Celebes Sea

Much of SE Asia, southern Borneo and western Sulawesi was emergent during the Palaeocene and the Early Eocene (Moss & Wilson 1998). Basin formation probably began in the Eocene around the margins of Sundaland, although the exact timing is still uncertain because the oldest parts of sequences in all basins are continental and devoid of fossils. Geological similarities between east Borneo and west Sulawesi suggest that they have moved apart (Katili 1978; Hamilton 1979), although the timing is not well constrained. There was extension and subsidence in the Makassar Straits (Burollet & Salle 1981; Situmorang 1982; Bergman *et al.* 1996) which occurred between early Palaeogene and Early Miocene (Situmorang 1982, 1987). Hamilton (1979) suggested the Makassar Strait had a basement of oceanic crust, whereas other workers have argued for attenuated continental crust (e.g. Dürbaum & Hinz 1982). There is now good evidence for oceanic crust in the northern part of the Makassar Straits based on gravity modelling by Cloke *et al.* (1999) and seismic profiles acquired across the northern Makassar Straits, which are interpreted to show a basement of oceanic crust beneath layered sediments (Baillie *et al.* 2000). To the north, the Celebes Sea formed in the Late Eocene (Weissel 1980; Silver & Rangin 1991) and received very little clastic detritus until the Early Miocene (Smith *et al.* 1990; Nichols & Hall 1999).

Sulu Sea

The Sulu Sea is thought to have opened as a backarc basin during the Early Miocene (Holloway 1982; Hinz *et al.* 1991; Rangin & Silver 1991). Its northern part is underlain by continental crust north of the Cagayan ridge. The Cagayan Ridge is interpreted as a volcanic arc, active for a short period in the Early Miocene, which collided with the south China margin at the end of the Early Miocene (Rangin & Silver 1991) following southward subduction of the proto-South China Sea between 20 and 15 Ma. This caused subduction to shift south leading to southward subduction of part of the Sulu Sea beneath the Sulu arc between 15 and 10 Ma (Rangin & Silver 1991). Thus, from the Early Miocene onwards it formed a deep marine basin to the NE of Borneo that received terrigenous clastic detritus (Nichols *et al.* 1990).

South China Sea

The central and deepest part of the South China Sea is floored by oceanic crust (Taylor & Hayes 1980), which was formed between the middle Oligocene and the Middle Miocene (Taylor & Hayes 1983; Briais *et al.* 1993). The area north of Borneo includes thinned continental crust of the Indochina and south China margin, which, during the Palaeogene, was separated from north Borneo by the proto-South China Sea ocean crust (Fig. 4). During the Palaeogene much of the north Borneo margin was an active margin with subduction beneath NE Borneo (Hamilton 1979; Taylor & Hayes 1983; Tan & Lamy 1990; Hall 1996). Sediment may have been supplied from SW Kalimantan and Indochina to turbidite fans in Sarawak and Sabah. In the Early–Middle Eocene turbidites in Sarawak were uplifted and deformed by the 'Sarawak orogeny' (Hutchison 1996*b*), reflecting regional plate reorganization at this time (Hall 1997). However, subduction continued in Sabah throughout the Oligocene forming the Crocker Formation (Hamilton 1979; Tan & Lamy 1990). By the Late Oligocene (Fig. 4) a deep marine basin remained to the north of Borneo with thick turbidites and a large accretionary complex on its south side.

Neogene sedimentation around Borneo

In the Early Miocene there was a significant change in the character of sedimentation around Borneo. Large amounts of clastic sediments began to pour into the deep basins to the north and east of the island, and major delta systems formed which prograded rapidly away from the island. The only possible source for the sediment is Borneo itself. Although some of the Palaeogene sediments in Borneo could have been supplied by river systems from Indochina (Hutchison 1989, 1996*a*; Hall 1996; Hutchison *et al.* 2000), by the Early Miocene the South China Sea would have been a barrier to sediment supply from the north, and sediment that was transported south from Asia was deposited in the Gulf of Thailand and the Malay basins.

The Neogene sediments in the basins around Borneo can provide some estimates of the amounts and timing of uplift in the mountains of the island. From published and unpublished information, including observations made by the SE Asia Research Group, it seems that much of the sediment in the north Borneo basins was derived initially from erosion of Palaeogene sediments and later from cannibalization of older Neogene sediments during inversion of the prograding deltas (e.g. van de Weerd & Armin 1992; Sandal 1996; Moss & Chambers 1999; Hutchison *et al.* 2000). No detailed studies of petrography or provenance have been made in most areas, but the Neogene sediments seem to be relatively mature, quartz-rich clastic rocks with some evidence of local clastic contributions from elevated nearby basement rocks. Hutchison (1996*b*) remarks that 'an interesting heavy mineral provenance study remains to be carried out' in Sabah and this is the case elsewhere in Borneo. In the Kutei Basin the only published provenance study (Tanean *et al.* 1996) record that all sediments of Early Miocene–Recent age have a recycled orogenic source. Tanean *et al.* report that Lower Miocene sandstones are moderately quartzose with a moderate lithic component, and a subsidiary but significant volcanic lithic component that records continuous volcanic activity during the Early Miocene. Uppermost Lower–lowermost Middle Miocene sandstones are volcanogenic, recording an increase in volcanic activity in western Kalimantan between 17 and 14.5 Ma. Middle and Upper Miocene sandstones are highly quartzose and are recycled products of basin inversion events to the west and record no evidence of volcanic activity.

A conservative estimate of the sediment volume around Borneo (Table 1) has been made using Hamilton's (1974, 1979) map which shows isopachs of sediment thickness in the basins (Fig. 5). Although there is a more recent compilation of sediment thickness data (CCOP 1991) this relies heavily on Hamilton's earlier maps for parts of the compilations and offers no more detail in this region. Although Hamilton produced his maps in the early 1970s, he had access to much unpublished industry data, and despite the limitations of the data set available at that time, Hamilton's observations and interpretations have been vindicated by subsequent work. However, his maps probably underestimate the thickness of Cenozoic sediment. For example, in the Makassar Strait deep seismic lines (Samuel *et al.* 1996) suggest the Neogene section alone is more than 9 km thick, whereas Hamilton shows a total Palaeogene and Neogene sediment thickness of 9 km. Ferguson & McClay (1997) and Moss & Chambers (1999) report the Tertiary Kutei Basin contains over 14 km of sediment, whereas the deepest depocentre shown by Hamilton is about 9 km. The contoured map produced by Moss & Chambers resembles that of Hamilton except that all isopachs are significantly deeper. It is suggested (J. L. C. Chambers pers. comm. 1997) that the sediment volume in the Kutei Basin may be overestimated because onland mapping in the Upper Kutei Basin indicates that part of the area shown by Hamilton as Cenozoic includes pre-Tertiary sedimentary rocks. However, even if this is the case, excluding the entire volume of sediments in the Upper Kutei Basin reduces the estimate of total volume of Neogene sediments in the Kutei Basin by *c.* 8%, and the total volume of Borneo Neogene sediments by *c.* 3%.

Beneath Brunei the pre-Neogene basement has not been drilled, but Sandal (1996) records that 'depths of up to 12 km have been reported (e.g. Taylor & Hayes 1980, 1983)'; their own cross-sections show up to 8 km of Middle Miocene and younger sediments from offshore Brunei. Offshore NW Borneo the base of the Neogene clastic sequence is at depths of 7–10 km (Hinz & Schluter 1985; Hazebroek & Tan 1993). Hazebroek & Tan (1993) report 'the sediment prism formed by the Inboard Belt and Baram Delta exceeds more than 12 km total thickness' in the Mio-Pliocene basins of NW Sabah. Hamilton indicates about 9 km of total Cenozoic sediment in this area. The Sandakan Basin is shown by Hamilton as containing little more than 7 km of sediment, whereas Swauger *et al.* (1995) report 10 km of Neogene sediments, and Graves & Swauger (1997) suggest that there may be as much as 15 km of sediment.

In the Tarakan Basin the Plio-Pleistocene sequence is about 4 km thick (H. Darman pers. comm. 1999) in areas contoured by Hamilton as containing 4–5 km of Cenozoic sediment. Wight *et al.* (1992) record that the deepest Pleistocene depocentre attains a thickness of up to 4.5 km and that Pleistocene depocentres do not coincide with Pliocene depocentres. However, in the area shown as the central depocentre with more than 5 km of Cenozoic sediment on Hamilton's map, Wight *et al.* map a similar pattern to Hamilton with the central depocentre at greater than 4 seconds two-way time (TWT), except that the sequence is shown to be entirely Plio-Pleistocene.

Some of the variation in estimates may be due to the methods adopted by some authors in calculating total thicknesses; for example, a few estimates may include summations of thicknesses from different areas and all others depend on values chosen for seismic velocities. No drilling has penetrated the deepest parts of any of these basins. Unfortunately, the published literature does not contain the information to produce new isopach maps; most of the data remain in proprietary form and oil companies are currently reluctant to release the detailed data required to improve estimates. It is currently impossible to map the Neogene–Palaeogene surface on a regional basis, and in the deeper parts of

Table 1. *Sediment volumes for Borneo basins, and for Alpine and Himalayan basins, based on sources discussed in the text*

	Sarawak	South Palawan	Sandakan	Tarakan	Kutei	Barito	Asem-Asem	Total Borneo
Total sediment volume (km^3)	1 939 981	100 415	157 493	553 438	1 027 043	213 136	34 404	4 025 910
Neogene sediment volume (km^3)	1 560 300	100 415	151 296	551 845	834 000	208 064	34 404	3 440 324
						Borneo	Himalayas	Alps
Neogene sediment volume (km^3)						3.44×10^6	1.29×10^7	1.38×10^6
Eroded rock volume (km^3)*						2.91×10^6	1.14×10^7	1.16×10^6
Present denudation area (km^2)						4.48×10^5	1.48×10^6	7.12×10^4
Denudation rate (m Ma^{-1}) based on present area of land above 100 m						326	383	817
Crust thickness (km) removed based on present area of land above 100 m						6.5	7.7	16.3
Min mountain area (km^2)†						1.51×10^5	6.00×10^5	
Denudation rate (m Ma^{-1}) based on minimum area of mountains						965	946	
Crust thickness (km) removed based on minimum area of mountains						19.3	18.9	

* Eroded rock volume calculated using Neogene sediment density of 2.2 Mg m^{-3} and original crust density of 2.6 Mg m^{-3} crust.

† For Borneo minimum mountain area is estimated area of emergent land in the Early Miocene and is approximately one third of the present-day area of land above 100 m. For the Himalayas this is the area of present high mountains from Einsele (1992).

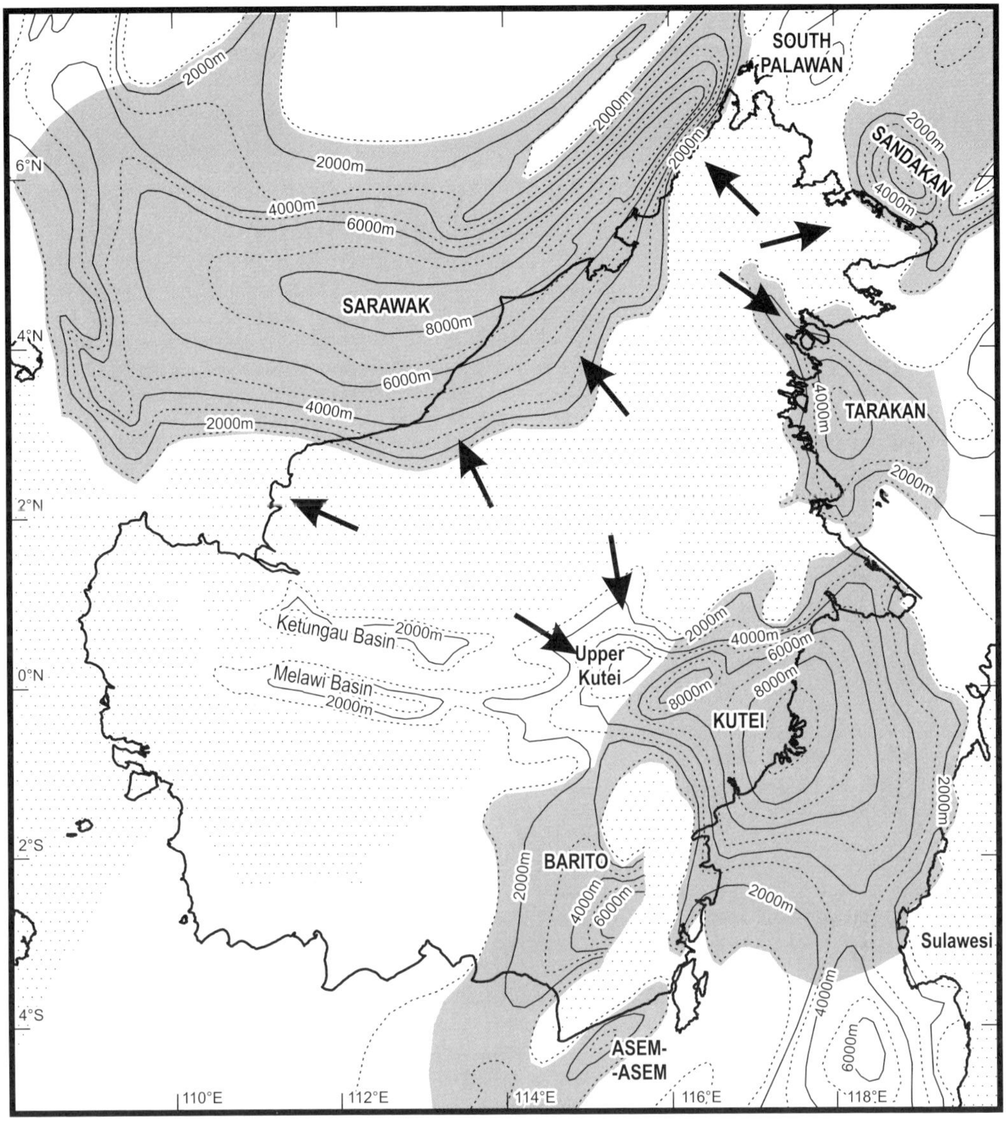

Fig. 5. Isopachs of sediment thicknesses used in the calculation of sediment volumes (based on Hamilton 1974, 1979). Stippled areas are principally areas without Cenozoic sediments. The grey shaded areas of the basins are those which have been included in the total volume of sediment in the circum-Borneo basins and used to calculate the value of 6 km of crust removed during the Neogene. The Ketungau and Melawi Basins contain pre-Neogene sediments. The Upper Kutei Basin is excluded because mapping in the Upper Kutei Basin suggests that part of the area shown by Hamilton as Cenozoic includes pre-Cenozoic sedimentary rocks (J. L. C. Chambers pers. comm. 1997). The South Palawan Basin is entirely excluded, although some of the sediment must have come from Borneo. The Tarakan Basin has been arbitrarily truncated close to Borneo, although some sediment in the Celebes Sea is thought to be derived from Borneo (see text). The arrows indicate the principal sediment pathways into the Borneo basins.

many basins identification of the Neogene–Palaeogene boundary is based on correlation of seismic horizons from the basin margins. However, it is very clear that: (1) the thicknesses of Cenozoic sediments estimated by Hamilton (1974, 1979) need to be increased as a result of more recent work; and (2) in many cases the thicknesses estimated for the Neogene sections alone are greater than Hamilton's estimates for the entire Cenozoic section. In many areas, such as the Kutei, Tarakan and Sandakan Basins, it would not be unreasonable to use Hamilton's isopachs as estimates of the Neogene thickness. It is probable that in most of the basins Neogene thicknesses are much greater than 5 km. However, in order to obtain a very conservative estimate of the volume of Neogene sediment, it has been arbitrarily assumed here that all sediment in each basin below 5 km is Palaeogene.

A further cause of uncertainty is in estimating the volume of non-terrigenous material in the total. Tropical areas are regions of high carbonate productivity, although, because of the amount of siliciclastic material being eroded from Borneo, there are no extensive areas of shallow water carbonates within the principal sedimentary basins around the island. Between the Kutai and Tarakan Basins is the Mangkalihat peninsula where there is a carbonate platform that may have contributed a small volume of clastic carbonates to the margins of both basins, although this is largely restricted to slopes close to the peninsula (Wilson *et al.* 1999). The total volume of carbonates now within the basins in and around Borneo based on Wilson *et al.* is less than 1% of the total sediment volume estimated. However, there may have been a contribution in deeper water due to pelagic background sedimentation. This contribution is impossible to quantify at present, although the limited information on sediment character, for example in the Celebes Sea (Smith *et al.* 1990) and the Sulu Sea (Nichols *et al.* 1990), suggests that carbonate contributes little to the volume in most areas. However, there is also an uncertainty in the amount of carbonate rock removed from the mountains of Borneo by solution during the Neogene. For the purposes of the calculations presented here, it is assumed the volume of eroded carbonate was equal to that of deposited carbonate, and the carbonate contribution to the total sediment volume has been ignored. A contribution from volcanic sources is also difficult to quantify: in ODP drill sites in the Celebes and Sulu Seas beyond the limits of the areas used in the calculation of sediment volumes, hemipelagic volcaniclastic material is recognized as forming a significant contribution to the sediment volume (Nichols *et al.* 1990; Smith *et al.* 1990), but in the parts of the succession that are interpreted as derived from Borneo, volcaniclastic detritus is largely absent (Nichols & Hall 1999).

In calculation of the sediment volumes derived from Borneo, the possibility of a contribution from outside Borneo must be considered. Sediment was certainly carried very far from Borneo during the Neogene. ODP drilling has shown that quartz-rich sandy turbidites were deposited in the Celebes Sea basin during a brief period in the Middle–Late Miocene (Smith *et al.* 1990; Nichols & Hall 1999) and Borneo was the source of this material. Nonetheless, in order to maintain a conservative estimate of sediment volumes, the basins have been cut at arbitrary limits relatively close to Borneo in order to include only sediment derived from Borneo. However, even excluding all the sediment in the South Palawan Basin and excluding all the sediment in the Celebes Sea forming part of the Tarakan Basin reduces the estimate of total sediment volume by only *c.* 6%. Seismic lines across the Makassar Strait (Bergman *et al.* 1996; Samuel *et al.* 1996) show that the primary source of sediment to the Kutei and Tarakan Basins was Borneo as the sediments thicken from Borneo and thin towards the central Makassar Straits. However, some of the sediment in the east Makassar Straits included in the Kutei Basin volume may have been derived from Sulawesi.

A final source of uncertainty is the neglect of onland Neogene sediments in NE Borneo. Large areas of Sabah and north Kalimantan are underlain by Neogene clastic sediments. The thickness of sediments onland is difficult to estimate because in many cases they are deformed or chaotic, but in Sabah there are several kilometres of section (Tongkul 1991; Swauger *et al.* 1995; Clennell 1996) and possibly even more (e.g. Balaguru 1997). These sediments have not been included in the estimates.

It is difficult to evaluate the uncertainties of these estimates. Einsele (1992) performed similar calculations to determine denudation rates (see below) and concluded that 'considering the many sources of error, all these values may deviate from the true rates by ± 20 to 30%'. Métivier *et al.* (1999) calculated volumes of rocks in Cenozoic basins of Asia. Only one of their basins (their Sarawak and Sabah Basin) is one of the Borneo basins (the Sarawak Basin) for which an estimate was calculated in this paper. Their estimate of total Cenozoic sediment volume was 2.1×10^6 km^3; our estimate is in good agreement but slightly lower at 1.94×10^6 km^3. In summary, the sediment volumes are considered to be as conservative as possible. The estimates might be reduced by around 10% if all the reservations listed above are accepted, although, as discussed above, there are also grounds for believing that the sediment volume is already substantially underestimated by the assumptions made.

Crustal thicknesses and denudation rates

The sediment volumes can be used to estimate the amount of crust removed by erosion using appropriate densities for sediments and consolidated crust (Table 1). By making assumptions about ages, these data can be used to estimate rates of uplift, erosion and deposition. It is inferred from the petrography of Neogene sandstones in Sabah that they are second-cycle deposits eroded from pre-existing sedimentary rocks rather than igneous basement (Hutchison *et al.* 2000), so a density of 2.6 Mg m^{-3} for the eroded source rocks has been used, which is the same value used by Einsele (1992) in estimating Himalayan denudation rates. Sediment densities in the Kutei Basin are 2.12 Mg m^{-3} for Upper–Middle Miocene, 2.07 Mg m^{-3} for overpressured Middle Miocene and 2.4 Mg m^{-3} for Lower Miocene (Cloke 1997): an average of 2.2 Mg m^{-3} has been chosen as a suitable average density for all sediment. Using these densities, a volumetric calculation indicates that during the Neogene the average thickness of crust removed from Borneo was 6.5 km. This estimate assumes an erosional area equal to the present-day area of the island above 100 m elevation. Even excluding sediments in the South Palawan Basin, and sediment in the Celebes Sea and Upper Kutei Basin, as discussed above, only reduces this value to 6 km.

This estimate of 6 km of crust removed is based on the assumption that the present denudation area in Borneo has remained constant. This is clearly unreasonable because at the beginning of the Miocene the area of the island above sea level was considerably smaller than at present, and marine deposits covered much of northern and eastern Borneo (Hamilton 1979; Moss & Wilson 1998). In east Kalimantan, deltas prograded rapidly east in the Early–Middle Miocene (van de Weerd & Armin 1992; Moss & Chambers 1999) and similar progradation occurred in north Borneo deltas somewhat later (Tan & Lamy 1990; Hazebroek & Tan 1993). In all areas progradation was accompanied by inversion of internal parts of the basins and reworking of the older parts of the deltaic sequences into younger parts. Thus, the denudation area must have increased since the Early Miocene. Using smaller values for the denudation area will increase the estimate of thickness of crust eroded.

There are few published estimates of the amount of overburden removed during the Neogene. Wain & Berod (1989) used data from one well in the Kutei Basin to estimate that 3 km of overburden has been removed. Van de Weerd & Armin (1992) reported that vitrinite reflectance data suggest 5 km of erosion of Lower and Middle Miocene strata in east Kalimantan. Ferguson & McClay (1997) note uplift and erosion of up to 1400 m of sediment in the western part of the Mahakam foldbelt. Cloke (1997) estimates 2.5 km of section has been removed based on balanced section modelling in the Upper Kutei Basin. For the NW margin of the Kutei Basin, Moss *et al.* (1998) used fission track data to estimate a mean denudation rate of *c.* 650 m Ma^{-1} and removal of *c.* 1.3 km of section over about 2 Ma in the Early Miocene, followed by a further 2 km of denudation at a lower rate. Vitrinite reflection and seismic velocity data from the Kutei Basin (D. Paterson pers. comm. 2000) suggests widespread removal of at least 2 km of Neogene section and local uplift of more than 4 km. Upper Miocene K–Ar ages recorded from the Mount Kinabalu granites (Jacobsen 1970) indicate rapid uplift, and fission track data indicate 4–8 km of Late Miocene denudation throughout the Crocker range (Swauger *et al.* 1995; Hutchison *et al.* 2000). Apatite fission track data from Miocene strata suggest that 2–3 km of section has been removed

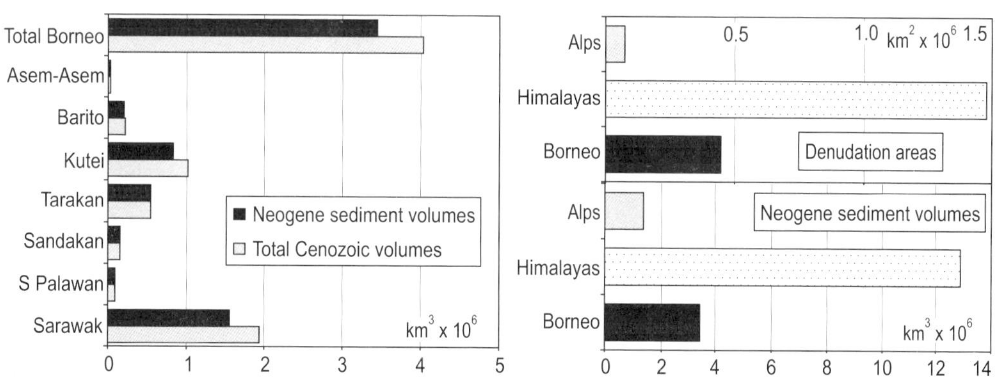

Fig. 6. Comparative sediment volumes for Borneo basins, and for Alpine and Himalayan basins, based on sources discussed in the text. Borneo and the Himalayas have supplied sediment at similar rates to adjacent sedimentary basins during the Neogene.

from eastern Sabah (Hutchison *et al.* 2000). Vitrinite reflection data suggest up to 4 km of section has been removed from south Sabah since the Late Miocene (Balaguru 2001).

Comparison with other orogenic belts

The volume of Neogene sediment in the Borneo basins can be compared to other areas in the world (Table 1, Fig. 6) using estimates from Neogene circum-Alpine basins compiled by England (1981), and for the basins of the Himalayas, including the foredeep and Bengal fan, compiled by Einsele (1992). It is clear from the estimates that the amount of sediment in the Borneo basins is very large, with the volume of sediment in one Borneo basin alone, the Sarawak Basin, larger than that in all the Neogene circum-Alpine basins. The area of land above 100 m in Borneo is much larger than the Alps, although the average height of the mountains is much lower, and the area of land above 100 m in Borneo is about one third of the area of the Himalayas, although once again considerably lower. Despite this, the amount of Neogene sediment in the Borneo basins is about one third of that eroded from the Himalayas for the same period. In other words, Borneo and the Himalayas have supplied similar amounts of sediment to adjacent sedimentary basins per unit area during the Neogene.

Using the present denudation area (the area of land above 100 m) implies an average denudation rate of 326 m Ma^{-1} over the last 20 Ma, again a minimum figure. For the Himalayas the comparable figure for the same period based on the present denudation area is 383 m Ma^{-1} (Einsele 1992), and for the Alps is 817 m Ma^{-1} (England 1981). Maximum denudation rates are more difficult to estimate: for Borneo, if it is assumed that sediment was supplied at a constant rate for the last 20 Ma and that the area of mountains has increased at a linear rate from one third of their present area, maximum denudation rates of 965 m Ma^{-1} are implied, comparable to the high Himalayas where Einsele (1992) calculates an average denudation rate of 950 m Ma^{-1}. The estimated value is well within the range of values from mountainous regions in the tropics such as New Guinea and Indonesia (Einsele 1992), and the minimum and maximum values are comparable with those derived from fission track studies of Moss *et al.* (1998) and Hutchison *et al.* (2000).

Causes of high erosion rates

There are problems accounting for the huge volume of Neogene sediments around Borneo. Many, if not most, published assertions about geomorphological and tectonic controls on sediment removal rates in the geological literature are questionable in their relevance to SE Asia. Most of the data were not obtained from SE Asia, those that were come from areas significantly modified by man, and conclusions are largely based on studies of much smaller, climatically different, regions of much younger uplift such as Taiwan and New Zealand.

Hydrologists have been collecting rainfall and erosion data in Borneo (e.g. Douglas *et al.* 1992; Douglas 1999) over periods of a few years and comparisons with their results are interesting. The sediment removed from Borneo estimated in this paper equates to *c.* 850 t km^{-2} a^{-1}, based on the present-day denudation area (the area of land above 100 m). This long-term average estimate is high in comparison with 'natural erosion' in Borneo but is comparable to man-induced 'accelerated erosion'. Natural erosion under undisturbed lowland rain forest in the Segama River of Sabah over a 2-year period was 312 t km^{-2} a^{-1} for a 1.7 km^2 catchment, but an adjacent 0.5 km^2 catchment subjected to selective logging generated an average of 1600 t km^{-2} a^{-1} (Douglas *et al.* 1992). Six–seven years after logging disturbance this catchment (N. A. Chappell pers. comm. 1998) generated 592 t km^{-2} a^{-1} and the period of this study included a 170 mm day^{-1} rainstorm, which demonstrated the importance of extreme events; 40% of the annual suspended-sediment flux was generated in a 1-day extreme event. Two other studies of Borneo indicate considerable variations. Malmer (1990) estimated that between 58 and 253 t km^{-2} a^{-1} was generated from < 1 km^2 catchments on sandstones and siltstones in SW Sabah disturbed by natural fires or cleared for plantation forestry. Murtedza (1990) recorded 448 t km^{-2} a^{-1} for the downstream Segama River and 718 t km^{-2} a^{-1} for the downstream Kinabatangan River in Sabah; both catchments had been disturbed by forestry operations. All these studies have problems: the catchment areas are small, the suspended sediment load was determined mainly from spot sampling and relatively simple flow gauging methods, there is little knowledge of bedload, and they are relatively short term and therefore have difficulty in considering the role of extreme events. Douglas *et al.* (1999) discuss the role of extreme events and their observations show that these need to be considered when attempting to translate rates measured during studies of a few years to geologically meaningful periods of time. For example, in the Segama River of Sabah the sediment yield on a single day in 1996 exceeded the annual sediment load in several previous years. Douglas (1999) estimated that in SE Asia sediment yields of the order of 10 000 t km^{-2} a^{-1} occur in the most tectonically active parts of island arcs, with around

1000 t km^{-2} a^{-1} on the weak Tertiary mud rocks of eastern Borneo.

There have been several attempts in recent years by geologists and geomorphologists to estimate sediment fluxes to the oceans (e.g. Milliman & Syvitski 1992; Summerfield & Hulton 1994; Hay 1998; Ludwig & Probst 1998). These studies are not entirely in agreement. Rainfall, runoff, elevation and relief may be factors. Runoff is related to precipitation, although seasonality of flow rather than total precipitation may be important for smaller basins. Elevation may have an indirect influence on precipitation through the orographic effect, but in the tropics this is not a simple relationship because maximum mean annual precipitation is reached at different elevations in different locations. Local relief may also be related to elevation, although elevation itself is not thought to have a direct effect on denudation rates. Milliman & Syvitski (1992) identified basin area and maximum elevation as the most important controls and concluded that other factors such as climate and runoff are of only secondary importance. However, their data show that runoff and sediment yield are poorly correlated because a few basins with high yield have low runoff (notably in China) and there is a strong positive correlation in their data between yield and runoff for basins with runoff greater than 100 mm a^{-1}. In tropical ever-wet areas such as Borneo (Morley & Flenley 1987) runoff is very much higher than this (Milliman & Syvitski 1992) as annual rainfall is typically greater than 2500 mm a^{-1} (e.g. Douglas *et al.* 1999; Staub *et al.* 2000). Other workers differ in their assessment of controls. Summerfield & Hulton (1994) identified runoff and local relief as most important, whereas basin area, runoff variability and mean temperatures were reported to be only weakly associated with denudation rates. Ludwig & Probst (1998) found that sediment yields were controlled principally by factors including runoff intensity, basin slope, rock hardness and annual rainfall variability. The Milliman & Syvitski maximum elevation factor is not quite the same as the local relief factor of other workers but is similar since they separate their basins into subgroups of different topographic categories with different regression values. Nonetheless, if basins with runoffs greater than 100 mm a^{-1} are considered, there is some agreement that sediment yields/denudation rates are a function of runoff and slope (local relief).

In general, tropical areas of SE Asia are almost completely neglected in the global analyses. The data for mountainous regions of South Asia/Oceania come from China, India, Indonesia, New Zealand, Philippines, Taiwan, Thailand and Vietnam. With the exception of small parts of New Zealand, only Indonesia includes ever-wet regions such as Borneo. Indonesia also occupies the whole circum-equatorial zone (5°N–5°S) between longitudes 95°E and 130°E, yet even the Milliman & Syvitski (1992) data set includes only seven rivers from Indonesia, all in Java. Java differs from Borneo in having a strongly monsoonal climate in contrast to ever-wet Borneo, an average rainfall which is significantly lower than Borneo, and a heavily populated, deforested and highly agricultural landscape substantially modified by man, compared to Borneo, which until recently was almost entirely rainforest with an average population density of 17 persons km^{-2}. During the Neogene Borneo had a similar ever-wet climate to the present, whereas Java and many other parts of SE Asia developed a monsoonal, highly seasonal climate by the Pliocene (Morley 1998).

Milliman *et al.* (1999) used modified forms of the Milliman & Syvitski (1992) algorithms, including data only from the East Indies (six Indonesian rivers in Java, two rivers in New Guinea and one river in the Philippines) to estimate sediment loads of SE Asian islands including Borneo. They concluded that six islands in SE Asia (Sumatra, Java, Borneo, Sulawesi, Timor and New Guinea) provide about 20–25% of the sediment discharge to the global ocean, even though these islands account for only about 2% of the land area draining into the oceans. For Borneo Milliman *et al.* (1999) calculated an average sediment yield of 1600 t km^{-2} a^{-1}, and although their data set includes no data from Borneo it is consistent with the yield estimated for the Rajang River of Sarawak (Staub *et al.* 2000), although much higher than that estimated for the Mahakam River of East Kalimantan (Allen & Chambers 1998).

It is important to note that estimates of sediment yields in Borneo are dependent on a very small amount of data. More studies are badly needed. However, the studies that have been reported suggest that: (1) the denudation rates indicated by estimates of sediment volumes from the circum-Borneo basins are compatible with estimates of sediment yields at the present day; (2) these yields are a function of several factors which may include runoff, relief and rock properties; (3) denudation rates estimated in this paper are closer to rates measured during extreme events rather than 'normal' conditions; and (4) that the sediment volumes in SE Asian basins may be a useful guide to the long-term erosion rates in the region. Like Milliman & Meade (1983) and Milliman *et al.* (1999), we suggest that SE Asia is exceptional in the large quantities of sediment being eroded and that climatic factors are important in determining the rates of denudation. If this is the case, climatic factors need to be included when considered the tectonic consequences of erosion and that tropical regions may develop in ways not typical of other orogenic regions in non-tropical settings.

Implications for the tectonic evolution of Borneo

These estimates indicate that Borneo mountains should be seen as comparable in importance to major mountain ranges such as the Himalayas. The Himalayas reach almost 9 km above sea level, and high erosion rates in the high mountain areas reflect relief. In contrast, Borneo is relatively low but the high erosion rates reflect climate, particularly very high rainfall and intense tropical weathering, as well as probably greater average heights and local relief early in the Neogene. The average thickness of crust removed is 6 km, and this is also comparable to the average thickness removed from the Himalayas for the same period. The average value implies that significantly more than 6 km has been removed from the highest mountainous parts of Borneo.

The structural style and development are likely to have been different in Borneo compared to the Alps or Himalayas. The critical wedge model (Davis *et al.* 1983) has been widely applied to orogenic belts and accretionary prisms, but for Borneo the model would appear to be inappropriate as usually presented. Numerical (e.g. Beaumont *et al.* 1992) and analogue modelling (e.g. Storti & McClay 1995; Mugnier *et al.* 1997) shows that erosion influences the structures that develop in orogenic belts. The high rate of erosion meant that as the Borneo mountains rose from the Early Miocene onwards, deformation did not propagate outwards in the form of large thrust sheets, and there would have been no requirement for the wedge to fail by faulting. Instead, the orogenic belt would have been denuded entirely by erosion. Unlike the Himalayan and Alpine belts, and most other orogenic belts in non-tropical regions, Borneo was surrounded by deep basins ready to receive sediment. But because there was no major thrusting, adjacent to Borneo there would have been no major lithospheric flexure due to thrust loading, and sediment was deposited adjacent to the orogenic belt in the deep oceanic areas of the Makassar Strait, Celebes Sea, Sulu Sea and the South China Sea. The sedimentary basins around Borneo are therefore not conventional foreland basins as they did not develop in response to loading by thrust sheets. The thicknesses of sediment deposited in these basins would have resulted in further subsidence due to sedimentary loading of continental margin and oceanic crust.

In Borneo, the most likely cause of the uplift in the Early Miocene was the arrival of thinned continental crust at the north Borneo subduction margin (Hamilton 1979; Taylor & Hayes 1983; Tan & Lamy 1990; Hazebroek & Tan 1993; Hutchison 1996*b*; Hall 1996; Hutchison *et al.* 2000). During the Oligocene the oceanic part of the proto-South China Sea was subducted, producing an accretionary complex (Hamilton 1979). The change in character of sedimentation in east Kalimantan implies that continental crust began to underthrust north Borneo at the end of the Oligocene, causing uplift of the region behind the Crocker accretionary prism (Hamilton 1979; Taylor & Hayes 1983; Tan & Lamy 1990; Hazebroek & Tan 1993; Hutchison *et al.* 2000). Underthrusting must have continued throughout the Early and Middle Miocene to cause the uplift and generate the large sediment volume. It has been suggested (Hall 1996) that South China Sea extension was driven by slab pull at the south side of the proto-South China Sea due to subduction beneath north Borneo and further east, and in this scenario the pull forces exerted by the subducted oceanic slab could have maintained underthrusting of thinned continental crust for a period after the beginning of continent–continent collision in Sabah (Hutchison *et al.* 2000), or arc–continent collision further east (Rangin *et al.* 1990). An additional cause for the uplift may be shortening due to the counter-clockwise rotation of Borneo, which is suggested to have occurred between about 25 and 10 Ma (Fuller *et al.* 1991, 1999; Hall 1996) as a consequence of Australia–SE Asia collision.

There is insufficient information from Borneo and elsewhere to examine all the consequences of the extremely high erosion rates in tropical regions. However, some speculation is justified. The Borneo mountains rose in the early Neogene as a result of collision. The elevation of mountains would also have influenced climate in the region and it is likely that increased elevation contributed to increased rainfall. The change from an Oligocene drier climate to the ever-wet climate of the Neogene in Borneo (Morley 1998) may therefore have been influenced by the elevation of mountains. As the major deltas prograded outwards during the Neogene the increased land area may have further influenced local climate and Borneo may have developed its own ability to cause tropical storms, thus further enhancing erosion rates. The increased loading of areas close to Borneo could then have contributed to rates of uplift of the central mountain ranges by flexural effects. Thus, climate changed at least in part as a result of tectonics, but the changing climate then caused the tectonic development of the island to diverge from that in non-tropical orogenic regions.

Conclusions

These results from Borneo support the view of Beaumont *et al.* (1992) that palaeoclimatic information may be required to understand the evolution of ancient orogens. In areas of low–moderate erosion rates a rising orogenic belt cannot be denuded fast enough by erosion and therefore has to be denuded mechanically by normal faulting, thrusting and nappe formation. In contrast, at high erosion rates, such as those experienced in the tropical climate of Borneo during the Neogene, an orogenic belt can be entirely denuded by erosion. In such cases, high-level nappes will not develop because thrust sheets are eroded as fast as they form. The absence of thin-skinned thrusting in regions of high erosion rates means that foreland basins will not develop in response to loading of the lithosphere by thrust sheets. More detailed work is needed to understand the geological evolution of Borneo, and to provide better data on the tectonic, thermal, denudational and sedimentary history of the area. However, these initial conclusions based on the estimates of sediment supply to adjacent basins indicate that a study of recent tropical orogenic belts might provide insights into the importance of climate in large-scale tectonic processes.

During the last one million years sea level and vegetation throughout SE Asia have varied dramatically with glacially-induced climate change. How have these changes influenced erosion and sedimentation rates? Are present rates representative? Logging, building and deforestation are activities which affect present erosion rates. Equally, climate change, forest fires and sea level changes are natural causes of variations in erosion rates. We require good estimates of longer-term erosion rates to assess anthropogenic effects and evaluate their impact. Improving estimates of sediment volumes in tropical regions such as Indonesia and New Guinea will improve our estimates of rates of processes, and will improve our understanding of the complex feedback system between climate and tectonics.

We are grateful to S. Baker, J. Chambers, N. Chappell, I. Cloke, J. Decker, M. Fuller, J. Granath, N. Harbury, C. Hutchison, S. Moss, R. Murphy, J. Noad, H. Schwing, M. Wilson, other members of the SE Asia Research Group, and many others for discussion.

References

ALLEN, G. P. & CHAMBERS, J. L. C. 1998. *Sedimentation in the Modern and Miocene Mahakam Delta*. Indonesian Petroleum Association.

AUDLEY-CHARLES, M. G. 1978. The Indonesian and Philippine archipelagos. *In*: MOULLADE, M. & NAIRN, A. E. M. (eds) *The Phanerozoic Geology of the World II. The Mesozoic, A*. Elsevier, Amsterdam, 165–207.

BAILLIE, P., GILLERAN, P., CLARK, W., MOSS, S. J., STEIN, A., HERMANTORO, E. & OEMAR, S. 2000. New insights into the geological development of the deepwater Mahakam delta and Makassar Straits. *Proceedings Indonesian Petroleum Association, 27th Annual Convention*, 397–402.

BALAGURU, A. 1997. Sedimentology and structural development of the Malibau basin, Sabah. *In*: *Tectonics, Stratigraphy and Petroleum Systems of Borneo*. Universiti Brunei Darussalam, Conference Abstracts, 65–66.

BALAGURU, A. 2001. *Tectonic Evolution and Sedimentation of the Southern Sabab Basin*. PhD Thesis, University of London.

BEAUMONT, C., FULLSACK, P. & HAMILTON, J. 1992. Erosional control of active compressional orogens. *In*: MCCLAY, K. R. (ed.) *Thrust Tectonics*. Chapman & Hall, London, 1–18.

BEN-AVRAHAM, Z. & EMERY, K. O. 1973. Structural framework of the Sunda Shelf. *AAPG Bulletin*, **57**, 2323–2366.

BERGMAN, S. C., COFFIELD, D. Q., GARRARD, R. A. & TALBOT, J. 1996. Tertiary tectonic evolution of western Sulawesi and the Makassar Strait, Indonesia. *In*: HALL, R. & BLUNDELL, D. J. (eds) *Tectonic Evolution of SE Asia*. Geological Society, London, Special Publications, **106**, 391–429.

BRIAIS, A., PATRIAT, P. & TAPPONNIER, P. 1993. Updated interpretation of magnetic anomalies and seafloor spreading stages in the South China Sea: implications for the Tertiary tectonics of Southeast Asia. *Journal of Geophysical Research*, **98**, 6299–6328.

BUROLLET, P. F. & SALLE, C. 1981. Seismic reflection profiles in the Makassar Strait. *In*: BARBER, A. J. & WIRYOSUJONO, S. (eds) *The Geology and Tectonics of Eastern Indonesia*. Geological Research and Development Centre, Bandung, Special Publications, **2**, 273–276.

CCOP. 1991. *Total sedimentary isopach maps offshore east Asia, 1:4 000 000*. Committee for Coordination of Joint Prospecting for Mineral Resources in Asian Offshore Areas, Bangkok. Technical Bulletin, **23**.

CLENNELL, B. 1996. Far-field and gravity tectonics in Miocene basins of Sabah, Malaysia. *In*: HALL, R. & BLUNDELL, D. J. (eds) *Tectonic Evolution of SE Asia*. Geological Society, London, Special Publications, **106**, 307–320.

CLOKE, I. 1997. *Structural Controls on the Basin Evolution of the Kutai Basin and Makassar Straits, Indonesia*. PhD Thesis, University of London.

CLOKE, I. R., MILSOM, J. & BLUNDELL, D. J. B. 1999. Implications of gravity data from east Kalimantan and the Makassar Straits: a solution to the origin of the Makassar Straits? *Journal of Asian Earth Sciences*, **17**, 61–78.

DALY, M. C., COOPER, M. A., WILSON, I., SMITH, D. G. & HOOPER, B. G. D. 1991. Cenozoic plate tectonics and basin evolution in Indonesia. *Marine and Petroleum Geology*, **8**, 2–21.

DAVIS, D. M., SUPPE, J. & DAHLEN, F. A. 1983. Mechanics

of fold and thrust belts and accretionary wedges. *Journal of Geophysical Research*, **88**, 1153–1172.

DOUGLAS, I. 1999. Hydrological investigations of forest disturbance and land cover impacts in South-East Asia: a review. *Philosophical Transactions of the Royal Society of London*, **354**, 1725–1738.

DOUGLAS, I., BIDIN, K., BALAMURUGAN, G., CHAPPELL, N. A., WALSH, R. P. D., GREER, T. & SINUN, W. 1999. The role of extreme events in the impacts of selective tropical forestry on erosion during harvesting and recovery phases at Danum Valley, Sabah. *Philosophical Transactions of the Royal Society of London*, **354**, 1749–1761.

DOUGLAS, I., SPENCER, T., GREER, T., BIDIN, K., SINUN, W. & MENG, W. W. 1992. The impact of selective commercial logging on stream hydrology, chemistry and sediment loads in the Ulu Segama rainforest, Sabah, Malaysia. *Philosophical Transactions of the Royal Society of London Series B*, **335**, 397–406.

DÜRBAUM, H.-J. & HINZ, K. 1982. SEATAR-related geophysical studies by BGR in the southwest Pacific. *Transactions of the Third Circum-Pacific Energy and Mineral Resources Conference*, 129–133.

EINSELE, G. 1992. *Sedimentary Basins*. Springer, New York.

ENGLAND, P. 1981. Metamorphic pressure estimates and sediment volumes for the Alpine orogeny: an independent control on geobarometers? *Earth and Planetary Science Letters*, **56**, 387–397.

ENGLAND, P. & HOUSEMAN, G. 1986. Finite strain calculations of continental deformation 2. Comparison with the India–Asia collision zone. *Journal of Geophysical Research*, **91**, 3664–3676.

FERGUSON, A. & MCCLAY, K. 1997. Structural modelling within the Sanga-Sanga PSC, Kutei basin, Kalimantan: its application to paleochannel orientation studies and timing of hydrocarbon entrapment. *In*: HOWES, J. V. C. & NOBLE, R. A. (eds) *Petroleum Systems of SE Asia and Australasia*. Indonesian Petroleum Association, Jakarta, 727–749

FULLER, M., ALI, J. R., MOSS, S. J., FROST, G. M., RICHTER, B. & MAHFI, A. 1999. Paleomagnetism of Borneo. *Journal of Asian Earth Sciences,* **17**, 3–24.

FULLER, M., HASTON, R., LIN, J.-L., RICHTER, B., SCHMIDTKE, E. & ALMASCO, J. 1991. Tertiary paleomagnetism of regions around the South China Sea. *Journal of SE Asian Earth Sciences,* **6**, 161–184.

GRAVES, J. A. & SWAUGER, D. A. 1997. Petroleum systems of the Sandakan basin, Philippines *In*: HOWES, J. V. C. & NOBLE, R. A. (eds) *Petroleum Systems of SE Asia and Australasia*. Indonesian Petroleum Association, Jakarta, 799–813.

HALL, R. 1996. Reconstructing Cenozoic SE Asia. *In*: HALL, R. & BLUNDELL, D. J. (eds) *Tectonic Evolution of SE Asia*. Geological Society, London, Special Publications, **106**, 153–184.

HALL, R. 1997. Cenozoic tectonics of SE Asia and Australasia. *In*: HOWES, J. V. C. & NOBLE, R. A. (eds) *Petroleum Systems of SE Asia and Australasia*. Indonesian Petroleum Association, Jakarta, 47–62.

HALL, R. 1998. The plate tectonics of Cenozoic SE Asia and the distribution of land and sea. *In*: HALL, R. & HOLLOWAY, J. D. (eds) *Biogeography and Geological Evolution of SE Asia*. Backhuys, Leiden, The Netherlands, 99–131.

HALL, R. 2001. Cenozoic reconstructions of SE Asia and the SW Pacific: changing patterns of land and sea. *In*: METCALFE, I., SMITH, J. M. B., MORWOOD, M. & DAVIDSON, I. D. (eds) *Faunal and Floral Migrations and Evolution in SE Asia–Australasia*. A. A. Balkema, Rotterdam, The Netherlands, 35–56.

HAMILTON, W. 1974. *Map of Sedimentary Basins of the Indonesian Region*. USGS Map-I-875-B. US Geological Survey, Reston, VA, USA.

HAMILTON, W. 1979. *Tectonics of the Indonesian Region*. U.S. Geological Survey, Professional Papers, **1078**.

HASAN, K. 1991. The Upper Cretaceous flysch succession of the Balangbaru Formation, Southwest-Sulawesi. *Proceedings of the Indonesian Petroleum Association 20th Annual Convention*, 183–208.

HAY, W. W. 1998. Detrital sediment fluxes from continents to oceans. *Chemical Geology*, **145**, 287–323.

HAZEBROEK, H. P. & TAN, D. N. K. 1993. Tertiary tectonic evolution of the NW Sabah continental margin. *Geological Society of Malaysia Bulletin,* **33**, 195–210

HINZ, K. & SCHLUTER, H. V. 1985. Geology of the Dangerous Grounds, South China Sea and the continental margin off southwest Palawan: results of Sonne Cruises SO23 and SO27. *Energy*, **10**, 297–315.

HINZ, K., BLOCK, M., KUDRASS, H. R. & MEYER, H. 1991. Structural elements of the Sulu Sea. *Geologische Jahrbuch,* **A127**, 883–506.

HOLLOWAY, N. H. 1982. The stratigraphic and tectonic evolution of Reed Bank, North Palawan and Mindoro to the Asian mainland and its significance in the evolution of the South China Sea. *AAPG Bulletin,* **66**, 1357–1383.

HUTCHISON, C. S. 1989. *Geological Evolution of South-east Asia*. Oxford Monographs on Geology and Geophysics, **13**.

HUTCHISON, C. S. 1996*a*. *South-east Asian Oil, Gas, Coal and Mineral Deposits*. Clarendon Press, Oxford.

HUTCHISON, C. S. 1996*b*. The 'Rajang Accretionary Prism' and 'Lupar Line' problem of Borneo. *In*: HALL, R. & BLUNDELL, D. J. (eds) *Tectonic Evolution of SE Asia*. Geological Society, London, Special Publications, **106**, 247–261.

HUTCHISON, C. S., BERGMAN, S. C., SWAUGER, D. A. & GRAVES, J. E. 2000. A Miocene collisional belt in north Borneo: uplift mechanism and isostatic adjustment quantified by thermochronology. *Journal of the Geological Society, London*, **157**, 783–793.

HUTCHISON, C. S., BERGMAN, S. C., SWAUGER, D. A. & GRAVES, J. E. 2001. Discussion of a Miocene collisional belt in north Borneo: uplift mechanism and isostatic adjustment quantified by thermochronology. *Journal of the Geological Society, London*, **158**, 396–400.

IOC, IHO, BODC. 1997. *GEBCO-97: The 1997 Edition of the Gebco Digital Atlas*. British Oceanographic Data Centre.

JACOBSEN, G. 1970. *Gunung Kinabalu area, Sabah, Malaysia*. Geological Survey of Malaysia Report, **8**.

KATILI, J. A. 1978. Past and present geotectonic position of Sulawesi, Indonesia. *Tectonophysics*, **45**, 289–322.

LEE, T. Y. & LAWVER, L. A. 1994. Cenozoic plate reconstruction of the South China Sea region. *Tectonophysics*, **235**, 149–180.

LEE, T.-Y. & LAWVER, L. A. 1995. Cenozoic plate reconstruction of Southeast Asia. *Tectonophysics*, **251**, 85–138.

LLOYD, A. R. 1978. Geological evolution of the South China Sea. *SEAPEX Proceedings*, **4**, 95–137.

LUDWIG, W. & PROBST, J. L. 1998. River sediment discharge to the oceans: present-day controls and global budgets. *American Journal of Science*, **298**, 265–295.

MALMER, A. 1990. Stream suspended sediment load after clear-felling and different forestry treatments in tropical rainforest, Sabah, Malaysia. *In*: *Research Needs and Applications to Reduce Erosion and Sedimentation in Tropical Steeplands*. International Association of Hydrological Sciences Publications, **192**, 62–71.

METCALFE, I. 1996. Pre-Cretaceous evolution of SE Asian terranes. *In*: HALL, R. & BLUNDELL, D. J. (eds) *Tectonic Evolution of SE Asia*. Geological Society, London, Special Publications, **106**, 97–122.

MÉTIVIER, F., GAUDEMER, Y., TAPPONNIER, P. & KLEIN, M. 1999. Mass accumulation rates in Asia during the Cenozoic. *Geophysical Journal International*, **137**, 280–318.

MILLIMAN, J. D. & MEADE, R. H. 1983. World-wide delivery of river sediment to the oceans. *Journal of Geology*, **91**, 1–21.

MILLIMAN, J. D. & SYVITSKI, J. P. M. 1992. Geomorphic/tectonic control of sediment discharge to the ocean: the importance of small mountainous rivers. *Journal of Geology*, **100**, 525–544.

MILLIMAN, J. D., FARNSWORTH, K. L. & ALBERTIN, C. S. 1999. Flux and fate of fluvial sediments leaving large islands in the East Indies. *Journal of Sea Research*, **41**, 97–107.

MORLEY, R. J. 1998. Palynological evidence for Tertiary plant dispersals in the SE Asian region in relation to plate tectonics and climate. *In*: HALL, R. & HOLLOWAY, J. D. (eds) *Biogeography and Geological Evolution of SE Asia*. Backhuys, Leiden, The Netherlands, 211–234.

MORLEY, R. J. & FLENLEY, J. R. 1987. Late Cenozoic vegetational and environmental changes in the Malay Archipelago. *In*: WHITMORE, T. C. (ed.) *Biogeographical evolution of the Malay Archipelago*. Oxford Monographs on Biogeography, **4**, 50–59.

MOSS, S. J. & CHAMBERS, J. L. C. 1999. Tertiary facies architecture in the Kutai Basin, Kalimantan, Indonesia. *Journal of Asian Earth Sciences*, **17**, 157–181.

MOSS, S. J. & WILSON, M. J. 1998. Tertiary evolution of Sulawesi–Borneo: implications for biogeography. *In*: HALL, R. & HOLLOWAY, J. D. (eds) *Biogeography and Geological Evolution of SE Asia*. Backhuys, Leiden, The Netherlands, 133–163.

MOSS, S. J., CARTER, A., HURFORD, A. & BAKER, S. 1998. A Late Oligocene tectono-volcanic event in East Kalimantan and the implications for tectonics and sedimentation in Borneo. *Journal of the Geological Society, London*, **155**, 177–192.

MUGNIER, J. L. , BABY, P., COLLETTA, R., VINOUR, P., BALE, P. & LETURMY, P. 1997. Thrust geometry controlled by erosion and sedimentation: a view from analogue models. *Geology*, **25**, 427–430.

MURTEDZA, M. 1990. Estimation of sediment yields of major watersheds in Sabah. *In*: *Workshop on Watershed Development and Management, 19–23 February 1990*. UPM/IHP/UNESCO.

NICHOLS, G. J. & HALL, R. 1999. Stratigraphic and sedimentological constraints on the tectonic history of the Celebes Sea Basin. *Journal of Asian Earth Sciences*, **18**, 47–59.

NICHOLS, G. J., BETZLER, C., BRASS, G. W., HUANG, Z., LINSLEY, B., MERRILL, D., MULLER, C. M., NEDERBRAGT, A., PUBELLIER, M., SAJONA, F. M., SCHERER, R. P., SHYU, J.-P., SMITH, R., SOLIDUM, R., SPADEA, P. & LEG 124 SCIENTIFIC PARTY. 1990. Depositional systems of the Celebes Sea from ODP sites 767 and 770. *Geophysical Research Letters*, **17**, 2065–2068.

PARKINSON, C. D. 1991. *The Petrology, Structure and Geologic History of the Metamorphic Rocks of Central Sulawesi, Indonesia*. PhD Thesis, University of London.

PELTZER, G. & TAPPONNIER, P. 1988. Formation and evolution of strike-slip faults, rifts, and basins during the India–Asia collision: an experimental approach. *Journal of Geophysical Research*, **93**, 15 085–15 117.

RANGIN, C. & SILVER, E. A. 1991. Neogene tectonic evolution of the Celebes–Sulu basins: new insights from Leg 124 drilling. *In*: SILVER, E. A., RANGIN, C., VON BREYMANN, M. T. *et al. Proceedings of the Ocean Drilling Program, Scientific Results*, **124**, 51–63.

RANGIN, C., BELLON, H., BENARD, F., LETOUZEY, J., MÜLLER, C. & SANUDIN, T. 1990. Neogene arc–continent collision in Sabah, Northern Borneo (Malaysia). *Tectonophysics*, **183**, 305–319.

SAMUEL, L., PURWOKO, H., PURNOMO, J., BERTAGNE, A. J. & SMITH, N. G. 1996. Results from interpretation of regional transects in central Indonesia. *The Leading Edge*, April, **15**, 261–266.

SANDAL, S. T. (ed.) 1996. *The Geology and Hydrocarbon Resources of Negara Brunei Darussalam*. Brunei Shell Petroleum Company, Syabas, Brunei.

SILVER, E. A. & RANGIN, C. 1991. Leg 124 tectonic synthesis. *In*: SILVER, E. A., RANGIN, C., VON BREYMANN, M. T. *et al. Proceedings of the Ocean Drilling Program, Scientific Results*, **124**, 3–9.

SITUMORANG, B. 1982. Formation, evolution, and hydrocarbon prospect of the Makassar basin, Indonesia. *Transactions of the Third Circum-Pacific Energy and Mineral Resources Conference*, 227–231.

SITUMORANG, B. 1987. *Seismic Stratigraphy of the*

Makassar Basin. Scientific Contribution on Petroleum Science and Technology Lemigas Sumbangan Ilmiah, **1/87**, 3–38.

SMITH, R., BETZLER, C., BRASS, G. W., HUANG, Z., LINSLEY, B., MERRILL, D., MULLER, C. M., NEDERBRAGT, A., NICHOLS, G. J., PUBELLIER, M., SAJONA, F. M., SCHERER, R. P., SHYU, J.-P., SOLIDUM, R., SPADEA, P. & LEG 124 SCIENTIFIC PARTY. 1990. Depositional systems of the Celebes Sea from ODP sites 767 and 770. *Geophysical Research Letters*, **17**, 2061–2064.

STAUB, J. R., AMONG, H. L. & GASTALDO, R. A. 2000. Seasonal sediment transport and deposition in the Rajang River delta, Sarawak, East Malaysia. *Sedimentary Geology*, **133**, 249–264.

STORTI, F. & MCCLAY, K. 1995. Influence of syntectonic sedimentation on thrust wedges in analogue models. *Geology*, **23**, 999–1002.

SUKAMTO, R. 1978. The structure of Sulawesi in the light of plate tectonics. *Proceedings of Regional Conference on Geology and Mineral Resources of SE Asia*, 121–141.

SUMMERFIELD, M. A. & HULTON, N. J. 1994. Natural controls of fluvial denudation rates in major world drainage basins. *Journal of Geophysical Research*, **99**, 13 871–13 883.

SWAUGER, D. A., BERGMAN, S. C., MARILLO, A. P., PAGADO, E. S. & SURAT, T. 1995. Tertiary stratigraphy and tectonic framework of Sabah, Malaysia: a field and laboratory study. *GEOSEA 95: 8th Regional Conference on Geology, Minerals, and Energy Resources of SE Asia, Manila*, 35–36.

TAN, D. N. K. & LAMY, J. M. 1990. Tectonic evolution of the NW Sabah continental margin since the Late Eocene. *Bulletin of the Geological Society of Malaysia*, **27**, 241–260.

TANEAN, H., PATERSON, D. W. & ENDHARTO, M. 1996. Source provenance interpretation of Kutei basin sandstones and the implications for the tectono-stratigraphic evolution of Kalimantan. *Indonesian Petroleum Association Proceedings 25th Annual Convention*, **1**, 333–345.

TAPPONNIER, P., PELTZER, G., LEDAIN, A., ARMIJO, R. & COBBOLD, P. 1982. Propagating extrusion tectonics in Asia: new insights from simple experiments with plasticine. *Geology*, **10**, 611–616.

TAYLOR, B. & HAYES, D. E. 1980. The tectonic evolution of the South China Basin. *In*: HAYES, D. E. (ed.) *The Tectonic and Geologic Evolution of Southeast Asian Seas and Islands*. American Geophysical Union Monographs, **23**, 89–104.

TAYLOR, B. & HAYES, D. E. 1983. Origin and history of the South China Basin. *In*: HAYES, D. E. (ed.) *The Tectonic and Geologic Evolution of Southeast Asian Seas and Islands, Part 2*. American Geophysical Union Monographs, **27**, 23–56.

TONGKUL, F. 1991. Tectonic evolution of Sabah, Malaysia. *Journal of SE Asian Earth Sciences*, **6**, 395–405.

UMBGROVE, J. H. F. 1938. Geological history of the East Indies. *AAPG Bulletin*, **22**, 1–70.

VAN BEMMELEN, R. W. 1949. *The Geology of Indonesia, Vol. 1a*. Government Printing Office, The Hague, The Netherlands.

VAN DE WEERD, A. A. & ARMIN, R. A. 1992. Origin and evolution of the Tertiary hydrocarbon bearing basins in Kalimantan (Borneo), Indonesia. *AAPG Bulletin*, **76**, 1778–1803.

VAN LEEUWEN, T. M. 1981. The geology of southwest Sulawesi with special reference to the Biru area. *In*: BARBER, A. J. & WIRYOSUJONO, S. (eds) *The Geology and Tectonics of Eastern Indonesia*. Geological Research and Development Centre, Bandung, Special Publications, **2**, 277–304.

WAIN, T. & BEROD, B. 1989. The tectonic framework and paleogeographic evolution of the Upper Kutei Basin. *Indonesian Petroleum Association Proceedings 18th Annual Convention*, **1**, 55–78.

WAKITA, K., SOPAHELUWAKAN, J., MIYAZAKI, K., ZULKARNAIN, I. & MUNASRI, O. 1996. Tectonic evolution of the Bantimala Complex, South Sulawesi, Indonesia. *In*: HALL, R. & BLUNDELL, D. J. (eds) *Tectonic Evolution of SE Asia*. Geological Society, London, Special Publications, **106**, 353–364.

WEISSEL, J. K. 1980. Evidence for Eocene oceanic crust in the Celebes Basin. *In*: HAYES, D. E. (ed.) *The Tectonic and Geologic Evolution of Southeast Asian Seas and Islands*. American Geophysical Union Monographs, **23**, 37–48.

WIGHT, A., HARE, L. H. & REYNOLDS, J. R. 1992. Tarakan Basin, NE Kalimantan, Indonesia: a century of exploration and future potential. *Bulletin of the Geological Society of Malaysia*, **33**, 265–288.

WILSON, M. E. J., CHAMBERS, J. L. C., EVANS, M. J., MOSS, S. J. & NAS, D. S. 1999. Cenozoic carbonates in Borneo: case studies from northeast Kalimantan. *Journal of Asian Earth Sciences*, **17**, 183–201.

Quantification of river-capture-induced base-level changes and landscape development, Sorbas Basin, SE Spain

M. STOKES[1], A. E. MATHER[2] & A. M. HARVEY[3]

Departments of [1]Geological Sciences and [2]Geographical Sciences, University of Plymouth, Drake Circus, Plymouth, Devon PL4 8AA, UK (e-mail: mstokes@plymouth.ac.uk)

[3]Department of Geography, University of Liverpool, Roxby Building, PO BOX 147, Liverpool L69 3BX, UK

Abstract: The Aguas/Feos river system of the Sorbas Basin, SE Spain was captured by an aggressive subsequent stream *c.* 100 ka. The consequence of the capture event was twofold: (1) basin-scale drainage reorganization via beheading of the southward flowing Aguas/Feos system and re-routing the drainage eastwards into the Vera Basin; and (2) the creation of a new, lower base level and associated upstream propagation of a wave of incision.

The sequence of pre- and post-capture events are well established from previous studies of the Quaternary terrace record. Using these studies, this paper makes the first attempt to quantify the impact of river capture in terms of spatial and temporal variations in rates of incision, sediment flux and surface lowering. This was carried out through construction of 43 valley cross-sections from the 'captured' (Upper Aguas), 'beheaded' (Feos) and 'capturing' streams (Lower Aguas) within the central–southern parts of the Sorbas Basin. Dated pre- and post-capture terrace and corresponding strath levels were plotted on to the valley cross-sections enabling incision amounts, rates and valley cross-sectional areas to be calculated. Sediment fluxes were calculated using a mean valley section method. Surface lowering calculations were made through reconstruction of the top basin-fill surface and subtraction from the modern contour values.

The lowering of base level has resulted in a dramatic increase in incision upstream of the capture site by a factor of 4 to 20. This in turn has been associated with significant pre- and post-capture changes in valley shape. The increased incision resulted in dramatic post-capture increases in valley erosion upstream and downstream of the capture site by a factor of 2 to 9 which can be related to changes in associated stream power as a function of increased gradient and discharge. In excess of 60% of the landscape change can be accounted for by valley-constrained erosion as opposed to overall surface lowering.

River capture is a commonly invoked process within landscape development models (see Bishop 1995 and references therein). A capture occurs when an actively incising and headward-eroding river system beheads and re-routes the drainage of a less active system. Such a capture may dramatically increase the catchment size of the active 'capturing' stream, and thus can affect stream power, flood magnitude and frequency. In some cases different bedrock lithologies may be added to the extended catchment area that can affect the sediment concentrations in flood events, reflected in aggradation or degradation below the capture point, depending on the direction of change (Shepherd 1982).

In order for capture to take place, the base level (*sensu* Leopold & Bull 1979) of the capturing stream must be lower than that of its 'victim' (Small 1978). The creation of a new lower base-level in the captured stream leads to disequilibrium in this captured system. The drainage network above the capture site will commonly respond to this disequilibrium by a wave of incision that migrates up the system with time (Germanoski & Harvey 1993; Mather 2000a; Florsheim *et al.* 2001).

Numerous examples of river capture effecting base-level changes within fluvial systems are documented (Bishop 1995; Harvey *et al.* 1995; Calvache & Viseras 1997). Most of this research is qualitative, although a few recent studies have attempted to quantify the spatial and temporal changes in incision resulting from capture-induced base level change (Harvey *et al.* 1995; Mather

From: JONES, S. J. & FROSTICK, L. E. (eds) 2002. *Sediment Flux to Basins: Causes, Controls and Consequences.* Geological Society, London, Special Publications, **191**, 23–35. 0305-8719/02/$15.00

2000*a*; Mather *et al.* 2000). This lack of quantification in part reflects the difficulty in establishing a chronology of events within a dominantly erosional record. River capture, however, has profound implications for sediment flux and routing between and within sedimentary basins (Mather 2000*b*; Mather *et al.* 2000). Thus, quantifying the direction and magnitude of change associated with river capture is an important issue for both basin and landscape modelling.

This paper takes a well-documented example of drainage rearrangement via river capture from the Sorbas Basin of SE Spain (Fig. 1) and makes the first attempt to quantify its basin-wide impact. Previous studies by Harvey & Wells (1987), Harvey *et al.* (1995), Mather & Harvey (1995) and Mather (2000*a*) have established a relative sequence of events associated with the capture event using a combination of geomorphic and geological evidence from the Quaternary river terrace record. Recently, a study by Kelly *et al.* (2000) has provided an absolute chronology for the river capture by establishing a U–Th isochron framework from river terrace soils. These data thus make the Sorbas Basin river capture an ideal candidate for quantifying the spatial and temporal impacts of capture-induced base-level changes on landscape development.

Regional setting

The Sorbas Basin is one of a series of small (30 × 20 km), intramontane sedimentary basins located within the Betic Cordillera, SE Spain (Fig. 1). The basin is bounded by mountain ranges to the north (Sierra de los Filabres) and south (Sierras Alhamilla-Cabrera) (Fig. 1). The eastern and western basin margins are poorly defined topographic highs. Since the formation of the Sorbas Basin during the Neogene an approximate N–S compressive regime has existed (Sanz de Galdeano & Vera 1992; Montenat & Ott d'Estevou 1999). Tectonic activity, dominated by strike- and oblique-slip faulting, has resulted in the development of a basin and range topography (Silva *et al.* 1993; Keller *et al.* 1995). This tectonic activity has had significant implications for the orientation of much of the drainage network (Mather & Westhead 1993; Mather & Harvey 1995; Mather 1999; 2000*a*, *b*).

Sedimentation within the Sorbas Basin occurred from the Mid-Miocene onwards (Ruegg 1964; Völk & Rondeel 1964). This basin-fill has a composite thickness of *c.* 1500 m. The Late Cenozoic palaeogeography of the Sorbas Basin has been reconstructed using fluvial sediments and landforms (Harvey 1987; Harvey & Wells 1987; Mather 1991, 1993*a*, *b*; 2000*a*, *b*; Mather & Harvey 1995). During the Pliocene the Sorbas Basin was infilled with sediments from aggrading fluvial systems of the Góchar Formation (Mather 1991, 1993*a*, 2000*b*; Mather & Harvey 1995). The uppermost level of sedimentary basin fill is represented by the morphological expression of the end Góchar Formation surface. During the Plio-Pleistocene the fluvial system began to incise into the underlying basin sediments resulting in the development of a series of inset river terraces (Harvey & Wells 1987; Harvey *et al.* 1995). The switch from deposition to erosion by the fluvial systems is primarily attributed to differential uplift (Mather & Harvey 1995). Calculations of crustal uplift for the Sorbas Basin during the Plio-Pleistocene reveal uplift rates in excess of 80–160 m Ma^{-1} (Mather 1991; Mather & Harvey 1995). It is these uplift rates that have increased regional surface gradients, culminating in river capture and drainage rearrangement during the Late Pleistocene (Mather & Harvey 1995).

The Río Aguas and Rambla de los Feos Catchments

Modern drainage record

The modern Río Aguas catchment covers an area of approximately 550 km^2. It is sourced in the west of the Sorbas Basin, and drains east through the southern parts of the Vera Basin, before entering the Mediterranean Sea near Mojácar (Fig. 1). Of its current catchment area, 50% (*c.* 225 km^2) was added by the river capture at *c.* 100 ka. The associated, beheaded Rambla de los Feos has a modern catchment area of *c.* 50 km^2 above its confluence with the Rambla Lucainena, the headwater to the Río Alías in the Carboneras Basin (Fig. 1), which drains east towards the Mediterranean Sea (Fig. 1).

Pleistocene terrace record

The river capture within the Sorbas Basin is recorded within the Pleistocene terrace record (Harvey & Wells 1987) (Fig. 2). This record consists of an inset sequence of five main river terrace levels (stages A–E, oldest–youngest; Harvey *et al.* 1995). Each terrace level comprises an unconformable/disconformable aggradation of up to 20 m of fluvial sediments (sandstones and conglomerates) which overlie basin-fill sediments or metamorphic basement rock. These terrace deposits suggest that the long-term erosion of the fluvial system was punctuated by periods of sedimentation, possibly determined by a number of factors including climatic change (Mather & Harvey 1995). The terrace levels can be correlated throughout much of the Sorbas Basin on the basis of their sedimentology, soil characteristics, lateral

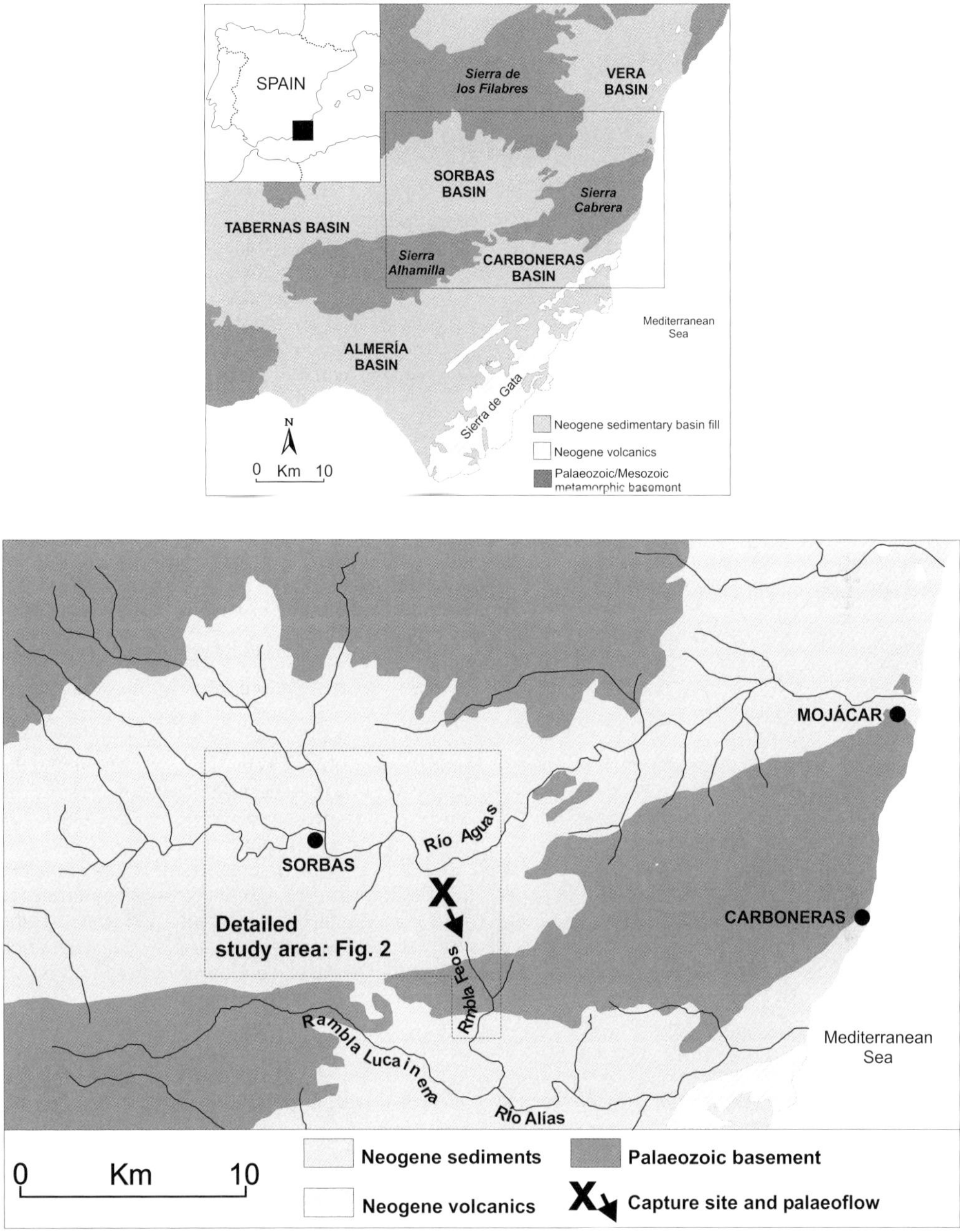

Fig. 1. (**a**) Study area location within SE Spain. (**b**) Location of detailed study area within the Aguas/Feos catchment of the Sorbas Basin. The arrow indicates the palaeoflow of the Early–Mid-Pleistocene Aguas/Feos drainage southwards into the Carboneras Basin. Note that the captured drainage is now routed eastwards into the Vera Basin.

continuity and height differences between horizontally eroded bases (Harvey & Wells 1987; Harvey *et al.* 1995).

The capture event involves two principal rivers, the Río Aguas and the Rambla de los Feos (Fig. 1) (Harvey & Wells 1987). Reconstruction of the Pleistocene terrace record reveals that a consequent fluvial system (Aguas/Feos) existed during the

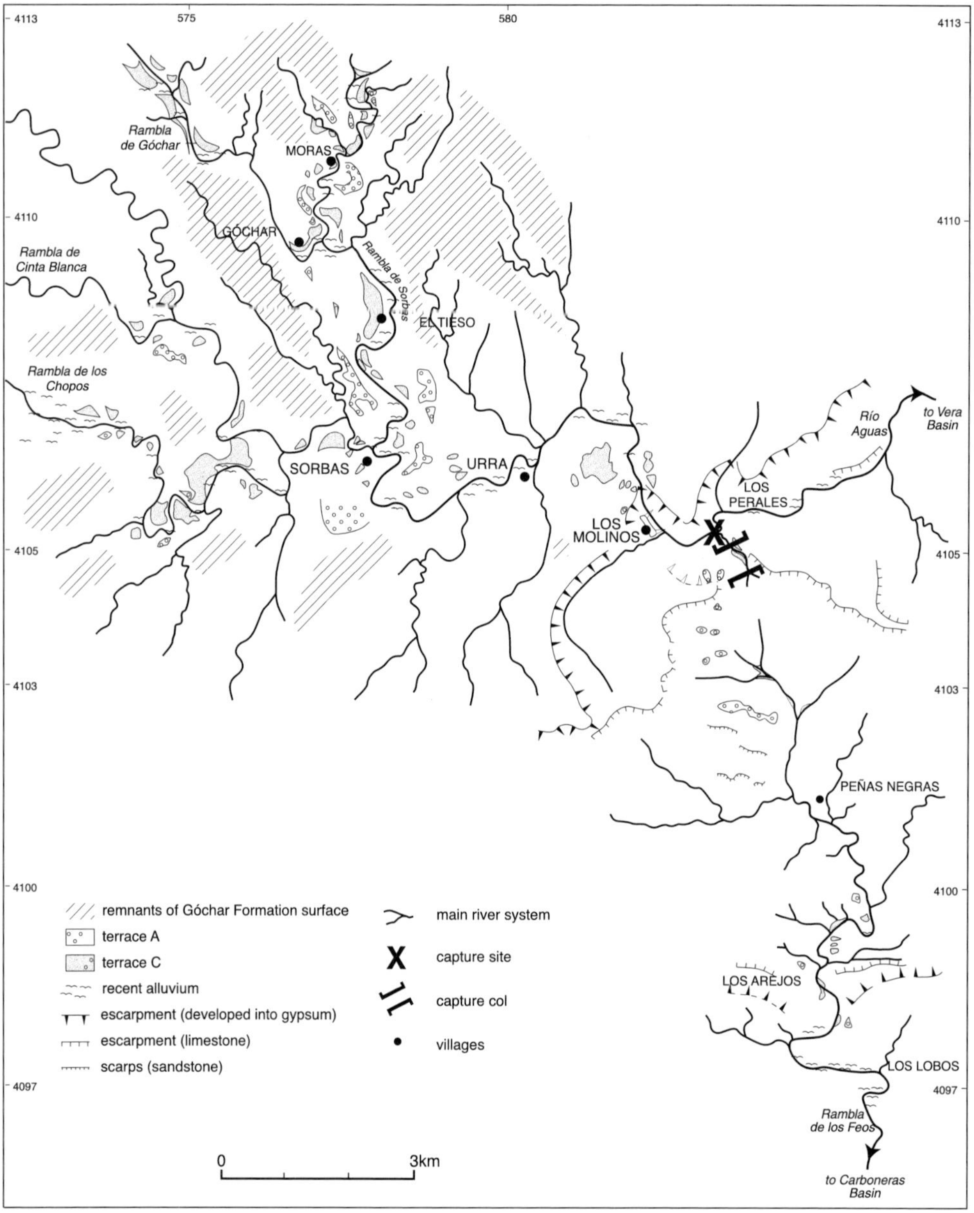

Fig. 2. Detail of the Aguas and Feos river systems illustrating the spatial relationships between the Góchar Formation surface, river terrace levels A and C, major morphological components (escarpments and scarps) and the modern drainage network (redrawn from Harvey *et al.* 1995).

Early and Mid-Pleistocene (Mather & Harvey 1995). Terrace levels A–C document this early pre-capture scenario and illustrate that the Aguas/Feos fluvial system flowed southwards following a discordant route across a structural low in the E–W-orientated axis of the Sierras Alhamilla–Cabrera range, exiting into the Carboneras Basin (Fig. 1). River capture took place during the Late Pleistocene as a consequence of Plio-Pleistocene tectonic elevation of the Sorbas Basin (Harvey & Wells 1987). This tectonic elevation lead to the steepening of regional surface gradients between the Sorbas Basin and its lower elevation neighbour, the Vera Basin, resulting in the development of an aggressive subsequent river originating in the Vera Basin (Río Aguas). The river cut back into the Sorbas Basin and captured the southwards flowing system (Aguas/Feos), beheading and re-routing it eastwards into the Vera Basin (Harvey & Wells 1987). The capture lead to a base-level lowering of 90 m at the capture site (Harvey *et al.* 1995; Mather 2000a). According to Mather (2000*a*), 73% of the original Sorbas Basin drainage was re-routed eastwards into the adjacent Vera Basin following the capture event. This stage of drainage development is documented by the valley evolution post-terrace C to the present.

Harvey & Wells (1987) and Harvey *et al.* (1995) have established a relative time framework for the capture event. They used the degree of soil development to suggest that the capture took place *c.* 100 ka. More recently, advances in absolute dating techniques have enabled Kelly *et al.* (2000) to apply U–Th isochron methods to carbonate soils from the terrace. Their results suggest that capture took place at 88 ka (68–104 ka). This date broadly accords with that of Harvey & Wells (1987). However, some problems arise with the compatibility of the relative and absolute dating of the older, pre-capture terraces, particularly with respect to the U–Th age of terrace level A (224 – >380 ka). This age range is at the limit of the dating technique (Kelly *et al.* 2000). Indeed, Harvey & Wells (1987) suggested a much older (relative) age for terrace level A of 700 – <1600 ka. Despite these uncertainties, the absolute and relative chronologies for the terrace record still provide a useful framework against which the base-level responses of the fluvial system to river capture can be explored.

Approach

In order to address the quantification of the base-level responses of the fluvial system to river capture this study focuses upon the central parts of the Sorbas Basin and its main trunk rivers, the Rambla de Sorbas, Río Aguas and Rambla de los Feos (Figs 1 and 2). This area contains the most complete terrace record for the Aguas/Feos system and has been mapped in detail by Harvey *et al.* (1995). The study area can be divided into three zones based upon those identified within Harvey et al. (1995):

- upstream of the capture site (Upper Aguas Valley: the captured stream);
- downstream of the capture site (Lower Aguas Valley: the capturing stream);
- downstream of the capture site (Feos Valley: the beheaded stream).

Throughout these zones a total of 43 valley cross-sections were constructed using 1:25 000 topographic maps (Mapa Topográfico Nacional de España 1985). On to each cross-section the position of key terrace levels or erosional features of suspected fluvial origin (strath lines) were plotted, together with the underlying bedrock geology. Representative valley cross-sections are presented in Fig. 3.

Valley incision

The height difference between the tops of terrace levels/strath lines and the modern river were used to calculate the total amounts of incision prior to capture (terraces A–C) and post-capture (post-terrace C to the modern river). Incision rates at each site were calculated by dividing the amount of incision for a given period of drainage evolution at each locality by the estimated age ranges of the terraces/straths. For terrace A, which has a wide possible age range on the basis of both absolute and relative dating techniques (380–1600 ka), maximum and minimum rates were calculated using these end-age values in order to show the probable scale of variation in incision rates.

Valley volumes

The drainage evolution documented by the terrace levels is an erosional record, demonstrating a net export of sediment by the fluvial system. A change in valley shape characterizes the pre-capture (broad and shallow) to post-capture (narrow and deep) valley evolution (Fig. 3). Cross-sectional areas from each of the sites provide an initial indication of sediment removal from each site. Volumes for the different stages of valley evolution (i.e. pre- and post-capture) can be calculated using the cross-sectional area of the A–C (pre-capture) and post- C to modern (post-capture) valley elements. The volume of sediment removed can be estimated using a mean valley section method:

sediment volume removed =

$$\sum_{n=1}^{n}\left[\frac{A_1 + A_2}{2} L_{(1-2)}, \frac{A_2 + A_3}{2} L_{(2-3)}, \ldots\ldots\ldots\ldots \frac{A_{n-1} + A_n}{2} L_{((n-1)-n)}\right] \quad (1)$$

where A_1 is the cross-sectional area of valley section 1 in m^2, A_2 is the cross-sectional area of valley section 2 in m^2, etc.; and L_{1-2} is the distance (m) between sections 1 and 2, L_{2-3} is the distance (m) between sections 2 and 3, and so on.

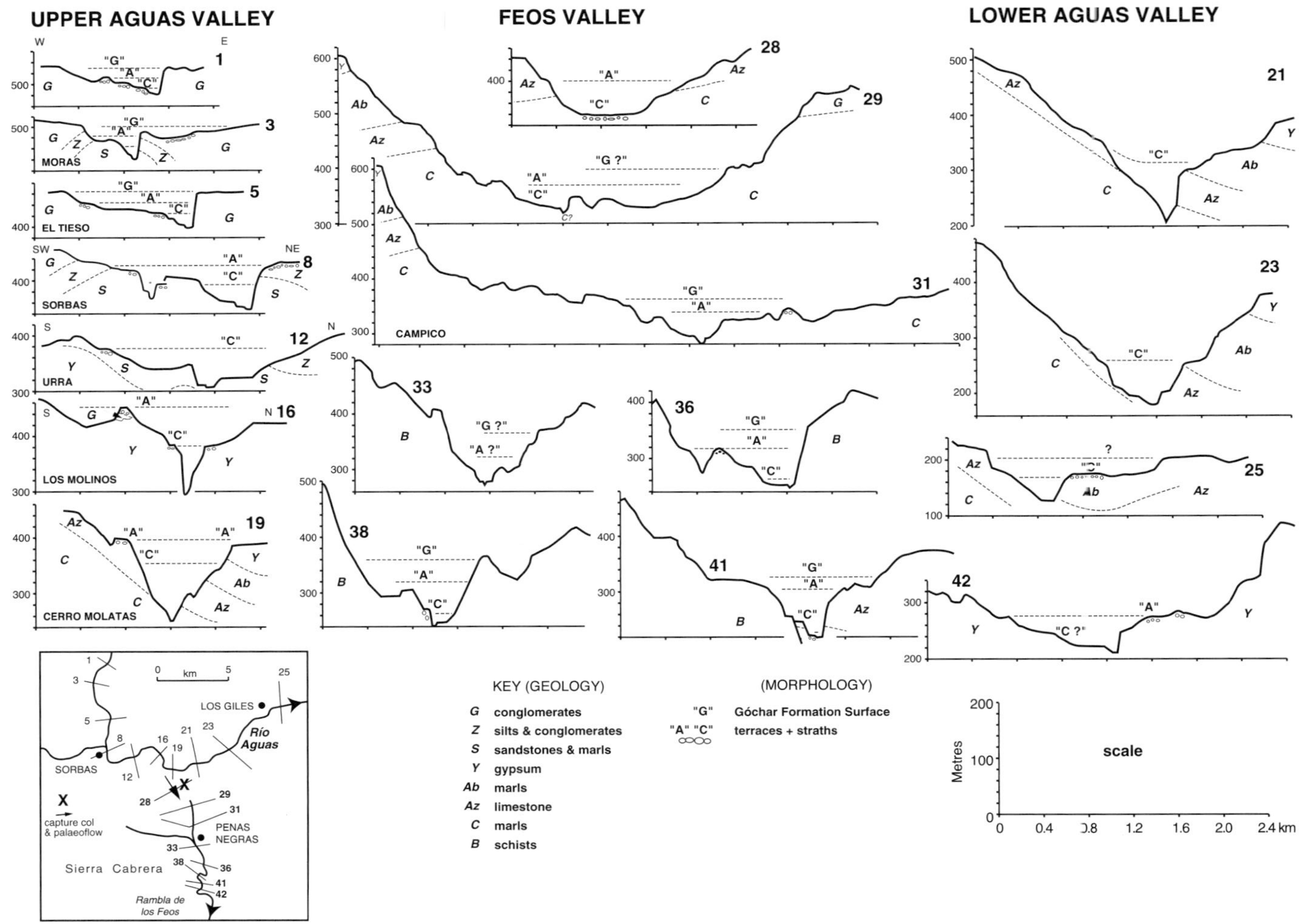

Fig. 3. Representative valley cross-sections from the 'captured' (Upper Aguas), 'capturing' (Lower Aguas) and 'beheaded' (Feos) drainage systems. The geology key corresponds to specific stratigraphic Formations where in stratigraphic order: B, basement; C, Chozas Formation; Az, Azagador Member; Ab, Abad Member; Y, Yesares Formation; S, Sorbas Member; Z, Zorreras Member; and G, Góchar Formation. See Völk & Rondeel (1964) for further details regarding these units.

Surface lowering

Surface lowering is an alternative estimate of dissection of the basin as a whole, as the switch from basin filling to dissection can be achieved by reconstructing the top basin-fill surface. This is represented by the top surface of the Góchar Formation, which is of approximate Early Pleistocene age (Mather 1991, 2000*a*, *b*; Mather & Harvey 1995). An evaluation of the form of the top Góchar Formation surface was made by constructing a generalized contour map of the basin using modern ridge top contours (which represent the pre-dissection surface) and projecting through distances of up to 1.5 km. Construction of the spatial patterns of dissection of this surface have been made by (1) using grid intersects of single points (spaced at 0.5 km) and (2) subtracting the modern elevation for these points from the reconstructed post Góchar Formation surface.

In some areas the Góchar Formation surface can clearly be identified in the form of a fan radiating from the northeast margin of the basin (Mather 1999; Mather *et al.* 2000). Elsewhere (southern basin margin, along the Rambla Lucainena, and southeastern area, along the Lower Aguas Valley) erosion has removed most of the relict surface. In areas to the south, east and northeast of Sorbas town, the surface has been significantly affected by left-lateral strike-slip fault activity and associated deformation along a prominent NE–SW- to NNE–SSW-orientated structural lineament (the Marchalico–Infierno lineament of Mather & Westhead 1993; Mather 2000b). As calculations of dissection can only be performed with any degree of confidence in areas of Góchar Formation outcrop affected by minimal erosion (i.e. north of Sorbas town), the volumes calculated are from the area of the basin encompassing the Ramblas de Góchar, Sorbas, Cinta Blanca and de los Chopos as far as the settlement at Urra (Fig. 2).

Valley incision record

Amounts of valley incision

The amounts of incision throughout the study area (Fig. 4) show spatial and temporal trends within individual valley zones and from the study area as a whole. Within the Upper Aguas Valley the amount of incision between the Góchar Formation surface and the first terrace level (A) is small (16 m average). The amount of incision recorded by the terrace sequence then progressively increases with time (pre-capture = 40 m average, post-capture = 59 m average). This pattern of incision can be traced upstream from the capture site, in the north to beyond the village of Moras and in the west to the distal parts of the Rambla de los Chopos (Fig. 2), a valley distance in each case of some 20 km. Within the Lower Aguas Valley no remnants of the Góchar Formation surface or terraces A–C are preserved. Despite this absence, large amounts of incision are recorded from the extrapolated Góchar Formation surface down to terrace C (78 m average). More importantly, post-capture incision is moderately high (66 m average). Within the Feos Valley the incision between the Góchar Formation surface and terrace A is moderate (26 m average).

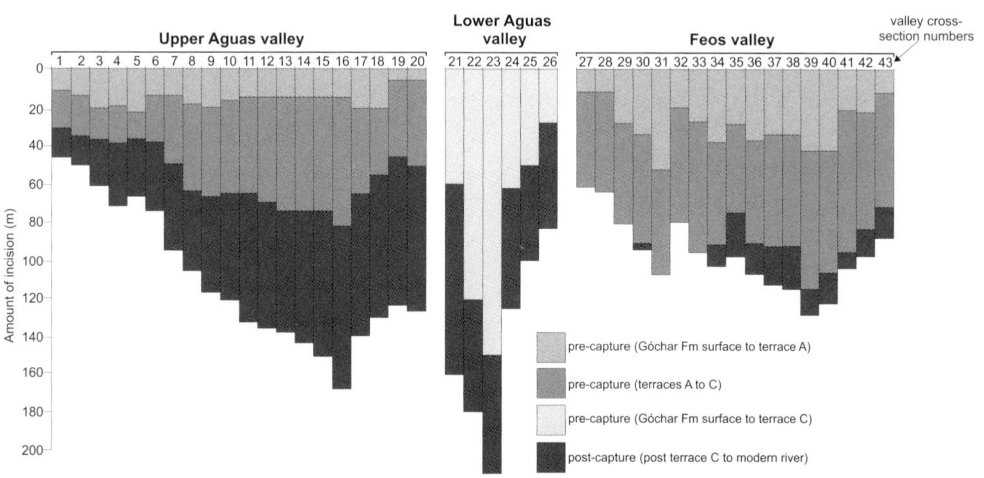

Fig. 4. Pre- and post-capture amounts of incision for the Aguas/Feos drainage systems. For location of sections see inset on Fig. 3.

An increase is then recorded within the pre-capture terrace sequence (60 m average). Incision post-capture is either absent or low (9 m average).

Two trends are evident from comparisons of the amounts of incision in each valley zone:

- the pre-capture terrace sequence is characterized by low to moderate amounts of incision throughout the entire study area;
- the post-capture terrace sequence typically shows higher amounts of incision and a greater degree of variance between valley zones – for example, incision amounts are moderate to high within the Upper and Lower Aguas Valleys but only negligible within the Feos Valley.

Rates of valley incision

Incision rates are presented in Fig. 5. Again, a series of spatial and temporal trends are apparent within individual valley zones and the study area as a whole. Pre-capture incision rates are characteristically low. The average maximum and minimum values range between 0.03 and 0.14 m ka^{-1} in the Upper Aguas Valley and are slightly higher (0.04–0.21 m ka^{-1}) in the Feos Valley. No calculations can be made from the Lower Aguas Valley due to the absence of level A. In complete contrast, the post-capture terrace sequence within the Upper Aguas Valley changes by a factor of 4 to 20. These values were calculated using 100 ka as the capture event date subtracted from the minimum and maximum possible ages for terrace A (380–1600 ka). Within the Feos Valley, incision rates show a significant change by a factor of –0.4 to –2.

Two trends are apparent when rates of incision are compared between the valley zones:

- the pre-capture terrace record is characterized by low–moderate rates of incision throughout the study area although there is some spatial variance, with the highest values recorded from the Feos valley;
- the post-capture terrace record shows higher rates of incision but with more spatial variance – the highest rates are recorded from the Lower Aguas Valley and contrast with the negligible rates recorded from the Feos Valley.

Rates of sediment removal

Valley erosion

Applying the mean section method (Equation 1, cross-sectional area data recorded in Fig. 6), the total volume of sediment removed from the end Góchar Formation surface to the present day is some 2 km^3. The Upper Aguas accounted for 0.9 km^3 of this, 0.5 km^3 was derived from the Lower Aguas and 0.6 km^3 came from the Feos Valley. Prior to river capture, the highest volume of sediment removed is recorded from the Upper Aguas Valley and the lowest from the Feos Valley. In contrast, within the Upper Aguas Valley the post-

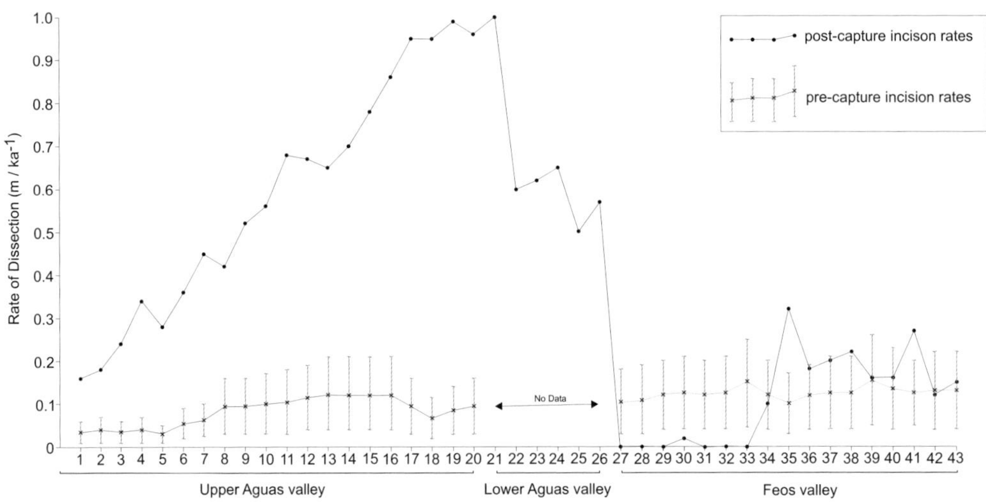

Fig. 5. Pre- and post-capture rates of incision for the Aguas/Feos drainage system. Pre-capture incision rates are plotted with a maximum/minimum range due to the errors involved with the absolute dating of terrace level A. For location of sections see inset on Fig. 3. The pre-capture 'no data' area between sites 21 and 26 located with the 'capturing' Lower Aguas Valley corresponds to the absence of terrace level A.

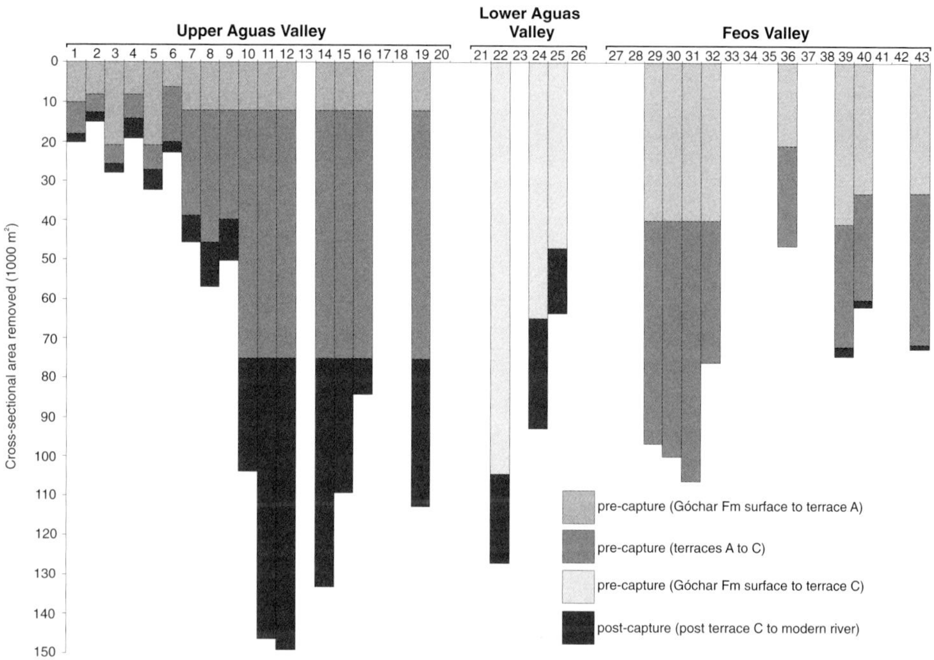

Fig. 6. Pre- and post-capture cross-sectional area removed for the Aguas/Feos drainage systems. No data sections correspond to the absence or unclear identification of terrace levels. For location of sections see inset on Fig. 3.

capture volumes of sediment removed show dramatic variations through time (pre- and post-capture) and space (maximum values concentrated around and upstream of the capture site). Post-capture, sediment flux changes by a factor of 2 to 9. In the Lower Aguas Valley the post-capture volumes of sediment removed change by a factor of –1 to 2 and in the Feos Valley by a factor of –6 to –34. The changes for the Upper Aguas and Feos valleys cannot be accounted for simply by errors in the dating methods and are therefore noteworthy.

Land surface lowering

The study area in which surface lowering was examined (the Upper Aguas Valley area only) covers 76 km^2 (Fig. 7). Within this area the mean lowering is 17.3 m. Thus, 1.29 km^3 of sediment have been removed from this area of the basin since the abandonment of the end Góchar Formation surface, i.e. 0.017 $km^3\ km^{-2}$. The results of the land surface lowering clearly show the concentration of erosion in areas close to the capture site (Fig. 7c).

Discussion

This study demonstrates a number of controlling factors for landscape development over Quaternary time scales. These include the primary controls of base-level change and its subsequent effects, on to which are superimposed anomalies arising from local variations in structure and lithology. The base-level change has had profound implications for the subsequent evolution of the fluvial system: (i) by stimulating a wave of incision that dominates valley evolution; and (ii) by producing a more spatially extensive phase of net surface lowering.

Wave of incision

The impact of the base-level change is of highest magnitude (some 90 m) around the capture site and decays spatially upstream to its limits at about 20 km. The headward propagation of a wave of incision stimulated by base-level change has been reported from both empirical and field-based studies (Begin *et al.* 1981; Gardiner 1983; Begin 1988; Goldrick & Bishop 1992, 1995). These studies support the spatial findings of this work and suggest that the wave of incision decays with time. A number of knick points can be identified within the Upper Aguas Valley (see Harvey *et al.* 1995) that are typically located within more resistant lithologies (e.g. cemented conglomerates and calcareous sandstones). These knick points are located some distance from lithological boundaries.

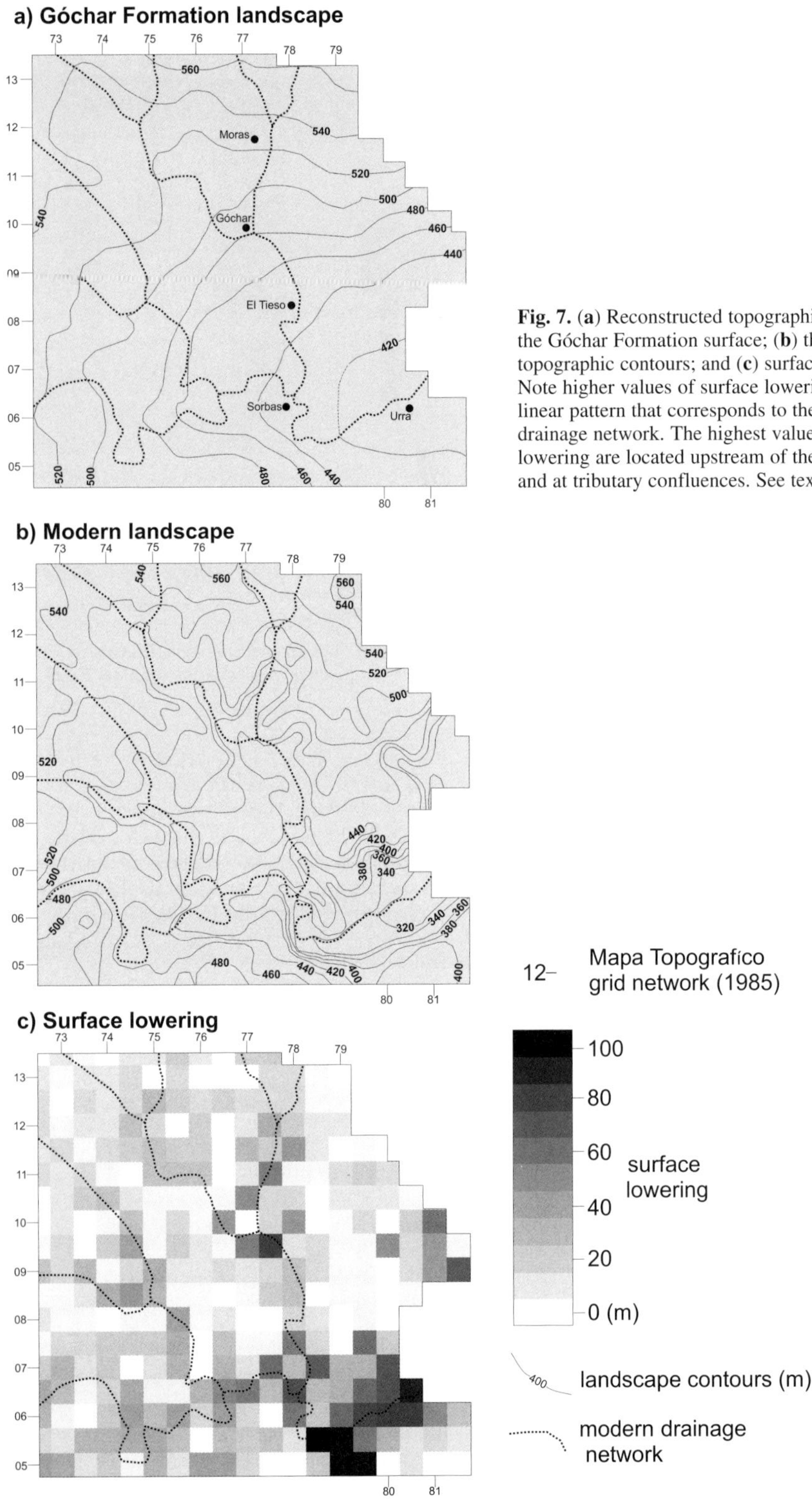

Fig. 7. (**a**) Reconstructed topographic expression of the Góchar Formation surface; (**b**) the modern topographic contours; and (**c**) surface lowering map. Note higher values of surface lowering show a linear pattern that corresponds to the modern drainage network. The highest values of surface lowering are located upstream of the capture site and at tributary confluences. See text for discussion.

In some cases they are of sufficient relief to generate dry waterfall features between 6 and 10 m in height. The knick points are thus still the focus of erosion during ephemeral stream flow events as a function of the associated high stream power they generate. These facts suggest that the knick points are still actively migrating upstream, and that the fluvial system is still in the process of adjusting to the capture-induced base-level change. The implications for sediment flux calculations are that the figures are still elevated in comparison to the pre-capture scenario (compare data in Figs 4–6). With time, however, these rates will probably decay.

The maximum impact of the capture-induced base-level change is evident around the capture site (Figs 1 and 2). Increased post-capture incision is recorded both upstream and downstream of this point (Upper and Lower Aguas Valleys: Figs 4 and 5). In the captured stream (Feos Valley), however, the proximal reaches (sites 27–33 on Fig. 5) demonstrate that post-capture incision rates are markedly lower than those of the pre-capture scenario. In both cases these probably reflect dramatic changes in stream power bought about by the river capture. In the case of the Lower Aguas Valley, stream power would have been amplified as a function of increased mean discharge *and* surface gradient. Conversely, the Feos Valley has experienced a change from occupying the pre-capture distal portion of a much larger catchment (the Upper Aguas Valley), to the headward reaches of a post-capture, much smaller catchment (Feos Valley). In this case it is evident that the reduction in discharge *alone* was sufficient to drive dramatic changes in stream power and associated incision. As both valleys are located on similar lithologies in this part of the river system we can speculate about the relative significance of discharge and gradient, to stream power. This can be achieved if we: (1) assume that the order of magnitude change in discharge was the same (although the direction was different) for both the Feos and Lower Aguas Valleys; and (2) use the data on incision rates in Fig. 5. This suggests that within the Lower Aguas Valley *c.* 80% of the experienced change in incision (and therefore stream power) can be attributed to gradient changes and 20% to discharge changes, whereas within the upper reaches of the Feos Valley 100% of the change is linked to changes in discharge alone. If, however, we examine the lower reaches of the Feos Valley (sites 35–43 on Fig. 5) we see a net increase of incision rates post-capture. This transition coincides with the Feos Valley entering the Sierra Cabrera (Fig. 1) and therefore could be related to a combination of the change in lithology (mica-schists) and the fact that this is an area of active uplift (Harvey & Wells 1987).

Another significant impact of the capture-induced base-level change and its associated accelerated incision is the change in valley shape (Fig. 3). Pre-capture, the valley sections have a mean width to depth ratio of 25. Post-capture this mean ratio changes to 3. This implies that following capture the typical valley side slopes will steepen, which in turn will have significant consequences for (a) the magnitude and frequency of slope-sediment transfer and (b) the dominant slope-sediment transfer process. This may also impinge upon the nature of sediment flux (Roering *et al.* 2001), although over these time scales it is difficult to quantify.

Surface lowering

Surface lowering data are illustrated in Fig. 7 and demonstrate that the highest values are located (a) around the capture site and (b) at channel confluences throughout the study area. These higher values can be explained by increases in stream power (see earlier discussion), although clearly in the case of channel confluences river capture is not a prerequisite. Away from these locations there are a number of other anomalies in the surface lowering pattern. These can be attributed to lithologies of low material strength (primarily marls) and structural lines of weakness (the Infierno–Marchalico lineament of Mather & Westhead 1993). For example, this applies to areas to the south of 070 and west of 780 (Fig. 7c, see Mather 2000*b* for more detail on this area). Many of the higher values are also focused, not surprisingly, adjacent to the main valleys (Fig. 7c).

We can compare the data generated on sediment volumes removed for the area covered by Fig. 7 from (a) valley erosion and (b) surface lowering. The volume removed by the main trunk river (Río Aguas) is 0.42 km^3, although this figure is underestimated since it is only calculated for a single trunk river and does not take into account tributary erosion. We can extrapolate sediment volumes removed by the major tributaries of the trunk drainage. This can be achieved by taking the volume of sediment calculated for the valley upstream of the tributary junction and assuming a similar volume for these adjoining tributaries (Ramblas de Góchar, Cinta Blanca and de los Chopos), which are all of a similar magnitude. This suggests a more reliable figure of 0.86 km^3 (67%), although again this must be severely underestimated as any valley below the scale of these drainages is ignored. From this we can postulate that the valley erosion accounts in excess of 60% of the landscape denudation within the study area. The significance of valley-constrained erosion as opposed to overall surface lowering has also been highlighted in adjacent study areas in work by Mather *et al.* (2000). These studies imply that the landscape of SE Spain in question has not yet fully adjusted to tectonically induced landscape changes.

Conclusions

This study has enabled us to quantify some elements of landscape change in relation to a river capture induced base level lowering. The lowering of base-level by some 90 m at the capture site has resulted in a dramatic increase in incision upstream (by a factor of 4 to 20). This in turn is associated with significant pre- and post-capture changes in valley shape (mean width to depth ratios of *c.* 25 pre-capture to *c.* 8 post-capture), which may have implications for the dominant slope-sediment transfer processes. The increased incision led to significant post-capture increases in valley erosion upstream and downstream of the capture site (by a factor of 2 to 9). These changes can be related to increases in associated stream power as a function of elevated gradient and discharge. In addition, the enlarged sediment flux was re-routed to an entirely different sedimentary basin, the Vera Basin to the east. The sediment flux entering the Carboneras Basin to the south (the original receiving basin) was changed by –6 to –34. Throughout the study area surface lowering accounted for 1.29 km^3 of sediment removed (i.e. a rate of 0.017 km^3 km^{-2}). This element of landscape change is outweighed by valley erosion that accounts for at least 0.86 km^3 of sediment removed.

The significance of this study is that it highlights the relative sensitivity of elements of the landscape to change over temporal scales of 10^4–10^6 years and spatial scales in excess of 10^2 km^2. Quantification of landscape changes over these temporal and spatial scales are uncommon, yet essential, for the calibration and development of basin and landscape models.

The research has been funded by the University of Plymouth (M. Stokes & A. E. Mather) and Liverpool staff development funds (A. M. Harvey). L. Walsh of Cortijo Urra Field Centre is thanked for her hospitality during fieldwork. P. Friend is acknowledged for his encouragement of the initial quantification idea. P. Bishop, L. E. Frostick and S. Jones are thanked for their constructive reviews and editing earlier versions of the manuscript.

References

BEGIN, Z. B. 1988. Application of a diffusion–erosion model to alluvial channels which degrade due to base-level lowering. *Earth Surface Processes and Landforms*, **13**, 487–500.

BEGIN, Z. B., MEYER, D. F. & SCHUMM, S. A. 1981. Development of longitudinal profiles of alluvial channels in response to base-level lowering. *Earth Surface Processes and Landforms*, **6**, 49–58.

BISHOP, P. 1995. Drainage rearrangement by river capture, beheading and diversion. *Progress in Physical Geography*, **19**, 449–473.

CALVACHE, M. & VISERAS, C. 1997. Long-term control mechanisms of stream piracy processes in Southeast Spain. *Earth Surface Processes and Landforms*, **22**, 93–105

FLORSHEIM, J. L., MOUNT, J. F. & RUTTEN, L. T. 2001. Effect of base level change on floodplain and fan sediment storage and ephemeral tributary channel morphology, Navarro River, California. *Earth Surface Processes and Landforms*, **26**, 219–232.

GARDINER, T. W. 1983. Experimental study of knickpoint and longitudinal profile evolution in cohesive, homogenous material. *Geological Society of America Bulletin*, **94**, 664–667.

GERMANOSKI, D. & HARVEY, M. D. 1993. Asynchronous terrace development in degrading braided channels. *Physical Geography*, **14**, 16–38.

GOLDRICK, G. & BISHOP, P. 1992. Morphology, processes and evolution of two waterfalls near Cowra, NSW. *Australian Geographer*, **23**, 116–121.

GOLDRICK, G. & BISHOP, P. 1995. Differentiating the roles of lithology and uplift in the steepening of bedrock river long profiles: an example from southeastern Australia. *Journal of Geology*, **103**, 221–227.

HARVEY, A. M. 1987. Patterns of Pleistocene aggradational and dissectional landform development in the Almería region, southeast Spain: a tectonically active landscape. *Die Erde*, **118**, 193–215.

HARVEY, A. M. & WELLS, S. G. 1987. Response of Pleistocene fluvial systems to differential epeirogenic uplift: Aguas and Feos river systems, southeast Spain. *Geology*, **15**, 689–693.

HARVEY, A. M., MILLER, S. Y. & WELLS, S. G. 1995. Pleistocene soil and river terrace sequences in the Aguas/Feos river systems: Sorbas basin, SE Spain. *In*: LEWIN, J., MACKLIN, M. G. & WOODWARD, J. C. (eds) Mediterranean Quaternary River Environments. A. A. Balkema, Rotterdam, The Netherlands, 263–282.

KELLER, J. V. A., HALL, S. H., DART, C. J. & MCCLAY, K. R. 1995. The geometry and evolution of a transpressional strike-slip system: the Carboneras fault, SE Spain. *Journal of the Geological Society, London*, **152**, 339–351.

KELLY, M., BLACK, S. & ROWAN, J. S. 2000. A calcrete-based U/Th chronology for landform evolution in the Sorbas basin, southeast Spain. *Quaternary Science Reviews*, **19**, 995–1010.

LEOPOLD, L. B. & BULL, W. B. 1979. Base level, aggradation and grade. *Proceedings of the American Philosophical Society*, **123**, 168–202.

MAPA TOPOGRÁFICO NACIONAL DE ESPAÑA. 1985. *Sorbas 1031-I, Turre 1031-II and Polopos 1031-III*, 1,25 000 Series. Mapa Topográfico Nacional de España.

MATHER, A. E. 1991. *Late Caenozoic Drainage Evolution of the Sorbas Basin, Southeast Spain*. Unpublished PhD Thesis, University of Liverpool.

MATHER, A. E. 1993*a*. Basin inversion: some

consequences for drainage evolution and alluvial architecture. *Sedimentology*, **40**, 1069–1089.

MATHER, A. E. 1993*b*. Evolution of a Pliocene fan delta: links between the Sorbas and Carboneras Basins, SE Spain. *In*: FROSTICK, L. & STEEL, R. (eds)*Tectonic Controls and Signatures in Sedimentary Successions*. IAS Special Publications, **20**, 77–290.

MATHER, A. E. 1999. Alluvial fans: a case study from the Sorbas Basin, southeast Spain. *In*: JONES, A. P., TUCKER, M. E. & HART, J. K. (eds) *The Description and Analysis of Quaternary Stratigraphic Sections*. Quaternary Research Association, Technical Guide **7**, 77–110.

MATHER, A. E. 2000*a*. Adjustment of a drainage network to capture induced base-level change. *Geomorphology*, **34**, 271–289.

MATHER, A. E. 2000*b*. Impact of headwater river capture on alluvial system development. *Journal of the Geological Society, London*, **157**, 957–966.

MATHER, A. E. & HARVEY, A. M. 1995. Controls on drainage evolution in the Sorbas Basin, southeast Spain. *In*: LEWIN, J., MACKLIN, M. G. & WOODWARD, J. C. (eds) *Mediterranean Quaternary River Environments*. A. A. Balkema, Rotterdam, The Netherlands, 65–75.

MATHER, A. E. & WESTHEAD, R. K. 1993. Plio/Quaternary strain of the Sorbas basin, SE Spain: Evidence from sediment deformation structures. *Quaternary Proceedings*, **3**, 57–65.

MATHER, A. E., HARVEY, A. M. & STOKES, M. 2000. Quantifying long term catchment changes of alluvial fan systems. *Geological Society of America Bulletin*, **112**, 1825–1833.

MATHER, A. E., GRIFFITHS, J. S. & STOKES, M. 2002. Anatomy of a Plio/Pleistocene landslip, SE Spain. *Geomorphology*, in press.

MONTENAT, C. & OTT D'Estevou, PH. 1999. The diversity of late Neogene sedimentary basins generated by wrench faulting in the Eastern Betic Cordillera, SE Spain. *Journal of Petroleum Geology*, **22**, 61–80.

ROERING, J. J., KIRCHNER, J. W., SKLAR, L. S. & DIETRICH, W. E. 2001. Hillslope evolution by non-linear creep and landsliding: an experimental study. *Geology*, **29**, 143–146.

RUEGG, G. J. H. 1964. *Geologische onderzoekingen in bet bekken von Sorbas, SE Spanje*. Internal Report, Geological Institute, Amsterdam.

SANZ DE GALDEANO, C. & VERA, J. A. 1992. Stratigraphic record and palaeogeographical context of the Neogene basins in the Betic Cordillera, Spain. *Basin Research*, **4**, 21–36.

SHEPHERD, R. G. 1982. River channel and sediment responses to bedrock lithology and stream capture, Sandy Creek drainage, Central Texas. *In*: RHODES, D. D. & WILLIAMS, G. P. (eds) *Adjustment of the Fluvial System*. Allen & Unwin, London, 255–275.

SILVA, P. G., GOY, J. L., SOMOZA, L., ZAZO, C. & BARDAJÍ, T. 1993. Landscape response to strike-slip faulting linked to collisional settings: Pleistocene tectonics and basin formation in the Eastern Betics, southeastern Spain. *Tectonophysics*, **224**, 289–303.

SMALL, R. J. 1978. *The Study of Landforms*, 2nd Edition. Cambridge University Press, Cambridge.

VÖLK, H. R. & RONDEEL, H. E. 1964. Zur glied erung des jungtertiars im becken von Vera, Sudostspanien. *Geologie en Mijnbouw*, **43**, 310–315.

WEIJERMARS, R. 1991. Geology and tectonics of the Betic zone, SE Spain. *Earth Science Reviews*, **31**, 153–236.

Tectonic control on changes in sediment supply: Quaternary alluvial systems, Körös sub-basin, SE Hungary

EDIT THAMÓ-BOZSÓ, ZSOLT KERCSMÁR & ANNAMÁRIA NÁDOR
Geological Institute of Hungary, Stefánia út 14, H-1143 Budapest, Hungary
(e-mail:bozso@mafi.hu)

Abstract: The Pannonian Basin of Hungary is Europe's largest inter-mountain basin, where an evolution in drainage development patterns during the Quaternary was caused by changes in sediment flux to the basin, the dynamics of basin morphology development and the uplift history of the Apuseni Mountains source area, all directly or indirectly related to the tectonic systems operating in the region. Micro-mineralogical data of detrital heavy minerals from modern rivers and two key boreholes covering a time span from the present back to 2.6 Ma have been grouped by statistical analysis into two main clusters and some sub-clusters. The samples within the same cluster have a similar composition, and originated from the same source area. Based on the similar palaeogeographical setting of the potential source areas during the Quaternary, it has been possible to extrapolate the present transport directions of the rivers with a well-known catchment area geology and the heavy mineral composition to Pleistocene borehole data. The changes in transport directions were clearly sharp and related to significant changes in the uplift history of the Apuseni Mountains catchment area. During the Pliocene and Early Pleistocene the compressional stress field operating in the East Carpathians region resulted in the thrust-driven uplift of the Apuseni Mountains and the formation of a syn-sedimentary trap at the western margin of the source area, which captured the sediments of short, transverse rivers. During this period the drainage of the study area was characterized by axial drainage parallel to this trap, and sediments were transported from the northeast, also shown by micro-mineralogical data of detrital heavy minerals. At about 1.95 Ma this trap stopped being active due to the release of the compressional stress field and the isostatic uplift of the Apuseni Mountains, together with the adjoining basin area. This resulted in the filling of the thrust-bounded trap, and the development of transverse drainage, which was shown by characteristic SE transport directions, based on micro-mineralogical data in the study area, and also by an increase of sediment flux. The axial capturing shifted further to the west, developing a configuration approaching to the present river course pattern. By analogy with the Himalayan foreland and retro-arc basins of the Cordilleran–Andean ranges, it was demonstrated that tectonic activity exerts a strong control on the drainage pattern through its influence on variations in sediment supply.

Variations in alluvial architecture and geomorphology of the rivers are determined by a complex interplay between climate, tectonics and base-level changes. These factors are usually interdependent, commonly involve feedback mechanisms and can rarely be unravelled (Jones *et al.* 1999; Vincent 1999). Most of the large alluvial plains in Europe are coastal plains, where base-level control through eustatic changes is emphasized when discussing fluvial sedimentology (e.g. Po Plain – Amorosi *et al.* 1999, River Meuse – Van den Berg 1994, southern Finland – Mansikkaniemi 1991). On the other hand, in continental interior settings eustasy is not relevant, and the major allogenic controls in determining fluvial architecture are basin subsidence, source area uplift and climate changes (Shanley & McCabe 1994). Climate change is an obvious factor that controls sediment supply to basins and may take the form of an increase or decrease in sediment input. Variations in precipitation directly affect discharges, and vegetation cover has an important role in trapping available sediments, while weathering processes control grain size and mineralogy. Climate change, especially during the Pleistocene has often been cited to explain changes in river aggradation and incision (Vandenberghe 1993, 1995; Blum 1994; Jones *et al.* 1999; Nádor *et al.* 2000). Limited vegetation favoured catchment erosion during glacial periods and, as a consequence, increased sediment supply to the rivers resulting in aggradation. More temperate and humid interglacials lead to extended vegetation, a reduction in sediment supply and a vertical incision of the rivers,

From: JONES, S. J. & FROSTICK, L. E. (eds) 2002. *Sediment Flux to Basins: Causes, Controls and Consequences.* Geological Society, London, Special Publications, **191**, 37–53. 0305-8719/02/$15.00

also as a consequence of the increased discharge. However, based on the study of NW European Pleniglacial to Late Glacial fluvial systems, Vandenberghe (1995) showed that a direct and linear response between climate and fluvial development does not exist. Incision takes place at climate transitions (from cold to warm, or from warm to cold), when the river system becomes unstable, compared to its environment due to climatic change. The effects of such delays, mostly caused by vegetation retreat or development, play a crucial role in the origin of these instability phases. During stable climate conditions, whether it is a glacial or interglacial period, the rivers are characterized by approximately balanced lateral sediment accretion and erosion.

In tectonically active areas, fluctuations in sediment supply are often linked to phases of episodic tectonic activity in the deforming catchment area (Seeber & Gornitz 1983; Burbank 1992; Hovius 1996; Vincent 1999). Since fluvial processes are responsible for the transport of most erosional material out of the mountain range, any changes in the drainage pattern are important for understanding not only the erosion of the orogens, but also of the infill of the adjacent sedimentary basins. Drainage network development is governed by a number of factors, the most important of which are slope, water discharge, lithology and rate of uplift (Seidl & Dietrich 1992; Howard *et al.* 1994). An additional control on sediment flux to a basin is drainage diversion in response to folding or thrusting of the catchment area, which can produce major shifts in the location and magnitude of sediment source.

The main aim of our research is the study of the long-term dynamics of the Quaternary fluvial systems in the southeastern part of the intermountain Pannonian Basin and their response to allogenic controls. This paper discusses the changes in catchment areas, river transport directions based on micro-mineralogical data and sediment supply to the basin. The conclusions drawn from this approach are used as supporting evidence for a model of the large-scale tectono-evolutionary history of the region in terms of

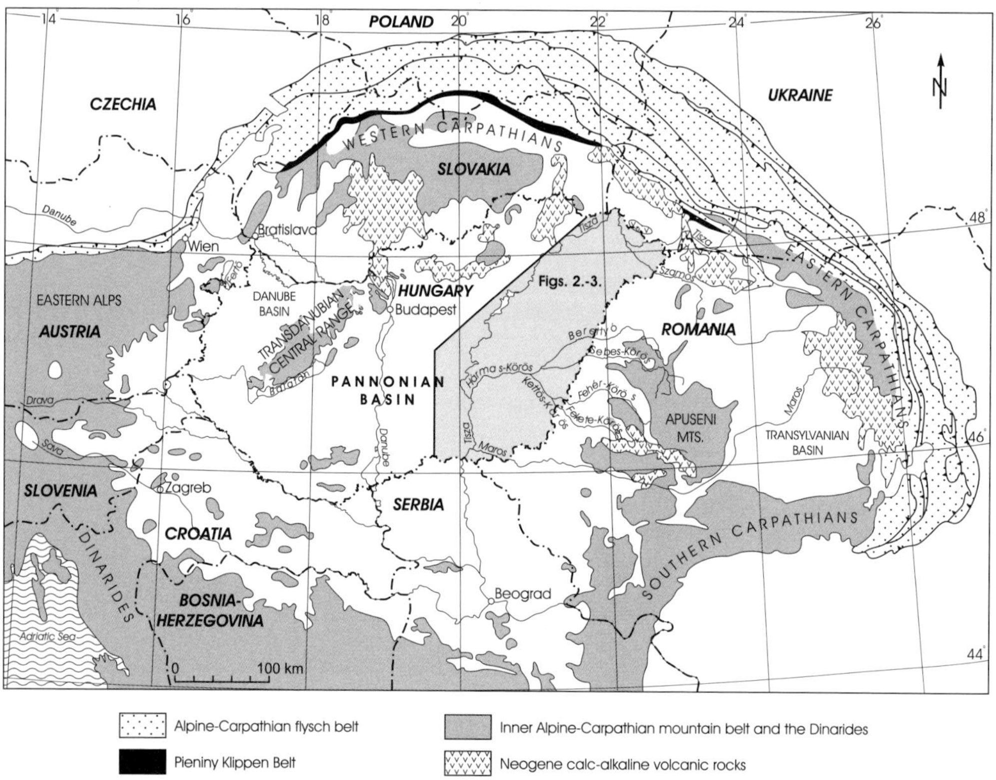

Fig. 1. Geological sketch of the Carpathian–Pannonian area. For a detailed map of the study area on the southeastern part of the Pannonian Basin see Figs. 2 and 3.

source area uplift and basin subsidence. A cyclostratigraphic model emphasizing Milankovitch-scale climate control on the Quaternary fluvial sequence of the area has been presented previously (Nádor *et al.* 2000) and is briefly summarized in this paper for the fuller evaluation of allogenic controls on sediment supply to the southeastern part of the Pannonian Basin during the Pleistocene.

Study area

The Pannonian Basin, located in Central Europe and surrounded by the Alps, Carpathians and Dinarides, is Europe's largest inter-mountain basin (Fig. 1). The Danube River, which is the major outflow is connected to the Black Sea only through a narrow opening in Romania.

The study area, the Körös sub-basin, is situated in the southeastern part of the Pannonian Basin, and currently the covers about 1600 km^2, although it cannot be clearly separated on the basis of surface topography. Its average height is 84–85 m above sea level, thus representing one of the lowest parts of the Pannonian Basin. The main rivers of the area are the Berettyó, the Sebes-, Fekete- and Fehér-Körös rivers, displaying a transverse pattern, orientated approximately E–W, draining the Apuseni Mountains and transporting sediment to the Pannonian Basin to the west. After the confluence of the Fekete- and Fehér- Körös rivers, they are called Kettös-Körös. The Berettyó and Sebes-Körös rivers to the north join the Kettös-Körös at the western part of the Körös sub-basin in a fan shape, and continue as the Hármas-Körös to their confluence with the low-gradient, southerly directed, axial Tisza river in the central part of the Pannonian Basin (Fig. 2). The lengths of the four transverse rivers are quite similar (Table 1), however there are differences in the size, geology and morphology of their catchment areas in the Apuseni Mountains in Romania, resulting in characteristic differences in the stream gradients (Fig. 2). Generally the large (>20 cm km^{-1}) stream gradient of the rivers coming from the mountain belt drops suddenly at the basin margin, close to the

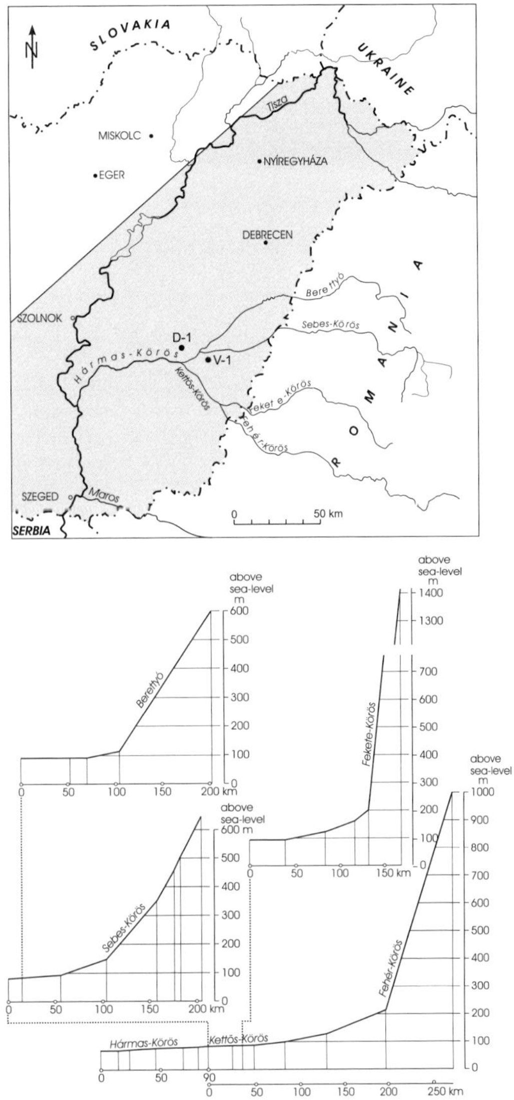

Fig. 2. Present river course pattern in the southeastern part of the Pannonian Basin with the location of the studied boreholes (D-1, V-1) and stream gradients.

Table 1. *Hydrological characteristics of the rivers of the Körös sub-basin (also see Fig. 2)*

Name of the river	Length (km)	Catchment area (km^2)	Highest point of the catchment area (m)	Average discharge at the border (m^3 s^{-1})	Average discharge at the Körös sub-basin (m^3 s^{-1})
Berettyó	203.8	6095	915	11	16
Sebes Körös	209.3	2788	1838	20	38
Fekete Körös	167.6	4729	1849	40	60
Fehér Körös	235.5	4275	1541	33	50

border of Hungary, and the rivers continue with a 1–8 cm km^{-1} gradient towards the axial Tisza river. In the studied Hungarian parts, the rivers now mostly carry suspended load and display no discernible systematic variations in alluvial architecture.

Regional geology

The formation of the Pannonian Basin started in the Early–Middle Miocene by back-arc style rifting, coeval with the late stages of thrusting of the Carpathian belt. Following the Middle Miocene rifting, characterized by two independent extensional phases, a post-rift thermal subsidence occurred during the Late Miocene–Pliocene (Horváth 1993; Horváth & Cloetingh 1996; Csontos & Nagymarosy 1999). The subsided basin was covered by the Pannonian lake, which was in contact with the former Paratethys up to the end of the Badenian. After the complete isolation of the Pannonian lake from the marine environment, it was filled up by prograded delta systems coming from the northwest and northeast (Bérczi & Phillips 1985; Pogácsás *et al.* 1988). Therefore, the Upper Miocene–Pliocene sequence represents a time-transgressive facies change from offshore basin sediments through basin slope and delta slope to delta front and delta plain sediments, passing up into the alluvial facies, which represent the latest stage of basin fill (Juhász 1992, 1994).

The latest phase of the multistorey development of the Pannonian Basin comprises a still active basin inversion, characterized by NW–SE and N–S compression, which resulted in significant uplift of the marginal parts and local subsidence of the basin centre during the Quaternary (Csontos *et al.* 1992; Horváth & Cloetingh 1996; Sanders 1998; Csontos & Nagymarosy 1999; Fodor *et al.* 1999; Horváth & Tari 1999; Müller *et al.* 1999; Van Balen *et al.* 1999). As a result of this varied morphology, the main rivers transported sediments from the north-west, north, northeast and east mountain regions towards the central part of the Pannonian Basin. The uninterrupted subsidence of the Körös sub-basin, which has been investigated here, was one of the largest subsiding areas, represented by a 400–500-m thick continuous Pleistocene fluvial record (Fig. 3) (Cooke *et al.* 1979; Rónai 1985; Jámbor *et al.* 1993; Nádor *et al.* 2000). The Körös sub-basin is bounded on the north and south by ENE–WSW transtensional faults (Csontos & Nagymarosy 1999; Fodor *et al.* 1999), and by a NNE–SSW fault on the east. Pleistocene fluvial sequences are best represented by two keybore-holes (Dévaványa D-1 and Vésztö V-1), the focus of this study, where palaeomagnetic measurements (Cooke *et al.* 1979; Rónai 1985) provide a precise chronostratigraphic framework.

The fluvial deposits of the Körös sub-basin are represented by fine-grained sediments, which are predominantly silt and clay with minor fine-grained sands, deposited from rivers coming from the Apuseni Mountains in Romania and the Carpathians (Rónai 1985). Quaternary transport directions and source areas of the Pleistocene fluvial sediments have been mostly modelled on the basis of the mineralogical composition of sands and silts (Molnár 1964, 1965*a*, *b*, 1966*a*, *b*, 1980; Gedeonné Rajetzky 1973, 1976*a*, *b*; Elek 1979, 1980), and on the basis of the position of alluvial fans (Borsy 1989, 1992). Previous large-scale mineralogical studies supposed that the Körös and Berettyó rivers, and their precursors, deposited their sediment load over the area of the Körös sub-basin (Elek 1980; Molnár 1980), changing their flow pattern over the area through time and space, as also did the precursor of the Tisza river (Elek 1980; Borsy 1989, 1992). However, the variations in transport directions have not been reconstructed in detail. The relationship between the uplift history of the hinterland area and its influence on the river course pattern had also not been studied.

Study methods

The Körös sub-basin area represents the downstream, lowland part of the Körös and Berettyó rivers. The Pleistocene fluvial strata under discussion do not crop out in the study area, but are known only from borehole data and well logs. Palaeocurrent direction indicators (planar cross-bedding, clast imbrications, etc.) could therefore not be evaluated.

Changes in catchment areas and transport directions to the basin have been studied on the basis of micro-mineralogical data. To be able to reconstruct more precisely the variations in transport directions, changes in the detrital heavy mineral composition of 123 samples from two key boreholes (Dévaványa D-1, Vésztő V-1) have been re-evaluated in detail, and these data have been compared with the composition of the present fluvial sediments of the main rivers in the Hungarian Plain. The mineralogical data of the studied boreholes comes from earlier examinations of the 0.1–0.2-mm fraction of sands (Elek 1980; Molnár 1980), determined by petrographic micro-scope. The heavy minerals have been quantified by point-counting. The mineralogical composition of the modern river sediments has been determined by P. Szabó (1955), Molnár (1964), D. Szabó (1967) and Gyuricza (in Kuti *et al.* 1987; Molnár *et al.* 1989, 1990).

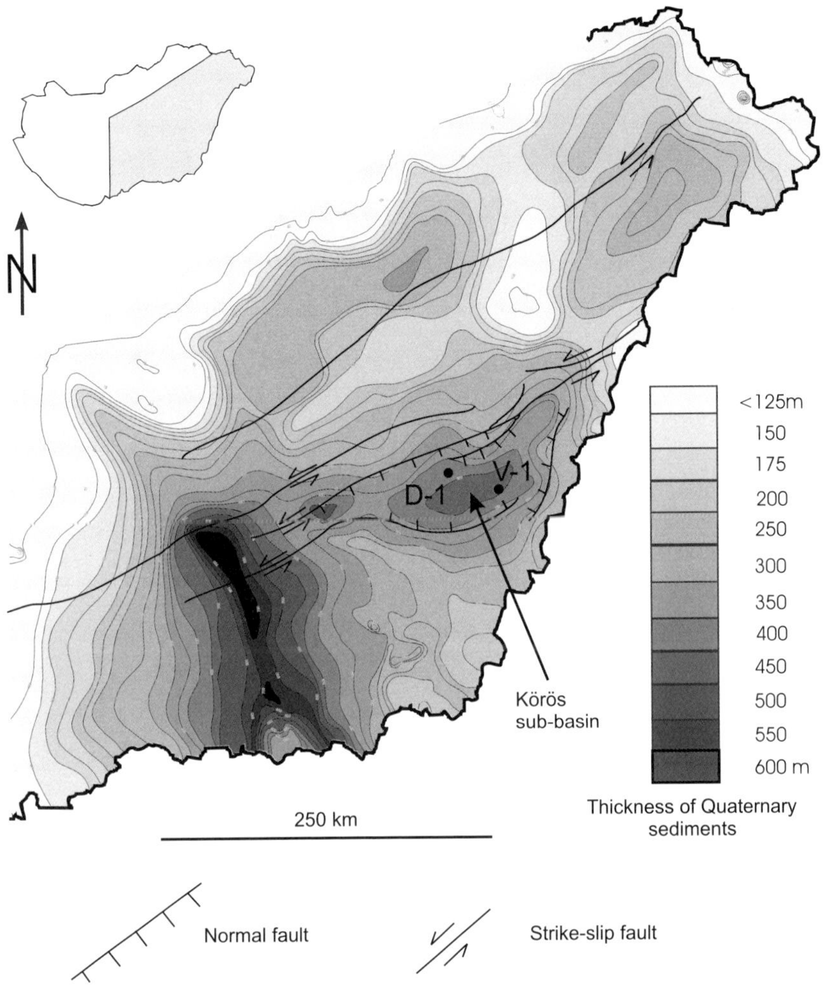

Fig. 3. Thickness of Quaternary sediments in the southeastern part of the Pannonian Basin, with the location of the studied boreholes (D-1, V-1) and the most important tectonic lines bounding the Körös sub-basin.

To be able to reconstruct transport directions of the palaeo-rivers, the detrital heavy minerals of the source rocks were initially considered, and the authigenic, diagenetic and secondary minerals were neglected. The frequencies of 22 detrital heavy mineral components of the sand samples from Pleistocene borehole data and modern river sediments have been compared using cluster analysis.

Cluster analysis is a standard technique of numerical classification that arranges samples into more or less homogeneous groups based on similarities among multiple variables. In this case, after several attempts of different methods of cluster analysis, a hierarchical classification was applied that clusters large numbers of samples by the weighted pair-groups method with arithmetic average, using Euclidian distance (Davis 1986). This method of cluster analysis was also used by Ibbeken & Schleyer (1991) to trace provenances of recent fluvial sediments in southern Italy, based on the light and heavy mineral composition and other sediment properties. In our case each sample, characterized by the frequency of their 22 detrital heavy mineral constituents, represented a point in the 22-dimensional space. The Euclidian distance between the samples in this space defined by 22 variables was used as a measure of similarity. The lower the distance is, the higher the similarity between two samples. This type of cluster analysis was performed using 'Deli-Clus-Dend' in-house software of Ó. Kovács and Kovács (Ó. Kovács 1987). The samples within the same group have a similar detrital heavy mineral composition, their main detrital heavy minerals are the same and, given the geological and geographical constraints of the region, it can be concluded that they originated from the same source area.

Some additional data, such as the distribution of fluvial sediments with contrasting grain size distribution and sedimentation rates in the studied boreholes, calculated on the basis of palaeomagnetic dating and thickness of sedimentary units, were also used as supplementary information in evaluating changes in sediment flux to the basin.

Micro-mineralogical investigations

To be able to reconstruct more accurately the variations in transport directions, the detrital heavy mineral composition of the D-1 and V-1 boreholes (for location see Figs 2 and 3), together with that of the sediments of the present river system, have been compared by cluster analysis. The results showed that the samples group into two major clusters and, within those, some sub-clusters (Fig. 4). Cluster I is characterized by hornblende, orthopyroxene, epidote, garnet and magnetite. Within this group, sub-cluster Ia is characterised mostly by hornblende, orthopyroxene and epidote, a characteristic assemblage for the current sediments of the Tisza river and its tributaries, with the exception of the Berettyó and the Sebes-Körös (Figs 5 and 6). Sub-cluster Ib is characterized by a high garnet content, which is characteristic for the source area of the current Berettyó and Sebes-Körös rivers (Figs 5 and 6). Cluster II is characterized by a high chlorite content, which is not known from the present river sediments, but it is considered to have a north-eastern origin since characteristic chlorite-rich Pliocene sediments of this area have been transported from the metamorphic rocks of the NE Carpathians (Molnár 1966*b*).

The two studied boreholes, representing the continuously cored Pleistocene sedimentary sequence of the Körös sub-basin with a thickness of 416 and 482 m respectively, are not identical in terms of their heavy mineral composition, as can be seen by their average values (Table 2). Of the main heavy minerals, chlorite, magnetite, garnet and pyroxene are more common in the D-1 borehole, while the V-1 borehole is typically characterized by amphibole and epidote. The characteristic detrital heavy mineral composition of the clusters and, accordingly, their distribution in the borehole logs is shown in Fig. 7.

In the two boreholes studied here, sediments with a similar detrital heavy mineral composition (chlorite, garnet) and similar age are known from the lower parts of the boreholes (Fig. 7). At about 1.95 Ma (known from the palaeomagnetic dating of the borehole sections: Cooke *et al.* 1979; Rónai 1985) there is a clearly defined change in the detrital heavy mineral composition of the two boreholes, characterized by the appearance of epidote–hornblende minerals (Fig. 7), indicating a change in provenance areas. The cluster analysis of the borehole sections showed that these sediments reached the area cut by the D-1 borehole only for a limited period, then most of the upper part is characterized by chlorite, pyroxene or garnet-rich sediments. Sands with different detrital heavy mineral contents vary frequently in the middle and upper part of the V-1 borehole, however hornblende and epidote remain dominant.

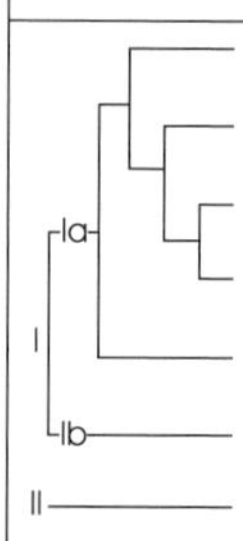

Cluster groups	most frequent detrial heavy minerals of the cluster-groups	sediments of modern rivers	transport direction from
am-e	hornblende + epidote, garnet, magnetite	–	SE
e-am	epidote, hornblende, magnetite, garnet, chlorite	Fekete-Körös	SE
px	orthopyroxene + clino-pyroxene, hornblende	Tisza, Szamos, Bodrog, Sajó, (Hernád)	NE (N)
am-px	hornblende + pyroxene, epidote	Fekete-, Fehér-, Hármas-Körös, Maros	SE
px-am	orthopyroxene + hornblende, epidote, magnetite	Fehér-Körös	SE
gr	garnet + magnetite, hornblende, pyroxene	Berettyó, Sebes-Körös, Zagyva	NE (NW)
kl	chlorite + magnetite, biotite, hornblende	–	NE?

Fig. 4. Composition of the detrital heavy mineral clusters and the sediments of modern rivers with similar mineralogical composition and their transport directions (for explanation see text).

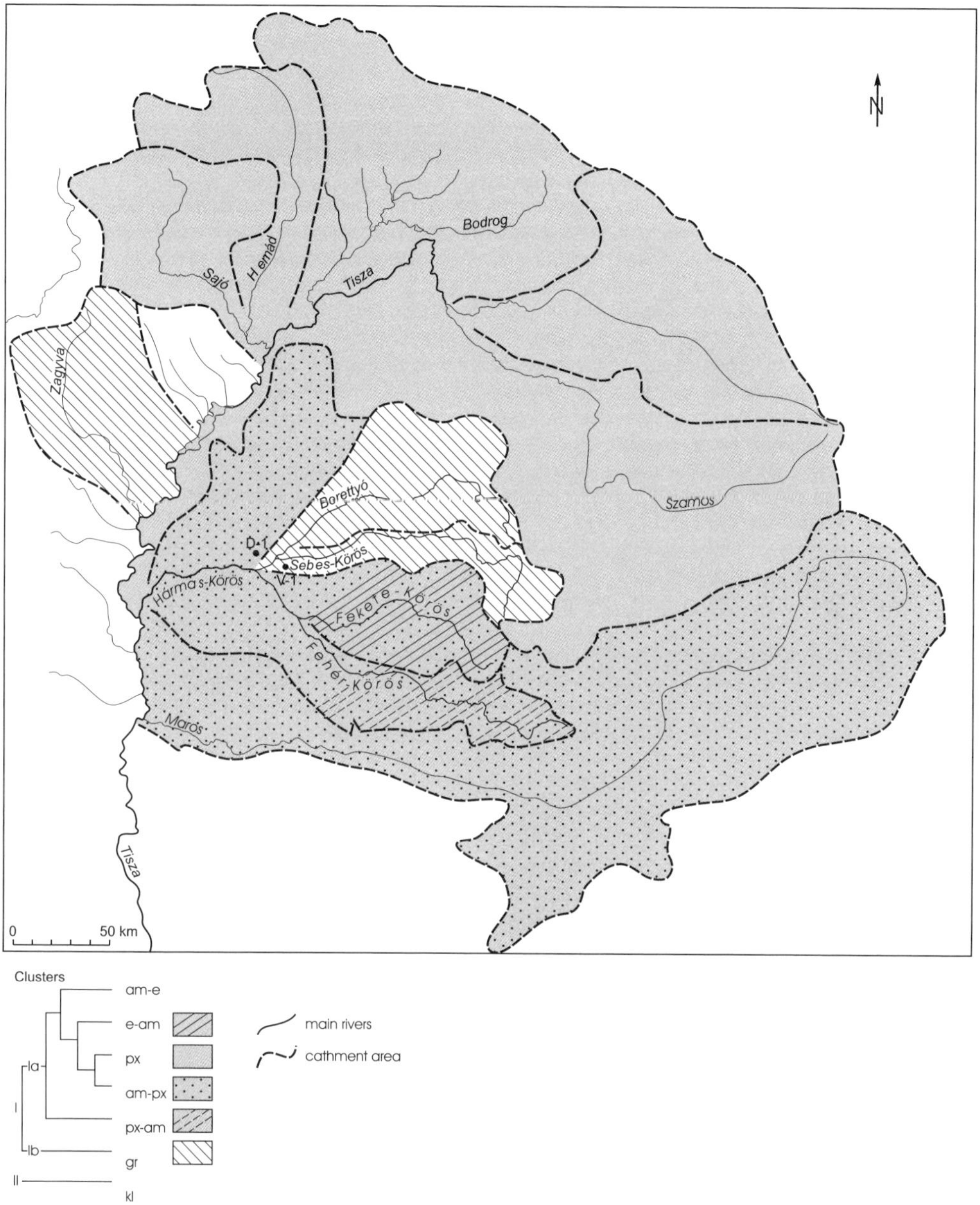

Fig. 5. Characteristic detrital heavy mineral clusters of the catchment area of the modern rivers in the study area, with the location of the studied boreholes (D-1, V-1). For the composition of the clusters see Fig. 4.

Additional sedimentary data

In order to get a more precise picture of the transport directions that are based on micro-mineralogical data, a few other observations were also considered during the interpretation of changes of sediment flux to the studied basin, summarized below.

A sharp contrast in grain size distribution of the alluvial strata has been shown by Sümeghy (1944), who found that coarse-grained fluvial sediments are missing from the area of the Körös sub-basin, and

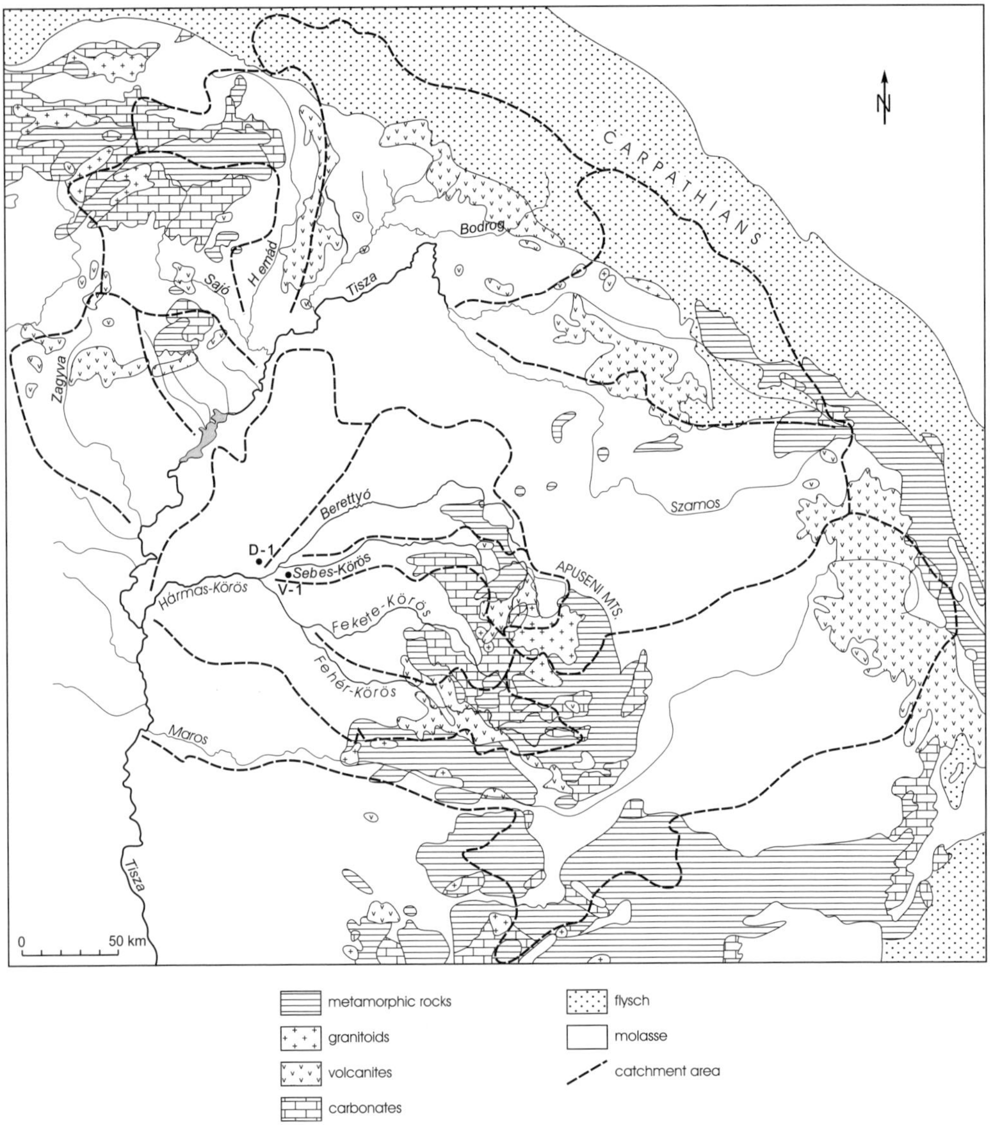

Fig. 6. Lithological sketch of the source area geology of the Tisza river and its tributaries (after Ianovici *et al.* 1976; Brezsnyánszky 1989; Fülöp 1989) with the location of the studied boreholes (D-1, V-1).

are known only along the western margin of the Apuseni Mountains where they form a great thickness at depth, bounded by a NNE–SSW-strike fault. Beyond this fault the continuation of the coarse-grained sediments towards the western basin areas is unknown, where only fine-grained sediments (mostly silt) are found.

An approximate sedimentation rate was calculated on the basis of ages inferred from palaeomagnetic data (Cooke *et al.* 1979; Rónai 1985) of the D-1 and V-1 boreholes, and the thickness of the sedimentary units that deposited at the corresponding time intervals (Fig. 7). As the regional subsidence of the area is considered to be uniform during the Quaternary (Rónai 1985), changes in sedimentation rates are possible indicators for changes in sediment flux to the basin.

Table 2. *Average values of heavy minerals in sand layers in the D-1 and V-1 boreholes. Detrital heavy minerals, which have been evaluated by cluster analysis are with fat letters (also see Fig. 7)*

Heavy minerals	Dévaványa D-1 43 samples (%)		Vésztő V-1 80 samples (%)	
Chlorite	**22.61**		**16.36**	
Oxyhornblende	**5.97**	**11.82**	**2.06**	**22.47**
Hornblende	**4.66**		**19.33**	
Other amphiboles	**1.19**		**1.08**	
Magnetite	**11.73**		**9.19**	
Epidote	**0.74**		**11.81**	
Garnet	**10.38**		**8.69**	
Orthopyroxene	**3.34**	**7.62**	**3.37**	**4.62**
Clino-pyroxene	**4.28**		**1.25**	
Biotite	**4.51**		**4.25**	
Apatite	**1.42**		**0.01**	
Tourmaline	**1.01**		**2.62**	
Rutile	**0.57**		**0.05**	
Clinozoisite	**0.54**		**0.77**	
Kyanite	**0.26**		**0.56**	
Staurolite	**0.18**			
Zoisite	**0.11**		**0.14**	
Topaz	**0.07**			
Sphene	**0.04**		**0.11**	
Zircon	**0.03**		**0.31**	
Olivine			**0.02**	
Piemontite			**0.01**	
Pyrite	3.03		1.76	
Carbonate	1.25			
Altered minerals	22.17		16.24	
Total	100.09		99.99	

Discussion

Changes in transport directions and source areas

If transport directions are to be reconstructed on the basis of the micro-mineralogical composition of the formations, then the potential source rocks have to be identified. This is a difficult task as there are only a few minerals that are uniquely characteristic for certain magmatic or metamorphic rocks. However, in the case of the Körös sub-basin, the situation is simplified by the fact that the Pleistocene palaeogeographical setting of the potential source areas was similar to the current one. Therefore, the present transport directions of the rivers with a well-known catchment area geology and detrital heavy mineral composition (cluster) can be extrapolated to the borehole data, and the transport directions for the Pleistocene sediments have been deduced from the recent ones on the basis of belonging to the same mineralogical cluster.

The most frequent detrital heavy minerals of the Pleistocene sands of the Körös sub-basin might have originated from several different rock types in the potential source areas, but the main source rocks of some minerals can be identified. Garnet and hornblende are characteristic for gneiss, mica schists and amphibolites. Epidote and chlorite might have been eroded from low-grade metamorphic rocks, while pyroxenes are characteristic for volcanic rocks of the source area (Figs 5 and 6).

Based on a similar detrital heavy mineral composition known from the lower parts of the boreholes (Fig. 7), it is concluded that chlorite-rich sediments have been transported by the same river, probably coming from the northeast, from the low-grade metamorphic units of the NE Carpathians. Garnet-rich sediments were deposited by another river, also coming from the northeast, mostly derived from Early Palaeozoic and older metamorphites of the Bihar autochthon of the Apuseni Mountains, as well as from Tertiary and Quaternary sedimentary rocks (molasse) (Fig. 6) similar to the present source area of the Berettyó and Sebes-Körös rivers (Fig. 5). The reconstruction of the river course pattern for the Early Pleistocene (Borsy 1992) also showed that axial rivers were flowing at this time from the northeast to the area of the Körös sub-basin.

The occurrence of epidote and hornblende-rich sediments indicate that SE transport directions, mostly from the sedimentary rocks and from Early Palaeozoic metamorphics of different thrusts of the Apuseni Mountains, as well as from Neogene volcanic rocks of the present source area of the Fekete-Körös river, were important (Figs 5 and 6). The cluster analysis of the borehole sections showed that these southeast-derived sediments reached the area of the distal D-1 borehole for only a short time. Subsequently, most of the upper part of the D-1 borehole is characterized by NE transport directions, inferred from chlorite, garnet or pyroxene-rich sediments (Fig. 7). Pyroxene is characteristic for the present source area of the Tisza and its northern tributaries, mostly derived from the north-northeast, from the Inner Carpathian Neogene volcanites. At the same time in the area of the V-1 borehole, which is closer to the Apuseni Mountains, SE and NE transport directions, inferred from the different clusters, varied in the central part of the borehole, while the SE transport direction became dominant in the upper part (Fig. 7). The orthopyroxene–hornblende-rich sediments indicate SE transport directions, characteristic for the sediments of the present Fehér-Körös river, coming from Tertiary and Quaternary sedimentary rocks (molasse) and Neogene volcanites. Hornblende–epidote-rich sediments, dominant in the upper part of the V-1 borehole, are most similar

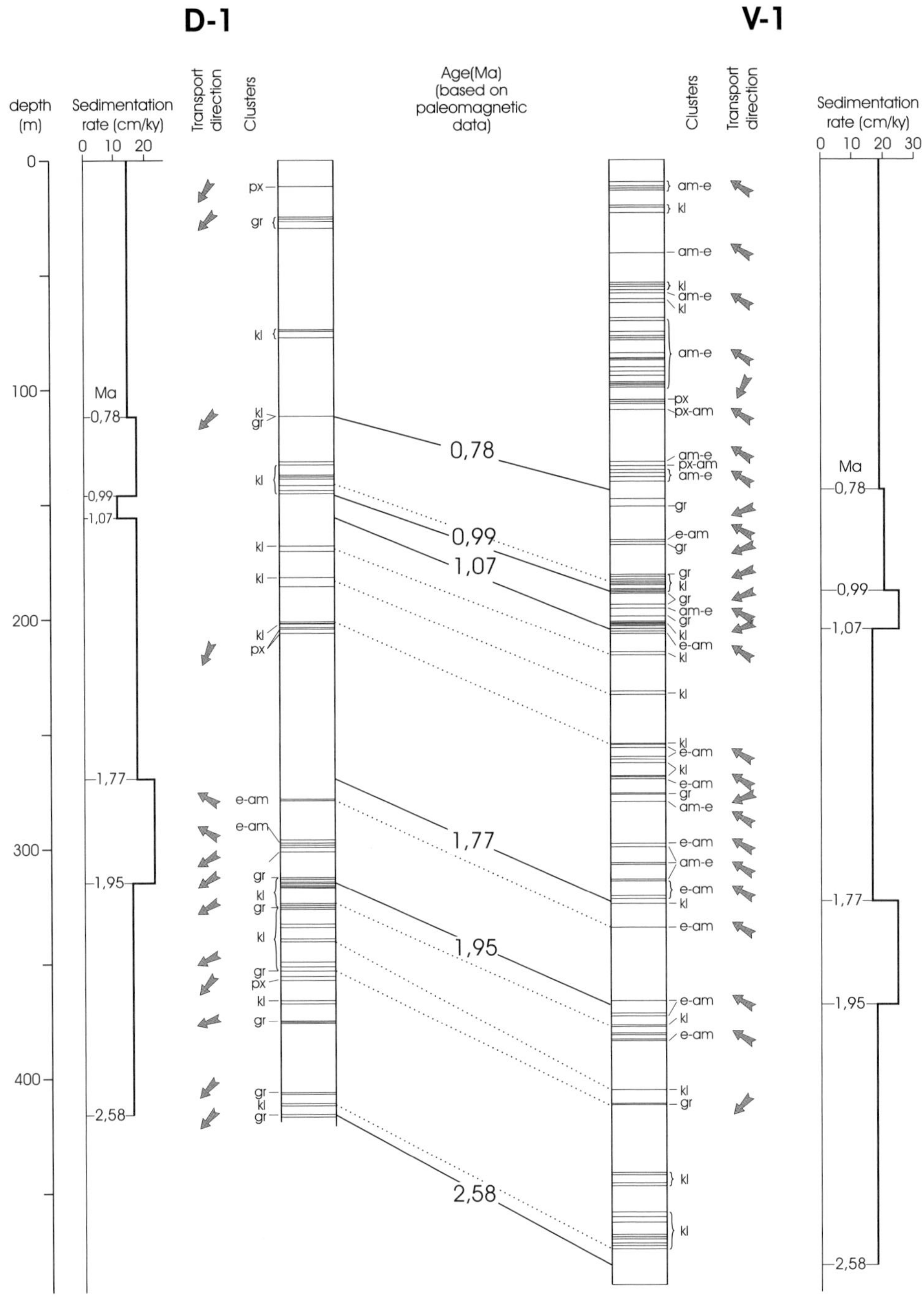

Fig. 7. Characteristic detrital heavy mineral clusters, and inferred transport directions and sedimentation rates, calculated on the basis of palaeomagnetic data in the D-1 and V-1 boreholes.

to the sediments of the Fekete-Körös river source area, and also indicate a SE transport direction (Figs 5 and 6). These last alternations in the mineralogical composition indicate that different rocks were sub-aerially exposed on this south-eastern source area due to tectonic and erosional unroofing.

Base-level and climate changes

The clear and abrupt change in transport directions from NE to SE at about 1.95 Ma ago, recorded in the boreholes (Fig. 7), was inferred from the cluster analysis of detrital heavy minerals. This switch in the provenance reflects shifts in transport directions. This time interval of the transport direction changes is also characterized by an increase in sediment flux, represented by a higher rate of sedimentation in the boreholes (Fig. 7).

Considering the possible causes responsible for the variations in sediment flux to the Körös sub-basin during the Pleistocene, eustasy controlled base-level change is an unlikely allogenic factor. The section of the Körös rivers that is the focus of this study is over many hundred kilometres from the Black Sea and its associated sea-level fluctuations, and therefore basin-scale, base-level changes are unlikely to have influenced variations in sediment flux. Large-scale base-level change (e.g. regional subsidence) would have been common to the system, and would not cause differences in transport directions to the boreholes, as was revealed on the basis of detrital heavy mineral data (Fig. 7).

The Quaternary fluvial systems of the south-eastern part of the Pannonian Basin were affected by the same climate histories. Based on high-resolution magnetic susceptibility measurements, correlated by pollen and gastropod data, as well as with the ODP 677 oxygen isotope proxy, cyclostratigraphical and palaeoclimate analyses showed that the Pleistocene fluvial sequences of the D-1 and V-1 boreholes represent fifth-order Milankovitch cycles. The cyclicity is 100 ka for the upper 1 Ma old section of the boreholes, while it is 40 ka for the older parts of the boreholes, which is also in good agreement with the global palaeoclimate records (Nádor *et al.* 2000). There was no difference in the climatic cycles between the two boreholes, which means that Milankovitch-scale climate changes exerted a regional influence on the river systems of the area. Intrinsic evolution within the system and local thresholds, which might have modified the global climate signal and could have influenced sediment flux, operate at scales of hundreds to thousands of years (Vandenberghe 1995), whose resolution was not obtained in the D-1 and V-1 boreholes.

Tectono-morphological model

As was discussed above, the Pleistocene fluvial deposits share the same regional subsidence and climate histories. Therefore, it is necessary to turn to other regional changes to explain the observed variations in sediment flux, particularly to differential rates of tectonic uplift and rates of denudation of the potential source areas. Such major shifts in transport directions, inferred from the cluster analysis of detrital heavy mineral data, are important indicators of tectonic control through its influence in topography. Plate tectonics play a predominant role in dividing land masses into sediment source regions, transport routes and sedimentary basins, and thus is primarily responsible for the accumulation of alluvial strata. The uplift style of the hinterland area strongly controls the associated basin morphology and drainage pattern (Burbank 1992; Jordan 1995; Hovius 1996).

To be able to understand the basin evolution and related change in drainage pattern in the study area throughout the Late Neogene and Quaternary, it is important to emphasis that the back-arc style, extensional formation of the Pannonian Basin occurred during the Middle Miocene and terminated in the Pliocene when it was infilled by prograding delta systems. It means that by the beginning of the period under investigation (Late Pliocene and Pleistocene), the basin was filled up and the Quaternary tectonics were superimposed on the earlier extensional back-arc tectonic style.

To reconstruct the changes in sediment flux to the Körös sub-basin throughout the Pleistocene, it is important to focus on the uplift history of the Apuseni Mountains within the geotectonic regime of the East Carpathians, as this was the key source area of the Körös fluvial sediments. The uplift history of the Apuseni Mountains was strongly controlled by the evolution of the subduction zone along the East Carpathians (Sanders 1998). Based on fission-track data, Sanders (1998) determined that the large-scale uplift of the Apuseni Mountains started in the Late Badenian–Early Sarmatian due to continent–continent collision along this subduction zone. East–west compression resulted in thrust-bounded uplift (Fig. 8a, constructive phase of Sanders 1988). Fission-track data (Sanders 1998) shows that the large-scale uplift of the Apuseni Mountains was waning throughout the Pleistocene, which has been related to the cessation of subduction at the East Carpathians and the release of the compressional stress field (Fig. 9a, destructive phase of Sanders 1998). However, these phases represent a continuous tectonic evolution, and no sharp boundaries in the processes can be defined.

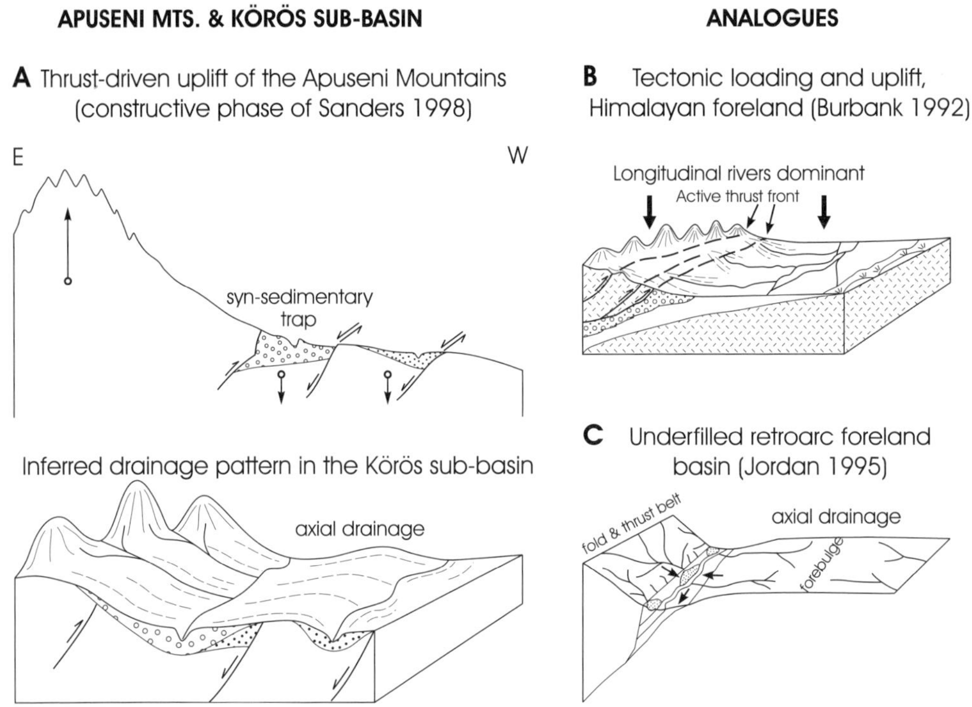

Fig. 8. Geotectonic model for the thrust-driven uplift of the Apuseni Mountains. A thrust-bounded syn-sedimentary trap captured the sediments of the transverse rivers and favoured the development of axial drainage.

The next step in reconstructing variations in transport directions and sediment flux to the basin is to envisage the palaeo-drainage pattern during the different tectono-evolutionary periods, described above. In a tectonically active area, the drainage pattern is predetermined by tectonics. River courses are often fault controlled, and differential subsidence due to faulting may control local base levels and flow directions.

The restricted occurrence of Lower Pleistocene coarse-grained fluvial sediments in a trough along the western margin of the Apuseni Mountains has been known for a long time (Sümeghy 1944). To explain this limited occurrence towards the basin area a sedimentary trap is envisaged, parallel to the strike of the mountain front, which captured the proximal coarse-grained sediments of the transverse rivers, draining the rapidly uplifting Apuseni Mountains. Such short transverse rivers are characteristic for actively uplifting mountain belts (Hovius 1996). The NNE–SSW-striking fault at the margin of the Apuseni Mountains, most probably functioning as a thrust fault, due to E–W compression during the Pliocene, might have been responsible for creating this sedimentary trap (Fig. 8a). Owing to this sedimentary trap at the western margin of the Apuseni Mountains, the sediments of the transverse rivers draining the Apuseni Mountains did not reach the area of the Körös sub-basin. Partly based on these tectonic considerations, and partly on the basis of the reconstruction of the river course pattern for the Early Pleistocene (Borsy 1989, 1992), it is concluded that during the Pliocene and Early Pleistocene axial rivers were flowing from the northeast along re-activated NE–SW-striking tectonic lines parallel to the mountain chain in the area of the Körös sub-basin (Fig. 8a), depositing their fluvial sediments. The transport directions, inferred from mineralogical composition, coincided with this predicted river course pattern, that is sediments have been deposited by rivers flowing from the northeast, carrying sediments from the source area of the present Berettyó and Sebes-Körös rivers to the northeast.

The present arrangement of the Körös rivers coming from the Apuseni Mountains shows a different drainage distribution. The cone-shaped alluvial fans extend as far as 50 km into the basin and indicate a well-developed transverse river course pattern with only a very distal axial capturing by the Tisza river (Fig. 2), quite different

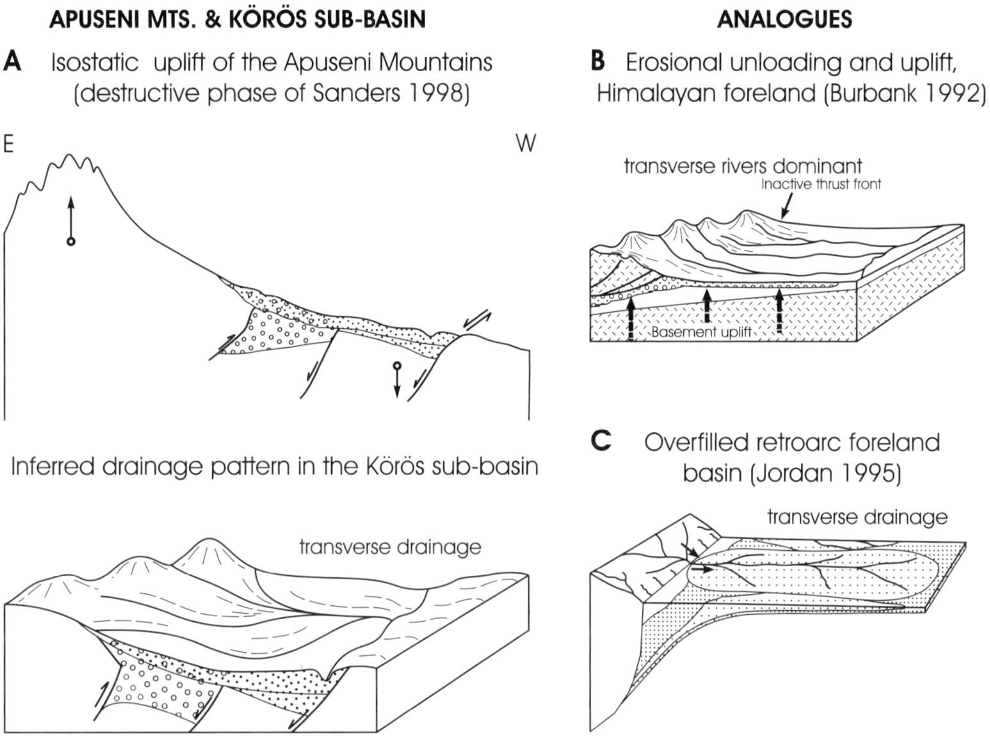

Fig. 9. Geotectonic model for the isostatic uplift of the Apuseni Mountains. The filling of the syn-sedimentary trap allowed the development of extended transverse rivers, while axial capturing shifted further to the west due also to the general uplift of the area.

from the Lower Pleistocene arrangement. This drainage reorganization could have been caused only by some significant change in allogenic controls, such as catchment area uplift linked to renewed sediment flux, basin subsidence or some major climate change, influencing sediment supply to the basin. Regional basin subsidence and climate changes have already been excluded as relevant factors for variations in sediment flux, therefore the changes in tectonic evolution of the area must be considered the dominant factor.

The development of a transverse river, extending far into the basin, presumes that the former sedimentary trap capturing these rivers at the western margin of the Apuseni Mountains was no longer functioning. Based on the changes in the tectonic regime of the studied region (Sanders 1998), it is concluded that the marginal fault that bounded the syn-sedimentary trap, characteristic for the first phase of uplift of the Apuseni Mountains, was not active by the Pleistocene, which contributed to the filling up of this tectonic trough (Fig. 9a). The disappearance of this trough was the main cause of drainage pattern changes, because it allowed the formation of well-developed transverse rivers, carrying sediments from the southeast from the Apuseni Mountains. This is supported by the occurrence of a SE transport direction in the borehole sections at about 1.95 Ma ago (Fig. 7). These SE-derived sediments reached the area of the distal D-1 borehole only for a short period, then most of the upper part is characterized again by NE transport directions, characteristic for the axial rivers. However, after an alternation with NE directions, the SE transport direction became dominant in the area of the V-1 borehole (Fig. 7), closer to the Apuseni Mountains, showing an active sediment supply from the southeastern source area towards the proximal parts of the Körös sub-basin.

Individual basins rarely follow the ideal behaviour of basin models, especially a basin with such a complicated and multiphase development like the Pannonian Basin. However, some analogues for basin evolution with similar stress fields help to unravel links of drainage development and large-scale tectonic evolution of the region, and might support further evidence for the tectono-morphological model described above. The following basin evolution models, as possible analogues, include the Himalayan foreland basin (Burbank 1992) and the retro-arc foreland basin model of Jordan (1995) of the Cordilleran–Andean ranges. Both models have genetic similarities in their stress field evolution history with the southeastern part of the Pannonian Basin and its hinterland area. However, the dimensions of the Himalayan foreland, or Cordilleran–Andean retro-arc foreland basins, are obviously different and plate tectonic settings cannot be fully compared. Despite these differences in reorganization of drainage pattern resulting from tectonic uplift, and related variations in sediment supply to the adjacent basins, these analogue areas show similarities with the Körös sub-basin.

Based on the study of Plio-Pleistocene foreland in the Himalayan massif, Burbank (1992) described the infilling of the Ganges Basin based on sediment infill and amount of sediment supplied to the basin, which controlled drainage patterns. Burbank (1992) identified two different scenarios in the tectonic evolution of the area:

- when the uplift of the hinterland is controlled by tectonic loading, then the associated basin morphology will be narrow and asymmetric. The major sediment accumulation is proximal to the mountain area, and the river course pattern is characterized by relatively short transverse streams, perpendicular to the strike of the mountain, depositing smaller alluvial fans, and by axial main river(s), parallel to the mountain front (Fig. 8b);
- when tectonic loading is replaced by erosional unloading and uplift, the formerly strongly subsiding basin also uplifts together with the isostatically uplifting hinterland area, and the area for fluvial sedimentation increases. This setting is characterized by the dominance of transverse rivers with large alluvial fans. Axial rivers are subordinate in this case (Fig. 9b).

The characteristic axial and transverse drainage patterns of the Himalayas are comparable with the under-filled and over-filled states of retro-arc foreland basin development (Jordan 1995), however plate tectonic settings are different. Retro-arc foreland basins form along the continental interior flanks of continental-margin orogenic belts due to the compression and tectonic loading resulting from the subduction (Jordan 1995). If sediment does not completely fill the tectonically subsiding basin, then a topographic trough develops between the thrust belt and the forebulge. This valley has a drainage parallel to the thrust front (under-filled basin), and the basin may receive sediment along strike, through axial drainage (Fig. 8c). This drainage scenario is comparable to the axial main river course pattern of the tectonic loading uplift of the Himalayas (Burbank 1992), as well as with the thrust-bounded syn-sedimentary trap and resulted axial drainage of the Körös sub-basin in Hungary.

When sediment supply exceeds subsidence, proximal alluvial fan facies propagate and the retro-arc foreland basin becomes 'over-filled', resulting in a sedimentary plain extending across the forebulge. In this case the sedimentary basin is a lowland region, and not restricted to an axial valley (Fig. 9c, Jordan 1995). This 'over-filled' scenario is comparable to the erosional unloading and uplift of the Himalayas and its related change in drainage pattern (Burbank 1992), that is the dominance of transverse rivers with extended alluvial fans propagating into the basin, as well as with the Körös sub-basin where the filling up of the syn-sedimentary trap allowed the formation of transverse drainage.

Both the Himalayan foreland and the retro-arc foreland basin models show that tectonic uplift has an important role in controlling drainage patterns through changes in amount and style of sediment supply to basins. By analogy with these models it is proposed that variations in sediment supply and changes in source areas, as well as in drainage pattern, are responsible for the alterations in transport directions inferred from detrital heavy minerals observed in the boreholes. These are likely to be caused by the changes in the style of tectonic uplift of the Apuseni Mountains and stress field, which exerted a strong control on the formation and subsequent cessation of a syn-sedimentary trap along the western margin of the catchment area.

Conclusions

Changes in transport directions in the Pleistocene fluvial sediments on the southeastern part of the Pannonian Basin have been studied by cluster analysis of detrital heavy minerals from modern rivers and two key boreholes covering the whole Pleistocene period. Samples within the same cluster have a similar composition, and they originated from the same source area. Based on a similar palaeogeographical setting of the potential source areas during the Quaternary, the present transport directions of the rivers with a well-known catchment area geology and heavy mineral composition have been extrapolated to the Pleistocene borehole data. The results of cluster analysis showed that the samples group into two major clusters and, within those, some sub-clusters. Cluster I is characterized by hornblende, orthopyroxene, epidote, garnet and magnetite. Within this group, sub-cluster Ia is characterized mostly by hornblende, orthopyroxene and epidote, indicating a SE transport direction, however the dominance of pyroxene indicates NE transport direction. Sub-cluster Ib is characterized by a high garnet content, referring to a NE transport direction. Cluster II is characterized by a high chlorite content, which is not known from the present river sediments but it is supposed to have a northeastern origin. In the D-1 and V-1 boreholes there is a clearly defined change in the detrital heavy mineral composition at about 1.95 Ma characterized by the occurrence of epidote and hornblende minerals indicating a switch in provenance areas and a change in transport direction from a NE to a SE one. This time coincides with an increased sediment flux to the basin, shown by the higher rate of sedimentation.

Owing to the inter-mountain basin setting, the Pleistocene fluvial sediments under discussion share the same regional subsidence, where base-level changes were unaffected by eustatic changes. The Pleistocene climatic signal is preserved in the sediment record as fifth-order Milankovitch cycles, identical in the two boreholes, therefore climate can also be considered as a regional influence on the river systems of the area. Variations in transport direction, determined on the basis of detrital micro-mineral composition as revealed by cluster analysis, were caused by changes in sediment supply, source areas and drainage pattern reorganization. These changes have been shown to be comparable to transport directions predicted on the basis of a theoretical tectono-morphological model, based on sedimentological observations and tectonic data, as well as on analogues for basin evolution with similar stress fields.

The tectonic model implies two phases of uplift of the Apuseni Mountains source area during the Late Neogene and Quaternary, which was strongly controlled by the evolution of the subduction zone along the East Carpathians. During the Pliocene and Early Pleistocene, due to continent–continent collision, a compressional stress field was operating in the East Carpathians region that resulted in thrust-driven uplift of the Apuseni Mountains and the formation of a syn-sedimentary trap at the western margin of the mountain chain. For this phase transverse drainage is envisaged, characteristic for actively uplifting orogens, whose sediments have been captured in the thrust fault bounded syn-sedimentary trap, parallel to the mountain front. In addition to capturing the sediments of the transverse rivers, this trap favoured the development of axial drainage, and sediments were transported from the northeast to the study area, also inferred from micro-mineralogical data of detrital heavy minerals. The second phase of uplift of the Apuseni Mountains was characterized by an erosion-driven, isostatic uplift due to the relief of the compressional stress field resulting from the waning collision. As a result the trap ceased to be active and was filled up rapidly by the sediments of the transverse rivers, and also allowed the spread of alluvial fans over the basin to the west. This is indicated by the occurrence of SE transport directions in the boreholes at about 1.95 Ma ago, which also gives the timing for the tectonic processes. These southeast-derived sediments reached the area of the distal D-1 borehole only for a short period, then most of the upper part is characterized again by NE transport directions, characteristic of axial rivers. However the SE transport direction became dominant in the area of the proximal V-1 borehole, showing an active sediment supply from the southeastern Apuseni Mountains source area. The isostatic uplift of the whole area was responsible not only for the development of transverse rivers, extending far to the basin, but also for gradual shift to the west of the axial capturing throughout the Pleistocene.

Analogues from the Himalayan foreland and Andean retro-arc basins also showed that tectonic activity exerts a strong control on the drainage pattern through its influence on variations in sediment supply.

This research were funded by the Hungarian National Scientific Research Fund (OTKA T-32956) and by the Bolyai János Research Fund of the Hungarian Academy of Sciences (A.N.). Helpful and constructive reviews from S. Jones, L. Frostick and O. Sztanó are gratefully acknowledged.

References

AMOROSI, A., COLALONGO, M. L., PASINI, G. & PRETI, D. 1999. Sedimentary response to Late Quaternary sea-level changes in the Romagna coastal plain (northern Italy). *Sedimentology*, **46**, 99–121.

BÉRCZI, I. & PHILLIPS, R. L. 1985. Processes and depositional environments within deltaic–lacustrine sediments, Pannonian Basin, Southeast Hungary. *Geophysical Transactions*, **31**, 55–74.

BLUM, M. D. 1994. Genesis and architecture of incised valley fill sequences: a Late Quaternary example from the Colorado river, Gulf Coastal Plain of Texas. *In*: WEIMER, P. & POSAMENTIER, H. (eds) *Siliciclastic Sequence Stratigraphy – Recent Developments and Applications*. American Association of Petroleum Geologists, Memoirs, **58**, 259–283.

BORSY, Z. 1989. Az Alföld hordalékkúpjainak negyedidőszaki fejlődéstörténete. *Földrajzi Értesítő*, **38**, 211–224.

BORSY, Z. 1992. Evolution of the alluvial fans of the Alföld. *In*: RACHOCKI, A. H. & CHURCH, M. (eds) *Alluvial Fans: A Field Approach*. John Wiley, Chichester, 229–246.

BREZSNYÁNSZKY, K. 1989. *Magyarország és környékének hegységszerkezete 1:2 000 000*. Magyar Állami Földtani Intézet kiadványa.

BURBANK, D. W. 1992. Causes of recent Himalayan uplift

deduced from deposited patterns in the Ganges Basin. *Nature,* **357**, 680–683.

COOKE, H. B. S., HALL J. M. & RÓNAI, A. 1979. Paleomagmatic, sedimentary and climatic records from boreholes at Dévaványa and Vésztő, Hungary. *Acta Geologica Academiae Scientiarum Hungaricae,* **22**, 89–109.

CSONTOS, L. & NAGYMAROSY, A. 1999. Late Miocene inversion versus extension in the Pannonian Basin. Abstracts of the 4th Workshop on Alpine Geological Studies, Tübingen. *Geowissenschaftlichte Arbeiten, Ser. A,* **52**, 132.

CSONTOS, L., NAGYMAROSY, A., HORVÁTH, F. & KOVAC, M. 1992. Tertiairy evolution of the Intra-Carpathian area: a model. *Tectonophysics,* **208**, 221–241.

DAVIS, J. C. 1986. *Statistics and Data Analysis in Geology*. Wiley, New York.

ELEK, I. 1979. A kunadacsi Ka-3, kerekegyházi Ke-3 és kecskeméti Kecs-3. sz. perspektívikus kutató fúrások mikromineralógiai vizsgálata. *Földtani Intézet Évi Jelentése,* **1977**, 113–120.

ELEK, I. 1980. A vésztői V-1. sz. Kutató fúrás mikromineralógiai eredményei. *Földtani Intézet Évi Jelentése,* **1978**, 167–172.

FODOR, L., CSONTOS, L., BADA, G., GYÖRFI, I. & BENKOVICS, L. 1999. Tertiary tectonic evolution of the Pannonian basin system and neighbouring orogens: a new synthesis of paleostress data. *In*: DURAND, B., JOLIVET, L., HORVÁTH F. & SÉRRANE, M. (eds) *The Mediterranean Basins: Tertiary Extension Within the Alpine Orogen.* Geological Society, London, Special Publications, **156**, 295–334.

FÜLÖP, J. 1989. *Bevezetés Magyarország geológiájába.* Akadémiai Kiadó, Budapest.

GEDEONNÉ RAJETZKY, M. 1973. A mindszenti és a csongrádi kutatófúrások mikromineralógiai vizsgálata, különös tekintettel az anyagszállítás egykori irányaira. *Földtani Intézet Évi Jelentése,* **1971**, 169–184.

GEDEONNÉ RAJETZKY, M. 1976*a*. Adatok az Észak-Alföld üledékösszletének ismeretéhez. *Földtani Intézet Évi Jelentése,* **1973**, 181–194.

GEDEONNÉ RAJETZKY, M. 1976*b*. *Pliocén végi negyedkori üledékciklusok mikromineralógiai spektruma a Szarvas 1. sz. fúrásban.* Magyar Geológiai Szolgálat Adattára, Kézirat.

HORVÁTH, F. 1993. Towards a mechanical model for the formation of the Pannonian basin. *Tectonophysics,* **226**, 333–357.

HORVÁTH, F. & CLOETINGH, S. 1996. Stress induced late-stage subsidence anomalies in the Pannonian Basin. *In*: CLOETINGH, S., BEN AVRAHAM, Z., SASSI, W. & HORVÁTH, F. (eds) *Dynamics of Extensional Basins and Inversion Tectonics. Tectonophysics,* **266**, 287–300.

HORVÁTH, F. & TARI, G. 1999. IBS Pannonian Basin project: a review of the main results and their bearings on hydrocarbon exploration. *In*: DURAND, B., JOLIVET, L., HORVÁTH F. & SÉRRANE, M. (eds) *The Mediterranean Basins: Tertiary Extension Within the Alpine Orogen.* Geological Society, London, Special Publications, **156**, 195–213.

HOVIUS, N. 1996. Regular spacing of drainage outlets from linear mountain belts. *Basin Research,* **8**, 29–44.

HOWARD, A. D., DIETRICH, W. E. & SEIDL, M. A. 1994. Modelling fluvial erosion on regional to continental scale. *Journal of Geophysical Research,* **99**, 13 971–13 986.

IANOVICI, V., BORCOS, M., BLEAHÚ, M., PATRULIUS, D., LUPU, M., DIMITRESCU, R. & SAVU, H. 1976. *Geologia Muntilor Apuseni.* Bukarest.

IBBEKEN, H. & SCHLEYER, R. 1991. *Source and Sediment. A Case Study of Provenance and Mass Balance at an Active Plate Margin (Calabria, Southern Italy).* Springer, New York.

JÁMBOR, Á., BIHARI, D., CHIKÁN, G., FRANYÓ F., KAISER M., RADÓCZ GY. & SIKHEGYI, F. 1993. *Magyarország pleisztocén tektonikai térképe.* Magyar Geológiai Szolgálat Adattára, Kézirat.

JONES, S. J., FROSTICK, L. E. & ASTIN, T. R. 1999. Climatic and tectonic controls on fluvial incision and aggradation in the Spanish Pyrenees. *Journal of the Geological Society, London,* **156**, 761–769.

JORDAN, T. 1995. Retroarc foreland and related basins. *In*: BUSBY, C. J. & INGERSOLL, R. V. (eds) *Tectonics of Sedimentary Basins.* Blackwell Scientific, Oxford, 331–362.

JUHÁSZ, GY. 1992. A pannóniai s.l. formációk térképezése az Alföldön: elterjedés, fácies és üledékes környezetek. *Földtani Közlöny,* **122**, 133–165.

JUHÁSZ, GY. 1994. Magyarországi neogén medencerészek pannóniai s. l. üledéksorának összehasonlító elemzése. *Földtani Közlöny,* **124**, 341–365.

KUTI, L., MOLNÁR, P., ELEK, I., VERMES, J., SALLAY, M. & GYURICZA GY. 1987. *Magyarország recens és fosszilis torlatainak kutatása.* Magyar Geológiai Szolgálat Adattára, Kézirat.

MANSIKKANIEMI, H. 1991. Regional case studies in Southern Finland with reference to glacial rebound and Baltic regression. *In*: STARKEL, L., GREGORY, K. J. & THORNES, J. B. (eds) *Temperate Paleohydrology.* John Wiley, Chichester, 79–104.

MOLNÁR, B. 1964. A magyarországi folyók homok üledékeinek nehézásvány összetétel vizsgálata. *Hidrológiai Közlöny,* **44**, 347–355.

MOLNÁR, B. 1965*a*. Ősföldrajzi vizsgálatok a Dél-Tiszántúlon. *Hidrológiai Közlöny,* **45**, 397–404.

MOLNÁR, B. 1965*b*. Changes in area and directions of stream erosion in the eastern part of the Hungarian basin (Great Plain) during the Pliocene and Pleistocene. *Acta Universitatis Szegediensis, Acta Mineralogica-Petrographica,* **17**, 39–52.

MOLNÁR, B. 1966*a*. Lehordási területek és irányok változásai a Dél-Tiszántúlon, a pliocénben és a pleisztocénben. *Hidrológiai Közlöny,* **46**, 121–127.

MOLNÁR, B. 1966*b*. Pliocén és pleisztocén lehordási területek változása az Alföldön. *Földtani Közlöny,* **96**, 403–413.

MOLNÁR, B. 1980. Changes of source areas as reflected by the depression of the Körös Rivers. *Acta Universitatis Szegediensis, Acta Mineralogica-Petrographica,* **24**, 339–353.

MOLNÁR, P., GYURICZA Gy. & TAMÓNÉ BOZSÓ E. 1990. *Jelentés a 'Magyarország recens és fosszilis torlatainak kutatása' c. Program 1989 évi munkálatairól*. Magyar Geológiai Szolgálat Adattára, Kézirat.

MOLNÁR, P., THAMÓNÉ BOZSÓ E. & GYURICZA Gy. 1989. *Jelentés a 'Magyarország recens és fosszilis torlatainak kutatása' c. Program 1988 évi teljesítéséről*. Magyar Geológiai Szolgálat Adattára, Kézirat.

MÜLLER, P., DANA, H. G. & MAGYAR, I. 1999. The endemic molluscs of the Late Miocene Lake Pannon: their origin, evolution, and family-level taxonomy. *Lethaia*, **32**, 47–60.

NÁDOR, A., MÜLLER, P., LANTOS, M., THAMÓNÉ BOZSÓ, E., KERCSMÁR, Z. S., TÓTHNÉ MAKK, Á., SÜMEGI, P., FARKASNÉ BULLA, J. & NAGY, T. 2000. A klímaváltozások és az üledékes ciklusok kapcsolata a Körös-medence negyedidőszaki folyóvízi rétegsoraiban. *Földtani Közlöny*, **130**, 623–645.

Ó. KOVÁCS, L. 1987. Cluster-analízis eljárások TPA/L számítógépen. *Magyar Állami Földtani Intézet Évi Jelentése*, **1985**, 571–582.

POGÁCSÁS, Gy., LAKATOS, L., UJSZÁSZI, K., VAKARCS, G., VÁRKONYI, L., VÁRNAI, P. & RÉVÉSZ, I. 1988. Seismic facies, electro facies and Neogene sequence chronology of the Pannonian basin. *Acta Geologica Hungarica*, **31**, 175–207.

RÓNAI, A. 1985. *Az Alföld negyedidőszaki földtana*. Geologica Hungarica series. Geologica, **21**.

SANDERS, C. 1998. *Tectonics and Erosion. Competitive Forces in a Compressive Orogene – A Fission Track Study of the Romanian Carpathians*. PhD Thesis, Vrije Universitiet, Amsterdam.

SEEBER, L. & GORNITZ, V. 1983. River profiles along Himalayan arc as indicators of active tectonics. *Tectonophysics*, **92**, 335–367.

SEIDL, M. A. & DEITRICH, W. E. 1992. The problem of channel erosion into bedrock. *Catena Supplement*, **23**, 101–124.

SHANLEY, K. W. & MCCABE, P. J. 1994. Perspectives on the sequence stratigraphy of continental strata. *AAPG, Bulletin*, **78**, 544–568.

SÜMEGHY, J. 1944. *A Tiszántúl. Magyar Tájak Földtani Leírása VI*. Magyar Királyi Földtani Intézet.

SZABÓ, D. 1967. *Összefoglaló földtani jelentés és készletszámtás a biharkeresztes-ártándi felderítőfázisú kavicskutatásról*. Magyar Geológiai Szolgálat Adattára, Kézirat.

SZABÓ, P. 1955. A Duna-Tisza közi felsőpleisztocén homokrétegek származása ásványos összetétel alapján. *Földtani Közlöny*, **85**, 442–456.

VAN BALEN, R. T., LENKEY, L., HORVÁTH, F. & CLOETINGH, S. A. P. L. 1999. Two dimensional modelling of stratigraphy and compaction-driven fluid flow in the Pannonian Basin. *In*: DURAND, B., JOLIVET, L., HORVÁTH, F. & SÉRRANE, M. (eds) *The Mediterranean Basins: Tertiary Extension within the Alpine Orogen*. Geological Society, London, Special Publications, **156**, 391–414.

VAN DEN BERG, M. W. 1994. *Fluvial Sequences of the Maas: a 10 Ma Record of Neotectonics and Climatic Change at Various Time-scales*. PhD Thesis, Universal Press, Veenendaal.

VANDENBERGHE, J. 1993. Changing fluvial processes under changing periglacial conditions. *Zeitschrift für Geomorphologie Neue Folge Supplementband* **88**, 17–28.

VANDENBERGHE, J. 1995. Timescales, climate and river development. *Quaternary Science Reviews*, **14**, 631–638.

VINCENT, S. J. 1999. The role of sediment supply in controlling alluvial architecture: an example from the Spanish Pyrenees. *Journal of the Geological Society, London*, **156**, 749–759.

Filling and cannibalization of a foredeep: the Bradanic Trough, Southern Italy

MARCELLO TROPEANO[1], LUISA SABATO[2] & PIERO PIERI[2]

[1]*Dipartimento di Scienze Geologiche, Università della Basilicata, Campus Macchia Romana, I-85100 Potenza, Italy (e-mail: tropeano@unibas.it)*

[2]*Dipartimento di Geologia e Geofisica, Università di Bari, Via Orabona 4, I-70125 Bari, Italy (e-mail: l.sabato@geo.uniba.it; p.pieri@geo.uniba.it)*

Abstract: The Bradanic Trough (southern Italy) is the Pliocene–present-day south Apennines foredeep. It is a foreland basin as subsidence due to westward subduction of the Adria Plate involves the continental crust of the Apulian domain. The infill succession of the Bradanic Trough is characterised by the presence of a long thrust sheet system (the so called 'allochthon') that occupied part of the accommodation space created on the foreland by subduction. The upper part of the infilling succession crops out along numerous sections. About 600 m of the 3–4 km basin-fill succession is exposed as the Bradanic Trough has experienced uplift during Quaternary times.

Outcropping successions are mainly characterized by shallow-marine deposits comprising carbonates of the Calcarenite di Gravina Formation, silty clay hemipelagites of the Argille subappennine Formation and coarse-grained bodies of the 'Regressive coastal deposits'.

The Calcarenite di Gravina Formation (Middle–Late Pliocene–Early Pleistocene in age) crops out in a backstepping configuration onto the flanks of the Apulian Foreland highs. It displays evidence of strong transgression onto a karstic region previously dissected in a complex horst and graben system.

The Argille subappennine Formation (Late Pliocene–Middle Pleistocene in age) succeeds the carbonate sedimentation on the foreland side of the basin and represents the shallowing of the basin in the other sectors of the Bradanic Trough. Toward the Apennines chain, in the wedge-top area of the foredeep, the Argille subappennine Formation covers the allochthon, while in the depocentre (in the foredeep *sensu stricto*) the same formation overlays turbidite deposits. The latter characterize the deeper part of the successions, and are mainly buried below the allochthon.

The Regressive coastal deposits (Early–Late Pleistocene in age) represent the upper part of the succession. They consist of coarse-grained wedges that lie on the hemipelagites of the Argille subappennine Formation in, alternatively, conformable or erosional contact. The wedges of the Regressive coastal deposits stack in a downward-shifting configuration, which indicates deposition during uplift.

The Quaternary development of the Bradanic Trough differs from that of the central and northern Apennines foredeep. The latter is characterized by aggradation of shallow-marine and alluvial sediments in a subsiding remnant basin, whose filling records a basin-scale depositional regression. In contrast, the Bradanic Trough is characterized by a basin-scale erosional regression and the last evolutive phase of this sector of the Apennines foredeep is best defined as a cannibalization phase rather than a filling or overfilling phase.

Foredeeps are basins that form due to flexural subsidence of a subducting plate. When foredeeps develop on continental crust they are called foreland basins (Dickinson 1974; Beaumont 1981; Allen *et al.* 1986; Doglioni 1993; De Celles & Giles 1996). The Apennines foredeep (Italy) is a Neogene–present-day foreland basin characterized during Miocene mainly by turbidite sedimentation in deep-marine environments (Mutti & Ricci Lucchi 1972; Ricci Lucchi 1975; 1986; Pescatore 1978).

The post-Messinian Apennines foredeep is represented by the Po Plain in the northern sector, the north-central Adriatic Sea in the central sector, and the Bradanic Trough and Taranto Gulf (Ionian Sea) in the south (Fig. 1a). Turbidite systems developed in these Apennines foredeep sectors also during the Pliocene (Cremonini & Ricci Lucchi

From: JONES, S. J. & FROSTICK, L. E. (eds) 2002. *Sediment Flux to Basins: Causes, Controls and Consequences.* Geological Society, London, Special Publications, **191**, 55–79. 0305-8719/02/$15.00

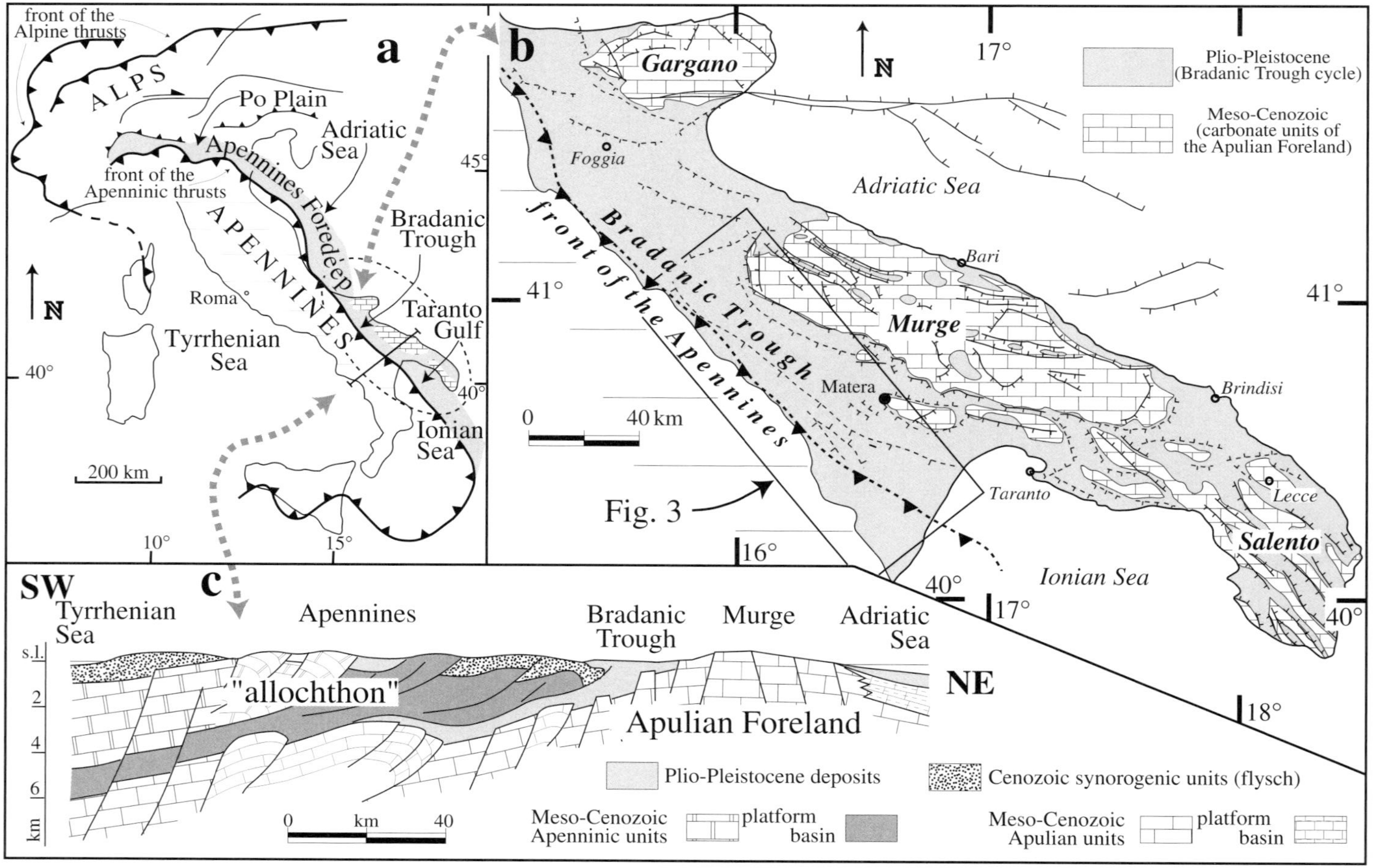

Fig. 1. (**a**) Schematic structural map of Italy showing the present-day location of the Apennines foredeep; (**b**) schematic geological map of the Bradanic Trough (modified from Pieri *et al.* 1997*b*); (**c**) geological cross-section showing the main structural elements of the southern Apennines orogenic system (modified from Sella *et al.* 1988).

1982; Casnedi 1988*a*, 1991; Bettazzoli & Visentin 1998). In contrast with the Neogene depositional history of the Apennines foredeep, the present-day Apennines foredeep is mainly overfilled (Po Plain and Bradanic Trough) or nearly filled (north-central Adriatic Sea) (Casnedi *et al.* 1982; Ricci Lucchi 1986; Ori *et al.* 1986) (Fig. 1a). Only the southernmost sector of the present-day foredeep (the Taranto Gulf in the Ionian Sea) is a deep basin and only here does turbidite sedimentation take place (Pescatore & Senatore 1986) (Fig. 1a). Therefore, these different sectors of the Apennines foredeep did not develop similarly during Quaternary times and the differences are mainly attributed to subduction of a segmented lithosphere with different thicknesses (Royden *et al.* 1987; Doglioni *et al.* 1994).

The evolution of a deep- to shallow-marine basin in the Apennines foredeep took place after the Pliocene and this work concentrates on the Plio-Quaternary history of the south Apennines foredeep (the Bradanic Trough, see Fig. 1). The upper part (mainly Quaternary) of the infilling marine succession of the Bradanic Trough is deeply incised and well exposed along numerous sections cut in the exposed part of the basin (about 200 km long and up to 50–60 km wide, Fig. 1b). This allows a study of the stratigraphy and sedimentology of these foredeep deposits. Collating field and subsurface published data with our new field studies we interpret filling of the basin in terms of migration of sedimentary systems caused by the interaction between inherited Meso-Cenozoic crustal and/or lithospheric structures, geodynamics and sedimentary processes, and suggest that uplift of the basin, recognized by neotectonic studies (Ciaranfi *et al.* 1979, 1983), accompanies sedimentation, at least from Early Pleistocene time, and induces cannibalization of the foredeep succession.

The Apennines chain

The growth of the Apennines chain started during late Oligocene times (Boccaletti *et al.* 1990) when the Adria Plate, characterized by both oceanic (Ionian sector) and continental (Apulia sector) lithosphere (Finetti 1982; Catalano *et al.* 2001), began to subduct back to the Alpine–Betic thrust belt (Gueguen *et al.* 1998 and references therein). The Alpine–Betic back-thrust belt and the hinterland of the same Alpine–Betic orogen (the undeformed Adria) was progressively involved in this westward subduction (Doglioni *et al.* 1998 and references therein). As a consequence, the south Apennines chain now incorporates: (i) terranes of the European continental plate (i.e. Alpine–Betic back-thrust belt fragments and/or related foredeep successions); (ii) successions derived from neo-Tethyan oceanic domains (i.e. Sicilide and Liguride complexes); (iii) large sequences of Mesozoic sedimentary cover delaminated from the subducted continental Adria Plate (i.e. south Apennines carbonate platform and Lagonegro basin successions); and (iv) Oligo-Miocene synorogenic successions (i.e. Irpinian flysches) (Casero *et al.* 1988; Roure *et al.* 1991; Pescatore *et al.* 1999 and references therein). All these units overthrust the Apulian platform, which is directly involved in thrusting only in its buried part (Mostardini & Merlini 1986; Patacca & Scandone 1989; Carbone & Lentini 1990) (Fig. 1c).

The Apennines foredeep (Fig. 1a) is the foreland basin related to the Apennines thrust belt; its subsidence rate is higher than 1 km Ma^{-1} (Doglioni 1994; Doglioni *et al.* 1999a) and is induced by the eastward roll-back of the Adria Plate (slab pull and sinking, and/or opposition to relative eastward mantel flow) rather than by thrust-sheet loading (Royden & Karner 1984; Malinverno & Ryan 1986; Royden *et al.* 1987; Doglioni 1991) (Fig. 2a). Apennines foredeep depocentres moved toward the ENE (Ricci Lucchi 1986; Boccaletti *et al.* 1990) following the fast Apennines arc migration, which has been calculated to be in the order of 700 km in southern Italy (i.e. Calabrian arc) (Patacca *et al.* 1990; Gueguen *et al.* 1997); this value linearly decreases moving toward the northern Apennines and, accordingly, both the amount and velocity of crustal shortening decrease from about 5–6 to < 1 cm $year^{-1}$ moving from south to north along the Apennines (Patacca & Scandone 1989; Doglioni *et al.* 1999*a*). The structural and sedimentary history of the Apennines foredeep records this progressive eastward migration of the orogenic system (Fig. 2a). Thrusting propagated within the foredeep, splitting it into an internal shallower basin (a piggyback area or a wedge-top basin) and an external deeper one (foredeep *sensu stricto*); high subsidence rates gradually involved more eastern sectors of the foreland (Ricci Lucchi 1986; Boccaletti *et al.* 1990; Ori *et al.* 1991; De Celles & Giles 1996) (Fig. 2b).

According to Patacca & Scandone (1989), two different styles of tectonic deformation characterize the central and southern Apennines. In the central Apennines, north of the Maiella Mountain and the Tremiti line, thrusts propagation mostly proceeded by imbricate fan development in piggyback sequences, which progressively involved the foreland (Bally *et al.* 1986) (Fig. 2a and b). In the southern Apennines, thrusts propagation mostly proceeded by long sheets (the ‘allochthon’) that overthrust the undeformed Apulian Foreland. Thrusts propagated in the foreland only after its underthrusting (Mostardini & Merlini 1986) (Fig. 2c and d).

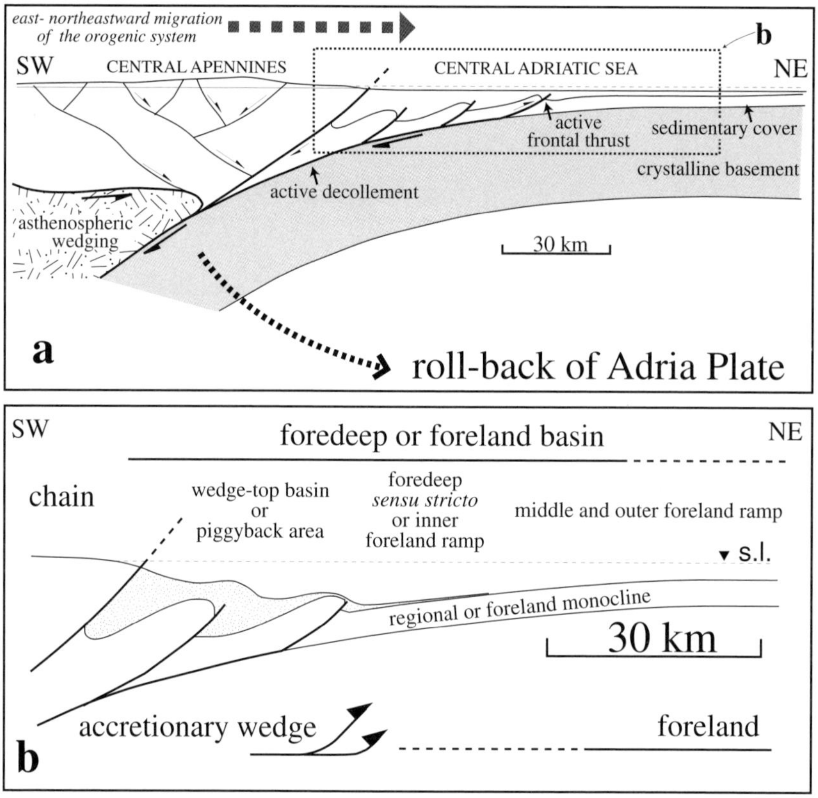

Fig. 2. (**a**) Structural and geodynamic sketches of a section of the central Apennines belt and its foreland basin (modified from Doglioni *et al.* 1994); the foreland subsidence is strictly related to the eastward roll-back of the Adria Plate, while thrust propagation follows basin subsidence; kinematics of these elements induce the eastward migration of the entire orogenic system; (**b**) particular of (a) in which is evidenced the terminology used in this paper (mainly from Boccaletti *et al.* 1990; De Celles & Giles 1996; Mariotti & Doglioni 2000); note the separation of the foredeep basin in three areas of sedimentation (wedge-top or piggyback basins; foredeep *sensu stricto* or inner foreland ramp; and middle–outer foreland ramp) due to the ENE propagation of both subsidence and frontal thrusts.

The Bradanic Trough

Stratigraphy of the basin

The Bradanic Trough (Migliorini 1937; Selli 1962; D'Argenio *et al.* 1973) is the Pliocene–Pleistocene southern Apennines foredeep sector, and it developed between the chain and an exposed and flexured karstic foreland (the Apulian Foreland) (Ricchetti 1980; Ricchetti & Mongelli 1980) (Fig. 1c). The in-fill of the Bradanic Trough began during late Early–Middle Pliocene times on a wide subsiding area of the Apulian Foreland (Ciaranfi *et al.* 1979); foredeep deposits lie on a carbonate bedrock mainly exposed since Late Cretaceous time (Crescenti 1975). According to Casnedi (1988*b*, 1991), the buried Pliocene–Early Pleistocene succession is mainly characterized by a turbidite complex consisting of slope, fan and basin plain deposits, while the outcropping Quaternary succession is basically made up of hemipelagic silty clay (Valduga 1973; Ciaranfi *et al.* 1979) and is overlain by coarse-grained shallow-marine siliciclastic or carbonate deposits attributed to the regressive stage of basin filling (Ricchetti 1967*a*) and to marine terrace development (Vezzani 1967*a*).

Following the descriptions of Ricci Lucchi (1975), Casnedi *et al.* (1982) recognized the three evolutionary stages of the northern Apennines Miocene foredeep (pre-turbidite, turbidite and post-turbidite stages) in the Pliocene–Pleistocene succession of the Bradanic Trough. Studies based only on drilling, log data and seismic profiles show that the stratigraphy of these deposits is made up of three different groups of sediments, the last of which represents the outcropping succession: a lower marl and clay group (shallow pre-turbidite hemipelagite); a clay and sand middle group (turbidite sand); and a clay and sand upper group (post-turbidite hemipelagite) (Balduzzi *et al.* 1982*a*, *b*; Casnedi *et al.* 1982). The whole in-fill succession is characterized by the presence, above

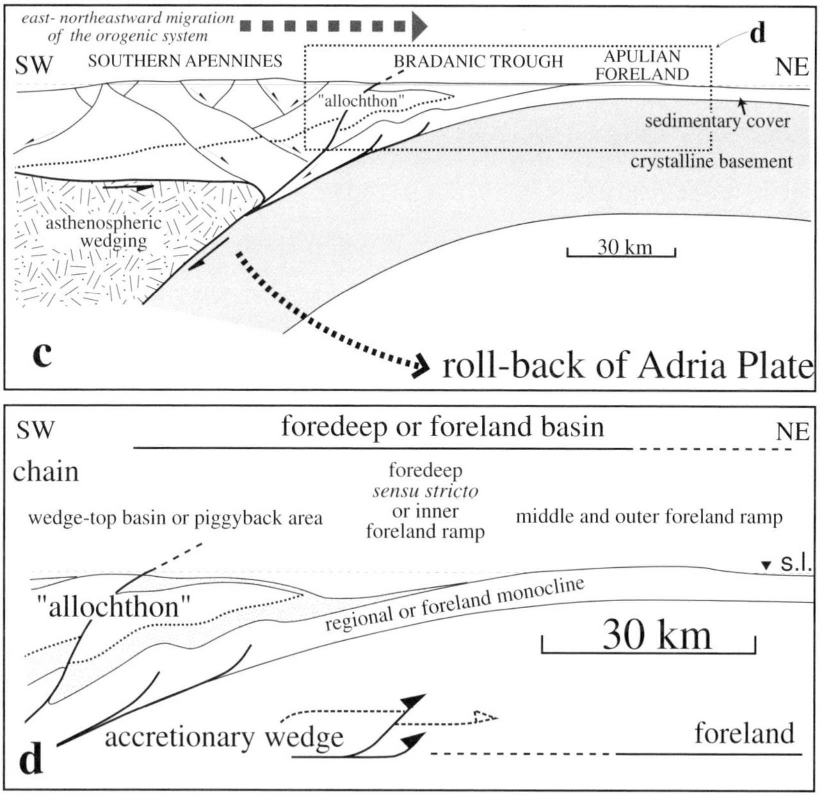

Fig. 2. cont. (**c**) structural and geodynamic sketches of a section of the southern Apennines belt and its foreland basin (modified from Doglioni *et al.* 1994); 'allochthon' propagation follows basin subsidence, while thrusts in the Apulian Foreland grow after underthrusting; (**d**) particular of (c) in which the same terminology used in (b) for the central Apennines foredeep is applied.

the turbidite sands, of the so-called 'allochthon', which emplaces predominantly pre-Pliocene deformed units (Mesozoic–Palaeogene pre-orogenic units passing to Miocene synorogenic ones) over Pliocene–Early Pleistocene foredeep deposits (Carissimo *et al.* 1962; Mostardini *et al.* 1966; Ogniben 1969; Ciaranfi *et al.* 1979; Lentini 1979; Sella *et al.* 1988) (Fig. 1b and c). The allochthon was considered a gravitational slide of the frontal part of the Apennines chain that moved toward the axis of the foredeep, but most recent data indicate that it represents a long thrust sheet system (as long along the dip as the Apennines chain is wide, Fig. 1c) which overthrusts the Apulian platform units (Mostardini & Merlini 1986) (Fig. 2c and d).

Following both the stratigraphical synthesis of Casnedi (1988*a*, *b*, 1991) and the structural bedrock data set collected by Sella *et al.* (1988), and using new field data, Pieri *et al.* (1996*b*, *c*) recognized two kinds of successions in the Bradanic Trough depending on the inclination of the Apulian Foreland ramp: thicker successions, associated with both the allochthon and turbidites, developed on the inner foreland ramp, a sector of the trough with a highly inclined bedrock surface; thinner successions, not associated with the allochthon and turbidites, developed on the outer, less inclined, sector of the foreland ramp (pre-Murge and pre-Gargano plateaux) and around the highs of the exposed foreland.

According to all published subsurface data and our field studies, three representative successions best describe the Bradanic Trough in-fill. They are measured along WSW–ENE transects, perpendicular to the elongation of the trough (Fig. 3). From the inner to the outer margins of the Bradanic Trough, these stratigraphic successions can be synthesized as follows:

- inner successions (wedge-top basin – succession A in Fig. 3b) are composed of hemipelagites which pass upward into a thick pile of turbidites; a thin accretionary wedge (the allochthon)

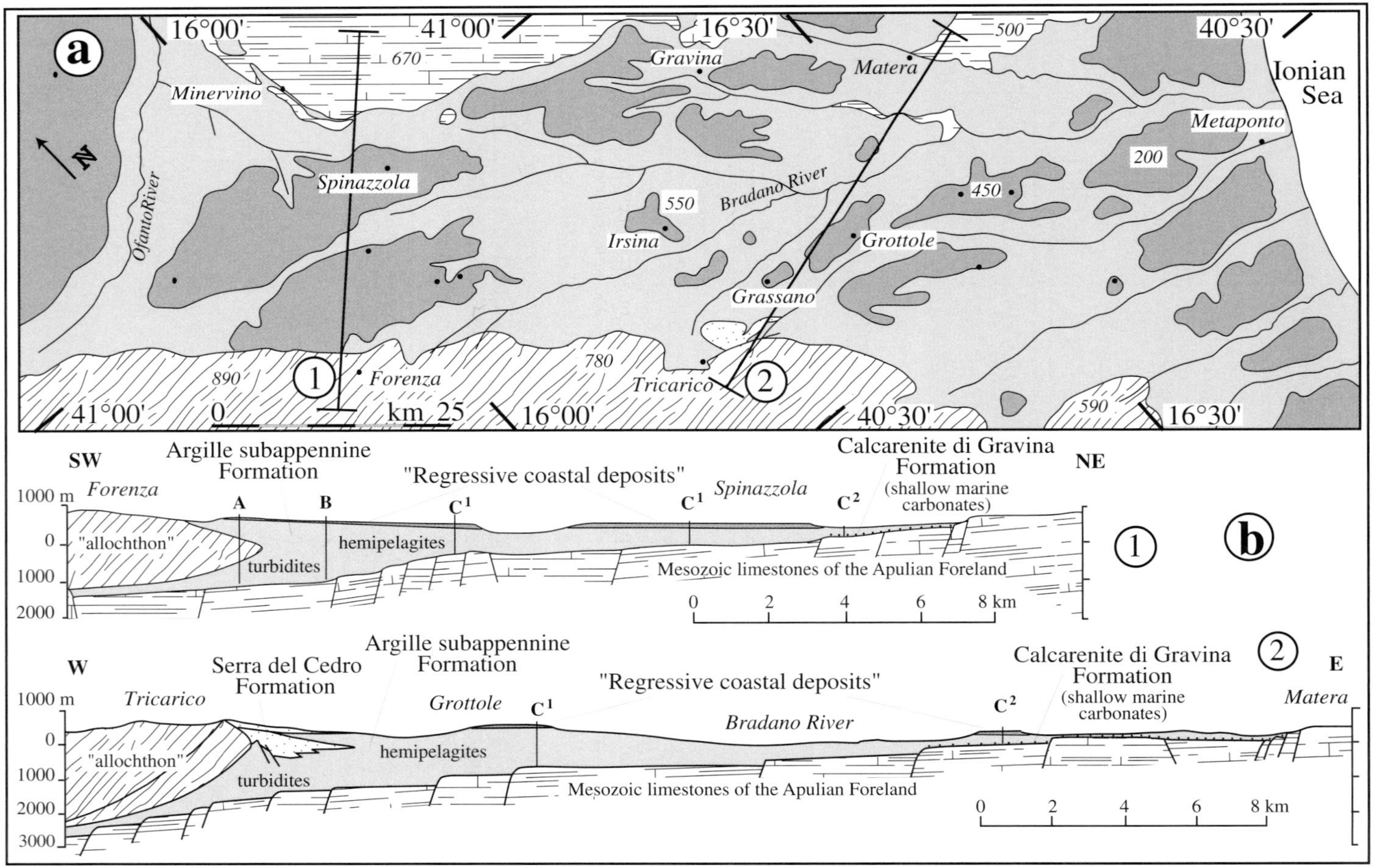

Fig. 3. (**a**) Schematic geological map of the deposits outcropping in the central and southern part of the Bradanic Trough (for location see Fig. 1b); (**b**) geological cross-sections in which the structural features of the substratum and the distribution of the deposits filling the Bradanic Trough are shown; A, B, C^1 and C^2 indicate the most representative successions of the Bradanic Trough in-fill (modified from Pieri *et al.* 1994, 1996*b*).

tectonically overlies the turbidite deposits. Hemipelagites and, subsequently, coarse-grained coastal and alluvial deposits cap the sequence on the wedge top. The whole thickness of this succession, including the allochthon, is in the order of 3–4 km. During the Pleistocene, only in the southern sector of the Bradanic Trough there was sedimentation on the wedge-top, whilst in the central and northern sectors of the basin the wedge top was exposed at least from Early Pleistocene times;

- thick depocentre successions (foredeep *sensu stricto* – succession B in Fig. 3b) are adjacent to the front of the allochthon, are up to 2 km thick and are composed of hemipelagites that pass upwards into a thick pile of turbidites that, in turn, are overlain by hemipelagites and, finally, coarse-grained coastal and alluvial deposits. Hemipelagites and coarse-grained deposits are laterally continuous with similar deposits formed both on the wedge top and on the outer margin of the trough;
- middle and outer margin successions (middle and outer foreland ramp) characterize a wide area of the presently exposed basin and have a variable thickness from a few tens of metres up to 1 km. They are mainly composed of hemipelagites passing to coarse-grained shallow-marine and alluvial deposits. The thicker successions (middle foreland ramp) are located in the central part of the basin (succession C^1 in Fig. 3b). The thinner successions (outer foreland ramp) are located in the eastern part of the basin (succession C^2 in Fig. 3b) with a basal unit of transgressive shallow-marine carbonates overlain by hemipelagites and finally coarse-grained deposits. On the foreland side of the trough, these thin successions are completely exposed from the bedrock and also characterize some depressions inside the Apulian Foreland (i.e. the Murge grabens and the Brindisi Basin). In the upper part of the flanks of the Apulian Foreland highs shallow-marine carbonates are erosionally covered by the coarse-grained deposits without the interposition of silty clays.

These stratigraphic features reflect the wedge-shape (asymmetry) of the basin with the location of its depocentre in a very narrow area, where the turbidites could develop. The upper part of the described successions is exposed (up to about 600 m in thickness) and is mainly characterized by silty clay hemipelagites (Argille subappennine Formation), which are transitional and/or in erosional contact with the overlying coarse-grained deposits ('Regressive coastal deposits') (Pieri *et al.* 1996*b*; Sabato 1996) (Fig. 3). As described above, a carbonate unit (the Calcarenite di Gravina Formation) is exposed in the outer margin successions below the Argille subappennine Formation.

The outcropping Pliocene–Quaternary successions

The Apulian Foreland and the Calcarenite di Gravina Formation. The Middle–Upper Pliocene–Lower Pleistocene Calcarenite di Gravina Formation (shallow-marine carbonates of Fig. 3b) crops out on the outer margin of the Bradanic Trough and on the Apulian Foreland, and unconformably overlies faulted strata of the Apulian carbonate platform (Azzaroli 1968; Iannone & Pieri 1979; Tropeano & Sabato 2000; Pomar & Tropeano 2001). The Apulian platform was a relic of Mesozoic rifting and passive margin development across the Adria, an African lithospheric promontory (D'Argenio 1974; Channel *et al.* 1979; Ricchetti 1980) that is considered a present-day separate plate (Ricchetti *et al.* 1988 and references therein). The Apulian platform became an emerged region at the end of the Mesozoic (Pieri 1980); only thin and discontinuous Cenozoic shallow-marine carbonates covered marginal areas of the wide Mesozoic platform (Ricchetti *et al.* 1988). During the Neogene, the Apulian platform became part of the foreland of the southern Apennines chain (Selli 1962; D'Argenio *et al.* 1973). As a consequence of the eastward roll-back of the subducting Adria Plate, the western part of the Apulian platform was overthrust by the allochthon and successively involved in thrusting (Mostardini & Merlini 1986; Patacca & Scandone 1989) whereas, from Early Pliocene time, the eastern part of the Apulian platform (the remaining part of the platform at that time not reached by thrusting) underwent a progressive increase in regional subsidence (Ciaranfi *et al.* 1979; Iannone & Pieri 1982). A thin (no more than a few tens of metres thick) Middle–Upper Pliocene–Lower Pleistocene mantle of bioclastic and/or lithoclastic carbonates deposited on the faulted rocks of the Apulian Foreland. The lower boundary of these carbonates is a long-term ravinement surface abraded onto the bedrock and regionally these rocks represent the Calcarenite di Gravina Formation, which was deposited in shallow-marine systems backstepping onto the flanks of the Apulian Foreland highs (Tropeano & Sabato 2000; Pomar & Tropeano 2001). From northwest to southeast and from west to east carbonate systems were diachronously covered by hemipelagites of the Argille subappennine Formation derived from the Apennines thrust belt at progressively higher levels on the flanks of the foreland-highs (Pieri *et al.* 1996*b*). A drowning unconformity (*sensu* Schlager

1989) bounds the top of the Calcarenite di Gravina Formation. In some places, glauconitic-rich deposits indicate a demise of carbonate production before the arrival of silty clay (drowning *sensu stricto*); in other localities a few thin carbonate strata alternate with silty clay layers before passing definitively into the Argille subappennine Formation (suffocation and drowning).

Marine carbonates of the Calcarenite di Gravina Formation crop out from the present-day sea level (along the Adriatic and Ionian cliffed coasts) up to more than 400 m in elevation (Ciaranfi *et al.* 1988; Tropeano & Sabato 2000). The highest outcrops are subaerially eroded, whilst lower outcrops on the flank of the foreland highs are at times either overlain by the Argille subappennine Formation or disconformably overlain by marine and/or alluvial terrace deposits. Moreover, some eolianites belonging to the Calcarenite di Gravina Formation very discontinuously surround the top of the Murge Alte plateau (filling joints or forming small bodies) (Iannone & Pieri 1979); these eolianites represent the highest outcrops of the Calcarenite di Gravina Formation (up to 500 m in elevation in the Matera, Minervino and Monte Trazzonard areas).

The Argille subappennine Formation. The hemipelagic outcropping succession represents the Argille subappennine Formation (hemipelagites of Fig. 3) (Ricchetti 1965, 1967*a*; Azzaroli *et al.* 1968*a*, *b*; Valduga 1973) which crops out extensively in subhorizontal arrangement or shows a monoclinal arrangement, with a dip of a few degrees toward the SE. Those hemipelagites that were deposited onto the top of the allochthon now dip at up to 10–15° towards the west or east recording, respectively, thrust or backthrust growth (Pieri *et al.* 1997*b*).

The Argille subappennine Formation shows a widespread distribution in the Bradanic Trough and it crops out from the present-day sea level up to more than 500 m in elevation onto the flanks of the main river valleys that deeply cut the basin succession. The cropping thickness is variable but can be up to a few hundreds of metres. The age becomes younger from NW to SE and from west to east. The silty clays (except for those older ones involved in the allochthon thrusting) are Late Pliocene(?)–Santernian in age in the northern part of the trough (Scalera 1986; Aucelli *et al.* 1997) and early to middle Pleistocene in the south (Verhallen 1991; Ciaranfi *et al.* 1994*a*, 1997).

In general, the depositional environment is interpreted to be a relatively deep-water environment and the deposits are described as massive or monotonous. Detailed sedimentological studies are very rare but they permit the recognition of the shallowing-upward trend that characterizes the Argille subappennine Formation. The macrofaunal associations in the 500-m thick outcropping succession of Montalbano area indicate a regressive trend from bathyal to shallow sublittoral environments, with a maximum palaeodepth of 500–600 m (D'Alessandro *et al.* 2000); furthermore, a palaeocommunities study has described fifth-order cyclicity probably related to eustatic causes (Ciaranfi *et al.* 1997). In the middle part of the Montalbano succession, detailed sedimentological analyses demonstrate a shorter cyclicity in thin facies sequences (Ciaranfi *et al.* 1996). These sequences are comprised predominantly of clayey silts and minor fine sands, have variable thicknesses from about 10 cm to over 1 m, are normally graded, bioturbated and are often separated by erosive surfaces. The finer sequences, made up exclusively by clayey silts, have a thickness of up to 1 m and are intensely bioturbated; the bottom is represented by ghosts of parallel laminations, with abundant plant remains and bioclastic detritus. There are no significant boundary surfaces within these sequences. In contrast, the coarser sequences, represented by fine sands and silts, are less thick, less bioturbated and are clearly separated by erosive surfaces marked by a reddish sandy layer with abundant plant remains and transported macrofossils (bryozoans, echinoids, lamellibranchs). From the base upward these sequences show a series of structures: parallel laminations to hummocky cross-laminations to wave ripple cross-laminations. The top of these sequences, when present, is massive or bioturbated. As a whole, the sedimentary features of the succession are related to storm–fair weather sequences in offshore environments (*sensu* Johnson & Baldwin 1986; Ciaranfi *et al.* 1996).

In other localities, the upper part of the Argille subappennine Formation is related to shelf environments (Caldara *et al.* 1979) or to shallow shelf passing to lower shoreface environments, essentially on the basis of palaeontological features (Caldara *et al.* 1989). In the Irsina area the facies analysis of the upper part of the Argille subappennine Formation suggests that these deposits should be referred to shelf passing to shoreface environments (Sabato 1996). The succession is characterized here by parallel laminated and ripple cross-laminated silts with numerous interlayered sandy beds, erosively based, parallel laminated and wavy cross-laminated, with abundant lamellibranchs.

According to lithostratigraphical and sedimentological data collected by Massari (1997), the Bradanic Trough was characterized by relatively steep shelves. These data agree with the morphological features of the basin margins observed by Pescatore & Senatore (1986) in the present-day southern Apennines foredeep. Therefore, we suggest that the term 'depositional ramp' (indicating a wide slope connecting nearshore zones to basinal areas) best describes the depositional system in which the Argille subappennine Formation was deposited. In this case: (i) storm-related processes and storm-induced currents could have reached relatively deep environments characterized by bathyal fauna (storm-dominated offshore facies); and (ii) a progressive upward-shallowing of the hemipelagitic facies (up to the transition to lower shoreface) within the formation can be recorded without sedimentary evidence of a significative break in the depositional profile.

The 'Regressive coastal deposits'. Hemipelagites of the Argille subappennine Formation are either transitionally or erosionally overlain by coarse-grained deposits. These deposits represent the top of the foredeep successions and are mainly shallow marine in origin. Thicknesses vary from a few metres up to over 100 metres, and facies belong to paralic, deltaic and/or alluvial environments (Massari & Parea 1988, 1990; Pieri *et al.* 1994*b*, 1996*b*; Sabato 1996; Lazzari 1998). In the central part of the Bradanic Trough, these deposits were interpreted as the Lower Pleistocene synchronous final in-fill of the basin before uplift of the region (i.e. the Sabbie di Monte Marano, the Calcarenite di Monte Castiglione and the Conglomerato di Irsina Formations by Ricchetti 1965, 1967*a*; Azzaroli *et al.* 1968*a*, *b*; Valduga 1973). These deposits, at present outcropping at about 600 m down to 350–400 m above sea level, were considered different from the southernmost Middle–Upper Pleistocene deposits, which more clearly represent marine terrace deposits (now about 350–400 m above sea level down to the present-day Ionian coastline) (Cotecchia & Magri 1967; Ricchetti 1967*b*; Vezzani 1967*a*; Boenzi *et al.* 1971, 1976; Brückner 1980; De Marco 1990). Also in the northern part of the Bradanic Trough, known as the Tavoliere Plain the Argille subappennine Formation top, from about 450 m above sea level down to the Adriatic coastal plain, is overlain by coarse-grained sediments. These deposits, no more than 50–70 m in thickness, may be either marine or continental in origin (i.e. Jacobacci *et al.* 1967; Boni *et al.* 1969; Caldara & Pennetta 1993; Capuano *et al.* 1996; Aucelli *et al.* 1997), but they were also interpreted as E-dipping alluvial fan deposits (Parea 1986).

As regards the central and southern sectors of the basin, Pieri *et al.* (1994*b*, 1996*b*) informally named all the coarse-grained deposits on top of the Argille subappennine Formation as 'Regressive coastal deposits' of the Bradanic Trough (Fig. 3). The highest and oldest Regressive coastal deposits are Sicilian in age and prograde towards the ENE, whilst the younger Middle and Late Pleistocene deposits prograde to the SE towards the Ionian coast (Pieri *et al.* 1994*b*, 1996*b*; Lazzari 1998). These deposits are mainly siliciclastic in origin and of Apenninic provenance, but locally, near the foreland highs, they may be carbonate or mixed in origin (Ricchetti 1965). According to Pieri *et al.* (1996*b*), the stratigraphic organization of these coarse-grained deposits is arranged into prograding wedges, often characterized by wave-dominated Gilbert-type bodies (*sensu* Massari 1996), whose top depositional surface progressively falls in elevation along progradation dip. The erosionally based wedges, commonly characterized by fan delta deposits (i.e. Massari & Parea 1990; Sabato 1996), alternate with transitionally based ones, characterized mainly by aggrading deposits which from bottom to top pass from offshore-transition facies of the Argille subappennine Formation up to shoreface, beach and, in some places, to alluvial facies (i.e. Massari & Parea 1988; Sabato 1996) (Fig. 4).

As regards the northern sector of the basin (the Tavoliere Plain), the coarse-grained deposits on top of the Argille subappennine Formation erosionally lie on the hemipelagites (Caldara & Pennetta 1991), except for the highest ones (Jacobacci *et al.* 1967; Boni *et al.* 1969; Caldara & Pennetta 1993). These latter, marine and/or transitional in origin, represent scattered remnants of the oldest Regressive coastal deposits of the Tavoliere Plain preserved from erosion only near the flanks of the emerging chain; other coarse-grained marine deposits erosionally lie on the hemipelagites of the Argille subappennine Formation near the present-day coastlines, north-westward and southward of the Gargano promontory (Caldara & Pennetta 1991, 1993) and represent the youngest Regressive coastal deposits formed in the Tavoliere Plain. Lithologies and facies of these coastal deposits are similar to those observed in the central-southern part of the Bradanic Trough. In contrast, in a wide part of the Tavoliere Plain (between the areas in which the oldest and the youngest Regressive coastal deposits crop out), alluvial terraces only (either alluvial fan or braided river in origin) erosionally lie on top of the Argille subappennine Formation (Pieri *et al.* 2001).

The present-day south Apennines foredeep basin (Taranto Gulf)

The Taranto Gulf is the present-day foredeep area of the south Apennines orogenic system where sedimentation is still active; it is a semi-enclosed basin that opens towards the southeast in the Ionian Sea (Fig. 5a). According to Pescatore & Senatore

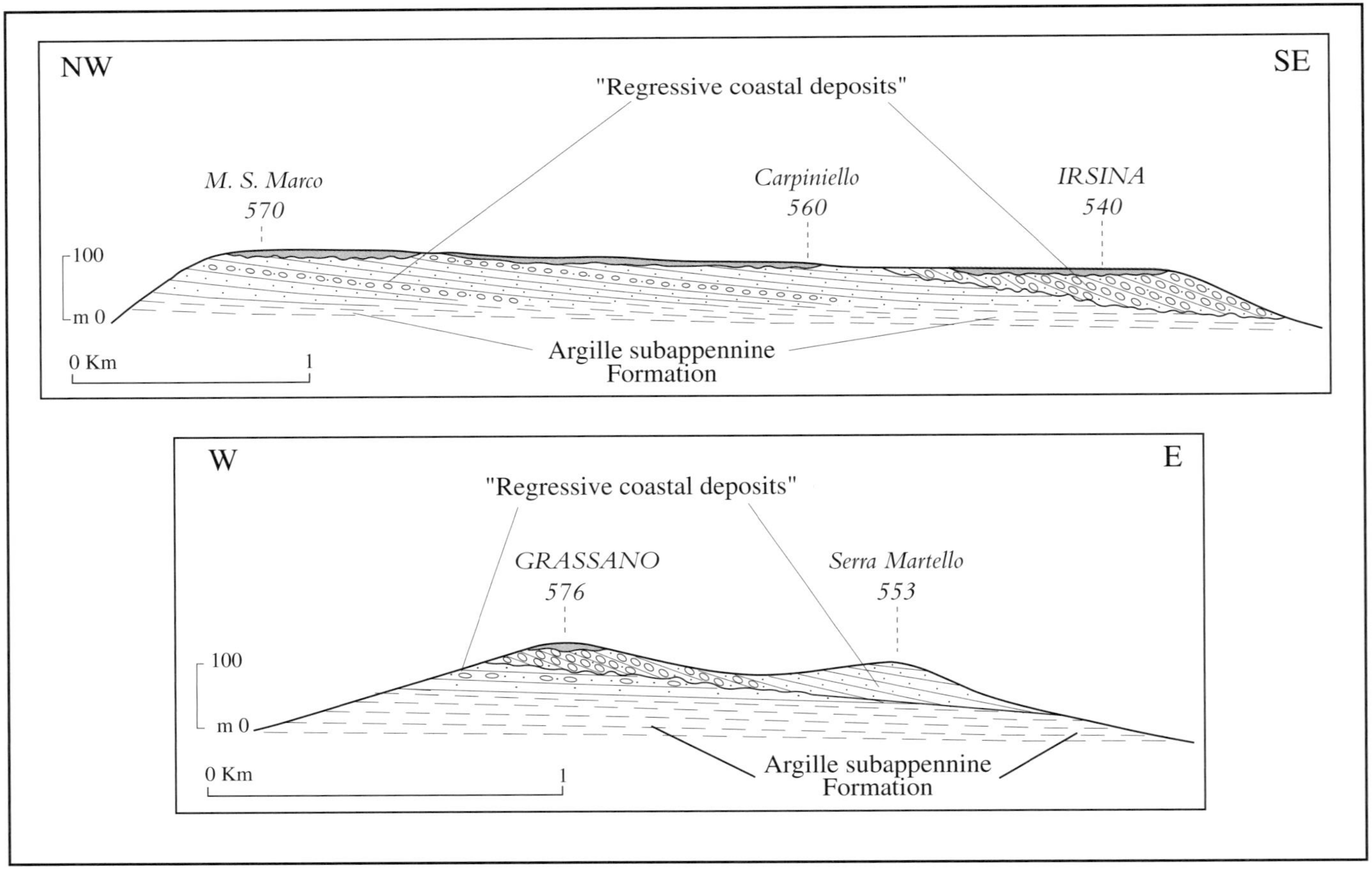

Fig. 4. Depositional schemes of the 'Regressive coastal deposits' outcropping in Irsina and Grassano localities (see Fig. 3). Modified from Sabato (1996).

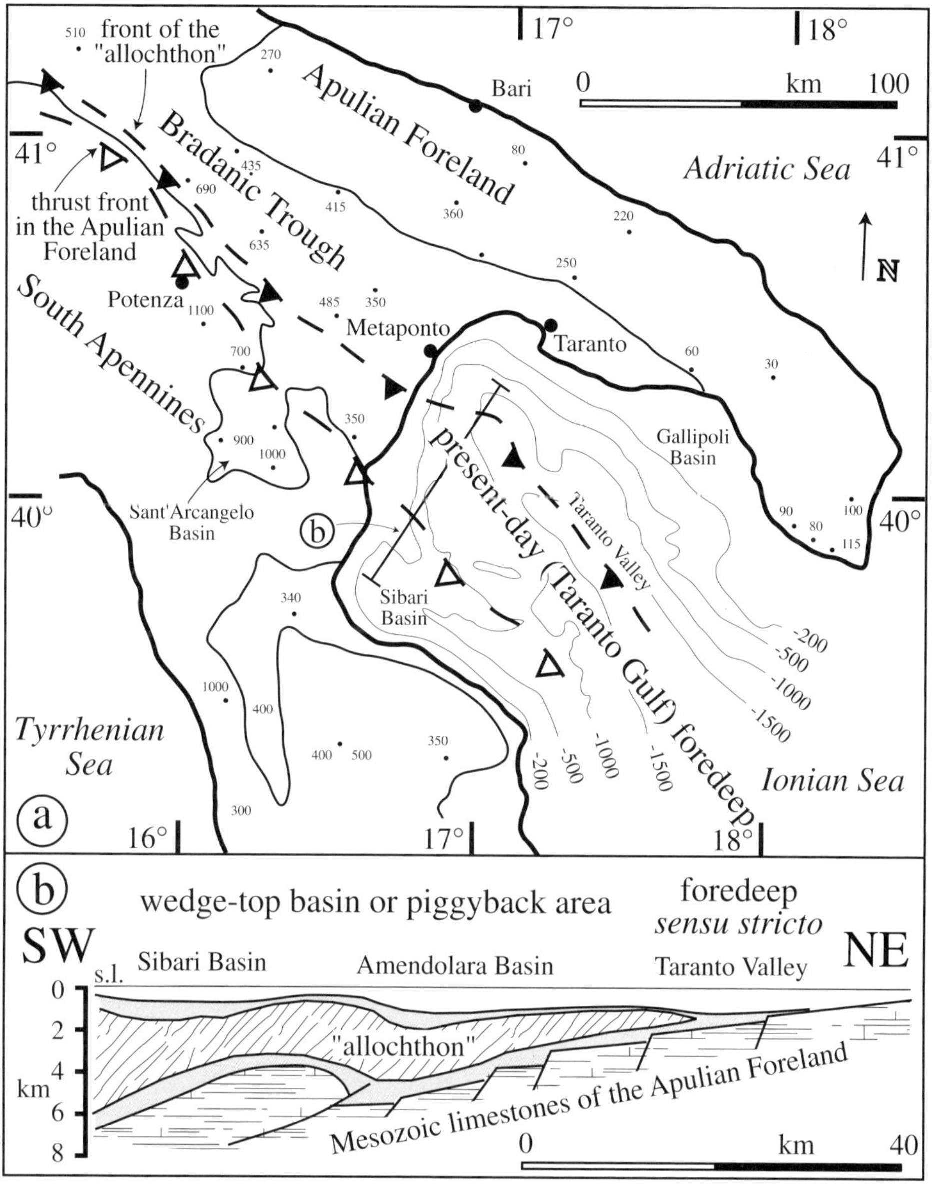

Fig. 5. (**a**) Prolongation of the Bradanic Trough in the Taranto Gulf. Bathymetric features and the main structural elements are shown (modified from Senatore *et al.* 1988); (**b**) schematic cross-section through the Taranto Gulf. Note the restricted area in which the present-day foredeep *sensu stricto* is located (modified from Vitale 1995).

(1986), three sectors can be distinguished in the Taranto Gulf: (i) the western sector (corresponding to the wedge-top basin of Fig. 2d); (ii) the central sector (corresponding to the foredeep *sensu stricto* of Fig. 2d); and iii) the eastern sector (corresponding to the middle and outer foreland ramp of Fig. 2d).

The western sector is characterized by two types of depositional ramp. The first type, here called the 'abrupt ramp', is from 1 to a few kilometres wide, is elongated SE–NW (adjacent to the exposed chain) and dips a few degrees towards the NE (towards the foredeep axis). The abrupt ramp passes laterally to another type of depositional ramp, here called the 'gentle ramp', which is up to 15 km wide, is oriented NW–SE and dips 1°

towards the SE (parallel to the foredeep axis). Only the gentle ramp shows active frontal progradation of the shallow marine systems. A significative change in the depositional profile is located at depths varying between 30 (near the chain) and 150 m. A wide slope connects the shallow-marine environments to a basin characterized by several depressions separated by submerged ridges mainly corresponding to thrust folds (Pieri *et al.* 1997*b*; Doglioni *et al.* 1999*b*) (Fig. 5b). Very small turbidite systems exist in these depressions and are linked to the main rivers (i.e. the Crati submarine fan in the Sibari Basin, see Fig. 5a and b: Ricci Lucchi *et al.* 1984; Colella & Di Geronimo 1987), but the wide wedge-top basin is basically underfed.

The central sector of the Taranto Gulf is the foredeep *sensu stricto* and corresponds to a very narrow submarine channel (the Taranto valley), up to 6–8 km wide, confined between the allochthon front and the highly inclined (up to 6° in dip) regional monocline (Tramutoli *et al.* 1984; Senatore *et al.* 1988; Doglioni *et al.* 1999*b*; Merlini *et al.* 2000) (Fig. 5b). It is a site of turbidite sedimentation (Pescatore & Senatore 1986).

The eastern sector of the Taranto Gulf occupies the regional monocline, corresponding to the subsided Apulian Foreland. The depositional ramp is oriented NW–SE, parallel to the Salento peninsula, and dips toward the Taranto valley. It is characterized by a series of terraces mainly caused by faulting and it is here called the 'faulted ramp'. A steep slope connects the faulted ramp to the Taranto valley. Carbonate sedimentation occurs along the coast (Dal Cin & Simeoni 1987), while depressions located seaward may be characterized by silty-clay deposits (i.e. the Gallipoli Basin) (Senatore *et al.* 1982) (Fig. 5a). Overall, there is little sedimentation on the Apulian Foreland and slumping exposes wide areas of bedrock on the faulted ramp and slope (Senatore *et al.* 1982).

In summary, the majority of the present-day foredeep (the Taranto Gulf) is occupied by the allochthon. A wide wedge-top basin characterizes the foredeep, whilst the foredeep *sensu stricto* is limited to a narrow channel. Shallow-marine depositional systems, either 'abrupt', 'gentle' or 'faulted' ramps, dip at least 1° seaward and represent the inner sector of major and more inclined sloping depositional ramps.

Pliocene–Pleistocene depositional history of the south Apennines foredeep

Middle and late Pliocene

Similar to the present-day Taranto Gulf basin, and in accordance with Patacca & Scandone (1989), the Pliocene south Apennines foredeep was characterized by an accretionary wedge formed by long thrust sheets (the allochthon) that progressively occupied accommodation space generated by roll-back of the Adria Plate (Fig. 1c).

In such a basin framework, coarse-grained sediments were deposited in narrow nearshore zones against the chain flank, while hemipelagites were deposited seaward on the wide wedge-top basin. These sediments, whose depositional systems were probably similar to the abrupt ramps of the present-day Taranto Gulf, were promptly deformed and cannibalized. Tectonized Pliocene marine deposits crop out discontinuously over the exposed chain, up to several tens of kms from its present-day front (Bonardi *et al.* 1988; Patacca *et al.* 1990; Amato & Cinque 1992; Amore *et al.* 1998; Matano & Staiti 1998). These Pliocene successions may reach more than 1 km in thickness and are often characterized by internal angular or progressive unconformities (Maggiore & Walsh 1975; Caldara *et al.* 1993; Pieri *et al.* 1994*a*; Sabato & Marino 1994; Riviello *et al.* 1997). Pliocene turbidite deposits on top of the allochthon were observed in the Baronia area only (Matano & Staiti 1998), and this situation could be considered an ancient counterpart of the present-day Crati submarine fan deposits. In contrast, turbidite systems described by Casnedi (1991) occupied the narrow axis of the foredeep *sensu stricto* between the allochthon and the regional foreland ramp (Fig. 6). These turbidites were mainly supplied from north along the axis of the Bradanic Trough (Casnedi 1988*b*), which at that time was a seaway between the Adriatic Sea, to the north, and the Ionian Sea to the south (Fig. 6b). Owing to their structural position, these turbidite deposits were underthrust by the allochthon. Later, younger thrusts which grew in the Apulian Foreland monocline involved the foredeep *sensu stricto* deposits previously buried below the allochthon (Fig. 2c and d).

As regards the Apulian Foreland ramp, this was characterized by two sectors with different degrees of dip, during Pliocene time. The more inclined sector (the inner foreland ramp) corresponded to the bedrock of the foredeep *sensu stricto*, whilst the lesser inclined sector corresponded to the pre-Gargano and pre-Murge plateaux (middle and outer foreland ramp) (Fig. 6c and d). Inner and middle foreland ramp sectors were connected by normal and/or transtensional faults, which in the space of a few kilometres vertically displaced the foreland ramp for about 1 km (Pieri *et al.* 1996*b*, *c*) (see the margins of the plateaux in Fig. 6c–e; see also Fig. 3b). Displacement took place mainly during the Pliocene since, after the Pliocene, in the southern part of the basin only there was space for turbidite

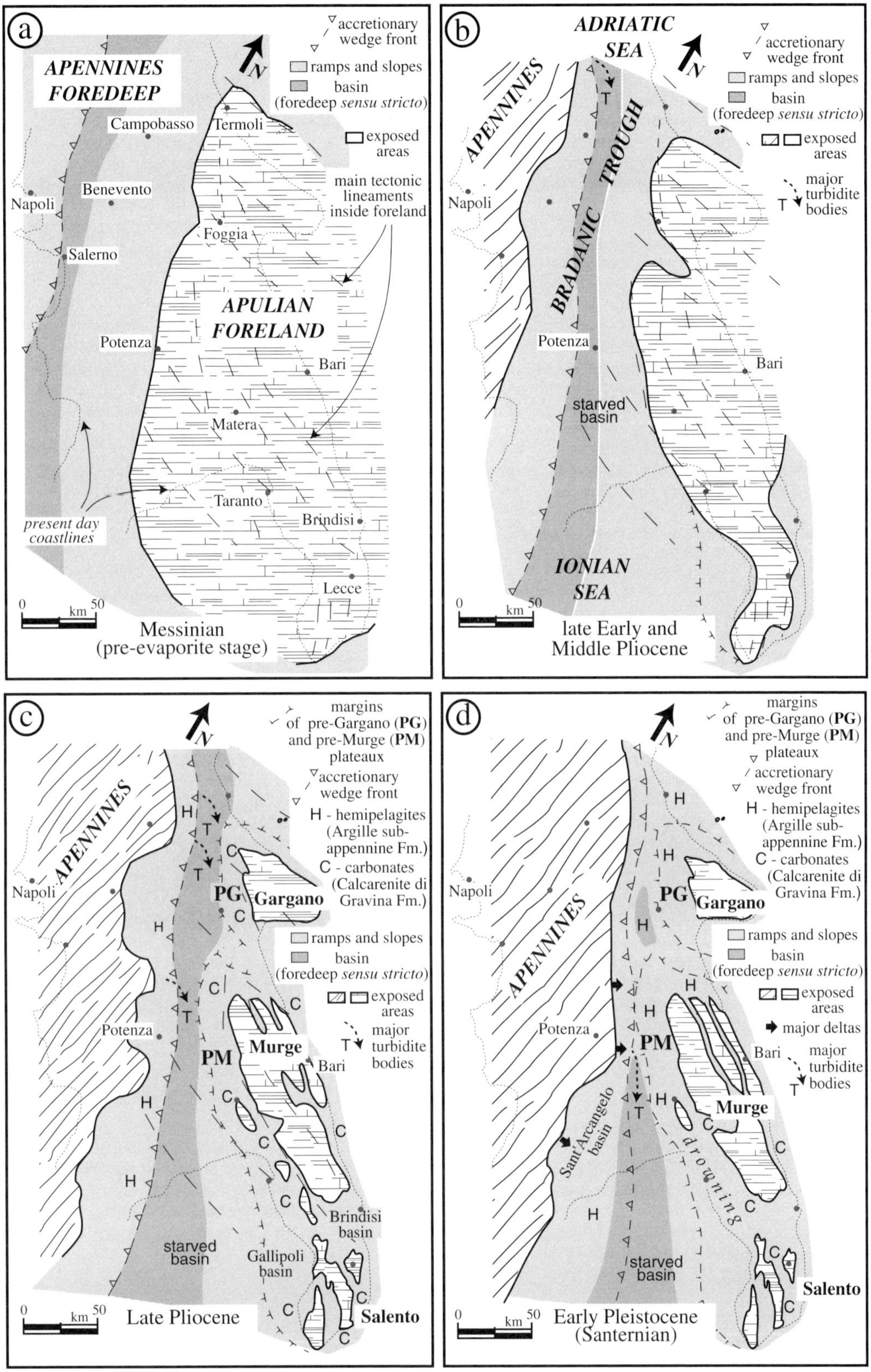

Fig. 6. Palaeogeographies of the southern Apennines foreland basin system from the Messinian to the Early Pleistocene.

sedimentation on the inner foreland ramp (Fig. 6d). Therefore, during Late Pliocene, the eastward migration of the allochthon was not accompanied by the eastward migration of the axis of the foredeep (the foredeep *sensu stricto*), since turbidites did not deposit onto the pre-Gargano and pre-Murge plateaux (Fig. 6c and d).

Early Pleistocene

The present-day distribution of Lower Pleistocene deposits reflects the palaeogegraphy of the Bradanic Trough during that time. The wedge-top basin was very narrow in the central and northern sectors of the Bradanic Trough (Fig. 6d and e) as: (i) practically no Pleistocene marine deposits crop out on the adjacent sectors of the Apennines chain; and (ii) some thick Lower Pleistocene delta bodies were located on the front of the allochthon, where rivers flowed into the basin (i.e. Serra del Cedro Formation: Gambassini 1967; Loiacono & Sabato 1987 – Figs 3b and 6d and e). Therefore, the front of the allochthon corresponded to the exposed front of the chain or was buried a few kilometres from it, and sediments coming from the chain were transported directly into the foredeep *sensu stricto*. In contrast, southward (along strike) the wedge-top basin remained wide, and a very large piggyback basin (the Sant' Arcangelo Basin) collected a large amount of sediment (Vezzani 1967*b*; Pieri *et al.* 1994*a*, 1996*a*) (Fig. 6d and e).

According to the distribution of Lower Pleistocene turbidite bodies described by Casnedi *et al.* (1982), the foredeep *sensu stricto* was present only in the southern part of the trough (Fig. 6d–f). It was characterized along strike by relative highs and depressions that controlled the thicknesses of sediment and determined the development of turbidite flows, which were restricted to very narrow and confined depressions (Fig. 6d). These morphostructural features were governed by faults of the foreland ramp (Sella *et al.* 1988).

In the central and northern part of the basin, since pre-Gargano and pre-Murge plateaux had not developed in the foredeep *sensu stricto*, the allochthon front reached the middle foreland ramp and effectively stopped against the fault system that displaced inner from middle foreland ramp (Pieri *et al.* 1996*c*) (Figs 3a and 6d and e). Lacking both the wedge-top basin and the foredeep *sensu stricto*, the siliciclastic depositional systems deriving from the chain prograded toward the foreland via eastward-dipping sedimentary gentle ramps (Fig. 6e and f). Where the front of the allochthon was contiguous with the middle foreland ramp, pre-existing structural features of the foreland became the main control on sedimentation of shallow-marine systems deriving from the Apennines chain. The Apulian Foreland was affected by E–W- and WNW–ESE-striking faults (Ciaranfi *et al.* 1983), which were mainly inherited from the Mesozoic development of the carbonate platform and reactivated during Neogene time (Festa 1999; Chilovi *et al.* 2000) (Fig. 6a and b). Faults dissected the foreland into a complex horst and graben structure (Iannone & Pieri 1983; Pieri *et al.* 1997*a*) and consequently carbonates of the Calcarenite di Gravina Formation covered narrow shore platforms around, and on top of, palaeo-islands (horsts), or were deposited into narrow straits (grabens), or onlapped degraded fault scarps (Tropeano & Sabato 2000; Pomar & Tropeano 2001) (Fig. 6c–e). Hemipelagites, coastal and, finally, alluvial deposits prograded onto the pre-Gargano and pre-Murge plateaux and then onto the outer foreland ramp, in places suffocating the remnant carbonate systems of the Calcarenite di Gravina Formation. After the chain and the foreland highs of Murge were connected (Fig. 6e and f), siliciclastic depositional systems prograded laterally (either north- or southward) progressively occupying the large accommodation space of the whole foreland ramp not filled by the preceding carbonate systems. Siliciclastic systems also filled small depressions inside the foreland (i.e. the Murge grabens and the Brindisi Basin) (Fig. 6c, d and g).

Middle and Late Pleistocene

After sedimentary infilling of that part of the basin between the chain and the Murge plateaux, two large gulfs became the only underfilled remnants of the Bradanic Trough (Fig. 6f and g). Lacking a foredeep *sensu stricto*, the northern gulf was progressively filled, whilst the southern one, which was deeper and longer than its northern counterpart, is still present in the Taranto Gulf today. Sediments progressively invaded shallow-marine environments of both gulfs and sedimentary closure of the strait between the chain and the Gargano island probably occurred soon after the strait adjacent to the Murge archipelago had infilled. Filling of the basin is recorded by the shallowing-upward trend of the Pleistocene successions and by the widespread sedimentation of the Regressive coastal deposits (Fig. 6g and h). Moreover, during Middle and Late Pleistocene time contrasting forms of erosion characterize deposits filling the two palaeogulfs. In the northern part of the Bradanic Trough, planar erosion took place and most of the Regressive coastal deposits were widely cannibalized. This erosion is recorded in the Quaternary succession of the central Adriatic Sea as a large sedimentary wedge which prograded from the south (Ciabatti *et al.* 1987; Trincardi & Correggiari

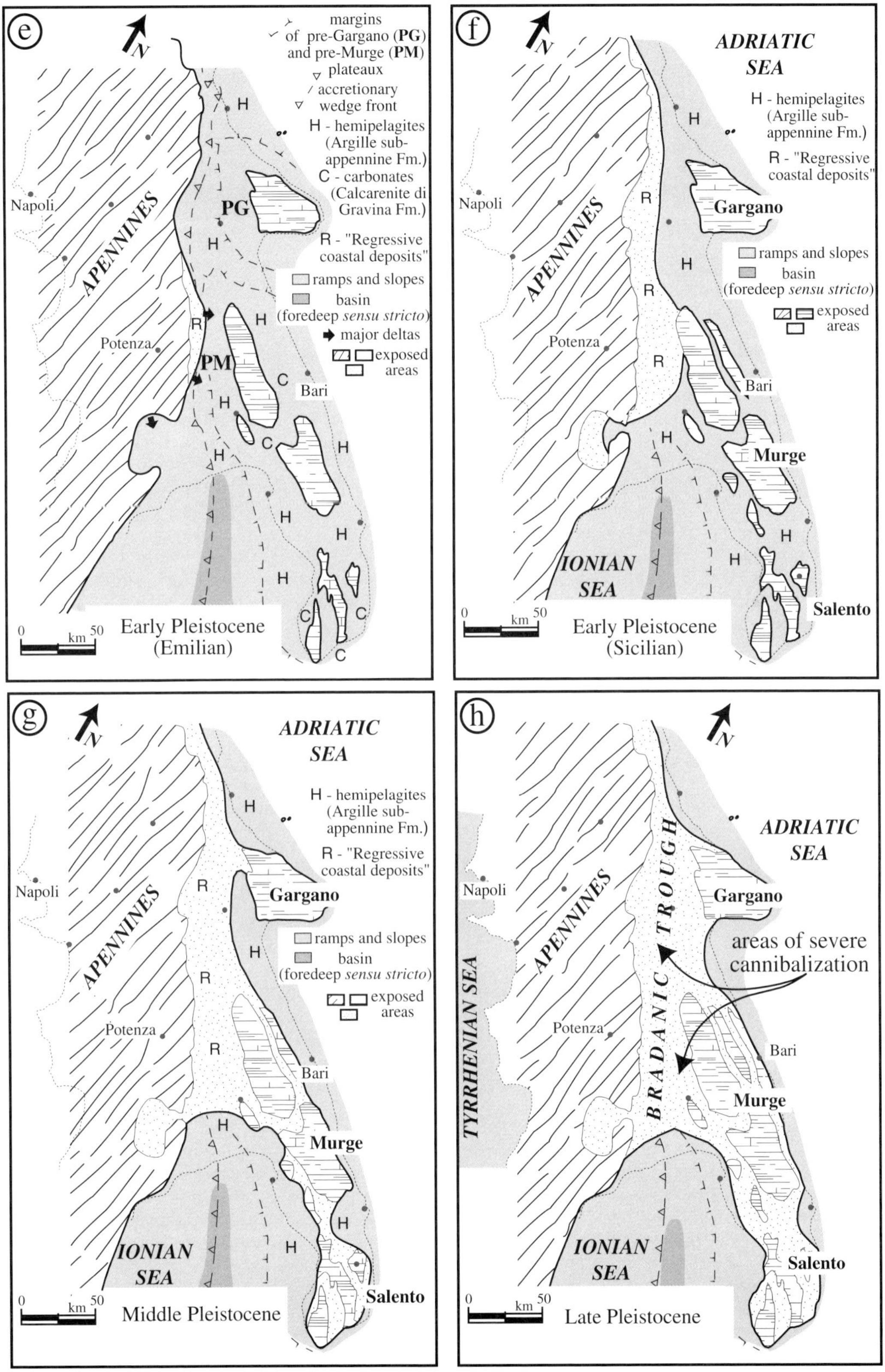

Fig. 6. cont. Palaeogeographies of the southern Apennines foreland basin system from the Early Pleistocene to the Late Pleistocene.

2000). In the southern part of the Bradanic Trough, river incisions deeply cut the succession and large remnants of the oldest Regressive coastal deposits still exist. Cannibalization of these deposits is the main source for the present-day depositional systems of the Taranto Gulf.

Discussion

Stratigraphic constraints on geodynamics in the southern Apennines foredeep

In the southern Apennines foreland, before sedimentation of hemipelagites, the backstepping configuration that characterizes shallow-marine carbonates of the Calcarenite di Gravina Formation records a severe transgression on karstic areas of the Apulian Foreland and indicates a relative sea-level rise that occurred up to Late Pliocene–Early Pleistocene time. During Middle Pliocene–Early Pleistocene time the eastward migration of flexural subsidence towards Apulia induced by the westward subduction of the Adria Plate was the only agent capable of producing this high-amplitude transgression on the Apulian Foreland. Therefore, nearshore sedimentation of the Calcarenite di Gravina Formation is indicative of the arrival of dynamic subsidence in the region and, as suggested by Tropeano (1994) and by Pieri *et al.* (1997*a*), the whole Apulian Foreland (exposed and submerged) should be considered part of the Pliocene outer foreland ramp. Moreover, sediments similar to the Calcarenite di Gravina Formation should have been deposited on the bedrock (below the pre-turbidite hemipelagites) at the base of the Bradanic Trough succession (as transgression on previously exposed lands should be accompanied by nearshore sedimentation), but they never appear in the subsurface description of the Bradanic Trough. The eolianites (backshore deposits), that characterize the highest outcrops only, record the maximum transgression around the structural highs of the Apulian Foreland. Beach facies (both foreshore and backshore ones) had a low preservation potential during transgression, being eroded by ravinement processes (Tropeano & Sabato 2000). Their preservation in outcrops more elevated than those characterized by marine facies indicates that ravinement processes never reached these eolian sedimentation areas, and records the end of the transgression during Late Pliocene–Early Pleistocene time. The maximum relative sea level on the Apulian Foreland was up to 450–500(?) m in elevation above present-day sea level, and this may be inferred from the elevation of the highest eolianites outcrops.

Contemporaneously to sedimentation of the Calcarenite di Gravina Formation, turbidite systems developed in the foredeep basin. In the central part of the Bradanic Trough the oldest turbidite bodies lie below the allochthon, whilst younger ones pass upward to Upper Pliocene–Lower Pleistocene hemipelagites, indicating a shallowing-upward trend in the basin that was not induced by thrust accretion. In this part of the basin the shallowing is contemporaneous to the end of the transgression on the Apulian Foreland, and the outcropping deposits belonging to the Argille subappennine Formation record this shallowing in the foredeep *sensu stricto* and the drowning of carbonates in the outer foreland ramp. Moreover, the upper part of the Argille subappennine Formation shows facies characteristics related to storm-dominated depositional ramp environments that pass to shoreface; this sedimentary passage also indicates a shallowing trend. The Argille subappennine Formation never covered the highest Plio-Pleistocene carbonates on the Apulian Foreland as most elevated hemipelagites in the outer foreland ramp successions (below the Regressive coastal deposits) outcrop at about 400 m above present-day sea level. Age of hemipelagites becomes younger from north to south and from west to east, indicating the progressive shallowing and filling of the basin in the same directions.

Progradation of the Regressive coastal deposits of the Bradanic Trough near the chain front is coeval to the sedimentation of the top of the Argille subappennine Formation in the basin. Regressive coastal deposits progressively replaced hemipelagites during shallowing and filling of the basin, and their altitude and age progressively decrease from the central part of the trough up to the present-day coasts. Regressive coastal deposits reach no more than 100 m in thickness and are mainly marine or transitional in origin. Alluvial facies may be present in the upper part of these regressive successions but they reach no more than a few tens of metres in thickness. Regressive coastal deposits consist of coarse-grained wedges that lie on the hemipelagites in, alternatively, conformable or erosional contact. In the first case the deposits represent the product of a depositional regression (*sensu* Curray 1964; Heward 1981), favoured by rates of sediment supply higher than rates of increase of accommodation space, while in the second case the deposits are the response of a forced regression (*sensu* Posamentier *et al.* 1992). This type of wedge stacking may be compared with the downward-shifting configuration, described by Pomar (1993), which indicates deposition induced by high-frequency relative sea-level changes during a long-term relative sea-level fall (or tectonic uplift). Neotectonic studies indicate that both the Bradanic Trough and the adjacent Apulian Foreland are affected by uplift (Ciaranfi *et al.* 1983; Doglioni *et al.* 1996; Pieri *et al.* 1997*a*). According to these studies, uplift acts at least from the Early–Middle Pleistocene time as Early Pleistocene coastal deposits crop out at about 400–450 m above sea level in the foreland and at more than 600 m in the trough. According to geochronological data collected in Late Pleistocene deposits by Cosentino & Gliozzi (1988), Dai Pra & Hearty (1988) and Hearty & Dai Pra (1992), uplift rates of the Apulian Foreland reached 0.2–0.3 mm year^{-1}, while, according to stratigraphical data collected in the Early, Middle and Late Pleistocene sedimentary units by Ciaranfi *et al.* (1988, 1994*b*), uplift rates of the Apulian Foreland are calculated at about 0.5–0.7 mm year^{-1}. As regards the Regressive coastal deposits, according to morphological evidence, their tops in the central part of the trough were interpreted to be a single depositional surface successively uplifted and eastward tilted (Patacca *et al.* 1990; Cinque *et al.* 1993; Amato & Cinque 1999). However, we suggest that the complex (downward shifting) configuration that characterizes the stratigraphic architecture of the Regressive coastal deposits indicates that they formed during uplift and that the decreasing elevation of the deposits should be considered an original feature rather than a post-depositional one. Moreover, the highest outcrops with marine facies of the Calcarenite di Gravina Formation are subaerially eroded while, lower down, the Regressive coastal deposits disconformably lie on the Calcarenite di Gravina Formation, indicating that uplift was active before the arrival of the coastal siliciclastic sediments. These data pre-date the beginning of uplift in the region suggesting that the upper part of the Argille subappennine Formation also probably formed during uplift or at least during relative tectonic still-stand of sea level in an unfilled basin. The last consideration is in accordance with the age of the end of transgression on the Apulian Foreland inferred by the presence of eolianites belonging to the Calcarenite di Gravina Formation.

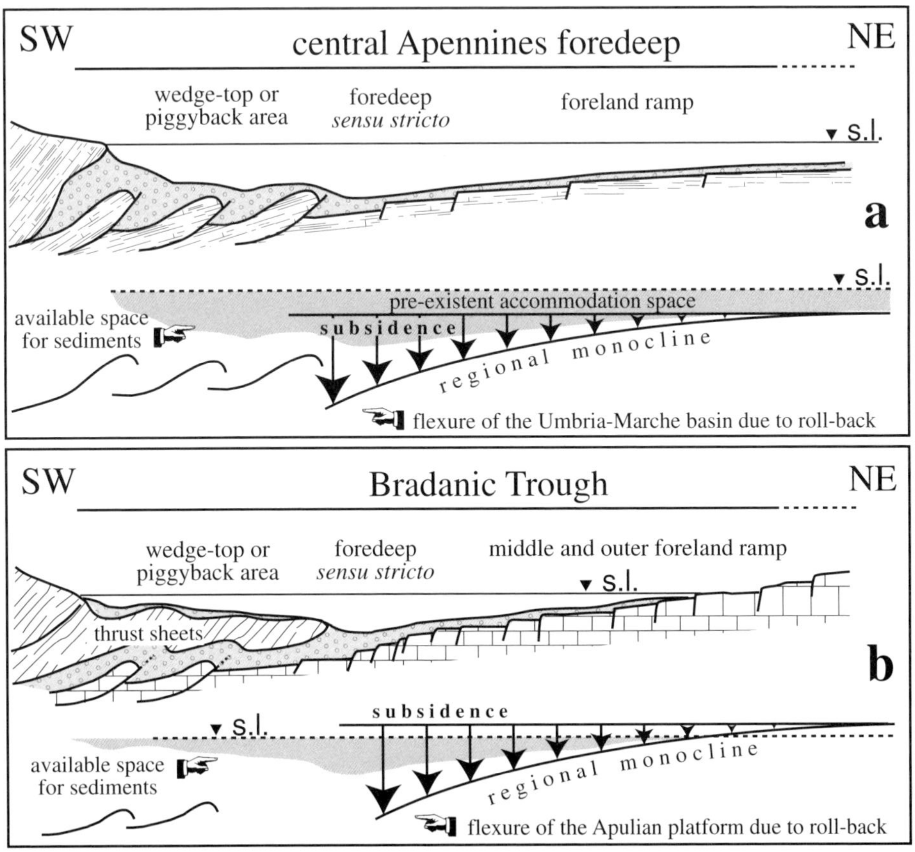

Fig. 7. Comparative sketches between the central Apennines foredeep (**a**) and Bradanic Trough (**b**) during the Pliocene.

Filling and overfilling vs cannibalization in the Plio-Pleistocene Apennines foredeep

In accordance with data from Patacca & Scandone (1989), during the Pliocene two different styles of foredeep migration characterized the central and southern Apennines (Fig. 7). In the central Apennines, the foreland sedimentary cover was directly involved in the thrust propagation and was predominantly added to the front of the accretionary wedge (Argnani *et al.* 1991) (Fig. 2a and b); in contrast, in the southern Apennines, foreland sedimentary cover was underthrust by the allochthon (Ciaranfi *et al.* 1979; Casnedi *et al.* 1982) (Figs 1c, 2c and d and 3b).

During the Pliocene, the central Apennines foreland was characterized by pelagic sedimentation on the Meso-Cenozoic Umbria–Marche Basin succession. When flexural subsidence reached the region it was added to the accommodation space formerly existing in these deep-marine environments (Fig. 7a). Here the hemipelagites, which were deposited in the basin before the turbidites, represented the foreland geodynamic state before the onset of the dynamic subsidence (pre-turbidite stage). In contrast, during the Pliocene the southern Apennines foreland was mainly an exposed area, characterized by limestones of the wide and thick Mesozoic Apulian platform. When flexural subsidence reached these continental areas it induced a transgression on a karstic land creating accommodation space only after submersion of the region (Fig. 7b). In this case nearshore carbonates and hemipelagites, which deposited on the bedrock (prior to the turbidites), were themselves indicative of the arrival of dynamic subsidence and so correlation between sedimentary stages (pre-turbidite, turbidite and post-turbidite) and geodynamic stages (foreland, foredeep and thrust accretion) cannot be correctly applied.

During the Pleistocene, the allochthon propagation in the southern Apennines slowed down and the front of the allochthon progressively stopped against steps of the Apulian Foreland ramp, whose middle–outer sector did not subside further. As a consequence, backthrusts developed and both wedge-top basin and foredeep *sensu stricto* did not migrate and did not propagate toward the east and this led to their infilling. In southern Italy, Pliocene and Pleistocene strike-slip structures affected either the chain (Catalano *et al.* 1993; Schiattarella 1998) or the foredeep and the foreland (Doglioni *et al.* 1994; Gambini & Tozzi 1996; Pieri *et al.* 1997*a*). According to Doglioni *et al.* (1994) all of these features are related to the arrival of a thicker lithosphere (the Apulian swell) in the south Adriatic subduction hinge, which lowered subduction rates and caused the slab to buckle. Uplift of the Apulian Foreland was the consequence of this buckling, while northward and southward the Apennines foreland continued to subside (Doglioni *et al.* 1996).

Owing to the eastward migration of the orogenic system, the northern-central Apennines belt approached other mountain chains (Alps and Dinarides) and its foredeep became a semi-enclosed epicontinental shallow-marine basin during Quaternary times (Ciabatti *et al.* 1986; Trincardi & Correggiari 2000). The Po Plain foredeep sector was overfilled by sediments, which predominantly originated in the Alps, and evolved from a marine into an alluvial basin (Cremonini & Ricci Lucchi 1982); river discharge produced more than 50 km of shelf progradation toward the central Adriatic Sea (Ciabatti *et al.* 1986, 1987; Ori *et al.* 1986; Roveri *et al.* 1986). The small amount of accommodation space precluded turbidite deposition in this remnant basin and prodelta mud accumulation occurs in the deeper part of north-central Adriatic Sea (no more than 200 m deep). Consequently, sedimentary accumulation rates compensate or exceed the foredeep subsidence rates (Ciabatti *et al.* 1986, 1987; Roveri *et al.* 1986; Trincardi & Correggiari 2000). This datum is well constrained in the Po Plain, as the top of the foredeep succession, no more than 100 m in elevation above sea level, is made up of Quaternary alluvial deposits that reach 450–500 m in thickness, indicating alluvial sedimentary aggradation during basin subsidence (Dondi *et al.* 1982; Ori 1993; Argnani *et al.* 1997; Regione Emilia Romagna, ENI-AGIP 1998).

The cause of the filling of the Bradanic Trough is very different from that of the northern part of the foredeep. Shallowing of the basin began in the central area of the trough in front of the western corner of Murge. Afterwards, depositional systems formed marine and/or continental terrace deposits that prograde and are progressively lower in elevation towards the present-day coastlines, and become younger towards the north (the central Adriatic Sea) east (the Manfredonia Gulf in the south Adriatic Sea), and south (the Taranto Gulf in the Ionian Sea). In contrast to the north-central Adriatic foredeep, lack of accommodation space for turbidite deposits was due mainly to uplift of the region rather than to overfilling of the remnant basin. Elevation up to more than 600 m of Quaternary marine deposits in the Bradanic Trough clearly demonstrates this uplift. Deposits corresponding to the outcropping Argille subappennine Formation lie several thousands of metres below sea level in the central and northern Apennines foredeep (Doglioni *et al.* 1994). Moreover, filling of the Po Plain is recorded by a thick pile of aggrading alluvial deposits, which practically never deposited in the Bradanic Trough.

Therefore, the terms 'filling' and 'overfilling' should be used only for: (i) the Pleistocene sedimentary development of the central and northern Apennines foredeep; and (ii) for the Late Pliocene up to Early Pleistocene (Santernian) sedimentary development of the Bradanic Trough; the post-Santernian development of the Bradanic Trough is better defined as a phase of 'cannibalization' induced by uplift. Also the use of the term 'post-turbidite deposits' (or stage) to indicate those deposits here called the Regressive coastal deposits should be abandoned as, in the case of the south Apennines foredeep, these deposits mainly cover areas that never reached deep-marine environments, and they did not form after filling and aggradation of turbidite deposits.

According to Trobeano *et al.* (2001) the Quaternary stratigraphic arrangement records these two different developments of the Apennines foredeep. The north-central Adriatic basin in-fill simulates the development of a shelf-margin systems tract (*sensu* Van Wagoner *et al.* 1988), while the Bradanic Trough simulates the development of a forced-regressive systems tract (*sensu* Hunt & Tucker 1992; Plint & Nummedal 2000).

Conclusions

The north (Po Plain) and the south (Bradanic Trough) Apennines foredeep were both considered overfilled (molasse) basins, but geological data reported in this work suggest that they differ greatly. Differences developed during Quaternary time, as the Quaternary history of the Bradanic Trough is characterized by erosional regression (cannibalization) in an uplifting basin while the Quaternary history of the north-central Apennines foredeep is characterized by depositional regression (filling and overfilling) in a subsiding remnant basin. In contrast, only the southernmost sector of the present-day foredeep (the Taranto Gulf in the Ionian Sea) is a deep basin in which turbidite sedimentation can take place.

Also, the Pliocene evolution is different as crustal features of the south Apennines foreland (Apulian Foreland) forced the growth of an accretionary wedge, characterized by long thrust sheets (the allochthon) which overthrust turbidites located along the axis of the foredeep and which occupied a large part of the accommodation space created on the foreland by flexural subsidence due to the roll back of the west-subducting Adria Plate. Starting from the Late Pliocene, the subsidence rate of the middle and outer Apulian Foreland ramp was reduced or practically ceased, while the allochthon continued to move toward the foreland. In such a context, during Early Pleistocene times the allochthon reached the foreland and physically connected depositional systems on the flank of the chain, with the shallow-marine areas of the Apulian Foreland. Accommodation space on the foreland was quickly infilled by siliciclastic deposits (mainly hemipelagites), which 'suffocated' the nearshore carbonate systems of the Apulian archipelago. The Early Pleistocene was also characterized by the onset of the Bradanic Trough uplift, which favoured the filling of the basin before its cannibalization. The upper part of the infilling succession is made up of the Regressive coastal deposits; they were previously interpreted to be aggrading and synchronous, at least in the central part of the basin, but they show a stratigraphic configuration that records deposition induced by high-frequency relative sea-level changes during uplift.

Careful reviews of M. Pedley and R. Robinson, and suggestions of M. Schiattarella improved the manuscript. Discussion with C. Doglioni, V. Festa, L. Pomar and G. Prosser were very useful. The editorial patience of S. Jones was greatly appreciated. The research was supported by CNR (grants 98.00273.CT05 and 99.00700.CT05 to P. Pieri) and MURST (grants 60% 1999 and 2000 to L. Sabato; grants PRIN 2000 to L. Simone and A. Laviano).

References

ALLEN, P. A., HOMEWOOD, P. & WILLIAMS, G. D. 1986. Foreland basins: an introduction. *In*: ALLEN, P. A. & HOMEWOOD, P. (eds) *Foreland basins*. International Association of Sedimentologists, Special Publications, **8**, 3–12.

AMORE, O., BASSO, C., CIARCIA, S., DI NOCERA, S., MATANO, F., TORRE, M., CIAMPO, G., DI DONATO, V., ESPOSITO, P. & STAITI, D. 1998. Nuovi dati stratigrafici sul Pliocene affiorante tra il Fiume Ufita e il Torrente Cervaro (Irpinia, Appennino meridionale). *Bollettino della Società Geologica Italiana*, **117**, 455–466.

AMATO, A. & CINQUE, A. 1992. Il bacino plio-pleistocenico di Calvello (Potenza): evoluzione geologica e geomorfologica. *Studi Geologici Camerti*, Volume Speciale **1992/1**, 181–189.

AMATO, A. & CINQUE, A. 1999. Erosional landsurface of the Campano-Lucano Apennines (S. Italy): genesis, evolution, and tectonic implications. *Tectonophysics*, **315**, 251–267.

ARGNANI, A., ARTONI, A., ORI, G. G. & ROVERI, M. 1991. L'avanfossa centro-adriatica: stili strutturali e sedimentazione. *Studi Geologici Camerti*, Volume Speciale **1991/1**, 371–381.

ARGNANI, A., BERNINI, M., DI DIO, G. M., PAPANI, G. & ROGLEDI, S. 1997. Stratigraphic record of crustal-scale tectonics in the Quaternary of the northern Apennines (Italy). *Il Quaternario*, **10**, 595–602.

AUCELLI, P. P. C., CINQUE, A. & ROBUSTELLI, G. 1997.

Evoluzione quaternaria del tratto di avanfossa appenninica compreso tra Larino (Campobasso) e Apricena (Foggia). Dati preliminari. *Il Quaternario*, **10**, 453–460.

AZZAROLI, A. 1968. *Studi illustrativi della Carta Geologica d'Italia – Formazioni Geologiche*. Servizio Geologico d'Italia, **I**, 183–185.

AZZAROLI, A. PERNO, U. & RADINA, B. 1968*a*. *Note illustrative della Carta Geologica d'Italia, F°188 'Gravina di Puglia'*. Servizio Geologico d'Italia.

AZZAROLI, A. RADINA, B., RICCHETTI, G. & VALDUGA, A. 1968*b*. *Note illustrative della Carta Geologica d'Italia, F°189 'Altamura'*. Servizio Geologico d'Italia.

BALDUZZI, A., CASNEDI, R., CRESCENTI, U., MOSTARDINI, F. & TONNA, M. 1982*a*. Il Plio-Pleistocene del sottosuolo del bacino lucano (Avanfossa appenninica). *Geologica Romana*, **21**, 89–111.

BALDUZZI, A., CASNEDI, R., CRESCENTI, U. & TONNA, M. 1982*b*. Il Plio-Pleistocene del sottosuolo del bacino pugliese (Avanfossa appenninica). *Geologica Romana*, **21**, 1–28.

BALLY, A. W., BURBI, L., COOPER, C. & GHELARDONI, R. 1986. Balanced sections and seismic reflection profiles across the Central Apennines. *Memorie della Società Geologica Italiana*, **35** (I), 257–310.

BEAUMONT, C. 1981. Foreland basin. *Geophysical Journal of the Royal Astronomical Society*, **65**, 291–329.

BETTAZZOLI, P. & VISENTIN, C. 1998. 3D Architecture of the Porto Garibaldi turbidite complex (northern Apennines foreland basin – Italy) (Abstract). *In*: *15th International Sedimentological Congress, Alicante, Spain, 12–17 April 1998*. Universidad de Alicante, Alicante, Spain, 192–193.

BOCCALETTI, M., CIARANFI, N., COSENTINO, D., DEIANA, G., GELATI, R., LENTINI, F., MASSARI, F., MORATTI, G., PESCATORE, T., RICCI LUCCHI, F. & TORTORICI, L. 1990. Palinspastic restoration and paleogeographic reconstruction of the pery-Tyrrenian area during the Neogene. *Palaeogeography, Palaeoclimatology, Palaeoecology*, **77**, 41–50.

BOENZI, F., PALMENTOLA, G. & VALDUGA, A. 1976. Caratteri geomorfologici dell'area del Foglio 'Matera'. *Bollettino della Società Geologica Italiana*, **95**, 527–566.

BOENZI, F., RADINA, B., RICCHETTI, G. & VALDUGA, A. 1971. Note illustrative della Carta Geologica d'Italia, F°201 'Matera'. Servizio Geologico d'Italia.

BONARDI, G., D'ARGENIO, B. & PERRONE, V. (eds) 1988. *Geological Map of Southern Apennines*. Memorie della Società Geologica Italiana, **41**.

BONI, A., CASNEDI, R., CENTAMORE, E., COLANTONI. P., CREMONINI, G., ELMI, C., MONESI, A., SELLI, R. & VALLETTA, M. 1969. *Note illustrative della Carta Geologica d'Italia F°155 'San Severo'*. Servizio Geologico d'Italia.

BRÜCKNER, H. 1980. Flussterrassen und flusstäler im küstentiefland von Metapont (süditalien) und ihre beziehung zu meeresterrassen. *Düsseldorfer Geographische Schriften*, **15**, 5–32.

CALDARA, M., CIARANFI, N. & MARINO, M. 1993. I depositi plio-pleistocenici di avanfossa al bordo dell'Appennino meridionale fra Oliveto lucano e Garaguso (Basilicata). *Bollettino della Società Geologica Italiana*, **112**, 893–908.

CALDARA, M., COLELLA, A. & D'ALESSANDRO, A. 1979. Studio sedimentologico e paleoecologico di alcune sezioni pleistoceniche affioranti presso Cerignola (FG). *Rivista Italiana di Paleontologia e Stratigrafia*, **85** (I), 173–242.

CALDARA, M., D'ALESSANDRO, A. & LOIACONO, F. 1989. Regressive Pleistocene sequence near Gravina in Puglia, southern Italy: sedimentological and palaeoecological analyses. *In*: *Atti 3° Simposio di Ecologia e Paleoecologia delle comunità bentoniche, Catania, Italy, 12–16 October*. Università di Catania, Catania, Italy, 417–469.

CALDARA, M. & PENNETTA, L. 1991. Pleistocenic buried abrasion platforms in southeastern "Tavoliere" (Apulia, South Italy). *Il Quaternario*, **4**(2), 303–310.

CALDARA, M. & PENNETTA, L. 1993. Nuovi dati per la conoscenza geologica e morfologica del Tavoliere di Puglia. *Bonifica*, **VIII**(3), 25–42.

CAPUANO, N., PAPPAFICO, G. & AUGELLI, G. 1996. Ricostruzione dei sistemi deposizionali plio-pleistocenici del margine settentrionale dell'avanfossa pugliese. *Memorie della Società Geologica Italiana*, **51**, 273–292.

CARBONE, S. & LENTINI, F. 1990. Migrazione neogenica del sistema Catena-Avampaese nell'Appennino meridionale: problematiche paleogeografiche e strutturali. *Rivista Italiana di Paleontologia e Stratigrafia*, **96**, 271–295.

CARISSIMO, L., D'AGOSTINO, O., LODDO, C. & PIERI, M. 1962. Le ricerche petrolifere dell'Agip mineraria e nuove informazioni geologiche nell'Italia centro-meridionale dall'Abruzzo al Golfo di Taranto. *In: proceeding of VI Congresso Mondiale del Petrolio*, Francoforte, Germany, **I**, AGIP Report, San Danato Milanese, Italy.

CASERO, P., ROURE, F., ENDIGNOUX, L., MORETTI, I., MULLER, C., SAGE, L. & VIALLY, R. 1988. Neogene geodinamic evolution of the Souhern Apennines. *Memorie della Società Geologica Italiana*, **41**, 109–120.

CASNEDI, R. 1988*a*. La Fossa bradanica: origine, sedimentazione e migrazione. *Memorie della Società Geologica Italiana*, **41**, 439–448.

CASNEDI, R. 1988*b*. Subsurface basin analysis of fault-controlled turbidite system in Bradano Trough, southern Adriatic foredeep, Italy. *AAPG Bulletin*, **72**, 1370–1380.

CASNEDI, R. 1991. Hydrocarbon accumulation in turbidites in migrating basins of the southern Adriatic foredeep (Italy). *In*: BOUMA, A. H. & CARTER, P. M. (eds) *Facies models in exploration and development of hydrocarbon and ore deposits*. VSP International Science Publisher, Zeist, the Netherlands.

CASNEDI, R., CRESCENTI, U. & TONNA, M. 1982. Evoluzione della avanfossa adriatica meridionale nel Plio-Pleistocene, sulla base di dati di sottosuolo. *Memorie della Società Geologica Italiana*, **24**, 243–260.

CATALANO, R., DOGLIONI, C. & MERLINI, S. 2001. On the

Mesozoic Ionian Basin. *Geophysical Journal International*, **144**, 49–64.

CATALANO, S., MONACO, C. & TORTORICI, L. 1993. Pleistocene strike-slip tectonics in the Lucanian Apennine (Southern Italy). *Tectonics*, **12**, 656–665.

CHANNEL, J. E. T., D'ARGENIO, B. & HORWATH, F. 1979. Adria, the African Promontory, in Mesozoic Mediterranean Palaeogeography. *Earth Science Reviews*, **15**, 213–292.

CHILOVI, C., DE FEYTER, A. J. & POMPUCCI, A. 2000. Wrench zone reactivation in the Adriatic Block: the example of the Mattinata Fault System (SE Italy). *Bollettino della Società Geologica Italiana*, **119**, 3–8.

CIABATTI, M., CURZI, P. V. & RICCI LUCCHI, F. 1986. Sedimentazione quaternaria nell'Adriatico centrale. *In:* NANNI, T. (ed.) *Atti Riunione Gruppo di Sedimentologia CNR, Ancona, Italy, 5–7 June 1986*, Università di Ancona, Ancona, Italy, 125–139.

CIABATTI, M., CURZI, P. V. & RICCI LUCCHI, F. 1987. Quaternary sedimentation in the central Adriatic Sea. *Giornale di Geologia*, **49**, 113–125.

CIARANFI, N., D'ALESSANDRO, A. & MARINO, M. 1997. A candidate section for the Lower-Middle Pleistocene Boundary (apennine Foredeep, South Italy). *In:* NAIWEN, W. & REMANE, J. (eds) *Proceeding of 30th Geological Congress, China*, **11**, VSP International Science Publisher, Zeist, the Netherlands, 201–211.

CIARANFI, N., D'ALESSANDRO, A., MARINO, M. & SABATO, L. 1994*a*. La successione argillosa infra e mediopleistocenica della parte sudoccidentale della Fossa bradanica: la sezione di Montalbano Jonico in Basilicata. *In: Guida alle escursioni, 77° Congresso della Società Geologica Italiana, Bari, Italy, 23 September-1 October 1994*. Quaderni della Biblioteca Provinciale di Matera, **15**, 117–156.

CIARANFI, N., GHISETTI, F., GUIDA, M., IACCARINO, G., LAMBIASE, S., PIERI, P., RAPISARDI, L., RICCHETTI, G., TORRE, M., TORTORICI, L. & VEZZANI, L. 1983. *Carta Neotettonica dell'Italia meridionale*. Progetto Finalizzato Geodinamica del CNR, **515**.

CIARANFI, N., MAGGIORE, M., PIERI, P., RAPISARDI, L., RICCHETTI, G. & WALSH, N. 1979. *Considerazioni sulla neotettonica della Fossa bradanica*. Progetto Finalizzato Geodinamica del CNR, **251**, 73–95.

CIARANFI, N., MARINO, M., SABATO, L., D'ALESSANDRO, A. & DE ROSA, R. 1996. Studio stratigrafico di una successione infra e mesopleistocenica nella parte sudoccidentale della Fossa bradanica (Montalbano Ionico, Basilicata). *Bollettino della Società Geologica Italiana*, **115**, 379–391.

CIARANFI, N., PIERI, P. & RICCHETTI, G. 1988. Note illustrative alla carta geologica delle Murge e del Salento (Puglia centromeridionale). *Memorie della Società Geologica Italiana*, **41**, 449–460.

CIARANFI, N., PIERI, P. & RICCHETTI, G. 1994*b*. Linee di costa e terrazzi marini pleistocenici nelle Murge e nel Salento: implicazioni neotettoniche. *In: Riassunti 77° Congresso della Società Geologica Italiana, Bari, Italy, 23 September-1 October 1994*, Università deglia Studi di Bari, Bari, Italy, 170-172.

CINQUE, A., PATACCA, E., SCANDONE, P. & TOZZI, M. 1993. Quaternary kinematic evolution of the Southern Apennines. Relationships between surface geological features and deep lithospheric structures. *Annali di Geofisica*, **XXXVI** (2), 249–260.

COLELLA, A. & DI GERONIMO, I. 1987. Surface sediments and macrofaunas of the Crati submarine fan (Ionian Sea, Italy). *Sedimentary Geology*, **51**, 257–277.

COSENTINO, D. & GLIOZZI, E. 1988. Considerazioni sulle velocità di sollevamento di depositi eutirreniani dell'Italia meridionale e della Sicilia. *Memorie della Società Geologica Italiana*, **41**, 653–665.

COTECCHIA, V. & MAGRI, G. 1967. Gli spostamenti delle linee di costa quaternaria del Mar Ionio fra Capo Spulico e Taranto. *Geologia Applicata e Idrogeologia*, **2**, 1–25.

CREMONINI, G. & RICCI LUCCHI, F. (eds) 1982. *Guida alla geologia del margine appenninico-padano*. Guide Geologiche Regionali della Società Geologica Italiana.

CRESCENTI, U. 1975. Sul substrato pre-pliocenico dell'Avanfossa appenninica dalle Marche allo Jonio. *Bollettino della Società Geologica Italiana*, **94**, 583–634.

CURRAY, J. R., 1964. Transgressions and regressions. *In*: MILLER, R. C. (ed.) *Papers in Marine Geology*, MacMillan Press, New York, 175–203.

DAI PRA, G. & HEARTY, P.J. 1988. I livelli marini pleistocenici del Golfo di Taranto. Sintesi geocronostratigrafica e tettonica. *Memorie della Società Geologica Italiana*, **41**, 637–644.

DAL CIN, R. & SIMEONI, U. 1987. Processi erosivi e trasporto dei sedimenti nelle spiagge pugliesi fra S.Maria di Leuca e Taranto (Mare Ionio). Possibili strategie di intervento. *Bollettino della Società Geologica Italiana*, **106**, 767–783.

D'ALESSANDRO, A., LA PERNA, R. & CIARANFI, N. 2000. Pleistocene deep-sea deposits in the Bradano Trough (Basilicata, South Italy). Macrobenthic associations and palaeoenvironmental evolution (abstract) *In: The Plio-Pleistocene boundary and the Lower/Middle Pleistocene Transition: Type Areas and Sections, SEQS Meeting, Bari, Italy, 25–29 September*, Università deglia Studi di Bari, Bari, Italy,.

D'ARGENIO, B. 1974. Le piattaforme carbonatiche periadriatiche. Una rassegna di problemi nel quadro geodinamico mesozoico dell'area mediterranea. *Memorie della Società Geologica Italiana*, **13**, 137–159.

D'ARGENIO, B., PESCATORE, T. & SCANDONE, P. 1973. Schema geologico dell'Appennino meridionale (Campania e Lucania). *Accademia Nazionale dei Lincei*, **182**, 49–72.

DE CELLES, P. G. & GILES, K. A. 1996. Foreland basin systems. *Basin Research*, **8**, 105–123.

DE MARCO, A. 1990. Rapporti fra geodinamica e sedimentazione nella Fossa bradanica durante il Pleistocene: testimonianze mineralogiche. *Bollettino della Società Geologica Italiana*, **109**, 313–324.

DICKINSON, W. R. 1974. Plate tectonics and sedimentation. *In*: DICKINSON, W. R. (ed.) *Tectonics and sedimentation*, SEPM Special Publication, **22**, 1–27.

DOGLIONI, C. 1991. A proposal for the kinematic modelling of W-dipping subduction – possible application to the Tyrrenian–Apennines system. *Terranova*, **3**, 423–434.

DOGLIONI, C. 1993. Some remarks on the origin of foredeeps. *Tectonophysics*, **228**, 1–20.

DOGLIONI, C. 1994. Foredeeps vs. subduction zones. *Geology*, **22**, 271–274.

DOGLIONI, C., HARABAGLIA, P., MERLINI, S., MONGELLI, F., PECCERILLO, A. & PIROMALLO, C. 1999*a*. Orogens and slabs vs. their direction of subduction. *Earth Science Reviews*, **45**, 167–208.

DOGLIONI, C., MERLINI, S. & CANTARELLA, G. 1999*b*. Foredeep geometries at the front of the Apennines in the Ionian sea (central Mediterranean). *Earth and Planetary Science Letters*, **168**, 243–254.

DOGLIONI, C., MONGELLI, F. & PIALLI, G. 1998. Boudinage of the Alpine Belt in the Apenninic back-arc. *Memorie della Società Geologica Italiana*, **52**, 457–468.

DOGLIONI, C., MONGELLI, F. & PIERI, P. 1994. The Puglia uplift (SE-Italy): an anomaly in the foreland of the Apenninic subduction due to buckling of a thick continental lithosphere. *Tectonics*, **13**, 1309–1321.

DOGLIONI, C., TROPEANO M., MONGELLI, F. & PIERI, P. 1996. Middle–Late Pleistocene uplift of Puglia: an "anomaly" in the Apenninic foreland. *Memorie della Società Geologica Italiana*, **51**, 101–117.

DONDI, L., MOSTRADINI, F. & RIZZINI, A. 1982. Stratigrafia dei pozzi nel bacino padano orientale. *In* CREMONINI, G. & RICCI LUCCHI, F. (eds) *Guida alla geologia del margine appenninico-padano*. Guide Geologiche Regionali della Società Geologica Italiana, 237–246.

FESTA, V. 1999. *L'avampaese apulo nel quadro dell'evoluzione dell'Appennino meridionale: la deformazione della piattaforma carbonatica apula nel settore murgiano (Puglia, Italia meridionale)*. PhD Thesis, University of Bari, Italy.

FINETTI, I. 1982. Structure, stratigraphy and evolution of central Mediterranean. *Bollettino di Geofisica Teorica ed Applicata*, **XXIV** (96), 247–312.

GAMBASSINI, P. 1967. Il Conglomerato di Serra del Cedro, presso Tricarico. *Atti Accademia Gioenia Scienze Naturali*, **18**, 153–157.

GAMBINI, R. & TOZZI, M. 1996. Tertiary geodynamic evolution of Southern Adria microplate. *Terranova*, **8**, 593–602.

GHISETTI, F. & VEZZANI, L. 1998. Segmentation and tectonic evolution of the Abruzzi-Molise thrust belt (central Apennines, Italy). *Annales Tectonicae*, **XII**, 97–112.

GUEGUEN, E., DOGLIONI, C. & FERNANDEZ, M. 1997. Lithospheric boudinage in the Western Mediterranean back-arc basin. *Terranova*, **9**, 184–187.

GUEGUEN, E., DOGLIONI, C. & FERNANDEZ, M. 1998. On the post-25 Ma geodynamic evolution of the western Mediterranean. *Tectonophysics*, **298**, 259–269.

HEARTY, P. J. & DAI PRA, G. 1992. The age and stratigraphy of middle Pleistocene and younger deposits along the Gulf of Taranto (Southeast Italy). *Journal of Coastal Researchs*, **8**, 882–905.

HEWARD, A. P. 1981. A review of wave-dominated clastic shoreline deposits. *Earth Science Reviews*, **17**, 223–276.

HUNT, D. & TUCKER, M. E. 1992. Stranded parasequences and the forced regressive wedge systems tract: deposition during base-level fall. *Sedimentary Geology*, **81**, 1–9.

IANNONE, A. & PIERI, P. 1979. Considerazioni critiche sui "Tufi calcarei" delle Murge. Nuovi dati litostratigrafici e paleoambientali. *Geografia Fisica e Dinamica Quaternaria*, **2**, 173–186.

IANNONE, A. & PIERI, P. 1982. Caratteri neotettonici delle Murge. *Geologia Applicata e Idrogeologia*, **XVIII**(II), 147–159.

IANNONE, A. & PIERI, P. 1983. Rapporti fra i prodotti residuali del carsismo e la sedimentazione quaternaria nell'area delle Murge. *Rivista Italiana di Paleontologia e Stratigrafia*, **88**, 319–330.

JACOBACCI, A., MALATESTA, A., MARTELLI, G. & STAMPANONI, G. 1967. *Note illustrative della Carta Geologica d'Italia F°163 "Lucera"*. Servizio Geologico d'Italia.

JOHNSON, H. D. & BALDWIN, C. 1986. Shallow clastic seas. *In*: READING, H. G. (ed.) *Sedimentary Environments: Processes, Facies and Stratigraphy*. Blackwell, Oxford, 232–280.

LAZZARI, M. 1998. *Evoluzione stratigrafica e paleoambientale della parte sommitale della successione della Fossa bradanica fra l'Appennino meridionale e le Murge settentrionali. Considerazioni sull'evoluzione tettonico-sedimentaria del bacino bradanico nel Pleistocene inferiore*. PhD Thesis, University of Bari, Italy.

LENTINI, F. 1979. Le Unità Sicilidi della Val d'Agri (Appennino lucano). *Geologica Romana*, **18**, 215–225.

LOIACONO, F. & SABATO, L. 1987. Stratigrafia e sedimentologia di depositi pleistocenici di fan-delta sul margine appenninico della Fossa bradanica (Tricarico, Basilicata). *Memorie della Società Geologica Italiana*, **38**, 275–296.

MAGGIORE, M. & WALSH N. 1975. I depositi plio-pleistocenici di Acerenza (Potenza). *Bollettino della Società Geologica Italiana*, **94**, 93–109.

MALINVERNO, A. & RYAN, W. B. F. 1986. Extension of the Tyrrhenian Sea and shortening in the Apennines as result of arc migration driven by sinking lithosphere. *Tectonics*, **5**, 227–245.

MARIOTTI, G. & DOGLIONI, C. 2000. The dip of the foreland monocline in the Alps and Apennines. *Earth and Planetary Science Letters*, **118**, 191–202.

MASSARI, F. 1996. Upper-flow-regime stratification types on steep-face, coarse-grained, Gilbert-type progradational wedges (Pleistocene, southern Italy). *Journal of Sedimentary Research*, **66**, 364–375.

MASSARI, F. 1997. High-frequency cicles within Pleistocene forced-regressive conglomerate wedge (Bradanic area, southern Italy) filling collapse scarps. *Sedimentology*, **44**, 939–958.

MASSARI, F. & PAREA, G. C. 1988. Progradational gravel beach sequences in a moderate- to high-energy, microtidal marine environment. *Sedimentology*, **35**, 881–913.

MASSARI, F. & PAREA, G. C. 1990. Wave-dominated Gilbert-type gravel deltas in the interland of the

Gulf of Taranto (Pleistocene, southern Italy). *In*: COLELLA, A. & PRIOR, D. B. (eds) *Coarse-grained Deltas*. International Association of Sedimentologists, Special Publications, **10**, 311–331.

MATANO, F. & STAITI, D. 1998. Studio stratigrafico della successione pliocenica affiorante nel settore meridionale della Baronia (unità di Ariano, Appennino campano). *Bollettino della Società Geologica Italiana*, **117**, 357–367.

MERLINI, S., CANTARELLA, G. & DOGLIONI, C. 2000. On the seismic profile Crop M5 in the Ionian Sea. *Bollettino della Società Geologica Italiana*, **119**, 227–236.

MIGLIORINI, C., 1937. Cenno sullo studio e sulla prospezione petrolifera di una zona dell'Italia meridionale. *In*: *2nd Petroleum world Congress*, Paris, AGIP Report, Roma, 1–11.

MOSTARDINI, F. & MERLINI, S. 1986. Appennino centro meridionale. Sezioni geologiche e proposta di modello strutturale. *Memorie della Società Geologica Italiana*, **35**, 177–202.

MOSTARDINI, F., PIERI, M. & PIRINI, C. 1966. Stratigrafia del Foglio 212, Montalbano Jonico. *Bollettino del Servizio Geologico d'Italia*, **87**, 57–143.

MUTTI, E. & RICCI LUCCHI, F. 1972. Le torbiditi dell'Appennino settentrionale: introduzione all'analisi di facies. *Memorie della Società Geologica Italiana*, **11**, 161–199.

OGNIBEN, L. 1969. Schema illustrativo alla geologia del confine calabro-lucano. *Memorie della Società Geologica Italiana*, **8**, 453–763.

ORI, G. G. 1993. Continental depositional systems of the Quaternary of the Po Plain (northern Italy). *Sedimentary Geology*, **83**, 1–14.

ORI, G. G., ROVERI, M. & VANNONI, F. 1986. Plio-Pleistocene sedimentation in the Apenninic–Adriatic foredeep (Central Adriatic Sea, Italy). *In*: ALLEN, P. A. & HOMEWOOD, P. (eds) *Foreland basins*. International Association of Sedimentologists, Special Publications, **8**, 183–198.

ORI, G. G., SERAFINI, G., VISENTIN, C., RICCI LUCCHI, F., CASNEDI, R., COLALONGO, M. L. & MOSNA, S. 1991. The Pliocene–Pleistocene Adriatic Foredeep (Marche and Abruzzo, Italy): An integrated approach to surface and subsurface geology. *In: 3rd European Association of Petroleum Geologists Conference, Adriatic Foredeep Field Trip Guide Book*. EAPG, Zeist, 1–85.

PAREA, G. C. 1986. I terrazzi marini tardo-Pleistocenici del fronte della catena appenninica in relazione alla geologia dell'avanfossa adroatica. *Memorie della Società Geologica Italiana*, **35**, 913–936.

PATACCA, E. & SCANDONE, P. 1989. Post-Tortonian mountain building in the Apennines. The role of passive sinking of a relic lithospheric slab. *In*: BORIANI, A, BONAFEDE, M., PICCARDO, G. B. & VAI, G. B. (eds) *The Lithosphere in Italy. Advances in Earth Science Research*. Atti dei Convegni dei Lincei, **80**, 157–176.

PATACCA, E., SARTORI, R. & SCANDONE, P. 1990. Tyrrhenian basin and Apenninic arcs: kinematic relations since late Tortonian times. *Memorie della Società Geologica Italiana*, **45**, 425–451.

PESCATORE, T. 1978. The Irpinids: a model of tectonically controlled fan and base-of-slope sedimentation in southern Italy. *In*: STANLEY, D. J. & KELLING, G. (eds) *Sedimentation in Submarine Canyons, Fans, and Trenches*, Hutchinson & Ross, Strandsbourg, Pennsylvania, USA, 325–339.

PESCATORE, T. & SENATORE, M. R. 1986. A comparison between a present-day (Taranto Gulf) and Miocene (Irpinian Basin) foredeep of the Southern Apennines (Italy). *In*: ALLEN, P. A. & HOMEWOOD, P. (eds) *Foreland Basins*. International Association of Sedimentologists, Special Publications, **8**, 169–182.

PESCATORE, T., RENDA, P., SCHIATTARELLA, M. & TRAMUTOLI, M. 1999. Stratigraphy and structural relationships between Meso-Cenozoic Lagonegro basin and coeval carbonate platforms in southern Apennines, Italy. *Tectonophysics*, **315**, 269–286.

PIERI, P. 1980. Principali caratteri geologici e morfologici delle Murge. *Murgia sotterranea, Bollettino Gruppo Speleo Martinense*, **2**, 13–19

PIERI, P., DE DONATO, G., FESTA, V., FIORE, A., GALLICCHIO, S., MORETTI, M., TILLI, A. & TROPEANO, M. 2001. I depositi continentali terrazzati medio pleistocenici della fascia pedemontana del Subappennino Dauno. *In*: *Riassunti della Riunione del Gruppo Informale di Sedimentologia del CNR, Geosed 2001, Potenza, 2–7 ottobre 2001*. Università della Basilicata, Potenza, Italy, 84–85.

PIERI, P., FESTA, V., MORETTI, M. & TROPEANO, M. 1997*a*. Quaternary tectonic activity of the Murge area (Apulian Foreland – southern Italy). *Annali di Geofisica*, **XL** (5), 1395–1404.

PIERI, P., SABATO, L., LOIACONO, F. & MARINO, M. 1994*a*. Il bacino di piggy-back di Sant'Arcangelo: evoluzione tettonico-sedimentaria. *Bollettino della Società Geologica Italiana*, **113**, 465–481.

PIERI, P., SABATO, L. & MARINO, M. 1996*a*. The Plio-Pleistocene piggyback Sant'Arcangelo Basin: tectonic and sedimentary evolution. *Notes et Mémoires du Service géologique du Maroc*, **387**, 195–208.

PIERI, P., SABATO, L. & TROPEANO, M. 1994*b*. Evoluzione tettonico-sedimentaria della Fossa bradanica a sud dell'Ofanto nel Pleistocene. *In*: *Guida alle escursioni, 77° Congresso della Società Geologica* Italiana, *Bari, Italy, 23 September–1 October 1994*, Quaderni della Biblioteca Provinciale di Matera, **15**, 35–54.

PIERI, P., SABATO, L. & TROPEANO, M. 1996*b*. Significato geodinamico dei caratteri deposizionali e strutturali della Fossa bradanica nel Pleistocene. *Memorie della Società Geologica Italiana*, **51**, 501–515.

PIERI, P., SABATO, L. & TROPEANO, M. 1996*c*. Evoluzione tettono-sedimentaria del segmento meridionale dell'avanfossa appenninica post-messiniana (Fossa bradanica). *In*: *Atti della riunione del Gruppo italiano di Sedimentologia del CNR, Catania, Italy, 10–14 October 1996*, 232–234.

PIERI, P, VITALE, G., BENEDUCE, P., DOGLIONI, C., GALLICCHIO, S., GIANO, S. I., LOIZZO, R., MORETTI, M., PROSSER, G., SABATO, L., SCHIATTARELLA, M., TRAMUTOLI, M. & TROPEANO, M. 1997*b*. Tettonica quaternaria nell'area bradanico-ionica. *Il Quaternario*, **10**, 535–542.

PLINT, A. G. & NUMMEDAL, D. 2000. The falling stage systems tract: recognition and importance in sequence stratigraphic analysis. *In*: HUNT, D. & GAWTHORPE, R. L. (eds) *Sedimentary Responses to Forced Regression*. Geological Society, London, Special Publications, **172**, 1–17.

POMAR, L. 1993. High-resolution sequence stratigraphy in prograding Miocene carbonates: application to seismic interpretation. *In*: LOUCKS, R. G. & SARG, J. F. (eds) *Carbonate Sequence Stratigraphy: Recent Development and Applications*. American Association of Petroleum Geologists Memoirs, **57**, 389–407.

POMAR, L. & TROPEANO, M. 2001. The Calcarenite di Gravina Formation in Matera (southern Italy): new insights for coarse-grained, large-scale, cross-bedded bodies encased in offshore deposits. *AAPG Bulletin*, **85**, 661–689.

POSAMENTIER, H. W., ALLEN, G. P., JAMES, D. P. & TESSON, M. 1992. Forced regressions in a sequence stratigraphic framework: concepts, examples, and exploration significance. *AAPG Bulletin*, **76**, 1687–1709.

REGIONE EMILIA ROMAGNA, ENI-AGIP. 1998. *Riserve idriche sotterranee della Regione Emilia Romagna*. DI DIO, G (ed.). S.EL.C.A., Firenze.

RICCHETTI, G. 1965. Alcune osservazioni sulla serie della Fossa Bradanica. Le 'Calcareniti di M. Castiglione'. *Bollettino della Società Naturalisti in Napoli*, **74**, 3–11.

RICCHETTI, G. 1967*a*. Lineamenti geologici e morfologici della media valle del fiume Bradano. *Bollettino della Società Geologica Italiana,* **86**, 607–622.

RICCHETTI, G. 1967*b*. Osservazioni preliminari sulla geologia e morfologia dei depositi quaternari nei dintorni del Mar Piccolo (Taranto). *Atti Accademia Gioenia di Scienze Naturali in Catania*, **18**, 123–130.

RICCHETTI, G. 1980. Contributo alla conoscenza strutturale della Fossa bradanica e delle Murge. *Bollettino della Società Geologica Italiana,* **49**, 421–430.

RICCHETTI, G. & MONGELLI, F. 1980. Flessione e campo gravimetrico della micropiastra apula. *Bollettino della Società Geologica Italiana*, **99**, 431–436.

RICCHETTI, G., CIARANFI, N., LUPERTO SINNI, E., MONGELLI, F. & PIERI, P. 1988. Geodinamica ed evoluzione sedimentaria e tettonica dell'avampaese apulo. *Memorie della Società Geologica Italiana*, **41**, 57–82.

RICCI LUCCHI, F. 1975. Miocene palaeogeography and basin analysis in the Periadriatic Apennines. *In*: SQUYRES, C. (ed.) *Geology of Italy*, **2**, Tripoli, Africa, 129–236.

RICCI LUCCHI, F. 1986. The Oligocene to Recent foreland basins of the northern Apennines. *In*: ALLEN, P. A. & HOMEWOOD, P. (eds) *Foreland Basins*. International Association of Sedimentologists, Special Publications, **8**, 105–139.

RICCI LUCCHI, F., COLELLA, A., GABBIANELLI, G., ROSSI, S. & NORMARK, W. R. 1984. The Crati submarine fan, Ionian sea. *Geomarine Letters*, **3**, 71–77.

RIVIELLO, A., SCHIATTARELLA, M. & VACCARO, A.M. 1997. Struttura geologica dell'area di Tolve (Basilicata) dedotta da dati di superficie e di sottosuolo. *Il Quaternario*, **10**, 557–562.

ROURE, F., CASERO, P.& VIALLY, R. 1991 Growth processes and melanges formation in the southern Apennines accretionary wedge. *Earth and Planetary Science Letters*, **102**, 395–412.

ROYDEN, L. & KARNER, G. D. 1984. Flexure of lithosphere beneath apennine and carpathian foredeep basins: evidence for an insufficient topographic load. *AAPG Bulletin*, **68**, 704–712.

ROYDEN, L., PATACCA, E. & SCANDONE, P. 1987. Segmentation and configuration of subducted lithosphere in Italy: an important control on thrust-belt and foredeep–basin evolution. *Geology*, **15**, 714–717.

ROVERI, M., ORI, G. G. & ZITELLINI, N. 1986. Sedimentazione plio-quaternaria nell'Adriatico centrale. *In*: *Atti Riunione Gruppo di Sedimentologia CNR, Ancona, Italy, 5–7 June 1986*, Università degli Studi di Ancona, Ancona, Italy, 141–146.

SABATO, L. 1996. Quadro stratigrafico-deposizionale dei depositi regressivi nell'area di Irsina (Fossa bradanica). *Geologica Romana*, **32**, 219–230.

SABATO, L. & MARINO, M. 1994. I depositi pliocenici del margine appenninico di Tricarico (Basilicata). *In: Guida alle escursioni, 77° Congresso della Società Geologica Italiana, Bari, Italy, 23 September-1 October 1994*. Quaderni della Biblioteca Provinciale di Matera, **15**, 87–104.

SABATO, L., PIERI, P. & TROPEANO, M. 2000. Morphostructural constraints on the Pleistocene depositional history of the Bradanic Trough (South-Apennines foredeep – Italy) (Abstract) *In*: *Millennium Flux: Sediment Supply to Basin Congress*, Southampton (UK), 41.

SCALERA, A. 1986. Il Pleistocene inferiore della Capitanata occidentale: analisi stratigrafico paleontologica. *Bollettino della Società Geologica Italiana,* **105**, 185–194.

SCANDONE, P. 1979. Origin of the Tyrrhenian sea and Calabrian arc. *Bollettino della Società Geologica Italiana*, **98**, 27–34.

SCHIATTARELLA, M. 1998. Quaternary tectonics of the Pollino Ridge, Calabria–Lucania boundary, southern Italy. *In*: HOLDSWORTH, R., STRACHAN, R. A. & DEWEY, J. F. (eds) *Continental Transpressional and Transtensional Tectonics*. Geological Society, London, Special Publications, **135**, 341–354.

SCHLAGER, W. 1989. Drowning unconformities on carbonate platforms. *In*: CREVELLO, P. D., WILSON, J. L., SARG, J. F. & READ, J. F. (eds) *Controls on Carbonate Platform and Basin Development*. Society of Economic Paleontologists and Mineralogists, Special Publications, **44**, 15–25.

SELLA, M., TURCI, C. & RIVA, A. 1988. Sintesi geopetrolifera della Fossa bradanica (avanfossa della catena appenninica meridionale). *Memorie della Società Geologica Italiana*, **41**, 87–107.

SELLI, R. 1962. Il Paleogene nel quadro della geologia

dell'Italia meridionale. *Memorie della Società Geologica Italiana*, **3**, 737–789.

SENATORE, M. R., DIPLOMATICO, G., MIRABILE, L., PESCATORE, T. & TRAMUTOLI, M. 1982. Franamenti sulla scarpata continentale pugliese del golfo di Taranto (alto Ionio-Italia). *Geologica Romana*, **21**, 497–510.

SENATORE, M. R., MIRABILE, L., PESCATORE, T. & TRAMUTOLI, M. 1980. La piattaforma continentale del settore nord-orientale del golfo di Taranto (piattaforma pugliese). *Geologica Applicata e Idrogeologia*, **XV**, 33–50.

SENATORE, M. R., NORMARK, W.R., PESCATORE, T. & ROSSI, S. 1988. Structural framework of the Gulf of Taranto (Ionian Sea). *Memorie della Società Geologica Italiana*, **41**, 533–539.

TRAMUTOLI, M., PESCATORE, T, SENATORE, M. R. & MIRABILE, M. 1984. Interpretation of reflection high resolution seismic profiles through the Gulf of Taranto (Ionian sea, eastern Mediterranean). The structure of Apennine and Apulia deposits. *Bollettino di Oceanografia Teorica ed Applicata*, **II** (1), 33–52.

TRINCARDI, F. & CORREGGIARI, A. 2000. Quaternary forced regression deposits in the Adriatic basin and the record of composite sea-level cycles. *In*: HUNT, D. & GAWTHORPE, R. L. (eds) *Sedimentary Responses to Forced Regression*. Geological Society, London, Special Publications, **172**, 245–269.

TROPEANO, M. 1994. *Sistemi costieri carbonatici nella Calcarenite di Gravina (Pliocene superiore – Pleistocene inferiore) nell'area delle Murge e della Fossa bradanica*. PhD Thesis, University of Bari, Italy.

TROPEANO, M. & SABATO, L. 2000. Response of Plio-Pleistocene mixed bioclastic–lithoclastic temperate-water carbonate system to forced regression: the Calcarenite di Gravina Formation, Puglia, SE Italy. *In*: HUNT, D. & GAWTHORPE, R. L. (eds) *Sedimentary Responses to Forced Regression*. Geological Society, London, Special Publications, **172**, 217–243

TROPEANO, M. & SABATO, L. & PIERI, P. 2001. The Quaternary "Post-turbidite" sedimentation in the South-Apennines Foredeep (Bradanic Trough-Southern Italy). *Bollettino della Società Geologica Italiana*, Volume Speciale, **1** in press.

TROPEANO, M. & SABATO, L., PIERI, P. & DOGLIONI, C. 1998. Geodynamic and stratigraphic history of the Bradanic Trough (south-Apennines foredeep – Italy) (Abstract) *In*: *15th International Sedimentological Congress, Alicante, Spain, 12-17 April 1998,* Universidad de Alicante, Alicante, Spain, 772–773.

VALDUGA, A. 1973. Fossa Bradanica. *In*: DESIO, A. (ed.) *Geologia dell'Italia*. UTET, Torino, Italy, 692–695.

VAN WAGONER, J. C., POSAMENTIER, H. W., MITCHUM, R. M., VAIL, P. R., SARG, J. F., LOUTIT, T. S. & HARDENBOL, J. 1988. An overview of the fundamentals of sequence stratigraphy and key definitions. *In*: WILGUS, C. K., HASTINGS B. S., KENDALL, C. G. ST. C., POSAMENTIER, H. W., ROSS, C. A. & VAN WAGONER, J. C. (eds) *Sea-level Changes: An Integrated Approach*. Society of Economic Paleontologists and Mineralogists, Special Publications, **42**, 39–45.

VERHALLEN, P. J. J. M. 1991. Late Pliocene to Early Pleistocene Mediterranean mud-dwelling Foraminifera; influence of a changing environments of community structure and evolution. *Micropaleontological Bulletin*, **40**, 1–219.

VEZZANI, L. 1967*a*. I depositi plio-pleistocenici del litorale ionico della Lucania. *Atti Accademia Gioenia Scienze Naturali,* **18**, 159–180.

VEZZANI, L. 1967*b*. Il bacino plio-pleistocenico di S. Arcangelo (Lucania). *Atti Accademia Gioenia Scienze Naturali*, **18**, 207–227.

VITALE, G. 1995. *Evoluzione tettonica e stratigrafica dal passaggio Miocene–Pliocene all'attuale del sistema catena-avanfossa lungo il fronte bradanico-ionico*. PhD Thesis, University of Bari, Italy.

Impact of periodicity on sediment flux in alluvial systems: grain to basin scale

LYNNE E. FROSTICK[1] & STUART J. JONES[2]

[1]*Department of Geography, University of Hull, Hull HU6 7RX, UK*

[2]*Department of Geological Sciences, University of Durham, South Road, Durham DH1 3LE, UK*

Abstract: Periodicity is a common component of many sedimentological processes, but seldom is it considered across all scales of fluvial processes in order to understand the complete impact on sediment supply to basins. Temporal changes in sediment supply within drainage systems and sedimentary basins are a consequence of the inherent instability in transport processes. The causes of fluctuations are of 2 main types: (i) changes in factors endemic to the supply of sediment but which are at least partly independent of erosive forces and (ii) changes in the magnitude of forces available to transport sediment. Fluctuations at spatial scales from grain – through reach – to basin – scales and at temporal scales from minutes to millennia are discussed and evaluated. All fluctuations are reflected in sedimentary deposits in some way. For example, irregular patterns of bed break-up during erosion can generate bedforms that are recorded in deposits, the passage of waves of sediment can cause cycles of incision and aggradation in a reach; large flood events will flush sediment into coastal regions and will be recorded as an identifiable 'package' in the deposits. Many models of basin processes and products assume a consistent supply of sediment which is far from the case in nature. One of the challenges in the coming decade is to move away from using long-term averages of sediment supply and to link models directly into geomorphic processes.

Modelling of sedimentary basins and the ways in which they fill with sediment is the focus of considerable research interest (e.g. Paola 2000). Many innovative numerical models have been developed to simulate landscape evolution, for example *ZSCAPE* the finite difference model of Densmore *et al.* (1998), which can be used to quantify such variables as water discharge, erosion, sediment flux and deposition in tectonically active landscapes. Other models (e.g. *DEMOSTRAT* Rivenaes 1992, 1997) attempt to link sediment flux from eroding river basins to the stratigraphy of deposits in the receiving basin. In Rivenaes (1997), the author tries to elucidate problems that relate to varying transport rates and efficiency in marine turbidite sequences. In many models denudation rates for drainage areas are quoted either as weight loss per unit area (e.g. tonnes per hectare) or as surface lowering (mm per year) often extrapolated from short-term monitoring at a few, isolated sites (Walling 1977; Powell *et al.* 1996). Sediment supply is also extrapolated, often from short-term experiments and for a limited range of environmental conditions which are presumed to be representative. Whatever the database used in the research, it is always projected to a long-term average, either as a volume or weight per unit time, where supply is presumed to approximate continuity with no major temporal changes (see Richards 2002). But this is very far from reality, where temporal changes in quantities of mobile sediment occur at all time scales and the physics of many geomorphic processes are still poorly known. The causes of fluctuations are many, but are of two main types:

(1) Changes in sediment-related factors endemic to the processes which bring about sediment transport that are independent or semi-independent of the erosive forces and often linked to the quality and quantity of sediment available for transport, for example the impact particle interactions of micro- and meso-scale bedforms, particle interactions kinematic waves and queues and sporadic supplies from hill slopes and banks.
(2) Changes in factors that relate to fluctuations in the magnitude of the forces available to

From: JONES, S. J. & FROSTICK, L. E. (eds) 2002. *Sediment Flux to Basins: Causes, Controls and Consequences.* Geological Society, London, Special Publications, **191**, 81–95. 0305-8719/02/$15.00

transport sediment, often expressed through magnitude-frequency diagrams, with larger but infrequent events stripping out large quantities of sediment from basins, e.g. flash floods, El Nino related floods (e.g. Honduras 1999), Monsoons, Tsunamis and earthquakes/ tectonics.

This paper explores some of the causes of temporal changes in sediment fluxes from the drainage to depositional basin, and investigates some of the important implications for basin modelling and sediment flux (Fig. 1).

Sediment related factors

These cause periodicity over time scales that vary from momentary to centennial and are spatially restricted in their impact, i.e. they cause fluctuations in a discrete section of the bed or over a river reach and add, incrementally, to the total catchment patterns.

Short time period fluctions–bed load

(a) *Grainscale.* Bedload transport in rivers has long been known to increase and decrease, or pulse, in a manner which is independent of the forcing variables and on timescales that vary from minutes to hours. The reasons for the pulses have never been fully explained and are interpreted variously as due to bedform migration (e.g. Einstein 1942) and kinematic waves (e.g. Reid & Frostick 1985). Recent research carried out by Allan & Frostick (1999) Brasington *et al.* (2000) and Middleton *et al.* (2000) has shown that the tendency of discrete sections of beds to break-up and dilate prior to commencing transport, almost instantaneously. This process may contribute to the sporadic nature of transport. It has been inferred that this failure is controlled by both the balance between fluid flow above and below the bed surface, and the particle interactions causing the bed to behave as a large aggregate, not as individual grains. This is inferred from the difference in behaviour between experiments carried out with unimodal and bimodal material, rapid and localized failure is more pronounced in mixed size sediment (Fig. 2a). From observations and analysis of individual grains, lift and not drag, has been identified as the important force during the early stages of entrainment and, as a result, fine matrix particles can be less mobile than those of the coarser framework. During entrainment, fine material can move down into subsurface pores, creating a reservoir of fine sediment which can be remobilized during larger floods.

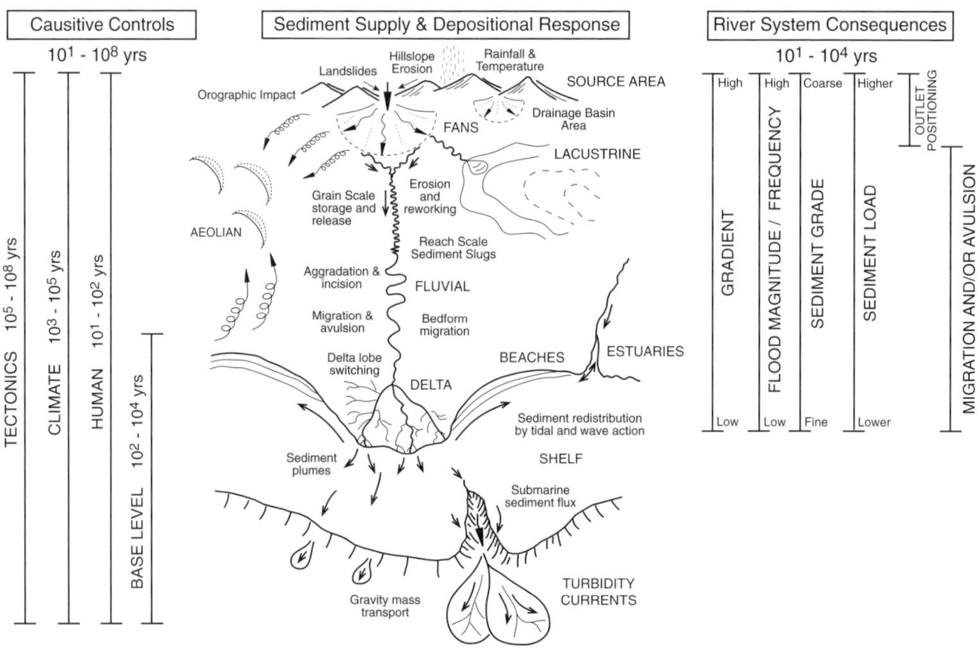

Fig. 1. Diagram illustrating the cause, response and consequences of variations in sediment supply to a depositional basin. The variety of controls all operating over different spatial and temporal time scales indicates the need for a holistic approach to understanding sediment flux.

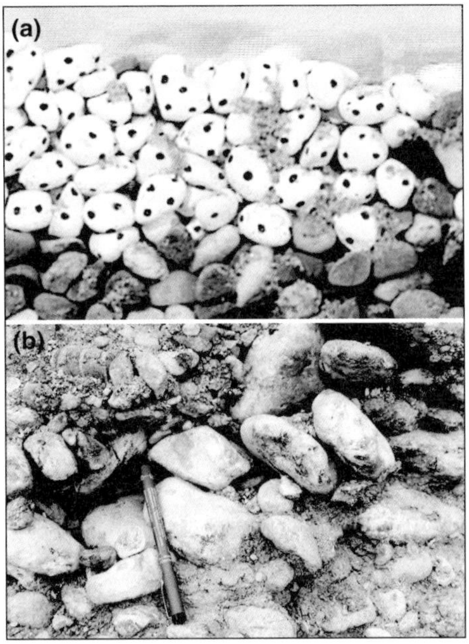

Fig. 2. (**a**) Photograph showing results of framework dilation during entrainment in a glass-sided tilting flume tank. Expansion of pore throats combines with suction of fluid and matrix particles into the bed to replace the moving framework, allowing matrix grains to penetrate deep into the gravel bed (see Allan & Frostick 1999; scale: white pebbles with black dots are 25 mm in diameter). (**b**) Plio-Pleistocene example of preserved open framework gravels, Rio Cinca Formation, South central Pyrenees, Spain (scale: pen is 150 mm long).

Thus, alternating layers of matrix-filled and open framework gravels can be created. By implication it may be possible to utilize texture to infer the palaeohydraulics of preserved fluvial deposits (Fig. 2b).

(b) *Reach scale.* Bedform migration is an obvious candidate for causing fluctuations in sediment transport in a reach. In gravel- and sand-bed rivers, the migration of ripples, dunes, bars, point bars and braids exert important controls on instantaneous transport rates at a point (Ashmore 1991). But, in the field, bedload transport is difficult to quantify due to logistical problems associated with point sampling in such a heterogeneous and episodic system. Studies of bedload transport rates in single thread channels are few, but all show pulses to a greater or lesser extent, with the size of pulses related to the magnitude of bed shear (Fig. 3; Reid & Frostick 1986). In braided systems the multi-thread nature of the channel make studies even more difficult and published results are few, but again a broad relationship with bed shear seems to exist, as long as sediment is available for transport and the bed is not armoured (Ashworth *et al.* 1992; Ferguson & Ashworth 1992). However, it has been observed (Ashmore 1991) that the passage of a bedload slug or pulse down one of the branches in a braided system can force major changes in confluence morphology – which might export another slug downstream as the balance between deposition and erosion shifts. However, on a channel-wide basis changes in local stream power per unit channel width may result only in larger material being transported rather than more (Bridge 1993). If grain size is kept constant, then an increase in stream power will result in an increase in time-averaged bedload transport and possibly a wider channel if the banks can be eroded to allow the channel pattern to adjust (Jones 2002). Any abrupt changes in sediment supply to a reach will result in changes in channel pattern – increasing supply causing bar development and braiding (Ashmore 1991) i.e. rivers adjust to the fluctuations delivered to them. Palaeohydraulic data collected from Oligo–Miocene and Plio–Pleistocene rivers of the Spanish Pyrenees allows the character of palaeorivers to be inferred, whether incising or aggrading (Jones *et al.* 1999). For a gravel-bed river with a low sediment supply, as during the Plio–Pleistocene, the bedload transport rates are

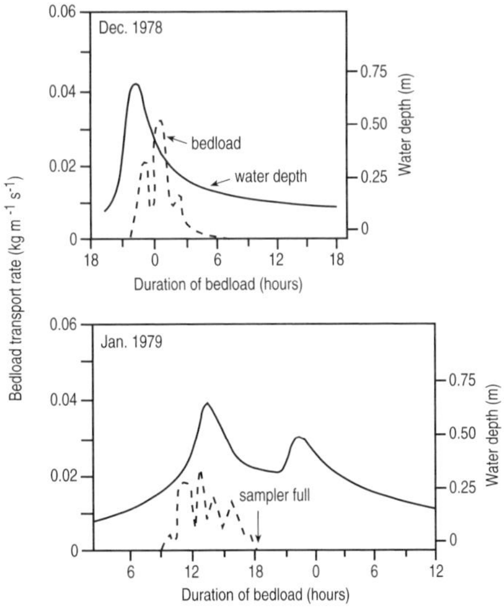

Fig. 3. Bedload transport per unit width and water depth during two selected floods on Turkey Brook, England. These results indicate that bedload transport is characterized by pulses of sediment (after Reid & Frostick 1986, 1987).

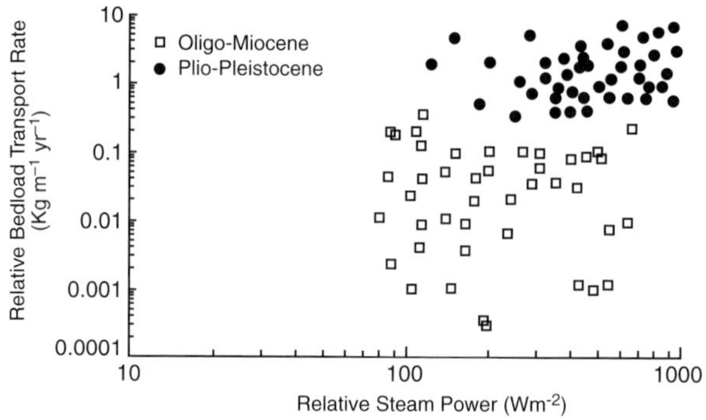

Fig. 4. Logarithmic plot of relative bedload transport and relative stream power for Oligo–Miocene and Plio–Pleistocene rivers of the south Central Pyrenees, Spain. The minimum estimates of sediment fluxes allows for comparison between different phases of a single river system (after Jones 1997; Jones *et al.* 1999).

high and the river incises (Fig. 4). Alternatively, when sediment supply is abundant, the bedload transport rates are low and the river system aggrades (Fig. 4; for methodology see Jones *et al.* 1999).

(c) *Supply-related fluctuations.* Sudden bank or slope failures are sporadic events (Brand *et al.* 1984; Crozier 1999) that can have catastrophic effects on river character and transport rates (Fig. 5). On a similar scale, a range of human-induced landscape disturbance, e.g. mining, urbanization, agriculture, have the potential for delivering large quantities of sediment to streams (Hooke 1994). Understanding how and when this material is translated downstream is important for sediment supply in the receiving basin as well as for predicting likely morphological changes within the drainage basin. These affect both bed- and suspended-load – the impact being more immediate for suspended load.

Some studies have emphasized the importance of temporary storage within the landscape (e.g. Magilligan 1985). Bedload material tends to be stored in river beds and banks where it is immediately available for re-erosion. Recent

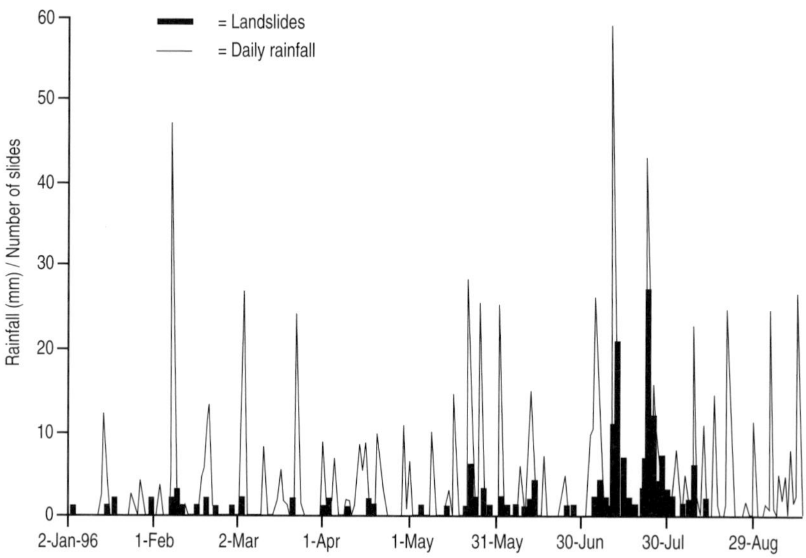

Fig. 5. Daily rainfall and landslide occurrence, Wellington City, New Zealand (after Crozier 1999).

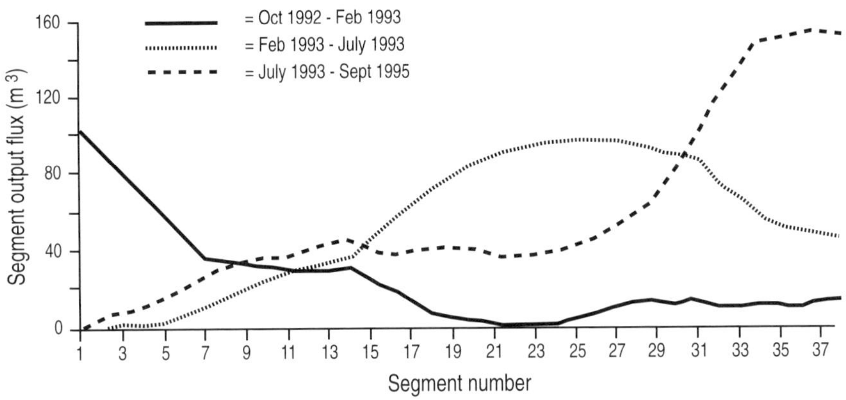

Fig. 6. Flux of sediment through a single reach, Allt Dubhaig, Scotland (after Wathen & Hoey 1998). The research suggests that sediment moves downstream as a discrete wave.

research has tended to suggest that when large quantities of coarse sediment are delivered to a reach the material moves downstream as a discrete wave (Fig. 6; Wathen & Hoey 1998; Jacobson & Bobbit Gran 1999), although experimental and theoretical studies of wave dispersal and translation downstream have suggested that dispersal may tend to dominate (Lisle *et al.* 1997). However, in natural rivers, observations have suggested that downstream translation does occur, following initial decay of the wave form and also that the rate of downstream translation depends on physiography, climate and the nature of the original disturbance, with scales involved ranging up to hundreds of years.

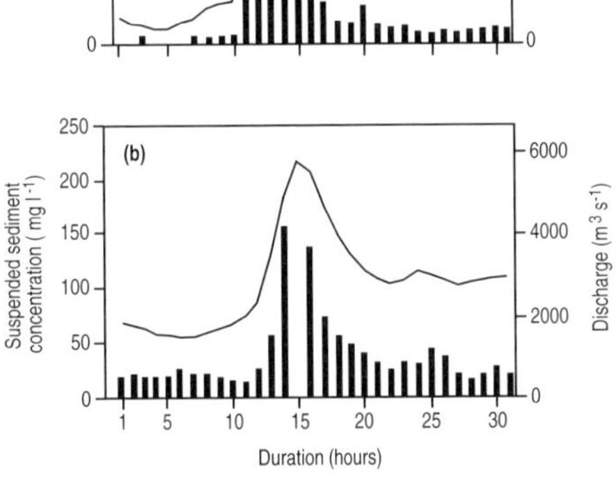

Fig. 7. Discharge and suspended sediment during a January 1993 flood in the rivers (**a**) Meuse and, (**b**) Rhine (adapted from Middlekoop & Asselman 1998).

Short timescale fluctuations: suspended load

Although the relationship between water and suspended sediment discharge is better than for bedload, problems of supply variations from tributaries and side slopes as well as temporary storage within, and release from, the bed and banks make prediction difficult. Additionally, because suspended sediment is closely controlled by the supply of material to the channel from the surrounding drainage basin, it is also necessary to look beyond the river channel into the drainage basin and floodplain in order to understand the changing patterns of suspended sediment yield. There are numerous studies of suspended sediment concentration and discharge many of which show a cloud of points and a hysteretic relationship generally, though not invariably, with high concentrations for the same discharge on the rising limb of a flood hydrograph due to erosion of readily available material from the bed (Fig. 7; Walling & Kleo 1979). Much of the variability in the relationships can be attributed to local bank erosion or localized soil erosion somewhere in the catchment (Fig. 7; Middelkoop & Asselman 1998). Storage within the floodplain occurs when floods over-top the riverbanks and suspended sediments are deposited from the flows. Deposits build up on the floodplain, with each layer of fining-upwards sand-silt-clay corresponding to a flood event. The deposits are safe from the erosive effects of flows in the channel until it shifts and the stored material is once more moved into the train of sediment and translated basinward. This can take from hundreds to tens of thousands of years (Walling & Bradley 1989) or may never happen so that the deposit can pass into the geological record. Frequently many of the models discussed above do not take into account for example, high-magnitude, low-frequency events such as storms that trigger intensive sheet and rill erosion or mass movements within a drainage basin, with the suspended sediment load associated with such events being substantially greater than for periods with only low magnitude events (Meade *et al.* 1990). For similar reasons, the suspended sediment loads of rivers will commonly be highly sensitive to changes in land use within a drainage basin, since, irrespective of whether or not the runoff regime is modified, these changes can influence the sediment supply to a river.

Volcanic activity in the catchment can provide vast reservoirs of easily erodable sediment in the form of volcanic ash. This is available for transport as suspended load and can cause dramatic increases in sediment transport in volcanically active areas. In response to eruptive events that have produced large volumes of unconsolidated volcaniclastic sediments and greatly disturbed local topographic equilibrium at Mount St. Helens, 10 m deep gullies were cut into the pyroclastic deposits within a few months of the 1980 eruption. At Mount Pinatubo, the 1991 pyroclastic flows had similar gullies within a matter of days (Davidson & De Silva 2000). In comparison pre-eruption and post-eruption rivers draining the flanks of Mount St. Helens indicate that mudflows significantly altered the channel configuration and hence flow capacity following the 1980 eruption. Measurements from the Cowlitz River revealed that sediment and debris raised the average elevation of the channel bottom by about 5 m (Cummans 1981). This dramatically affected the carrying capacity of the Cowlitz River channel to less than 10% of its former load without overbank flooding.

Flow and force-related factors

Hovius & Leeder (1998) comment that sedimentary basins record episodes of erosion, whereas the drainage basin and its networks are the 'negative imprint' of the same record, and their very existence needs to be inferred from stratigraphic successions. It is particularly important that the allogenic controls on sediment erosion from the drainage networks and the affects these have on the depositional basin are fully appreciated and understood. These include tectonic uplift (topography), climate, land use and other human-related factors (dam breaks etc). Sediment erosion and deposition both within the river catchment and in the receiving basin ultimately rely on the magnitude and frequency of flood events. Both suspended and bedload transport will generally increase with increasing stream power, all else being equal. As flood waves pass through a river system, the movement and delivery of sediment to an awaiting basin is episodic in a way that reflects hinterland geomorphology and climate.

Aspects of climate

Variation in sediment flux is greatest where the events (e.g. rainfall, temperature, erosion rates) are largest and most episodic and where there is little vegetation to prevent erosion, in deserts or semi deserts and high mountain ranges (Fig. 1). Recent comparisons between temperate and arid zone river bedload and suspended load transport (Powell *et al.* 1996) have shown that both bedload and suspended sediment yields are an order of magnitude higher in arid zones, that bedload is quantitatively more important than in temperate zones (up to 85% of the total load in some streams, compared with 2–10% in most UK streams), and that the relationship between bed shear and transport rates is clearer and more direct than for temperate streams (Powell *et al.* 1998). Variability is more predicatable in sub-arctic conditions but still extreme, where seasonal snowmelt is the trigger for sudden increases in discharge and sediment transport (Hodson & Ferguson 1999).

Attempts to recognize climate control in ancient sediments are often thwarted by feedback between climate change and tectonism (Shanley & McCabe 1994). Despite these difficulties, river character can be a valuable and sometimes a unique guide to climate-induced environmental change. The main climatic control on fluvial architecture is discharge variability. It is to be expected that lithofacies heterogenity will increase with increasing seasonal or longer-term variability in discharge and that the frequency of the more minor bounding surfaces (i.e. second and third-order surfaces) will also increase (Miall 1996). Recently, Jones *et al.* (2001) concluded the Rio Vero Formation deposits, consisting of braided fluvial facies, in the south central Spanish Pyrenees were deposited by perennial streams. The basis for this interpretation is that, in large perennial to intermittent rivers, channel macroform assemblages dominate the architecture. Under sustained discharge conditions the well-developed macroforms show an upward decrease in grain size and scale of bedding. Additionally, large dune fields, well-defined channels several metres deep with deep scours and lateral or downstream accretion are typical of perennial rivers interspersed with seasonal flow stage fluctuations and poorly developed flood plain facies. The interpretation of this sequence as a perennial stream deposit will allow geologists to predict the location of the associated sediment flux into the basin. The mouths of such streams will be the loci for long-lived deltas – areas where the sediment flux into the basin is of a high magnitude compared with the interspersed, slowly accreting, coastal areas.

Climate change

It is inevitable that any fluctuations in climate within a drainage system will have a consequent effect on sediment delivery to associated sedimentary basins. Research into Quaternary and Holocene fluvial deposits worldwide has yielded plentiful evidence of major changes in fluvial sediment storage and release in the form of aggradation and incision (e.g. Fryirs & Brierley 1998; Huisink 1997, 1999; Jones *et al.* 1999). These changes reflect altering patterns of discharge and sediment availability in response to the climatic fluctuations of the glacial, interglacial and post-glacial periods (Fig. 8). The quantities of sediment that can be involved in changes in a single drainage system are shown by Huisink (1999) where amounts of reworked sediment for the Maas valley, southern Netherlands, exceeds 1.8×10^8 m^3 over a 4 ka timescale in the Late-Glacial. To the extent that drainage basin area, tectonic uplift and relief remain more-or-less constant for a river system over a time scale of 10^3–10^5 years, then Hovius (1998) has demonstrated, using an empirical derived equation, that significant changes in sediment yields can be accounted for by changes in

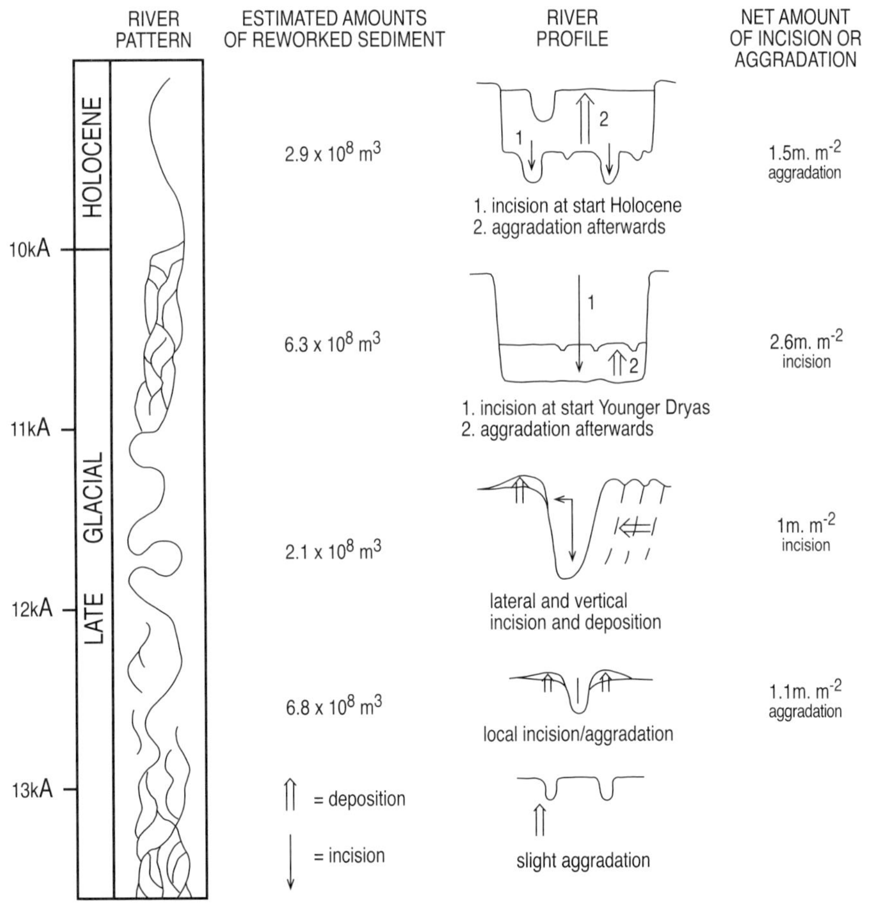

Fig. 8. River pattern, sediment reworking and patterns of incision /aggradation in the Maas valley, Netherlands (adapted from Huisink 1999).

climate parameters such as runoff and temperature fluctuations. These changing parameters can be recognized in ancient deposits through the affects on lithofacies architecture. One example is the change observed in gravel-bed river deposits of the Uncastillo (Oligo–Miocene age) compared with the Rio Cinca Formations (Plio–Pleistocene age) in the south central Pyrenees, Spain (Fig. 9; Jones 1997). This is thought to reflect a change in climate from semi-arid to temperate-humid conditions exerting a controlling influence on the behaviour of the gravel-bed rivers through variations in sediment supply. This change in climate led to a decrease in sediment supply and an accompanying increase in vegetation causing the Plio–Pleistocene rivers to incise (Jones *et al.* 1999). Changes in major facies attributes can be utilized as climatically sensitive criteria and applied to other clastic fluvial lithofacies (Table 1). Fluvial lithofacies must reflect the direct control of climate on sediment supply and the impacts of periodic climate change are especially evident in long-lived river systems.

Human impacts

Human intervention in river catchments can have a profound influence on temporal and spatial patterns of sediment erosion and deposition. Important human activities include dam construction and failure, flow diversion, river regulation and embankment, gravel mining, urbanization and land-use change. All of these change the patterns of water distribution and delivery in the catchment and therefore impact on sediment transport and all have been studied widely (e.g. Knighton 1989; Billi & Rinaldi 1997). Research carried out in the Piave River over the period 1928–1995 has shown how some of the impacts of river regulation may be reflected in changes in the river system that, in turn, are a response to changing patterns of sediment

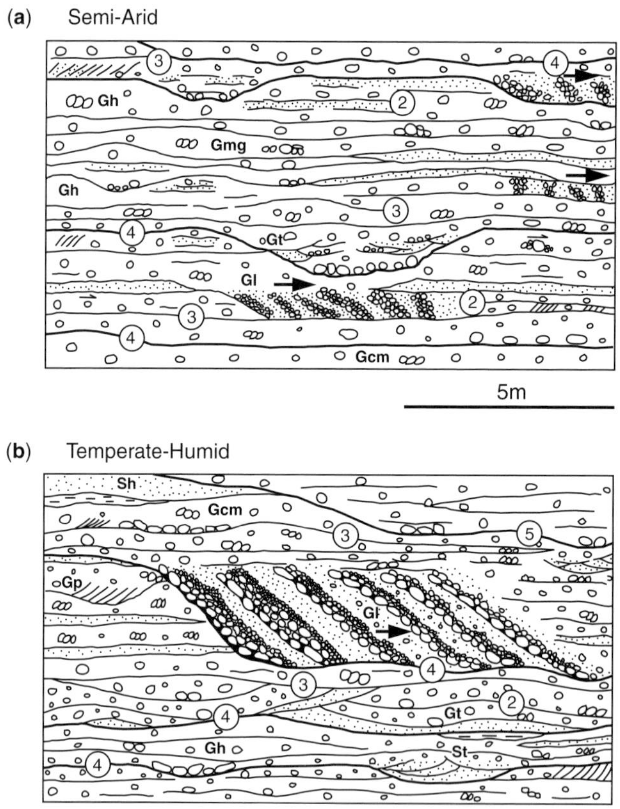

Fig. 9. Architectural sketches of gravel-bed river conglomerates from the South Central Pyrenees, Spain. (**a**) Typical gravel assemblage in a semi-arid climatic setting from the Oligo-Miocene. (**b**) Gravel assemblage in a temperate-humid climatic setting, with deeper channels and larger bedforms from the Plio–Pleistocene. Numbers in circles refer to bounding surfaces and letters to lithofacies classification (classification adapted from Miall 1996; Jones *et al.* 2001).

Table 1. *Field criteria for the interpretation of climate control from clastic fluvial lithofacies*

Semi arid systems	
Typical clastic lithofacies:	Mass flow and gravel sheet deposits
Typical architecture:	Sheets, gravel bars, accretionary units, complex channel fills and sandstone lenses
Major bounding surfaces:	Fifth and more rarely sixth-order
Petrology	Carbonate cements, some weathering of unstable grains
Autocyclic features:	Many truncated erosional tops of units, debris-flow events.
Other structures:	Characteristic paleosols, calcretes, evaportites and evaporitic cements. Lenses of erosional debris and plant debris lags.
Temperate-humid systems	
Typical clastic lithofacies:	Stream flow gravels, gravel sheets. Mass flow deposits rare.
Typical architecture:	Widely developed sandy channels, with gravel bars and fine-grained lithologies.
Major bounding surfaces:	Channel bases of fourth or fifth order.
Petrology:	No distinctive features, unless recent deposits that are poorly cemented and gravelly. Overall, increased clay content.
Autocyclic features:	Macroform, channel fill and gravel bars.
Other structures:	Seasonal variations in sediment accretion units. Increased vegetational debris.

storage (Fig. 10; Surian 1999). In this period dominant discharges halved in magnitude from a mean of 316 m^3/s to 165 m^3/s. The result is a

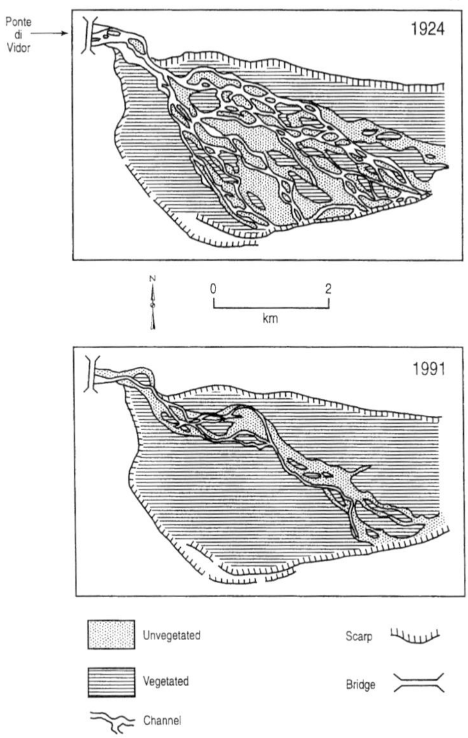

Fig. 10. Channel changes for the Piave River related to Human impacts (dams, gravel extraction, bank protection). Adapted from Surian (1999).

decrease in active channel area and an increase in the quantities of sediment in medium term storage (i.e. vegetated) situations. Similar research in the Kyparissiakos Gulf, Ionian Sea (Poulos *et al.* 2002) has shown that the construction of two dams along the Alfios River significantly reduced the normal sediment flux by about 30%.

More extreme impacts of human interference occur when engineering goes wrong and dams fail. Under these conditions huge volumes of water are released into the system instantaneously and the resulting floods can carry more sediment in a matter of hours and days than has been transported in the previous century (Scott & Gravlee 1968; Bathurst & Ashiq 1998). They also leave a heritage of increased sediment availability that can affect sediment yields for years after the flood. One example of this is the Roaring River, Colorado which was ravaged by the Lawn Lake dam failure in July 1982. Data collected in 1995 showed bedload transport rates remained at enhanced levels 13 years after the flood, with significant impact for sedimentation and flux through the system.

Tectonic activity

On a longer time scale of 10^4–10^8 years, changes in the morphology and topography of the drainage basin will have an impact on sediment supply. Tectonics controls the magnitude, position and evolution of a drainage basin and thus, with climate and local geology controls, the flux of sediment into a basin. Rising topography both within and at the margins of a basin will influence the quantity of sediment which is supplied and the main locations at which it enters the basin along-strike. Several authors (e.g. Oberlander 1985; Hovius 1996;

Talling *et al.* 1997) have commented upon the regularity of transverse drainage spacing from actively uplifting mountain belts based upon Hack's scaling law. It has been proposed that, once fixed by structure and bedrock, that the regular spacing of catchment outlets will have major implications for the infilling of adjacent depositional basins. However, even in mountain belts where the main rivers are regularly spaced, it does not necessarily equate to a regular distribution of sediment input into the adjacent depositional basin. For example, in the Spanish Pyrenees during the Tertiary the supply of sediment to the Ebro foreland basin was controlled by discrete palaeovalleys formed by lateral and oblique ramp anticlines (see Hirst & Nichols 1986; Vincent & Elliott 1997; Jones *et al.* 2001). Such tectonic controls are long-lived and continue to influence the modern transverse drainage patterns in the area today (Fig. 11).

In comparison, the transverse fluvial systems of the Zagros Mountains of Iran form a more minor component of the drainage network (Fig. 12a; Jones 1997). This may reflect stage in the evolution of the mountain belt whereby structural complications and halokinesis have caused the drainage patterns to become less regular, introducing non-transverse stream elements. However, it is evident from the Zagros mountains that, where transfer zones and/or structural re-entrants exist, then transverse drainage systems can evolve as important sediment supply pathways to the depositional basin (Fig. 12b; Baltzer & Pursor 1990).

Thrust movement disturbs gradients particularly on transverse rivers and can induce local or regional cycles of incision (e.g. Jones *et al.* 1999). Although some of the derived material may be stored temporarily within the system it is reasonable to assume that, as with climate change, the increase in gradient and stream power and bed shear stress will increase sediment excavation from the system, and control the patterns of deposition and erosion (Jones 2002). Importantly, once a river incises through bedrock it becomes fixed in

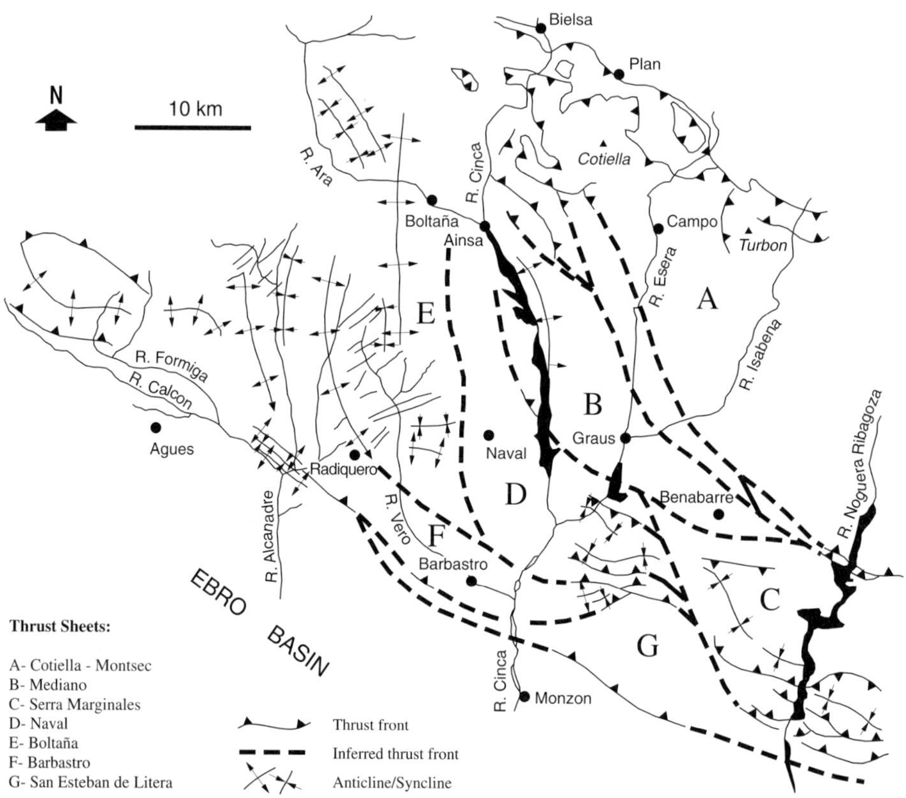

Fig. 11. The Spanish Pyrenees modern transverse drainage pattern indicating the continued influence of lateral and oblique ramp anticlines, especially influencing long-lived rivers, for example the Rio Cinca.

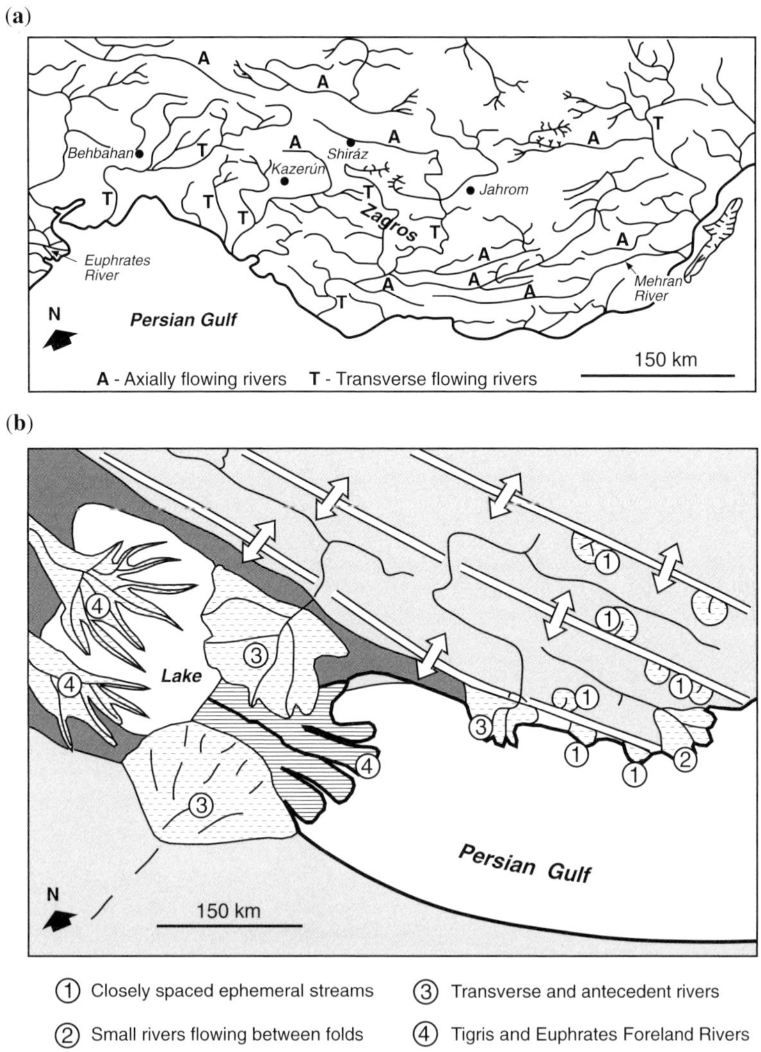

Fig. 12. (**a**) Modern drainage pattern of the Zagros Mountains, Iran. (**b**) Where transfer zones exist then important transverse drainage systems develop (adapted from Baltzer & Pursor 1990).

position, dominating sediment supply to a basin and becoming a long-lived system.

Changes in topography can deflect drainage networks either into or away from a basin and can prevent or encourage sediment supply to the basin. A classic example of drainage deflection can be seen in the East African Rift System where initial doming diverted rivers away from the opening rift valleys leading to many basins becoming starved of sediment (Fig. 13a; Frostick 1997; Frostick & Reid 1989). In addition, in some parts the development of transfer faults directly controls local drainage patterns and sediment supply, usually leading to increased sediment flux (Fig. 13a). In a similar way to the East African Rift System the Dead Sea strike-slip fault system controls sediment distribution through river drainage diversion and deflection (Fig. 13b). In the central Arava Valley, along the Dead Sea transform fault, near to Petra (see Fig. 13b), there is a notable discrepancy between the number and location of the feeding drainage basins from the eastern margins and those of the alluvial fans and the cross-rift large streams. Some of these alluvial fans lack any feeding drainage basin. These discrepancies can be explained by 15 km of left-lateral movement since the Late Pliocene or Early

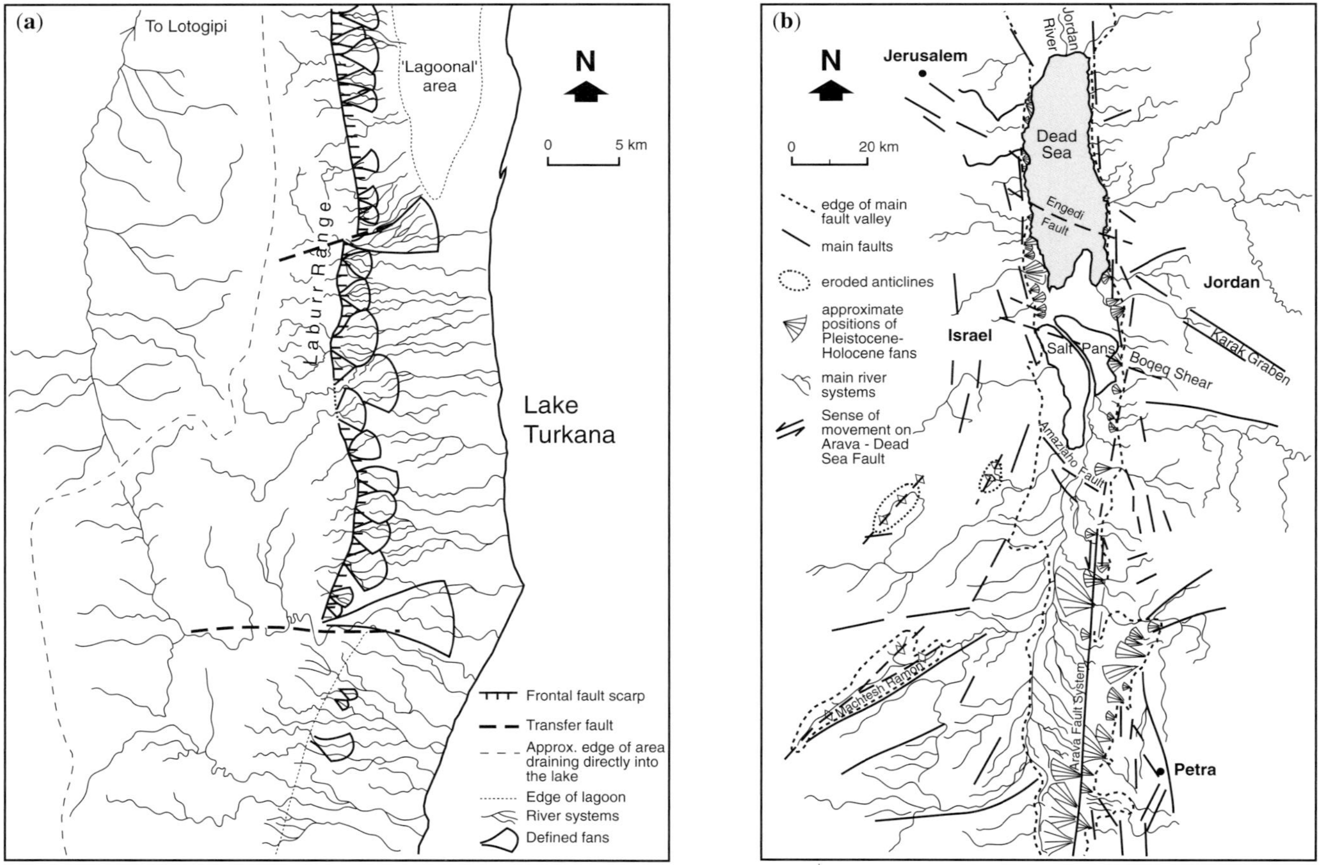

Fig. 13. (**a**) River drainage and structure around Lake Turkana, part of the Eastern Branch of the East African Rift system. (**b**) Simplified map illustrating the structure, distribution of Pleistocene to recent alluvial fan deposits, and pattern of river drainage along the Dead Sea transform zone.

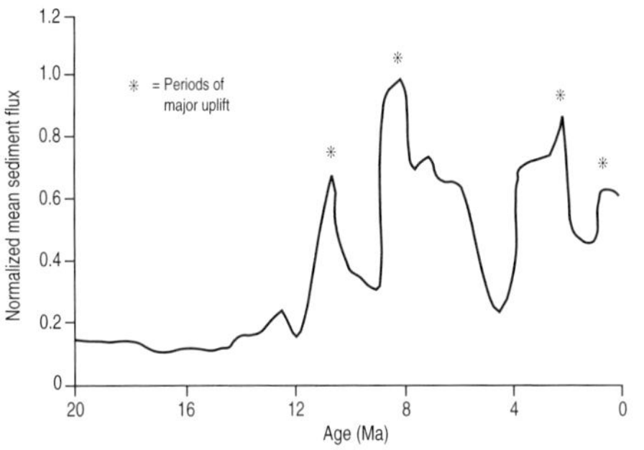

Fig. 14. Periods of renewed sediment flux coincide with periods of major Himalayan uplift recording a strong cyclicity over a 12 million year time period (after Hovan & Rea 1992).

Pleistocene along the Arava-Dead Sea segment of the transform significantly redistributing the sediment flux pathways.

On a larger scale, evidence from a study carried out of the Indian Ocean Deep Sea drilling cores by Hovan & Rea (1992) provides a close linkage between tectonics and sediment flux. The core material contains a useful record of Himalayan derived sediment deposition that can be biostratigraphically dated. Sediment flux rates derived from these data show peaks of sediment delivery at 10.5, 8, 2.5, and 0.3 Ma that do not correspond to any known climatic and eustatic fluctuations (Fig. 14). These are interpreted as reflecting cycles in orogenic activity and thrust emplacement accompanied by weathering and erosion.

Discussion and conclusion

It is evident that sediment delivery to basins is subject to temporal fluctuations at a range of scales. Examples of this phenomenon range from bedload pulses that pass in a few seconds to the impacts of climatic and tectonic fluctuations that may last millennia. The longer time scale and larger scale phenomena will have obvious and direct impacts on sediment supply to basins and can easily be incorporated into basin models. But variations that occur over shorter timescales may also be significant in influencing basin deposit architecture. The passage of pulses and slugs of sediment can cause threshold conditions for change to be exceeded causing knock-on changes in channel morphology accompanied by the release of large quantities of stored material; the big flood may strip out floodplain deposits; slope failure can flood a reach with sediment and the attenuated wave of increased sediment transport will then travel to the receiving basin. The net effect of all of these processes is that large scale sediment inputs to the receiving basin are episodic. The intervening periods of lower supply provide the opportunity for material to be reworked and deposits to be modified. Under these conditions, the large scale episodic deposits will be capped by thinner reworked sediments and erosive surfaces that define the fundamental 'parcels' of delivery. Examination of most deltaic deposits in the stratigraphic record supports this contention. The definition of individual delta top and front strata owes its existence primarily to episodic sediment supply, and sea level change. The challenge now is to build these observations into basin modelling and establish the scale of temporal fluctuations that will yield results that stand validation.

We would like to thank Martyn Pedley and Adrian Hartley for reviewing the manuscript, Kate Davis for drawing the figures and our many colleagues in geology and geography who inspired us to write this paper.

References

Allan, A. F. & Frostick, L. E. 1999. Framework dilation, winnowing and matrix particle size: the behaviour of some sand-gravel mixtures in a laboratory flume. *Journal of Sedimentary Research,* **69,** 21–26.

Ashmore, P. E. 1991. Channel morphology and bed load pulses in braided, gravel-bed streams. Geografiska Annaler Sweden, **73A**, 37–52.

Ashworth, P. J., Ferguson, R. I., Ashmore, P. E. Paola, C., Powell, D. M. & Prestegaard, K. L. 1992. Measurements in a braided river chute and lobe: sorting of bed load during entrainment, transport and deposition. *Water Resources Research*, **28**, 1887–1896.

Baltzer, F. & Purser, B. H. 1990. Modern alluvial fan and deltaic sedimentation in a foreland tectonic setting: the Lower Mesopotamian Plain and the Arabian Gulf. *Sedimentary Geology*, **67**, 175–197.

Bathurst, J. C. & Ashiq, M. 1998. Dambreak flood impact on mountain stream bedload transport after 13 years. *Earth Surface Processes and Landforms*, **23**, 643–649.

Billi, P. & Rinaldi, M. 1997. Human impact on sediment yield and channel dynamics in the Arno River basin (central Italy). *In*: Walling, D. E. & Probst, J. L. (eds) *Human impact on erosion and sedimentation*, International Association of Hydrological Sciences Publication, **245**, 301–311.

Brasington, J., Middleton, R., Frostick, L. E. & Murphy, B. J. 2000. Evaluating sediment transport dynamics in a laboratory flume using digital video image analysis. *Earth Surface Processes and landforms*, **25**, 191–196.

Bridge, J. S. 1993. The interaction between channel geometry, water flows, sediment transport and deposition in braided rivers. *In*: Best, J. L. &

BRISTOW, C. S. (eds) *Braided rivers*. Geological Society, London, Special Publications, **75**, 13–71.

BRAND, E. W., PREMCHITT, J. & PHILLIPSON, H. B. 1984. Relationship between rainfall and landslides in Hong Kong. *In*: *IV international symposium on landslides*, University of Toronto, Canada, 377–384.

CROZIER, M. J. 1999. Prediction of rainfall-triggered landslides: a test of the antecedent water status model. *Earth Surface Processes and landforms*, **24**, 825–833.

CUMMANS, J. 1981. Chronology of mudflows in the South Fork and North Fork Toutle River following the May 18 eruption. *In*: LIPMAN, P. W. & MULLINEAUX, D. R. (eds) *The 1980 eruptions of Mount St. Helens, Washington*, Geological Survey Professional Paper **1250**, 479–486.

DAVIDSON, J. P. & DE SILVA, S. 2000. Composite volcanoes. *In*: SIGURDSSON, H. (ed.) *Encyclopedia of volcanoes*, 663–681.

DENSMORE, A. L., ELLIS, M. A. & ANDERSON, R. S. 1998. Landsliding and the evolution of normal-fault-bounded mountains. *Journal of Geophysical Research, B. Solid Earth Planets*, **103**, 15203–15219.

EINSTEIN, H. A. 1942. Formulas for the transportation of bedload. *Transactions of the American Society of Civil Engineers*, **107**, 561–573.

FERGUSON, R. I. & ASHWORTH, P. J. 1992. Spatial patterns of bedload transport and channel change in braided and near-braided rivers. *In*: BILLI, P., HEY, R. D., THORNE, C. R. & TACCONI, P. (eds) *Dynamics of gravel-bed rivers*, 477–496.

FROSTICK, L. E. 1997. The East African rift basins. *In*: SELLEY, R. C. (ed.) *African Basins*. Sedimentary basins of the world, **3**, 187–209.

FROSTICK, L. E. & REID, I. 1989. Is structure the main control of river drainage and sedimentation in rifts? *Journal of African Earth Sciences*, **8**, 165–182.

FRYIRS, K. & BRIERLEY, G. 1998. The character and age structure of valley fills in upper Wolumla Creek catchment, south coast, New South Wales, Australia. *Earth Surface Processes and landforms*, **23**, 271–287.

HIRST, J. P. P. & NICHOLS, G. 1986. Thrust tectonic controls on Miocene alluvial distribution patterns, southern Pyrenees. *In*: ALLEN, P. A. & HOMEWOOD, P. (eds) *Foreland basins*. International Association of Sedimentologists, Special Publication, **8**, 247–258.

HODSON, A. J. & FERGUSON, R. I. 1999. Fluvial suspended sediment transport from cold and warm-based glaciers in Svalbard. *Earth Surface Processes and Landforms*, **24**, 957–974.

HOEY, T. 1992. Temporal variations in bedload transport rates and sediment storage in gravel-bed rivers. *Progress in Physical Geography*, **16**, 319–338.

HOOKE, R. 1994. On the efficacy of humans as geomorphic agents. *Geological Society of America Today*, **217**, 224–225.

HOVAN, S. A. & REA, D. K. 1992. The Cenozoic record of continental mineral deposition on Broken and Ninetyeast ridges, Indian Ocean; southern African aridity and sediment delivery from the Himalayas. *Paleoceanography*, **7**, 833–860.

HOVIUS, N. 1996. Regular spacing of drainage outlets from linear mountain belts. *Basin Research*, **8**, 29–44.

HOVIUS, N. 1998. Controls on sediment supply by large rivers. *In*: SHANLEY, K. W. & MCCABE P. J. (eds) *Relative role of Eustacy, climate and tectonics in continental rocks*. Society of Economic Paleontologists and Mineralogists, Special Publication, **59**, 3–16.

HOVIUS, N. & LEEDER, M. R. 1998. Clastic sediment supply to basins. *Basin Research*, **10**, 1–5.

HUISINK, M. 1997. Late-glacial sedimentological and morphological changes in a lowland river in response to climatic change: the Maas, southern Netherlands. *Journal of Quaternary Science*, **12**, 209–223.

HUISINK, M. 1999. Late-glacial sediment budgets in the Maas Valley, The Netherlands. *Earth Surface Processes and Landforms*, **24**, 93–109.

JACOBSON, R. B. & BOBBIT GRAN, K. 1999. Gravel sediment routing from widespread, low-intensity landscape disturbance, Current River basin, Missouri. *Earth Surface Processes and Landforms*, **24**, 897–917.

JONES, S. J. 1997. *The evolution of alluvial systems in the south central Pyrenees, Spain*. PhD Thesis, University of Reading.

JONES, S. J., FROSTICK, L. E. & ASTIN, T. R. 1999. Climatic and tectonic controls on fluvial incision and aggradation in the Spanish Pyrenees. *Journal of the Geological Society*, **156**, 761–769.

JONES, S. J., FROSTICK, L. E. & ASTIN, T. R. 2001. Braided stream and flood plain architecture. The Río Vero Formation, Spanish Pyrenees. *Sedimentary Geology*, **139**, 229–260.

JONES, S. J. 2002. Transverse rivers draining the Spanish Pyrenees: large scale patterns of sediment erosion and deposition. *In*: JONES, S. J. & FROSTICK, L. E. (eds) *Sediment Flux to Basins: Causes, Controls and Consequences*. Geological Society, London, Special Publications, **191**, 171–186.

KNIGHTON, A. D. 1989. River adjustment to changes in sediment load; the effects of tin mining on the Ringarooma River, Tasmania, 1875–1984. *Earth Surface Processes and Landforms*, **14**, 333–359.

LISLE, T. E., PIZZUTO, J. E., IKEDA, H., ISEYA, I. & KODOMA, Y. 1997. Evolution of a sediment wave in an experimental channel. *Water Resources Research*, **33**, 1971–1981.

MAGILLIGAN, F. J. 1985. Historical floodplain sedimentation in the Galena River basin, Wisconsin and Illinois. *Annals of the Association of American Geographers*, **75**, 583–594.

MEADE, R. H., YUZYK, T. R. & DAY, T. J. 1990. Movement and storage of sediment in rivers of the United states and Canada. *In*: WOLMAN, M. G. & RIGGS H. C. (eds) *The geology of North America*, Geological Society of America, 255–280.

MIALL, A. D. 1996. *The geology of fluvial deposits: sedimentary facies, basin analysis and petroleum geology*. Springer-Verlag.

MIDDELKOOP, H. & ASSELMAN, N. E. M. 1998. Spatial

variability of floodplain sedimentation at the event scale in the Rhine-Meuse delta, The Netherlands. *Earth Surface Processes and landforms*, **23**, 561–573.

MIDDLETON, R., BRASINGTON, J., MURPHY, B. J. & FROSTICK, L. E. 2000. Monitoring gravel framework dilation using a new digital particle tracking method. *Computers and Geosciences*, **26**, 329–340.

OBERLANDER, T. M. 1985. Origin of drainage transverse to structures in orgens. *In*: MORISAWA, M. & HACK, J. T. (eds) *Tectonic geomorphology*. Allen & Unwin, Boston, 155–182.

PAOLA, C. 2000. Quantitative models of sedimentary basin filling. *Sedimentology*, **47**, 121–178.

POULOS, S. E., VOULGARIS, V., KAPSIMALIS, V., COLLINS, M. B. & EVANS, G. 2002. Sediment fluxes and the evolution of a riverine-supplied tectonically active coastal system: Kyparissiakos Gulf, Ionian Sea (eastern Mediterranean). *In*: JONES, S. J. & FROSTICK, L. E. (eds) *Sediment Flux to Basins: Causes, Controls and Consequences*. Geological Society, London, Special Publications, **191**, 247–266.

POWELL, D. M., REID, I., LARONNE, J. B. & FROSTICK, L. E. 1996. Bedload as a component of sediment yield from a semi-arid watershed of the Northern Negev. *In*: *Erosion and sediment yield: global and regional perspectives*. International Association of Hydrological, Sciences Publication, **236**, 389–397.

POWELL, D. M., REID, I., LARONNE, J. B. & FROSTICK, L. E. 1998. Cross stream variability of bedload flux in narrow and wider ephemeral channels during desert flash floods. *In*: *Proceedings of the fourth International Workshop on Gravel Bed Rivers*.

REID, I. & FROSTICK, L. E. 1985. Role of settling, entrainment and dispersive equilivance and of interstice trapping in placer formation. *Journal of the geological Society of London*, **142**, 739–746.

REID, I. & FROSTICK, L. E. 1986. Dynamics of bedload transport in Turkey Brook, a coarse-grained alluvial channel. *Earth Surface Processes and Landforms*, **11**, 143–155.

REID, I. & FROSTICK, L. E. 1987. Flow dynamics and suspended sediment properties in arid zone flash floods. *Hydological Processes*, **1**, 239–253.

RICHARDS, K. 2002. Drainage basin structure, sediment delivery and the response to environmental change. *In*: JONES, S. J. & FROSTICK, L. E. (eds) *Sediment Flux to Basins: Causes, Controls and Consequences*. Geological Society, London, Special Publications, **191**, 149–160.

RIVENAES, J. C. 1992. Application of a dual-lithology, depth-dependent diffusion equation in stratigraphic simulation. *Basin Research*, **4**, 133–146.

RIVENAES, J. C. 1997. Impact of sediment transport efficiency on large-scale sequence architecture: results from computer simulation. *Basin Research*, **9**, 91–106.

SCOTT, K. M. & GRAVLEE, G. C. 1968. *Flood surge on the Rubicon River, California; hydrology, hydraulics and boulder transport*. United States Geological Survey Professional Paper, M1–M40.

SHANLEY, K. W. & MCCABE, P. J. 1994. Perspectives on the sequence stratigraphy of continental strata. *American Association of Petroleum Geologists Bulletin*, **78(4)**, 544–568.

SURIAN, N. 1999. Channel changes due to river regulation: the case of the Piave River, Italy. *Earth Surface Processes and Landforms*, **24**, 1135–1151.

TALLING, P. J., STEWART, M. D., STARK, C. P., GUPTA, S. & VINVENT, S. J. 1997. Regular spacing of drainage outlets from linear fault-blocks. *Basin Research*, **9**, 275–302.

VINCENT, S. J. & ELLIOTT, T. 1997. Long-lived transfer-zone palaeovalleys in mountain belts: an example from the Tertiary of the Spanish Pyrenees. *Journal of Sedimentary Research*, **B67**, 303–310.

WALLING, D. E. 1977. Limitations on the sediment rating curve technique for estimating suspended sediment loads, with particular reference to British rivers. *International Association of Hydrological Sciences Publication*, **122**, 34–48.

WALLING, D. E.& KLEO, A. H. A. 1979. Sediment yields of rivers in areas of low precipitation; a global view. *International Association of Hydrological Sciences Publication*, **128**, 479–493.

WALLING, D. E. & BRADLEY, S. B. 1989. Rates and patterns of contemporary floodplain sedimentation: a case study of the River Culm, Devon, UK. *GeoJournal*, **19**, 153–162.

WATHEN, S. J. & HOEY, T. B. 1998. Morphological controls on the downstream passage of a sediment wave in a gravel-bed stream. *Earth Surface Processes and Landforms*, **23**, 715–730.

Sediment budgeting techniques in gravel-bed rivers

P. A. BREWER[1] & D. G. PASSMORE[2]

[1]*Institute of Geography and Earth Sciences, University of Wales, Aberystwyth, Ceredigion SY23 3DB, UK*

[2]*Department of Geography, University of Newcastle upon Tyne, Newcastle upon Tyne NE1 7RU, UK*

Abstract: The difficulties in obtaining reliable sediment transfer data from direct field measurement or from sediment transport formulae are widely recognized by geomorphologists and river engineers. Quantifying morphological changes (erosion and deposition) in rivers by the analysis of archive data or by field survey, however, can overcome many of these difficulties and provide a mechanism by which sediment budgets can be calculated over a variety of spatial and temporal scales. This paper applies three sediment budgeting methods based on morphological changes in a hypothetical braided reach. These methods range from a simple two-dimensional planform budget to more sophisticated three-dimensional cross-profile and morphological budgets. The development of each budget technique is described and the limitations and applicability of each identified. The three methods are then used to calculated sediment transfer rates in a multi-thread reach on the River Severn in mid-Wales, UK. Results show that across four budget periods spanning 2.5 years the reach was a net exporter of sediment. Application of the planform budget to eight time periods since 1836 shows a similar pattern of net sediment export in the nineteenth century, but during the majority of the twentieth century the reach was a net sediment sink. Finally, the applicability of applying budgeting techniques to extended centennial and millennial timescales is discussed and an assessment made of the role they might play in advancing our understanding of Holocene river dynamics and informing sustainable river management practices.

It is widely understood that the stability of alluvial channels is dependent on the patterns and processes of fluvial sediment transfer. However, despite over a century of investigations, geomorphologists and river engineers are still seeking solutions to the problems of reliably monitoring and quantifying sediment transfers in natural and semi-natural rivers, where practical and (or) financial constraints preclude its direct measurement (e.g. Church 1985; Bradley *et al.* 1998). This is particularly true of gravel-bed rivers where bed-load transport and channel geometry are both spatially and temporally variable. These problems have tended to thwart attempts to develop bed-load formulae that can accurately predict sediment transport from measures of streamflow hydraulics and sediment characteristics (e.g. Gomez & Church 1989; Bradley *et al.* 1998; Reid *et al.* 1999). One means of overcoming these difficulties, however, is to assess the *morphological* changes that reflect erosion, transport and deposition of sediment over a known period. This approach has been identified as a valuable means of linking hydraulic processes and river geomorphology (Ashmore & Church 1998) and has been shown to yield credible sediment transport data even in comparatively complex anabranched and braided river environments (e.g. Goff & Ashmore 1994; Lane *et al.* 1995; Martin & Church 1995; Macklin *et al.* 1998; Wathen & Hoey 1998; McLean & Church 1999; Ham & Church 2000).

Sediment budgeting using the morphological approach in gravel-bed rivers has focused on a variety of spatial and temporal scales. Most studies have evaluated sediment transfers at the scale of discrete bars or bar assemblages extending over a few hundred metres, and over timescales ranging from annual to discrete events (e.g. Goff & Ashmore 1994; Lane *et al.* 1994, 1995). Other studies have focused on extended reaches of several or tens of kilometres (McLean 1990; Martin & Church 1995; Ham & Church 2000) and combine channel profiling and planform monitoring data (e.g. via aerial photographs or direct survey) taken at approximately decadal intervals or longer. Rather less attention has been focused on sediment

From: JONES, S. J. & FROSTICK, L. E. (eds) 2002. *Sediment Flux to Basins: Causes, Controls and Consequences.* Geological Society, London, Special Publications, **191**, 97–113. 0305-8719/02/$15.00

budgeting at intermediate reach scales, with Wathen *et al.* (1997) conducting one of the few UK-based studies.

However, valley reaches of 1–2 km in many anabranching upland British gravel-bed rivers are typically the setting for the development and subsequent recovery of discrete instability zones that are associated with complex bar assemblages and bifurcated or multi-thread channels; these reaches act to control reach-scale sediment transfers and may see complete transformation of channel and bar environments in as little as 50 years (Macklin & Lewin 1989; Passmore *et al.* 1993; Brewer & Lewin 1998; Macklin *et al.* 1998; Passmore & Macklin 2000). Channel widening and/or migration in these instability zones may be associated with rapid rates of bank erosion over annual or event timescales (Brewer & Lewin 1998; Macklin *et al.* 1998). A better understanding of the rates and patterns of sediment transfer and storage at these reach scales is not only of interest to geomorphologists, but also to river engineers who are responsible for managing river channels. The ability to develop generic sediment budgeting techniques, over a variety of timescales and using rapidly acquired data from field survey or archive sources, will enable more accurate assessments of the influence that intrinsic and extrinsic controls have in governing reach-scale sedimentation and erosion.

This paper assesses three sediment budgeting techniques that are intended to facilitate the monitoring and evaluation of long-term channel instability and sediment transfers at reach scales in upland gravel-bed river environments. The first technique analyses planform changes to estimate two-dimensional sediment budgets. This may employ primary field-based data, while the use of paired map and (or) aerial photograph overlays may extend budgets to decadal or century timescales. A second technique, intended to assess contemporary sediment budgets over defined events or longer intervals, combines data from repeat planform and channel profile surveys. This approach is refined in a third technique (referred to below as the 'morphological budget' technique) to integrate channel planform and cross-profile data at the spatial scale of discrete morphological units that sub-divide channel and bar complexes. The latter approach has previously been used to develop annual sediment budgets for a 2.75-km reach of the River South Tyne in northern England (Macklin *et al.* 1998), and more recently has also been employed by Fuller *et al.* (this volume) in a 3-year study of channel and bar development in an 850-m reach of the River Coquet, northern England.

The data requirements, and data processing routines, for each of the budgeting methods are initially described. All three methods are then applied here to a 500-m reach of the River Severn at Llandinam in mid-Wales, UK with the aim of assessing medium- (the last 150 years) and short-term (the last 2.5 years) sediment transfer and storage. The budgets generated by the three methods are finally compared and an assessment made of their usefulness in quantifying reach-scale sediment transfer and storage over different timescales.

Sediment budgeting techniques

Figures 1–3 illustrate the three budgeting techniques with reference to a hypothetical braided channel reach with a single braid bar. All budgets calculate the change in area or volume of morphological units within the sub-reach 'VW' over an arbitrary budget period ('T'). Morphological units comprise either channel areas (units 'b' and 'g'), bar forms (units 'c', 'd', 'e' and 'f') or areas of bank erosion (unit 'a'); in reality, these areas can be sub-divided on the basis of morphology, topographic elevation or grain-size characteristics. Channel change over the budget period in the sub-reach 'VW' is represented by two areas of sediment deposition on the braid bar (units 'c' and 'd') and one area of bank erosion (unit 'a'). Although the following descriptions of the budgeting procedures only refer to changes in one sub-reach over one budget period, the methods are readily applicable to multiple sub-reaches and multiple budget periods.

All of the sediment budgeting techniques outlined below are subject to a number of limitations. First, they only record sites of sediment erosion and deposition that are evident between surveys, and hence will fail to accommodate sediment transfers that occur during a budget period (e.g. localized scour and fill) and that are not manifested in net morphological changes. Second, the budgeting techniques only calculate gross/net sediment gain/loss from reaches, and do not accommodate sediment throughput in the reach as a whole. As such they will underestimate sediment transport rates. Finally, none of the described techniques differentiate between fine-grained and coarse-grained sediments within the active channel environment or adjacent fluvial terraces. Channel and bar morphology in the River Severn at Llandinam, as is typical of gravel-bed rivers elsewhere in upland Britain, is primarily controlled by the erosion, transport and deposition of coarse bed sediments. Finer material comprising the wash load (predominantly silts and clays in the River Severn), by contrast, forms only thin localized deposits on bar surfaces and comprise relatively thin cappings of overbank sediment on terraced alluvial fills in the study reach. We do make an attempt to estimate the wash load in this study, and it is recognized that interpretation of sediment budgets must account for the potential of fine sediments, once entrained, to be transported rapidly downstream of the study reach (Ham & Church 2000).

Finally, the budgeting techniques outlined here offer a means of establishing regular monitoring of field sites that may be widely adopted without recourse to the relatively high-cost option of obtaining topographic data via, for example, dedicated aerial photograph or Light Detection and Ranging (LiDAR) techniques (e.g. Ritchie 1995) sorties. It should also be noted that the latter solutions, while offering a rapid means of deriving potentially high-resolution topographic data, are problematic with respect to yielding information from below vegetation canopies and the water surface.

The planform budget

This is a simple two-dimensional budget derived solely from planform data collected from paired field-based planform surveys or from paired map and/or aerial photograph overlays. Planform data sources have been widely used in reconstructing past channel change (e.g. Lewin 1977, 1983, 1987; Hooke & Kain 1982) and offer the opportunity to extend budget analysis, albeit with decreasing resolution, to decadal or even century timescales – see Hooke & Kain (1982) and Passmore *et al.* (1993) for reviews on the accuracy and utility of cartographic sources. GIS-based analyses of cartographic and aerial photograph data has already proved a useful means of quantifying localized changes in active gravel area in the Rivers Severn and Tyne over the past 150 years (Passmore *et al.* 1993) and over a 40-year period in the Chilliwack River (Ham & Church 2000).

The planform budget developed in this study estimates two-dimensional net sediment losses (through bank erosion) or gains (through colonization of bar surfaces by vegetation) in the sub-reach 'VW' by assessing the change in area of the 'total channel environment' (the combined area of the channel and active bars) over the budget period (δA_{VW}). In digitized data bases this may be readily achieved by calculating the difference between the area of all morphological units at the start of the budget period (A^0_{VW}) and the area of all morphological units at the end of the budget period (A^1_{VW}) (Fig. 1). This technique is well suited to analyses that are solely based on historic maps and which lack data on the relative height of channel banks and channel/bar relief. However, it takes no account of differential changes in the morphology of channel and active bars, vertical changes resulting from scour or aggradation and variability in the thickness of eroding channel banks.

The channel profile budget

This is a relatively simple three-dimensional budget that combines data from repeat planform surveys (giving lateral sediment exchanges) and channel profile surveys (giving vertical sediment exchanges). Channel profile data, as with planform data, have been used extensively in monitoring channel change and bank erosion (Werritty & Ferguson 1980; Everitt 1993; Hickin 1995), and more recently have been used successfully in budgeting sediment movement through reaches (Martin & Church 1995; Wathen & Hoey 1998). Channel profile budgets may be derived by calculating net changes in cross-sectional area (m^2) at each channel profile (δA_V and δA_W) for the budget period, and then dividing these values by the profile width (L_V and L_W) to give a standardized net sediment gain/loss value per metre (Fig. 2). The active channel zone area of the sub-reach at the end of the budget period is then halved, and each 'half' area then multiplied by the unit gain/loss value for the upstream and downstream cross-profiles, respectively. Summing the resulting two values gives the net sediment gain/loss (m^3) for the sub-reach (δV_{VW}: Fig. 2).

Channel profile budgets thus accommodate vertical changes that are neglected by the planform budget, but are restricted in their spatial resolution to cross-profile sites and hence cannot account for morphological changes occurring in intervening reaches.

The morphological budget

The morphological budget offers a relatively sophisticated means of integrating both planform and channel cross-profile data. Here we develop a morphological budget at the resolution of discrete morphological units that sub-divide the channel and composite bar environments into geomorphologically-defined elements (e.g. bar platform, chute channel). The budget is calculated in two stages. First, the net volumetric change of each morphological unit (e.g. unit 'a' = δV_a) over the budget period is calculated (Fig. 3). Second, the volumetric change for all morphological units is summed to yield a net sediment gain/loss (m^3) for the sub-reach (δV_{VW}).

Application and appraisal of budgeting techniques

The three budgeting techniques were applied to a 500-m multi-thread reach on the River Severn at

Planform Budget:

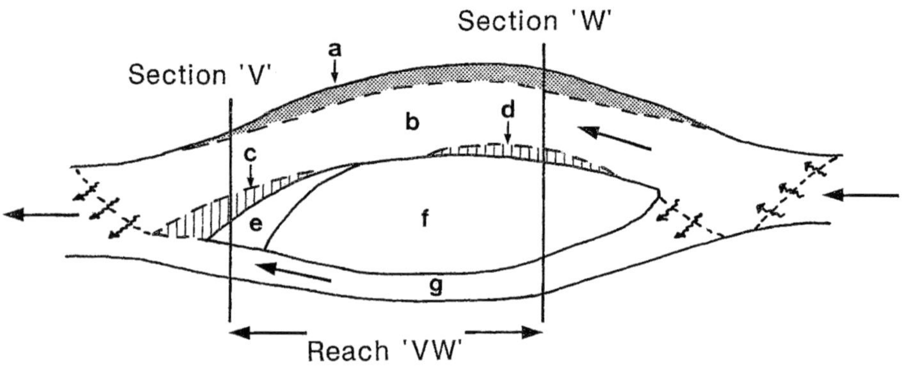

$$\delta A_{vw} = A^1_{vw} - A^0_{vw}$$

where:

$$A^0_{VW} = (b_0 + c_0 + d_0 + e_0 + f_0 + g_0) \qquad A^1_{VW} = (a_1 + b_1 + c_1 + d_1 + e_1 + f_1 + g_1)$$

δA_{VW} = net sediment gain/loss for sub-reach 'VW' (m^2)

A^0_{VW} = area of all morphological units at beginning of budget period(m^2)

A^1_{VW} = area of all morphological units at end of budget period (m^2)

$b_0 \rightarrow g_0$ = area of morphological units at beginning of budget period (m^2)

$a_1 \rightarrow g_1$ = area of morphological units at end of budget period (m^2)

Fig. 1. Calculation of the planform budget for a single sub-reach 'VW'.

Llandinam (National Grid Reference SO 022872) with the aim of examining medium- (last 150 years) and short-term (last 2.5 years) sediment transfer and storage. Historically, the reach has exhibited lateral instability associated with repeated switching between meandering (1836–1840 and 1948–1963) and bifurcating (1884–1903 and 1975–present) phases (see Brewer & Lewin 1998 for a detailed discussion of channel change at Llandinam). Although presently a bifurcating reach (Fig. 4), it does not exhibit large, unstable, overlapping, mid-channel gravel bars as found, for example, in many high-latitude environments (Rust 1972). The divided trunk stream channel more closely resembles the 'wandering gravel rivers' as described by Neill (1973) and Church (1983) or the 'sinuous braided channel' as described by Brice (1984).

Data sources

Data for the medium-term sediment budget were acquired from four historical (Ordnance Survey) 1:10 560-scale map editions (1836, 1840, 1884, 1903) and five *c.* 1:10 000-scale air photograph sorties (1948, 1951, 1975, 1981, 1984). Data for the short-term sediment budget were acquired from two principal field techniques. First, 31 monumented channel profiles were established at approximately 50 m intervals throughout the reach producing 30 sub-reaches for budgeting purposes (Fig. 5). The channel profile for each monumented section was determined by total station survey, with readings taken every 1 m across the channel. Additional points were surveyed at the bank top, bank base, water's edge and at any significant breaks of slope. Second, a planform survey of the

Channel Profile Budget

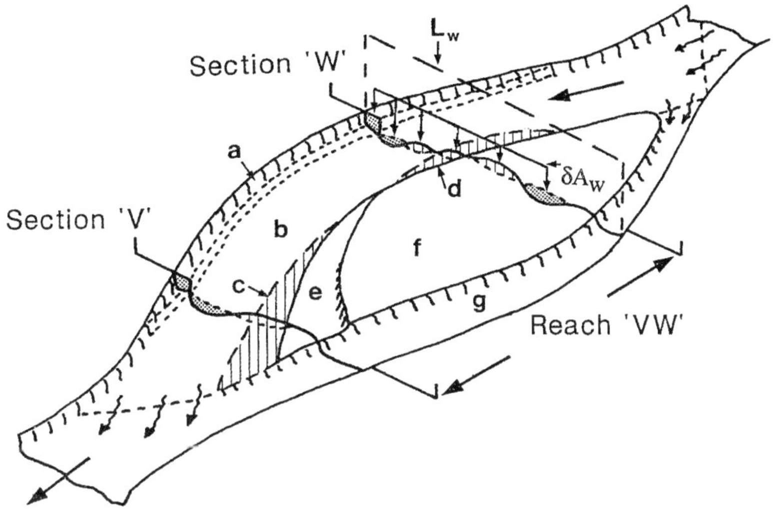

$$\delta V_{VW} = \frac{\Sigma(a+b.....+g)}{2}\left[\frac{\delta A_v}{L_v}+\frac{\delta A_w}{L_w}\right]$$

Where: δV_{VW} = net sediment gain/loss for sub-reach 'VW' (m^3)

δA_V = net change in cross-sectional area for section 'V' (m^2)

δA_W = net change in cross-sectional area for section 'W' (m^2)

L_V = width of active channel zone at section 'V' (m)

L_W = width of active channel zone at section 'W' (m)

a →g = morphological unit areas in sub-reach 'VW' (m^2)

Fig. 2. Calculation of the cross profile budget for a single sub-reach 'VW'.

study reach was undertaken using a combination of field mapping on enlarged aerial photographs (at a scale of *c.* 1:3000) and total station survey, recording channel margins, bar forms, islands, pools and riffles. Mapping was undertaken at low flows, and large compound bars were sub-divided on the basis of morphology, elevation and grain size. Here we focus on short-term budgets derived from five surveys conducted between August 1990 and March 1993.

Data processing

All field data (river channel cross-sections and planform maps) were digitized and subsequently analysed within the vector-based Arc/Info™ GIS. Transformation of archive maps to National Grid Coordinates requires points of known position (tic-marks) to be identified on each map. For maps with National Grid Lines, grid-line intersections provide ideal tic-mark locations generating a regular grid of tic-marks. However, for maps and aerial photographs which do not have grid lines, as used in this study, alternative tic marks need to be identified.

Three prerequisites were established for the location of tic-marks on these archive sources. First, tic-marks had to be present not only on the archive source being digitized but also present on the 1:10 000 metric map edition with National Grid lines. Second, the tic-marks had to be located on features that were stable throughout the survey period (e.g. corners of buildings and field boundary junctions). Finally, tic-marks had to be located as

Morphological Budget

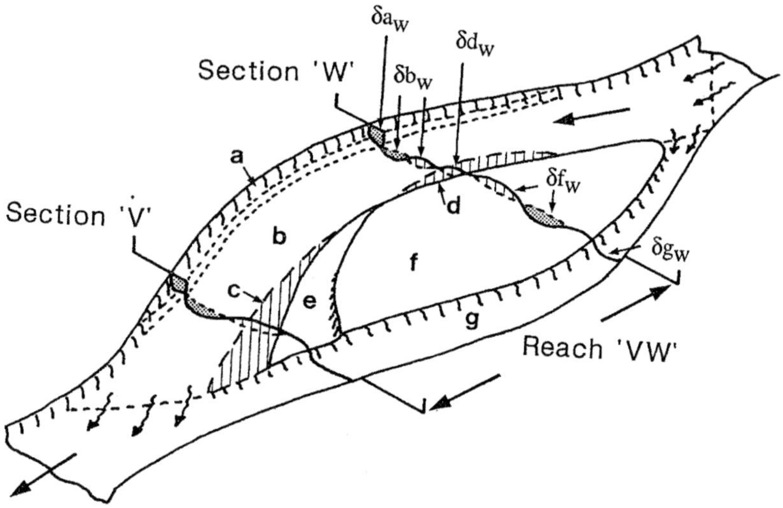

Fig. 3. Calculation of the morphological budget for a single sub-reach 'VW'.
Stage 1: *calculation of volumetric change for each morphological unit.*

for morphological units traversed by:

1 channel profile
(e.g. unit 'd' on profile 'W')

$$\delta V_{\mathrm{d}} = \left(\frac{\delta A_{\mathrm{dw}}}{L_{\mathrm{dw}}} \times A_{\mathrm{d}} \right)$$

2 channel profiles
(e.g. unit 'a' on profiles 'W' and 'V')

$$\delta V_{\mathrm{a}} = \left(\frac{\delta A_{\mathrm{aw}}}{L_{aw}} + \frac{\delta A_{\mathrm{av}}}{L_{\mathrm{av}}} \right) \times \frac{A_{\mathrm{a}}}{2}$$

where:

δV_{d} = volumetric change for morphological unit 'd' (m^3)
δA_{dw} = change in cross-sectional area of morphological unit 'd' (m^2)
δL_{dw} = length of morphological unit 'd' across profile 'W' (m)
A_{d} = planform area of morphological unit 'd' at end of budget period (m^2)
δV_{a} = volumetric change for morphological unit 'a' (m^3)
δA_{aw} = change in cross-sectional area of morphological unit 'a' across profile 'W' (m^2)
δA_{av} = change in cross-sectional area of morphological unit 'a' across profile 'V' (m^2)
L_{aw} = length of morphological unit 'd' along cross profile 'W' (m)
L_{av} = length of morphological unit 'd' along cross profile 'V' (m)
A_{a} = planform area of morphological unit 'a' at end of budget period (m^2)

Stage 2: *calculation of volumetric change for each sub-reach*

e.g. for sub-reach 'VW': $\delta V_{\mathrm{VW}} = \Sigma(\delta V_{\mathrm{a}} + \delta V_{\mathrm{b}} ... + \delta V_{\mathrm{g}})$

near as possible to the channel features being digitized, and certainly be located on the valley floor to avoid introducing topography-related distortion errors. A minimum of four tic-marks were necessary for digitizing, but in practice as many as possible of these irregularly spaced tic-marks were identified (>10) to ensure that transformation errors were reduced.

Once tic-marks were identified they were digitized, in conjunction with grid-line intersection tic-marks from the 1:10 000 metric map, into a pre-National Grid transformed Arc/Info™ coverage. This process of digitizing irregular tic-marks ensured that they were automatically converted to National Grid coordinates. The Arc/Info™ coverage was then reopened, but this time the irregular tic-marks were digitized from the archive source (e.g. aerial photograph). Channel features and lines demarcating the budget sub-reaches were then digitized as before with individual morphological units (polygons) given a unique label identifier. Digitizing errors within the coverage were corrected and, finally, point, line and polygon topology built. This approach to data capture

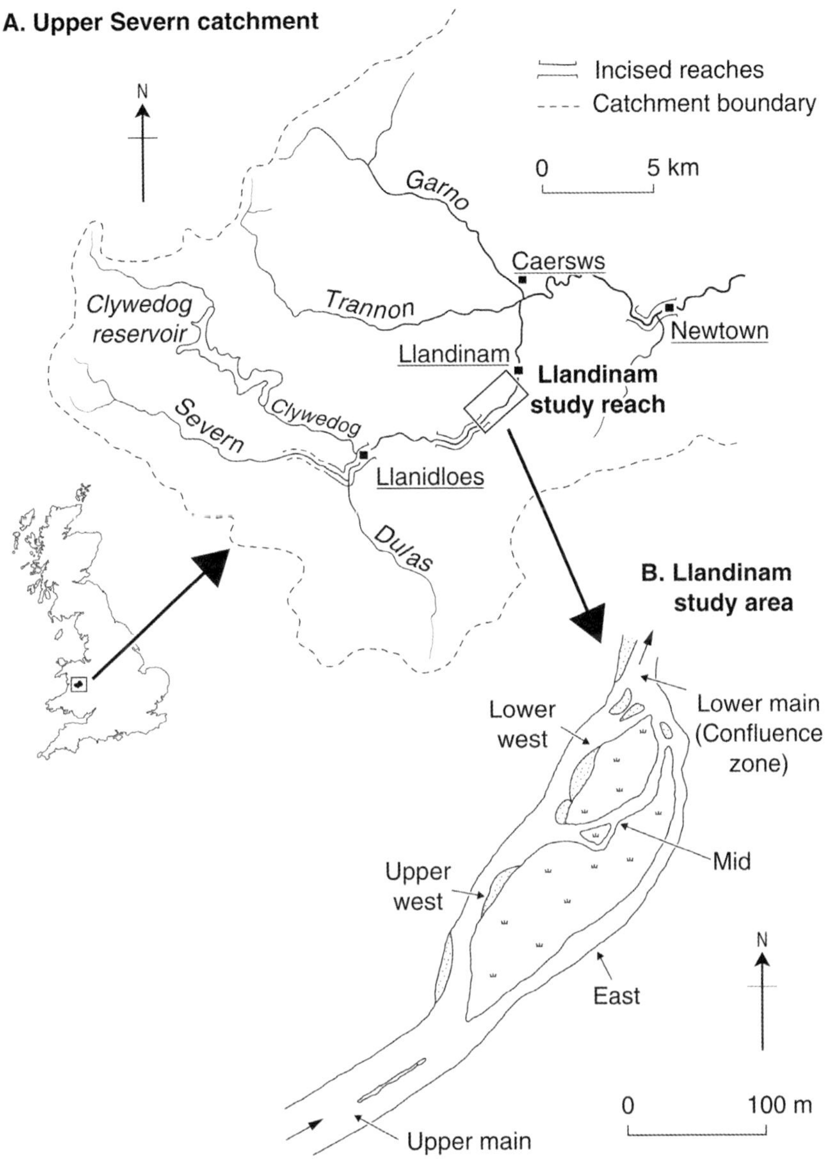

Fig. 4. The Llandinam study site.

ensured that digitized features were, again, automatically transformed into National Grid coordinates, thus reducing the risk of later data transformation introducing errors.

The precision and accuracy of the digitizing process was confirmed through repeat digitizing of selected river reaches for all archive sources. Displacement errors were typically less than 1.00 m for the 1:10 560 scale maps and *c.* 1:10 000 aerial photographs, and less than 0.1 m when digitizing from 1:2000 surveyed planform maps. In the context of channel change at Llandinam, these displacements were considered acceptable because they typically represent less than 1% of the observed channel movement between available archive sources.

Budget results

Table 1a summarizes the net and, in order to accommodate the different number of days in each

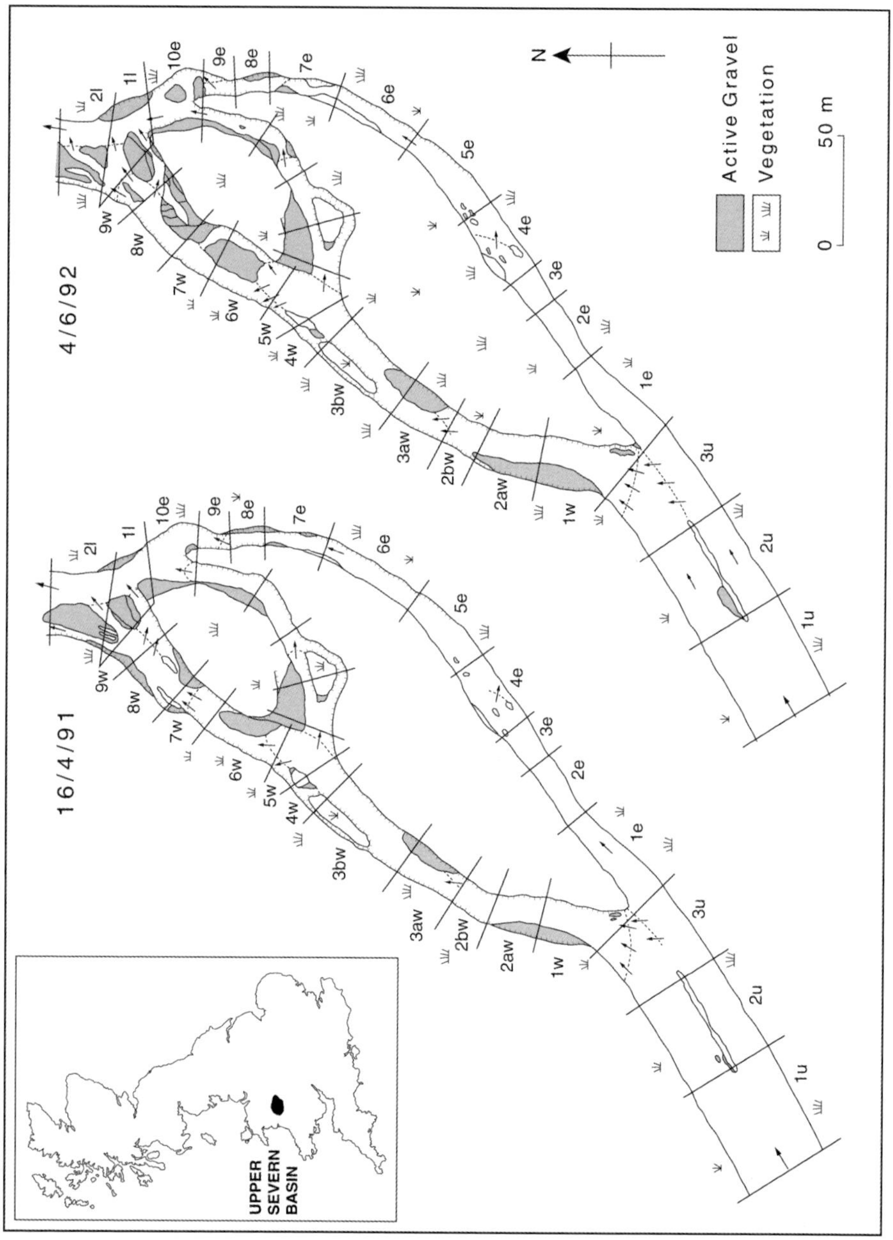

Fig. 5. Morphological maps of the Llandinam study reach for two survey dates (budget period 3) illustrating sub-reach locations and channel change.

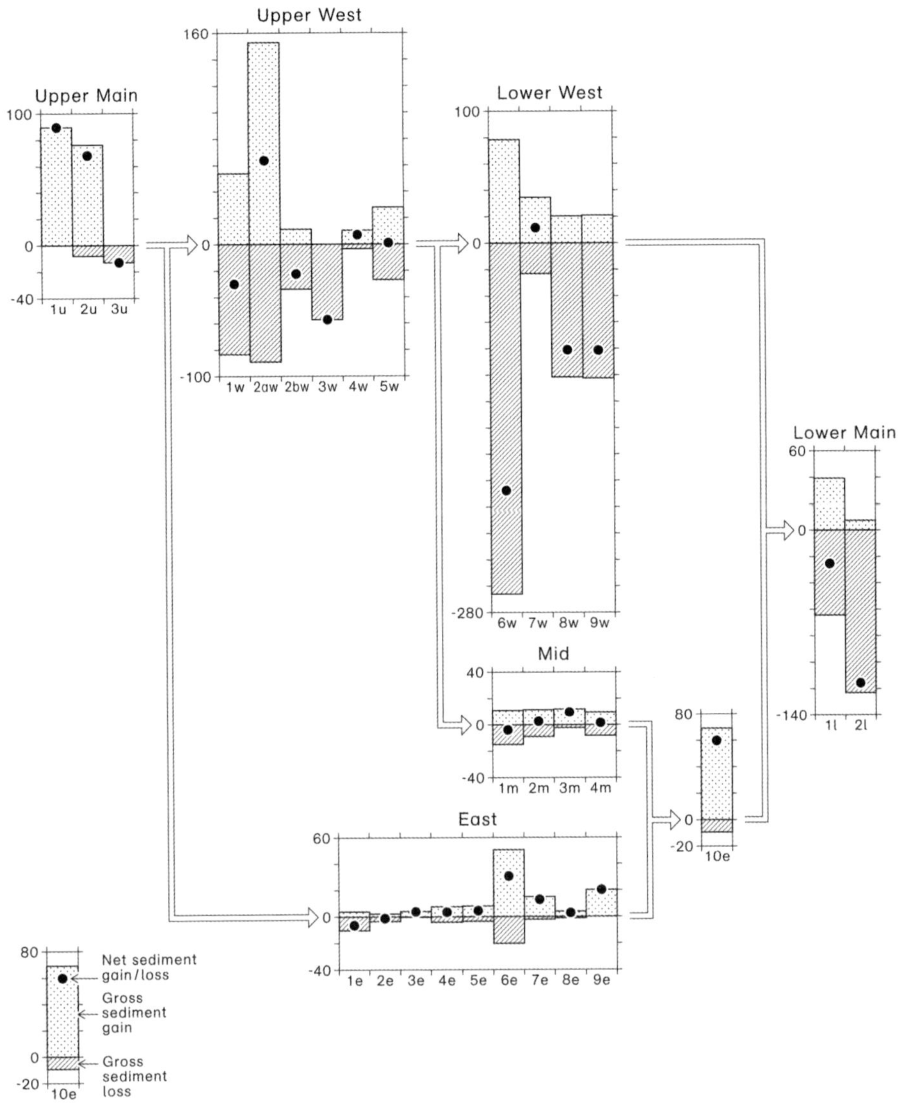

Fig. 6. Sub-reach scale morphological budget for budget period 3 (see Fig. 5 for the location of sub-reaches).

budget period, the average daily short-term sediment budgets for the Llandinam study reach for each budgeting technique over four time periods. The majority of the budget calculations (10 out of 12) indicate that there was net sediment loss in the reach (sediment export) over the four budget periods; only the cross-profile and morphological budgets for period 2 show net sediment gain in the reach. Net sediment export from monitored unstable reaches has also been recorded by these methods on the River South Tyne (Macklin *et al.* 1998; Passmore & Macklin 2001) and here, as at Llandinam, this pattern most probably includes a relatively high proportion of fine-grained sediment eroded from channel banks. The channel banks at Llandinam are composite with fine-grained silts and clays, comprising between 10 and 25% of bank height, overlying coarser gravels. Eroded basal gravels may be quickly deposited on bars within the reach, whereas the silts and clays are readily transported from the reach in the suspended load. The cross-profile and planform budgets have been recalculated to accommodate variable bank composition within the reach (Table 1b). This refinement reduces the amount of gravel input to the calculations and therefore has the effect of reducing

Table 1. *Gross short-term sediment budgets for the Llandinam study reach.(**a**) Net sediment erosion (–)/deposition (+) for the budget period; values in brackets are average daily figures. (**b**) Recalculated cross-profile and morphological sediment budgets accounting for bank composition; values in brackets are average daily figures*

Budget method	Period 1 2/8/90–5/2/91	Period 2 6/2/91–16/4/91	Period 3 17/4/91–4/6/92	Period 4 5/6/92–15/3/93
(a)				
Planform (m^2)	–158 (–0.84)	–125 (–1.81)	–611 (–1.44)	–210 (–0.74)
Cross-Profile (m^3)	–82 (–0.44)	+33 (+0.48)	–450 (–1.06)	–319 (–1.12)
Morphological (m^3)	–79 (–0.42)	+53 (+0.77)	–233 (–0.55)	–240 (–0.85)
(b)				
Cross-profile (m^3) (bank composition)	–48 (–0.26)	+63 (+0.91)	–348 (–0.82)	–284 (–1.00)
Morphological (m^3) (bank composition)	–56 (–0.30)	+83 (+1.20)	–179 (–0.42)	–219 (–0.77)

the gravel export from the study reach by between 9 and 40%. The sensitivity of the cross-profile and morphological budget calculations to the proportion of gravels in eroding banks demonstrates the importance of accounting for this sediment source in reach-scale sediment budgets.

Budget period 2 is unusual in that it is the only period where the three budgets do not consistently generate either net sediment gain or net sediment loss values for the reach (Table 1). This inconsistency is, in part, a function of the underlying assumptions and different data requirements of the budgets (as discussed above), but it is also a result of the nature of bar growth in the budget period. During period 2, channel cross-profiles showed that bars, particularly in the west channel, generally developed by vertical rather than lateral accretion of gravels, resulting in an increase in bar volume but only a modest increase in planform area. Although the cross-profile and morphological budgets can account for such vertical sediment deposition, the planform budget is solely driven by changes in surface area of morphological units. In general, the cross-profile and morphological budgets yield comparable volumetric estimates of reach-scale sediment storage/export, particularly when bank composition is factored into the calculations (Table 1), although the cross-profile budget tends to overestimate sediment export and underestimate sediment storage with respect to the more sophisticated morphological budget.

It is not surprising that gross sediment budgets vary between the four budget periods, since each period covers a different number of days. However, there is also significant variation between the average daily budget and this suggests that rates of sediment transfer in the study reach are not merely a function of time (Table 1). Mean daily discharge hydrographs at Llandinam for each budget period are plotted in Fig. 7, which shows that during each budget period there were a number of high flows potentially capable of mobilizing sediment. Using threshold shear stress, and DuBoys and Darcy Weisbach equations, threshold discharges for sediment entrainment at the Llandinam reach have been calculated as lying between 30.5 and 36.8 $m^3 s^{-1}$ (see Fuller *et al.* this volume, for a detailed description of the calculations). Taking the lowest threshold discharge value and comparing it with the discharge hydrographs (Fig. 7) suggests that there were competent flows for bed-load transport on 3 days in period 1, 1 day in period 2, 3 days in period 3 and 6 days in period 4. The general pattern that emerges from this analysis is that the more days that there are above threshold discharges in a budget period, the more sediment export from there is from the reach. However, the sediment budgets are not only sensitive to bed-load movement, but also to the supply of sediment from bank erosion. Given the complex nature and spatial variability of bank erosion processes, it is not appropriate to determine empirically a threshold discharge for bank erosion at the Llandinam study reach. However, because the river banks are composite and erode by cantilever failure, it is reasonable to assume that bank erosion will be initiated at discharges below those required for bed-load entrainment. Accepting this assumption, analysis of the discharge hydrographs (Fig. 7) and the sediment budget data (Table 1) illustrates the importance of the timing of bank erosion with respect to the timing of competent flows for bed-load transport.

In the first 150 days of budget period 1 there were three periods (14 days in total) where

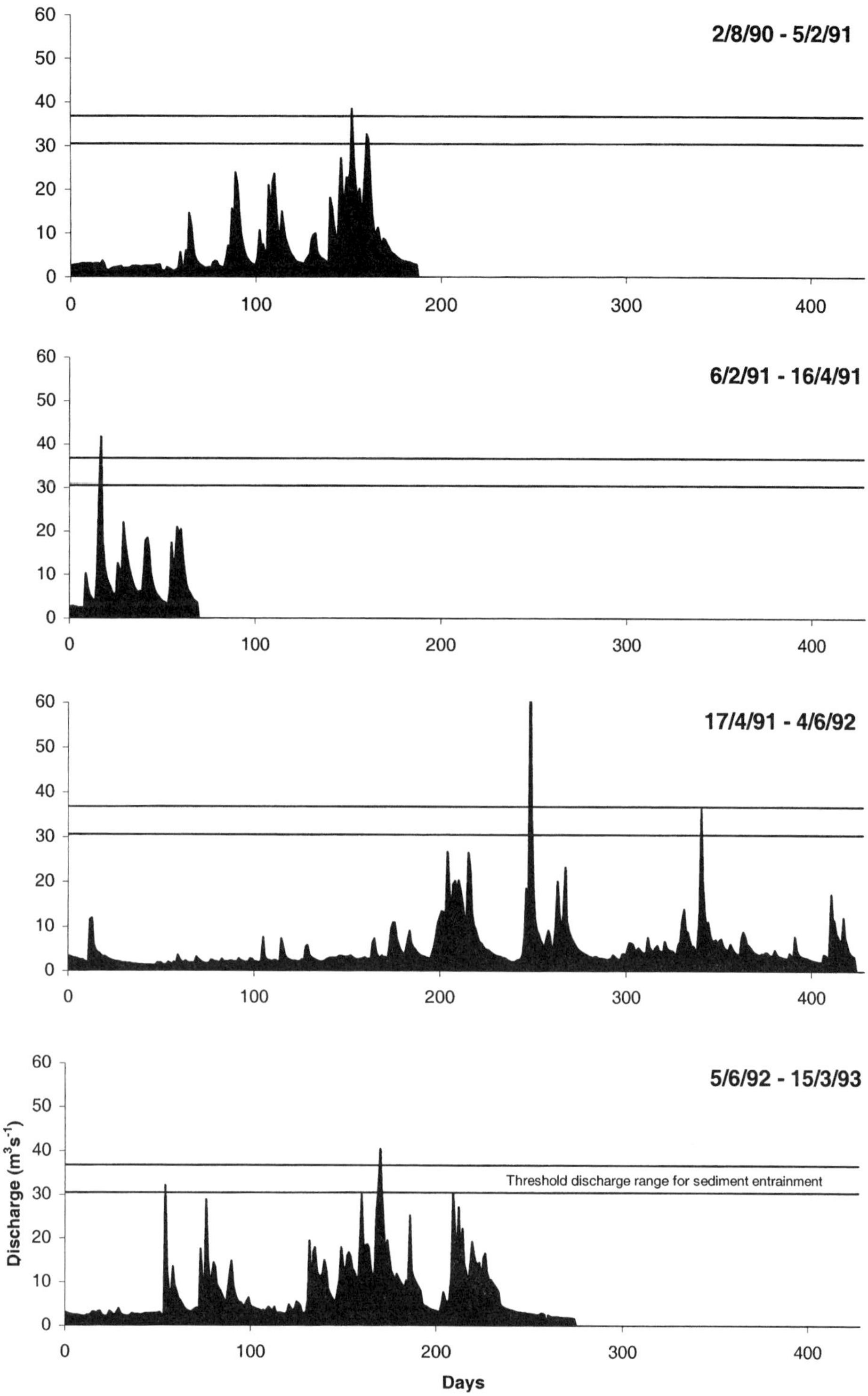

Fig. 7. Mean daily discharge hydrographs for the four budget periods (see Brewer & Lewin 1998 for a description of discharge calculations at Llandinam).

discharges exceeded 20 $m^3 s^{-1}$ but where threshold discharges were not exceeded. It is probable that during these periods coarse and fine-grained sediment was supplied to the channel from bank erosion but the coarse sediment became lodged at the bank base because discharges were not sufficiently competent to mobilize and transport bed material. However, towards the end of budget period 1 (days 153, 161 and 162) threshold discharges for bed-load entrainment were exceeded and the previously eroded bank material was entrained from the channel bed and contributed to the net sediment export from the reach. In budget period 2 the timing of sediment supply from bank erosion and sediment transport by competent flows was reversed. Early in the budget period there was 1 day when the threshold discharge for sediment entrainment was exceeded (day 18). However, later in the budget period there were three periods where flows were probably of sufficient magnitude to cause bank erosion but insufficient for bed-load transport. Therefore, the majority of budget period 2 was characterized by sediment supply from bank erosion without discharges of sufficient magnitude to entrain the gravel component from the channel bed; this led to a net storage of sediment in the reach (Table 1). In budget periods 3 and 4 there were 3 and 6 days, respectively, when the threshold discharge for bed-load entrainment was exceeded, and day 249 in budget period 3 had the highest discharge of all four budget periods (70.5 $m^3 s^{-1}$). In both of these budget periods, the threshold discharge for bed-load entrainment was exceeded on at least one occasion towards the end of the budget period and thus provided an opportunity for bed-load transport in, and hence sediment export from, the study reach (Table 1). One final implication of the *timing* of bank erosion and sediment transport is that it calls into question the value of using average daily figures in sediment budgets. For example, for the first 200 days of sediment budget flows were relatively low and consequently there was little or no coarse-grained sediment erosion or transport in the reach. Probably all the erosion and sediment transport occurred during the second half of the budget period and, hence, the averaged daily sediment budget is significantly overestimated for the first half of the budget period and is significantly underestimated in the second half of the budget period. To allow direct comparison of the geomorphological effectiveness of floods it would be better to constrain budget periods to identifiable events. However, this is not possible where discharge data are unavailable or where surveys have not been undertaken between each flood event.

Table 1 conceals a considerable amount of within-reach sediment transfer because it only provides for each budget period a single net sediment gain or net sediment loss value for the study reach. The degree of within-reach sediment transfer activity is shown in Fig. 6 where sediment gain and loss values for individual 50 m reaches, as calculated by the morphological budget for period 3, are plotted. In general, the upper, east and mid channels (Fig. 1) are dominated by sediment gain during the budget period, principally through bar and bed sedimentation. However, the west and lower channels are characterized by sediment loss, principally through channel scour in sub-reaches 1w, 2aw, 6w, 8w, 9w and 2l (Fig. 6). In addition, the amount of sediment gain and loss in the west channel is considerably higher than in the east channel. Indeed 25% of the sediment loss for the whole study reach is accounted for by a single sub-reach in the west channel (sub-reach 6w, Fig. 5). In budgets 1, 2 and 4 the greatest sediment losses were located in sub-reaches 2l, 7w and 1u, respectively (Fig. 5), demonstrating the spatial variability of sediment loss and gain within the reach. This variability is further highlighted when the behaviour of sub-reaches is examined over the four budget periods. Out of the 29 sub-reaches only three consistently lost sediment and only two consistently gained sediment across all four budget periods. The remaining 24 sub-reaches, particularly those located in the west channel, alternated between losing and gaining sediment over the four budget periods.

All three of the budgeting techniques described and implemented above are subject to a number of assumptions, which if not satisfied can lead to significant errors being introduced into the budgeting calculations. The assumptions and limitations of the planform budget have been outlined above and error analysis, through multiple

Table 2. *Error analysis for the cross-profile budget (adjusted for bank composition)*

	Period 1 2/8/90–5/2/91	Period 2 6/2/91–16/4/91	Period 3 17/4/91–4/6/92	Period 4 5/6/92–15/3/93
Original cross-profile budget (m^3)	–48	+63	–348	–284
Cross-profile budget using profile subset 1 (m^3)	–39	+71	–338	–279
Cross-profile budget using profile subset 2 (m^3)	–146	+65	–114	–226

digitizing of maps and aerial photographs, has shown displacement errors to be less than 1% of observed channel movement during the budget period. The accuracy of the cross-profile and morphological budgets is principally affected by how well the cross-profiles reflect changes in the adjacent sub-reaches. If cross-profiles are located in unrepresentative parts of the channel they may grossly over- or underestimate sediment erosion/ deposition occurring in adjacent sub-reaches. To assess the sensitivity of the cross-profile budget (and implicitly the morphological budget) to profile location, the budget was recalculated twice using two different sub-sets of cross-profiles. The first profile sub-set was generated by joining adjacent sub-reaches starting at sub-reach 1u (i.e. joining 1u with 2u, joining 3u with 1e and 1w, etc.) and then calculating the budget using the profiles at the upstream and downstream limits of the enlarged sub-reach. The second profile subset was generated by joining adjacent sub-reaches starting at sub-reach 2u (i.e. joining 2u with 3u, joining 1w with 2aw, etc.), thereby using all the cross-profiles not used in subset 1. Table 2 shows the results of this profile location-sensitivity analysis and compares the recalculated budgets with the original cross-profile budget). The recalculated cross-profile budget using profile subset 1 shows good agreement with the original budget, with only the budget for period 1 more than 10% different from the original value. However, the recalculated budgets using profile subset 2 show considerably more variability, over- and underestimating sediment export by 204% (budget 1) and 67% (budget 3), respectively. Virtually all the discrepancy between the budgets calculated by profile subset 2 and the original data set is introduced in the dynamic lower main (Fig. 4) section of the study reach; both profile subsets showed good agreement with the original budget in other parts of the study reach. This suggests that the cross-profile budget (and, by implication, the morphological budget) is generally robust in terms of cross-profile location and spacing; however, in reaches where there is rapid channel change (bank erosion and sedimentation) the budget can be extremely sensitive to the positioning of cross-profiles.

Applicability of budgeting techniques over different spatial and temporal scales

The three sediment budgeting techniques presented here are applicable over a range of time and space scales, but the choice of which budgeting technique to use in a particular study will often be constrained by available data and logistical constraints. The planform budget, although only generating areal estimates of change in fluvial sediment storage, does have the advantage of having the least demanding data requirements of all three budgets. Because it principally uses data from maps and aerial photographs, it can be applied to entire river systems and often extended to decadal and even century timescales. For example, Fig. 8 plots *per annum* values (to accommodate the different number of years in each budget period) of gross sediment loss and gross sediment gain in the study reach over eight budget periods since 1836. A cumulative budget curve is also plotted showing the net change *per budget period* in sediment storage in the reach over time. The highest rates of sediment loss and gain in the reach occurred between 1840 and 1884 and 1903 and 1948. These were both periods of rapid channel change when the planform switched from a single-thread meandering pattern to a braided pattern (1840–1884) and back again (1903–1948). A second switch to a braided channel pattern occurred between 1951 and 1975 but this was not associated with such high rates of sediment loss or gain (see Brewer & Lewin 1998 for a detailed discussion of the intrinsic and extrinsic controls on channel change at Llandinam).

The balance between sediment loss and sediment gain in the reach has also varied in the reach since 1836. From 1846 to 1903 and from 1981 to 1984, sediment gain in the reach exceeded sediment loss, but from 1903 to 1981 sediment loss exceeded sediment gain (Fig. 8). Sediment loss in the reach was also recorded in all four of the planform budgets calculated between 1990 and 1993 (Table 1). This changing pattern of sediment gain and loss at the reach scale over time is comparable with that observed at the sub-reach scale and highlights the variable nature of sediment transfer in space and time. Two-dimensional planform budgeting techniques may also be configured to yield long-term estimates of bank erosion rates, and thereby extend the timescales over which centennial and millennial rates and patterns of sediment transfers from alluvial storage may be assessed. This can be achieved by standardizing unit area estimates of bank erosion to give per annum bank retreat rates (Passmore & Macklin 2001).

The cross-profile and morphological budgets have more demanding data requirements and as such will often be restricted in their scale and applicability, particularly where direct field measurements are precluded by logistical and safety constraints. Nevertheless, field survey of channel planform and cross-profile data in 1–3 km reaches of smaller gravel-bed rivers (with catchment areas of less than *c*. 500 km^2) has proved practicable, as has been demonstrated in this study and by recent application of morphological budgeting techniques over a 3-year timescale in the

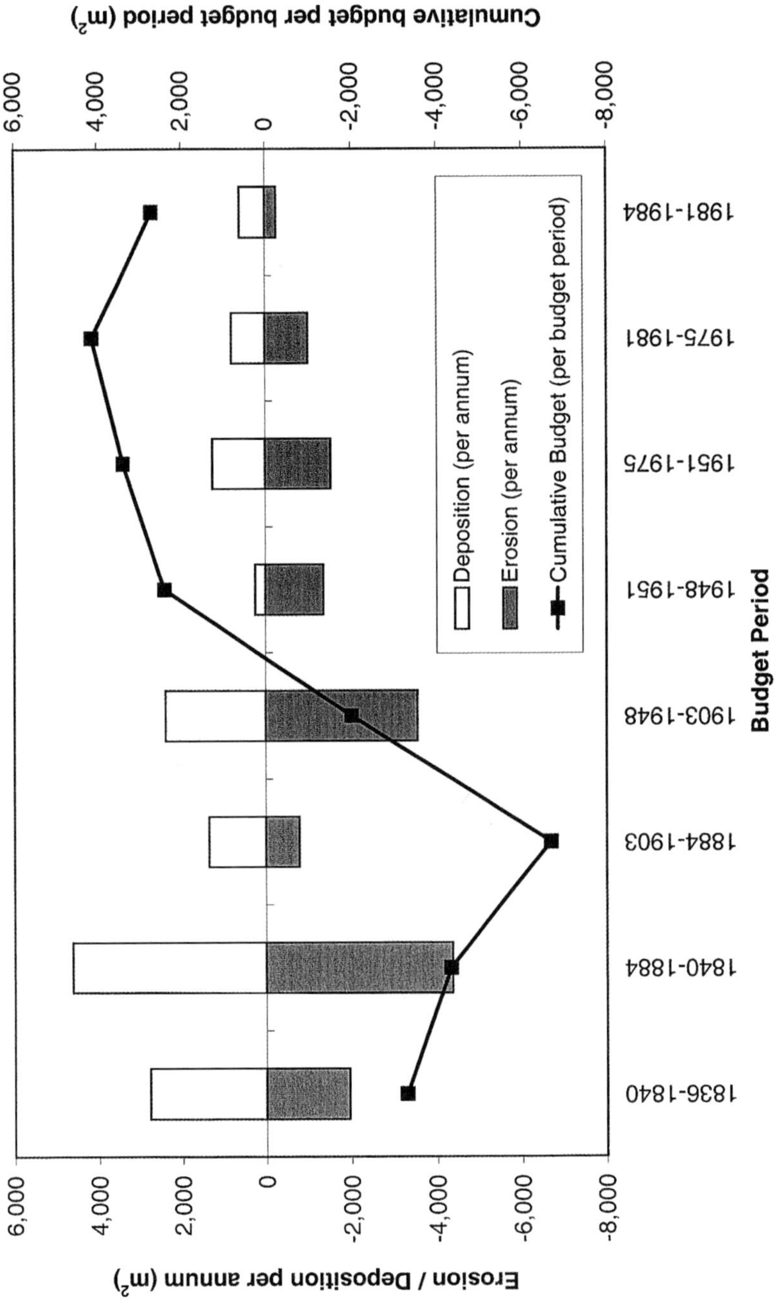

Fig. 8. Medium-term morphological budget for the Llandinam study reach.

River Coquet, northern England (Fuller *et al.* this volume). Historic maps, aerial photographs and channel profiling programmes undertaken for river engineering and management concerns do, however, offer a means of extending the spatial and temporal scale of budgeting analyses provided due consideration is paid to the timing of available surveys and/(or) the interval spacing of cross-profiles, neither of which may be optimal for the purposes of geomorphological investigation. Recent examples of progress in this regard are the mesoscale analyses of sediment transfers in 40–50 km reaches of the Vedder River (Martin & Church 1995) and Chilliwack River (Ham & Church 2000), western Canada.

Morphological approaches to sediment budgeting may also be applied to fluvial sedimentary sequences to yield a long-term (10^2–10^4 years) perspective on the residence time of stored fluvial sediment and the rates and patterns of sediment transfer, and thereby enhance our understanding of fluvial response to environmental changes of differing scale and character beyond those of the instrumental and documentary record. Although long-term sediment budgeting studies that are derived from field evidence are faced with problems of fragmentary survival of sedimentary sequences and establishing dating control for periods of incision and alluviation, recent work by Huisink (1999) and Passmore & Macklin (2001), for example, has demonstrated that morphological approaches to budgeting at extended reach scales (using digitized data bases combining the results of geomorphological mapping, lithostratigraphic analyses, topographic survey and geochronological analyses) can establish upper and lower bound rates of fluvial sediment fluxes over timescales of the order of 10^2–10^4 years.

Conclusions

Sediment budgeting techniques based on morphological approaches are likely to attract increasing attention from geomorphologists and river engineers seeking to investigate the patterns and rates of fluvial sediment erosion, transfer and deposition in gravel-bed rivers. Over timescales ranging from individual events to annual–decadal intervals, the three approaches to budgeting presented here offer a means of overcoming the inherent difficulties of directly measuring reach-scale sediment transport rates in dynamic river settings, and also the limitations of bed-load transport formulae.

Of these approaches, it is the three-dimensional morphological budget that represents the most robust, albeit relatively field- and computationally-intensive solution; a detailed example of the application of this technique is illustrated in Fuller *et al.* (this volume) for an unstable upland reach of the River Coquet, Northumberland, UK for the period between 1997–1999. This study reveals considerable variability in rates and patterns of sediment transfer that are related to reach-scale geomorphological and hydrological controls. It should be noted that the advent of widely available differential GPS technology allows field survey to be accomplished at greater resolutions that those presented in these studies, and with a corresponding increase in the reliability of sediment budgets. However, the combination of planform and cross-profile survey is likely to continue to be a cost-effective means of establishing budgets over reach lengths in the order of several kilometres.

Over centennial and millennial timescales, sediment budgets using two- or three-dimensional morphological approaches also represent a means of estimating the long-term residence time of stored sediment and the rates and patterns of reach-scale sediment transfer, and hence will enhance our understanding of fluvial response to environmental changes of differing scale and character beyond those of the instrumental and documentary record. These approaches can therefore be expected to complement advances in modelling of Holocene fluvial systems, and may also provide the long-term perspective that is necessary to inform the development of sustainable management frameworks for river valley environments.

Research was conducted as part of NERC Grant GR3/7617A. The authors are grateful for helpful comments on the manuscript provided by T. Hoey and S. McLelland.

References

ASHMORE, P. E. & CHURCH, M. A. 1998. Sediment transport and river morphology: a paradigm for study. *In*: KLINGERMAN, P. C., BESCHTA, R. L., KOMAR, P. D. & BRADLEY, B. (eds) *Gravel-bed Rivers in the Environment*. Water Resources Publications, LLC, Colorado, 115–148.

BRADLEY, J. B., WILLIAMS, D. T. & WALTON, R. 1998. Applicability and limitations of sediment transport modelling in gravel bed rivers. *In*: KLINGERMAN, P. C., BESCHTA, R. L., KOMAR, P. D. & BRADLEY, B. (eds) *Gravel-bed Rivers in the Environment*. Water Resources Publications, LLC, Colorado, 543–569.

BREWER, P. A. & LEWIN, J. 1998. Planform cyclicity in an unstable reach: complex fluvial response to environmental change. *Earth Surface Processes and Landforms*, **23**, 989–1008.

BRICE, J. C. 1984. Planform properties of meandering rivers. *In*: ELLIOT, C. M. (ed.) *Proceedings of the Conference – Rivers '83*. American Society of Civil Engineers, New Orleans, 1–15.

CHURCH, M. 1983. Pattern of instability in a wandering gravel bed channel. *In*: COLLINSON, J. D. & LEWIN, J. (eds) *Modern and Ancient Fluvial Systems.* IAS Special Publications, **6**, 169–180.

CHURCH, M. 1985. Bed load in gravel-bed rivers: observed phenomena and implications for computation. *In*: *Proceedings of the Annual Meeting of the Canadian Society for Civil Engineering, 7th Hydrotechnical Conference, Saskatoon*, Canadian Society for Civil Engineers, 17–37.

EVERITT, B. 1993. Channel responses to declining flow on the Rio Grande between Ft. Quitman and Presidio, Texas. *Geomorphology*, **6**, 225–242.

FULLER, I. C., PASSMORE, D. G., HERITAGE, G. L., LARGE, A. R. G., MILAN, D. J. & BREWER, P. A. 2001. Annual sediment budgets in an unstable gravel bed river: the River Coquet, northern England. *In*: JONES, S. J. & FROSTICK, L. E. (eds) *Sediment Flux to Basins: Causes, Controls and Consequences.* Geological Society, London, Special Publications, **191**, 115–132.

GOFF, J. & ASHMORE, P. E. 1994. Gravel transfer rates and morphological changes in braided Sunwapta River, Alberta, Canada. *Earth Surface Processes and Landforms*, **19**, 195–212.

GOMEZ, B. & CHURCH, M. 1989. An assessment of bed load sediment transport formulae for gravel bed rivers. *Water Resources Research*, **25**, 1161–1186.

HADLEY, R. F. 1986. *Drainage Basin Sediment Delivery*. IAHS Publication, 159.

HAM, D. G. & CHURCH, M. 2000. Bed-material transport estimated form channel morphodynamics: Chilliwack River, British Columbia. *Earth Surface Processes and Landforms*, **25**, 1123–1142.

HICKIN, E. J. 1995. Hydraulic geometry and channel scour, Fraser River, British Columbia, Canada. *In*: HICKIN, E. J. (ed.) *River Geomorphology*. Wiley, Chichester, 155–167.

HOOKE, J. M. 1979. An analysis of the processes of river bank erosion. *Journal of Hydrology*, **42**, 39–62.

HOOKE, J. M. & KAIN, R. J. P. 1982. *Historical Changes in the Physical Environment.* Butterworths, London.

HUISINK, M. 1999. Late glacial river sediment budgets in the Maas valley, the Netherlands. *Earth Surface Processes and Landforms*, **24**, 93–109.

LANE, S. N., CHANDLER, J. H. & RICHARDS, K. S. 1994. Developments in monitoring and modelling small-scale river bed topography. *Earth Surface Processes and Landforms*, **19**, 349–368.

LANE, S. N., RICHARDS, K. S. & CHANDLER, J. H. 1995. Within-reach spatial patterns of process and channel adjustment. *In*: HICKIN, E. J. (ed.) *River Geomorphology*. Wiley, Chichester, 105–130.

LEWIN, J. 1977. Channel pattern changes. *In*: GREGORY, K. J. (ed.) *River Channel Changes.* Wiley, Chichester, 167–184.

LEWIN, J. 1983. Changes of channel patterns and floodplains. *In*: GREGORY, K. J. (ed.) *Background to Palaeohydrology.* Wiley, Chichester, 303–319.

LEWIN, J. 1987. Historical channel changes. *In*: GREGORY, K. J., LEWIN, J. & THORNES, J. B. (eds) *Palaeohydrology in Practice.* Wiley, Chichester, 161–175.

MACKLIN, M. G. & LEWIN, J. 1989. Sediment transfer and transformation of an alluvial valley floor: the River South Tyne, Northumbria, U.K. *Earth Surface Processes and Landforms*, **14**, 233–246.

MACKLIN, M. G., PASSMORE, D. G. & NEWSON, M. D. 1998. Controls of short and long term river instability: processes and patterns in gravel-bed rivers, the Tyne basin, northern England. *In*: KLINGERMAN, P. C., BESCHTA, R. L., KOMAR, P. D. & BRADLEY, B. (eds) *Gravel-bed Rivers in the Environment.* Water Resources Publications, LLC, Colorado, 257–278.

MARTIN, Y. & CHURCH, M. 1995. Bed-material transport estimated from channel surveys: Vedder River, British Columbia. *Earth Surface Processes and Landforms*, **20**, 347–361.

MCLEAN, D. G. 1990. *Channel stability on Lower Fraser River*. PhD Thesis, Department of Geography, University of British Colombia.

MCLEAN, D. G. & CHURCH, M. 1999. Sediment transport along lower Fraser River 2. Estimates based on the long-term gravel budget. *Water Resources Research*, **35**, 2549–2559.

NEILL, C. R. 1973. *Hydraulic and Morphologic Characteristics of Athabasca River near Fort Assiniboine*. Highway River Engineering Division Report, REH/73/3, Alberta Research Council, Edmonton.

PASSMORE, D. G. & MACKLIN, M. G. 2000. Late Holocene floodplain and channel development in a wandering gravel-bed river: The River South Tyne at Lambley, northern England. *Earth Surface Processes and Landforms*, **25**, 1237–1256.

PASSMORE, D. G. & MACKLIN, M. G. 2001. Holocene sediment budgets in an upland gravel bed river: the River South Tyne, northern England. *In*: MADDY, D., MACKLIN, M. G. & WOODWARD, J. (eds) *River Basin Sediment Systems: Archives of Environmental Change*. Balkema, Rotterdam, The Netherlands, 423–444.

PASSMORE, D. G., MACKLIN, M. G., BREWER, P. A., LEWIN, J., RUMSBY, B. T. & NEWSON, M. D. 1993. Variability of late Holocene braiding in Britain. *In*: BEST, J. L. & BRISTOW, C. S. (eds) *Braided Rivers.* Geological Society, London, Special Publications, **75**, 205–230.

REID, I., LARONNE, J. B. & POWELL, D. M. 1999. Impact of major climate change on coarse-grained river sedimentation: a speculative assessment based on measured flux. *In:* BROWN, A. G. & QUINE, T. A. (eds) *Fluvial Processes and Environmental Change*. Wiley, Chichester, 105–115.

RITCHIE, J. C. 1995. Laser altimeter measurements of landscape topography'. *Remote Sensing of Environment*, **53,** 91–96.

RUST, B. R. 1972. Structure and process in a braided river. *Sedimentology*, **18**, 221–245.

WATHEN, S. J. & HOEY, T. B. 1998. Morphological controls on the downstream passage of a sediment wave in a gravel-bed stream. *Earth Processes and Landforms*, **23**, 715–730.

WATHEN, S. J., HOEY, T. B. & WERRITTY, A. 1997. Quantitative determination of the activity of within-reach sediment storage in a small gravel-bed river using transit time and response time. *Geomorphology*, **20**, 113–134.

WERRITTY, A. & FERGUSON, R. I. 1980. Pattern changes in a Scottish braided river over 1, 30 and 200 years. *In*: CULLINGFORD, R. A., DAVIDSON, D. A. & LEWIN, J. (eds) *Timescales in Geomorphology*. Wiley, Chichester, 53–68.

Annual sediment budgets in an unstable gravel-bed river: the River Coquet, northern England

I. C. FULLER[1], D. G. PASSMORE[2], G. L. HERITAGE[3], A. R. G. LARGE[2], D. J. MILAN[4] & P. A. BREWER[5]

[1]*Division of Geography and Environmental Management, University of Northumbria at Newcastle, Newcastle upon Tyne NE1 8ST, UK (e-mail Ian.Fuller@unn.ac.uk)*

[2]*Department of Geography, University of Newcastle, Newcastle upon Tyne NE1 7RU, UK*

[3]*Division of Geography, School of Environment and Life Sciences, University of Salford, Manchester M5 4WT, UK*

[4]*Geography and Environmental Management Research Unit, University of Gloucestershire, Cheltenham GL50 4AZ*

[5]*Institute of Geography and Earth Sciences, University of Wales, Aberystwyth, Ceredigion SY23 3DB, UK*

Abstract: Sediment budgeting procedures based on analysis of three-dimensional morphological change provide a useful mechanism by which rates and patterns of fluvial sediment erosion, transfer and deposition can be monitored. This paper presents results from an annual sediment budgeting programme established in a 1-km long piedmont reach of the gravel-bed River Coquet in Northumberland, northern England. The study reach has a locally braided channel planform and has experienced lateral instability over at least the past 150 years. Annual sediment budgets for 1997–1998 and 1998–1999 have been based on tacheometric survey of: (i) 15 monumented channel cross-profiles; and (ii) channel margins and gravel-bar morphology. Survey data have been analysed for each discrete morphological unit (differentiating channel and complex bar assemblages) within 17 sub-reaches of the study reach using Arc/Info™ GIS. The morphological sediment budgeting techniques used to generate these reach-scale budgets may be particularly valuable in unstable gravel-bed rivers due to the inherent difficulties in measuring bed-load transport. The results show considerable variability in rates and patterns of within-reach sediment transfer between the successive surveys. The channel at Holystone is characterized by substantial within-reach sediment transfer, with minimal net export downstream. This behaviour appears to be characteristic of UK gravel-bed piedmont rivers.

Sediment flux in rivers is a critical component determining channel stability (Schumm 1977; Werritty 1997). In gravel-bed rivers, changes in channel morphology reflect the movement of bed material (Leopold 1992; Martin & Church 1995), that is the sediment comprising the bed and lower banks of a river, which, once entrained, moves short distances by traction, saltation or suspension (Martin & Church 1995). However, the flux in bed load is notoriously difficult to measure (Reid *et al.* 1999). Bed-load transport is random and discontinuous (Einstein 1937), and often occurs in a series of pulses, reflecting the downstream passage of sediment waves (Gomez *et al.* 1989; Hoey 1992; Nicholas *et al.* 1995; Paige & Hickin 2000) and variable bed patchiness during initiation of motion (Garcia *et al.* 2000). Bed-load formulae consistently fail to predict transport rates and thresholds of entrainment accurately (Gomez & Church 1989). This is a product of the wide range of flow, fluid and sediment properties that condition sediment mobilization, together with the inherent natural variability of bed sediment and flow structure (Milan *et al.* 1999; Buffin-Bélanger *et al.* 2000).

Sediment budgeting provides a means by which bed-load flux can be monitored, by quantifying the changes in channel morphology that take place in response to sediment erosion, transfer and deposition (Davies 1987; Ashmore & Church 1998). This avoids some of the complications caused by spatial and temporal variability of bed-load

From: JONES, S. J. & FROSTICK, L. E. (eds) 2002. *Sediment Flux to Basins: Causes, Controls and Consequences.* Geological Society, London, Special Publications, **191**, 115–131. 0305-8719/02/$15.00

transport (Hubbell 1987). However, as Martin & Church (1995) note, few studies have investigated the use of changes in channel morphology to evaluate sediment transport. Most research into morphological adjustment of channels has been confined to an assessment of planform change, rather than a quantification of sediment volumes moving through a reach (e.g. Lewin 1976; Werritty & Ferguson 1980; Passmore *et al.* 1993; Brewer & Lewin 1998; Macklin *et al.* 1998).

Investigation of three-dimensional morphological change in gravel-bed rivers requires repeat survey of bed elevation and planform adjustments. The majority of research in this field has focused on the bar scale (Neill 1987; Lane *et al.* 1994). In relation to larger-scale investigations, Goff & Ashmore (1994) sought to determine bed-load transport rates between specific areas of the bed at a scale (*c.* 60-m reach length) extending beyond the individual bar complex in a pro-glacial braided river using field-based sediment budgeting. Channel planform in their study was simply derived from photography, however, rather than high-resolution topographic survey, and budgets were derived exclusively from repeat cross-profile data. In addition Goff & Ashmore's (1994) research was limited to a 2-month (summer melt) period, albeit providing a high-resolution data set for a short period. By comparison, a much lower-resolution, but more spatially and temporally extensive, study was conducted by Martin & Church (1995). This employed a series of channel surveys based on cross-profiles and planform maps to estimate bedload transport on the Vedder River, British Columbia, in an 8-km reach over 10 years. Their measurements examined sediment transfer through 800-m sub-reaches, which exceeded average bar spacing in the Vedder River, suggesting that the patterns of sediment transfer identified were at a coarse resolution only. Furthermore, although survey data were collected over a 10-year period, some surveys were separated by as much as 3 years making an accurate assessment of annual flux problematic. Similar approaches estimating sediment flux over the longer term, using changes in channel morphology over tens of years, and at a relatively coarse scale, have been used by McClean & Church (1999) and Ham & Church (2000). However, there remains a need to assess annual sediment transfers in the context of a spatially extensive reach.

Brewer & Passmore (this volume) present three budgeting techniques designed to evaluate sediment transfers at intermediate reach scales (1–3 km), which are more typical of major instability zones in UK piedmont rivers. Furthermore, in assessing sediment transfers at this higher reach-scale resolution, it is possible to identify more accurately sediment fluxes and storage in these instability zones, which may be overlooked by larger scales of study. Wathen *et al.* (1997) note that reach-scale sediment storage is rarely quantified in sediment budget studies, but recognize that it has a significant effect on the accuracy of morphological methods of bed-load estimation at the reach scale. This paper presents an application of a morphological budgeting technique developed by Brewer & Passmore (which is described in detail in this volume) to an unstable reach of the River Coquet, Northumberland, UK, and quantifies annual reach-scale sediment flux and storage for a 3-year period (1997–1999).

Study site

The River Coquet rises in the Cheviot Hills (776 m) in northern England (Fig. 1). The focus of this research has been an instability zone located at Holystone (National Grid Reference NY 958027), a piedmont setting at the upland fringe in the catchment, where the Coquet drains an area of *c.* 255 km^2 (Fig. 1). Historic (Ordnance Survey) maps show the Coquet at Holystone has been characterized by a high degree of lateral instability and channel avulsion over the last 150 years (Fig. 2). The contemporary active channel at Holystone comprises features common to both braided and meandering rivers, being locally divided by expanses of bare gravel, but having well-defined pool–riffle units (Fig. 3); as such it fits Ferguson & Werritty's (1983) classification as a wandering river. According to Nanson & Croke's (1992) classification, the 400–500-m wide valley floor at Holystone is classified as a medium-energy non-cohesive, wandering gravel-bed river floodplain. The valley as a whole is similar to other gravel-bed rivers in northern England, in that it displays a characteristic 'hourglass' valley morphology, with alternating confined and unconfined sections (Macklin 1999).

Methodology

The channel planform (channel boundaries, barforms and major chute channels) was surveyed on an annual basis between 1997 and 1999 using theodolite-EDM (electromagnetic distance measuring) ground survey with a total station (Fig. 4). Surveys were conducted during low-flow stage (*c.* 2 m^3 s^{-1}). The planform maps were then digitized using the Arc/Info^{TM} GIS. Repeat-channel cross-profiles were surveyed from monumented pegs using theodolite-EDM survey with a total station. Measurement of cross-profiles coincided with the planform surveys. The spacing of the cross-profiles along the study reach was designed to ensure that

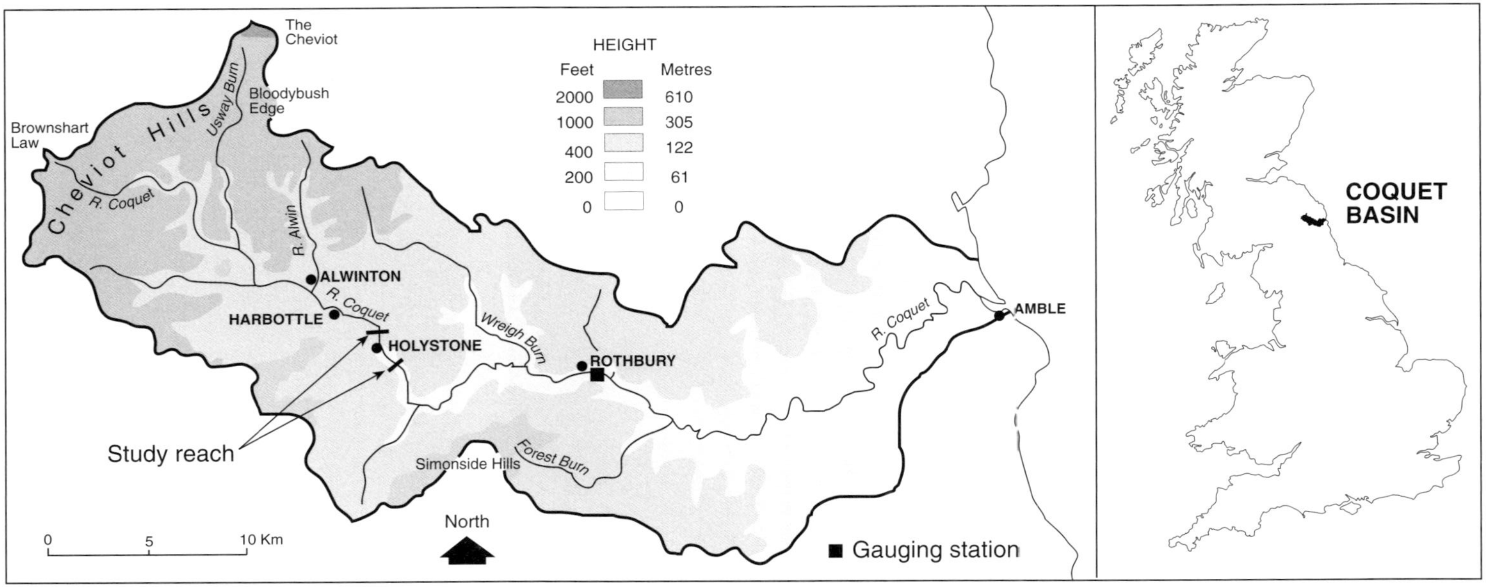

Fig. 1. Location and catchment physiography of the River Coquet, identifying the location of the study reach at Holystone and gauging station at Rothbury. Adapted and reproduced by kind permission of David Archer.

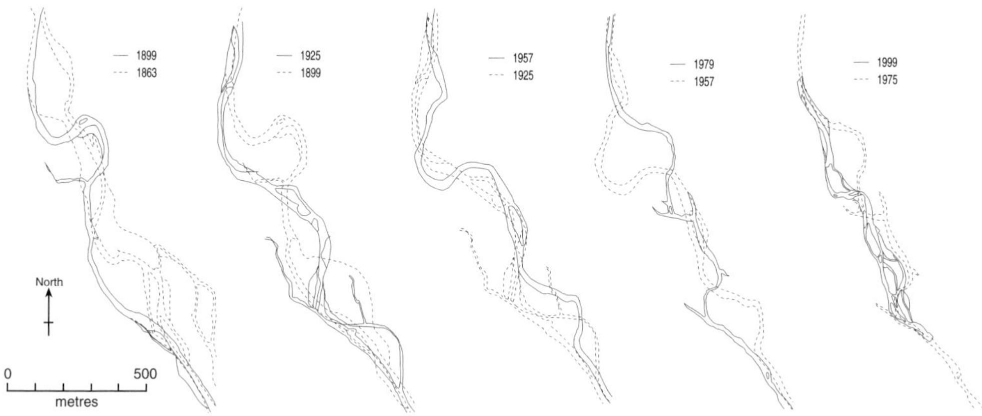

Fig. 2. Historic channel change at Holystone. Note the reduction in sinuosity in the contemporary channel compared with previous surveys.

channel-bed elevation was measured at regular intervals (not exceeding *c.* 75 m) and that each bar in the reach was crossed by at least one profile. This provided the means of assessing the vertical changes, produced by sediment gain or loss, that occurred across the morphological units in the reach. The locations of the cross-profiles along the reach are shown overlaying the channel planform (Fig. 3). The 13 cross-profiles surveyed in 1997 initially divided the reach into 15 sub-reaches. Two additional cross-profiles extended the reach downstream in 1998 (dividing the reach into 17 sub-reaches). Cross-profile data for each survey were incorporated into Arc/Info™.

Sediment budgets were calculated using the morphological budget described by Brewer & Passmore (this volume), and illustrated in Fig. 5. This approach integrates both planform and cross-profile data at the resolution of the discrete morphological unit within each sub-reach. Within Arc/Info™, vertical changes in area along each cross-profile were calculated and standardized to net gain/loss values per m^2. These values were then multiplied by the corresponding planform area values to give a net gain/loss value (m^3) for each morphological unit within the sub-reach. Where morphological units are bounded by two cross-profiles (e.g. large bar and main channel units), the planform area was halved and multiplied by the upstream and downstream cross-profile data, respectively. Where cross-profile data were missing from a monumented section (due to loss of monumented pegs that affected the 1998 re-survey), changes in unit elevation were derived from the nearest available section crossing the morphological unit. These data were then consolidated for each sub-reach into three volumetric measures (m^3): gross sediment gain; gross sediment loss; and net sediment gain/loss (Fig. 6). This approach also identifies the contribution of discrete morphological units to sub-reach sediment gains and/or losses (Fig. 6).

The accuracy of the budgeting approach used at Holystone is largely contingent upon the accuracy of the cross-profiles, as inaccuracies here will be compounded in the budget, generating potentially large errors in the calculated sediment volumes. Cross-profiles were generated from a Digital Elevation Model (DEM) (point density 0.033 m^{-2}, grid interval 0.25 m) constructed from planform survey data in 1999 (independent from the profile data) within sub-reach 2. This abstraction was then compared with the surveyed profiles bounding this sub-reach (cross-profiles 1 and 2). The average difference between the DEM and surveyed profiles was 0.073 m m^{-1}, which compares with the D_{50} of 0.051 m and D_{84} of 0.083 m (derived from sediment sampling described below). In other words, the errors associated with the cross-profile surveys do not exceed errors induced by grain-size characteristics, which suggests that the profiles are an accurate measure of the channel topography and should not constitute a significant error in the calculation of sediment budgets.

However, this approach to budgeting cannot account for vertical changes in topography (accretion/scour) that may occur between cross-profiles, and it assumes that profiles are an adequate representation of any changes in unit height between surveys. However, to assess the overall accuracy of this method requires a more detailed and extensive DEM, which is beyond the scope of this paper. As intimated by Brewer & Passmore (this volume), the main advantage of this

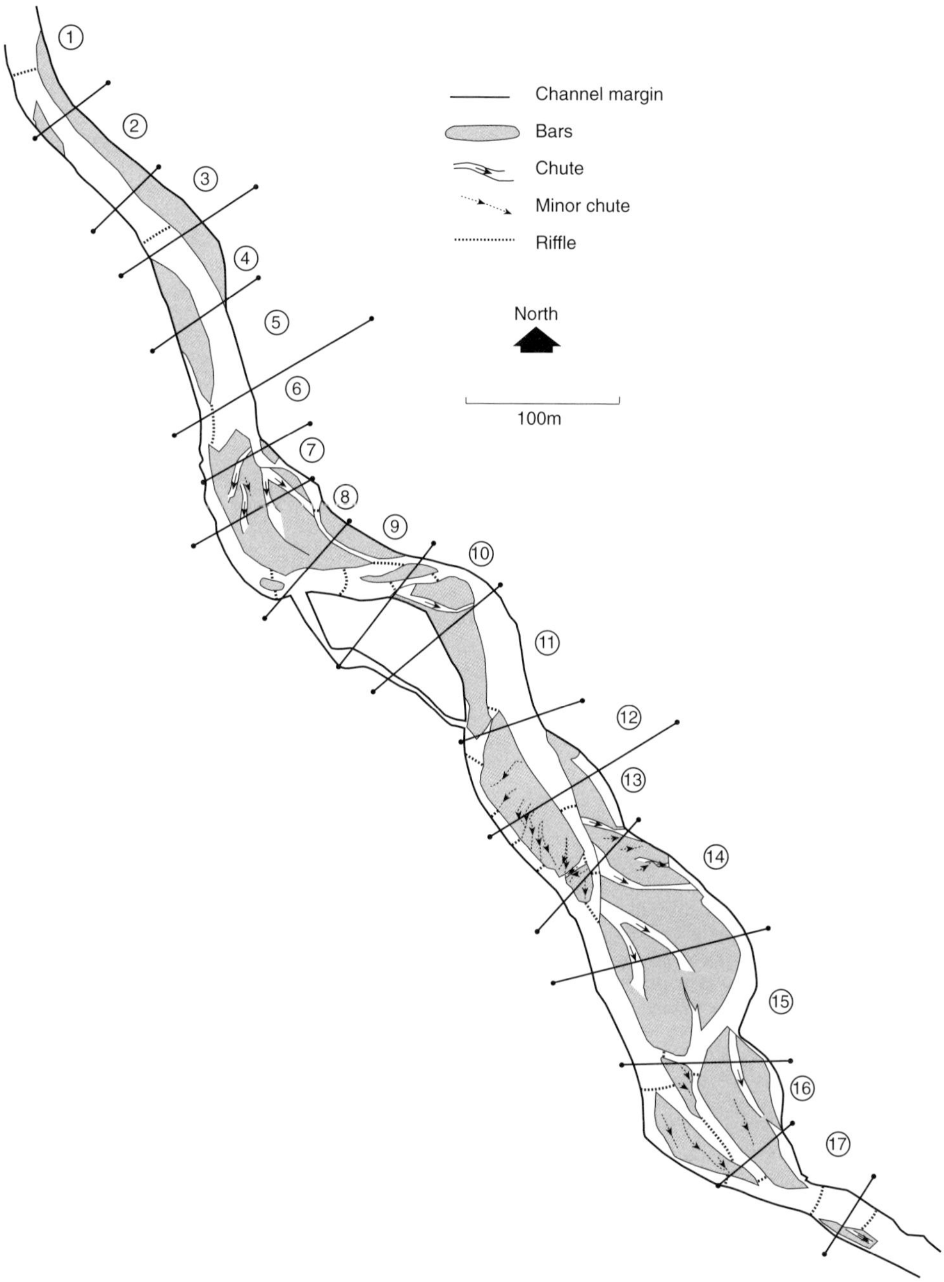

Fig. 3. Detailed channel morphology at Holystone, as mapped in March 1999. The sub-divisions of the reach used for sediment budgeting are numbered 1–17.

approach is that it should provide a reasonably reliable, rapid assessment of within-reach sediment transfers.

In order to characterize the size of sediment being transported through/within the reach at Holystone, the grain size of surface sediment was measured from each major bar unit throughout the reach, collecting a total of nine 100-clast samples

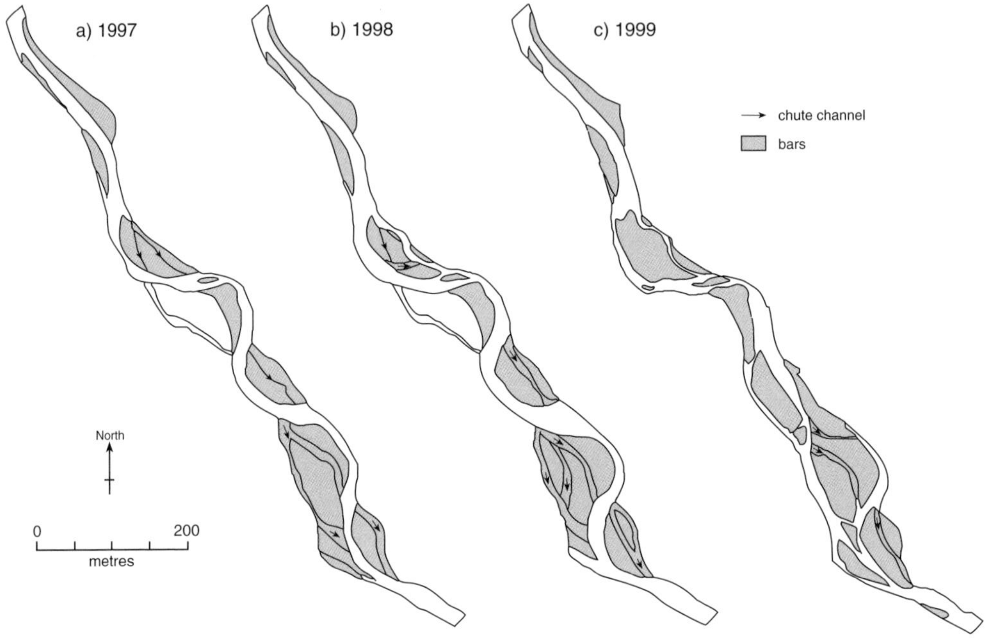

Fig. 4. Planform maps of the active channel at Holystone, defining units used to derive sediment budgets: (**a**) 1997, (**b**) 1998, (**c**) 1999. Arrows denote chute channels and identify the flow direction.

(as recommended by Wolman 1954). The errors associated with this sample size are in the order of 12% (using Hey & Thorne's 1983 equation). Samples were collected across the entirety of each bar complex. The resulting grain-size distribution is shown in Fig. 7.

Limitations

The budgets presented are derived on an annual basis, and do not account for changes occurring between surveys that do not result in elevation changes (such as scour and re-filling). As such, the budgets represent a lower-bound estimate of sediment flux. There is no means of measuring sediment throughputs that are not expressed in morphological changes within the active channel at the time of survey. Although this is potentially a major component of any sediment budget, it has, to date, not proven feasible to measure sediment throughputs at Holystone. This study considers only the net change of sediment storage *within* the reach, which in itself is of importance as a constituent (albeit small) part of the fluvial sediment transfer system. Supporting the concept that it is within-reach sediment transfers that are the focus of attention is the likely step-length of gravel in the reach. Although it has not been feasible to measure step lengths at Holystone, average displacements of the order of 22.8 and 18.8 m for moving particles (Hassan & Church 1992) have been measured in other UK gravel-bed river environments. A similarly short step length for the Coquet would imply that changes in the downstream end of the Holystone reach are largely locally derived. Without direct measurements however, such a conclusion remains tentative. A further limitation concerns the absence of measurement for the finer, wash load passing through the reach. This often constitutes a substantial portion of the river load (e.g. Leopold 1992), although at Holystone fine members in floodplain stratigraphy are only thinly developed. The budgets presented here focus on the gravel component of the bed-load material, which is considered to have the greatest impact on channel morphology (Leopold 1992; Martin & Church 1995). However, the volumetric estimates of sediment in these budgets will be crude, as no accounting for variation in grain packing is made, and volumetric changes may therefore not necessarily reflect changes in mass, although in the clast-supported gravels comprising the morphological units here a reasonable correspondence between volume and mass may be assumed.

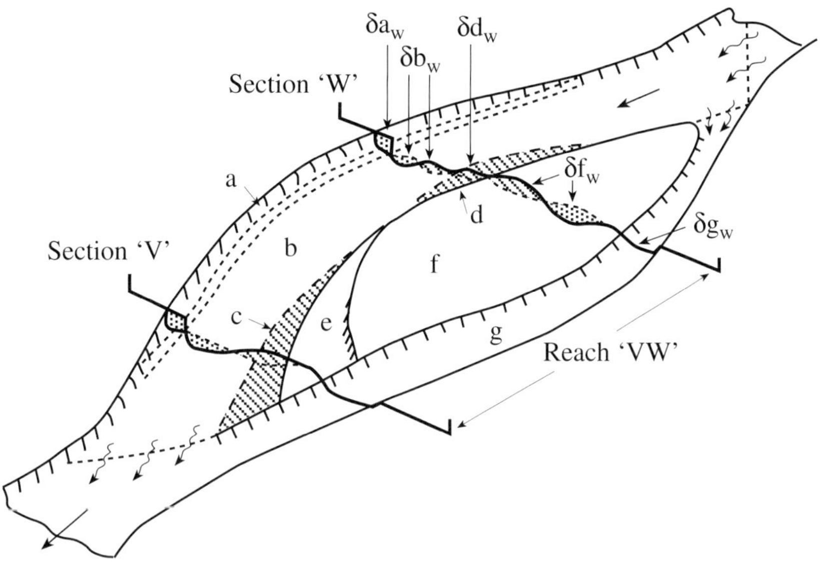

Fig. 5. Derivation of morphological budgets in this study (after Brewer & Passmore this volume).
Stage 1: calculation of volumetric change for each morphological unit. For morphological units traversed by: (i) one channel profile (e.g. unit 'd' on profile 'W'):

$\delta V_d = \left(\frac{\delta A_{dw}}{L_{dw}} \times A_d\right)$; (ii) two channel profiles (e.g. unit 'a' on profiles 'W' and 'V'): $\delta V_a = \left(\frac{\delta A_{aw}}{L_{aw}} + \frac{\delta A_{av}}{L_{av}}\right) \times \frac{A_a}{2}$;

where:
δV_d = volumetric change for morphological unit 'd' (m^3); δA_{dw} = change in cross-sectional area of morphological unit 'd' (m^2); L_{dw} = length of morphological unit 'd' across profile 'W' (m); A_d = planform area of morphological unit 'd' at end of budget period (m^2); δV_a = volumetric change for morphological unit 'a' (m^3); δA_{aw} = change in cross-sectional area of morphological unit 'a' across profile 'W' (m^2); δA_{av} = change in cross-sectional area of morphological unit 'a' across profile 'V' (m^2); L_{aW} = length of morphological unit 'a' across profile 'W' (m); L_{av} = length of morphological unit 'a' across profile 'V' (m); A_a = planform area of morphological unit 'a' at end of budget period (m^2).

Stage 2: calculation of volumetric change for each sub-reach, e.g. for sub-reach 'VW':

$$\delta V_{VW} = \Sigma(\delta V_a + \delta V_b ... + \delta V_g).$$

Patterns of sediment transfer

1997–1998

Figure 6a and c demonstrate the patterns of sediment transfer that occurred between 1997 and 1998. The upstream sub-reaches (1–3) show a sediment gain during this period, derived partly from bank erosion contributing sediment to the channel. However, this process delivered only 16 m^3 of sediment in sub-reach 1 and no evidence of substantial bank erosion occurred in sub-reaches 2 and 3. This suggests that bar and channel aggradation in the upstream parts of the reach was largely derived from upstream sediment inputs. This contrasts with the overall loss of sediment from both channel and bar units recorded in sub-reaches 4–6. The greatest sediment losses in these sub-reaches (389 m^3) occurred from the channel (Fig. 6a), while local bar units remained relatively stable.

Most of the subsequent sub-reaches (7–15) show both sediment gain and loss, reflecting a greater complexity of change. Sub-reaches 7 and 8 gained sediment from both bank erosion (87 m^3) and channel aggradation (65 m^3) (Fig. 6a). The losses of sediment in these sub-reaches were due largely to surface lowering of the large left-bank lateral bar complex in these reaches (Fig. 4) rather than chute dissection. Sediment losses from reach 9 were due to surface lowering and dissection of the left-bank lateral bar tail, while the channel in this reach gained a small amount (7 m^3) of sediment. Most sediment gain (*c.* 65 m^3) was associated with

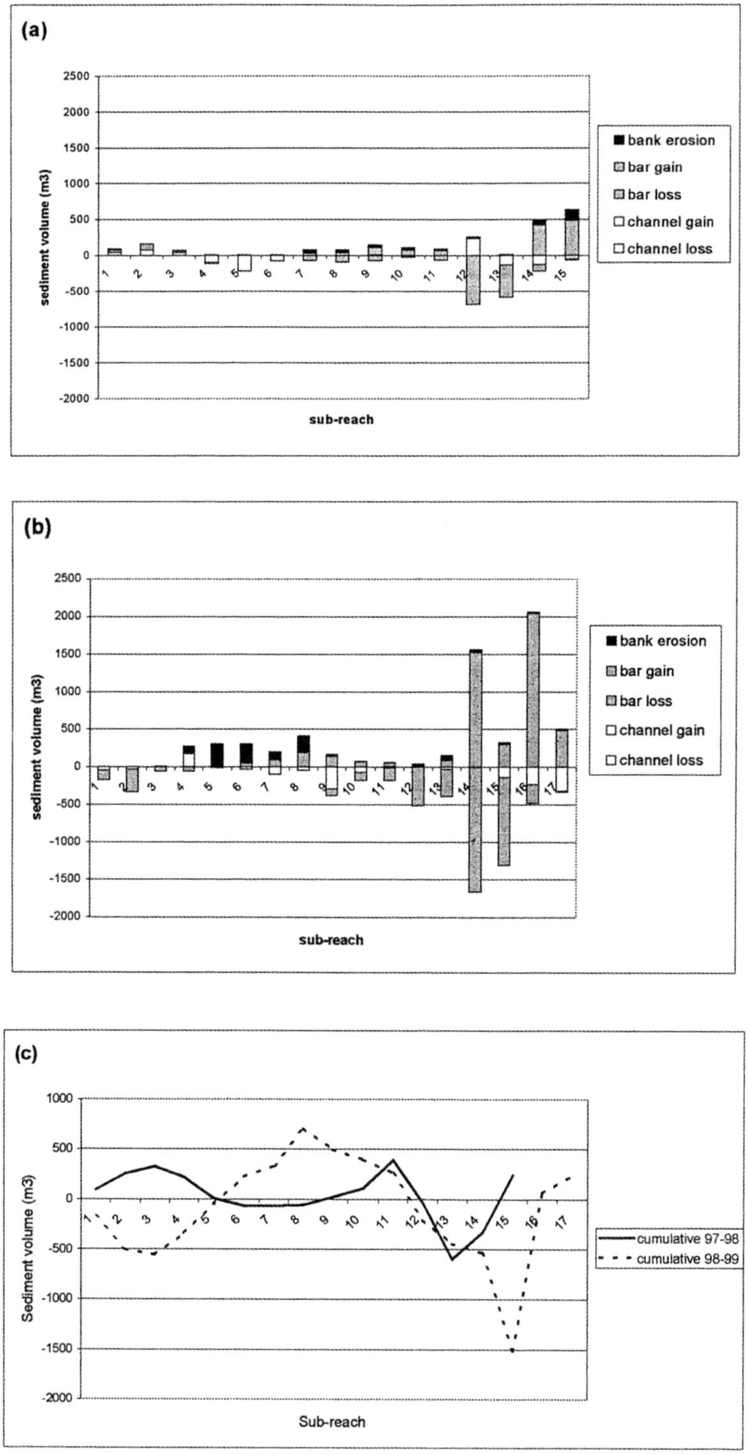

Fig. 6. Annual morphological sediment budgets, Holystone, showing contributions of specific morphological units to the active channel sediment gains/losses: (**a**) 1997–1998, (**b**) 1998–1999, (**c**) cumulative net sediment gain/loss through the instability zone for 1997–1998 and 1998–1999.

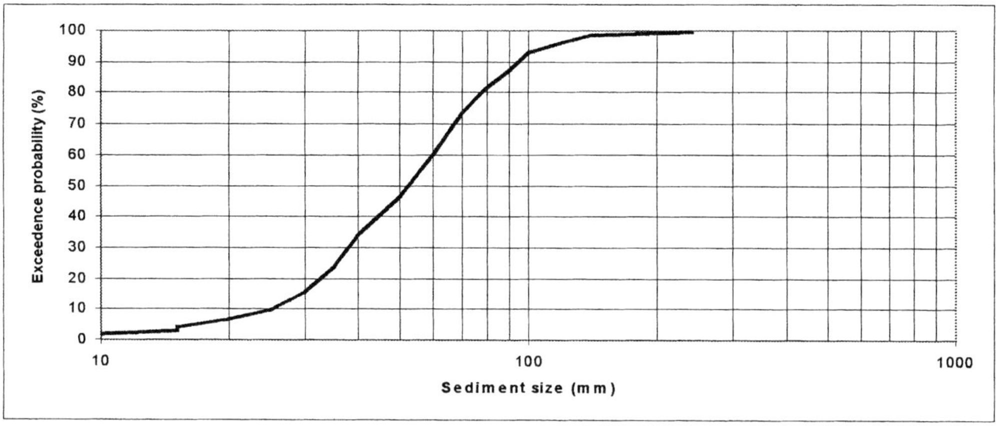

Fig. 7. Grain size frequency plot based on 900 clasts sampled from bar complexes throughout the length of the study reach at Holystone, River Coquet.

deposition across the surface of a right-bank lateral bar and chute channel in this reach, with bank erosion contributing 36 m^3 to the channel (Fig. 6a). Draping of sediment across the bar surfaces in sub-reach 10 also accounted for most of the sediment gain in the sub-reach, whilst bank erosion contributed 29 m^3 of sediment to the channel. Small-scale sediment loss (21 m^3) measured from sub-reach 10 took place from the channel. In sub-reach 11 the channel unit showed larger-scale aggradation (279 m^3), accounting for the majority of sediment gain in the sub-reach (Fig. 6a), the remainder being derived from inputs from banks and vertical accretion of the right-bank lateral bar and chute channel. Sediment losses were largely the product of left-bank lateral bar-head dissection at the downstream end of the sub-reach. Erosion from this bar also contributed most of the sediment losses in sub-reaches 12 and 13, and was associated with the development of a chute channel across the bar (Fig. 4) and overall surface lowering of the unit. The channel in sub-reach 12 was again characterized by aggradation of 239 m^3, accounting for the majority of sediment gain in the sub-reach (Fig. 6a). This contrasts with the sediment loss of 136 m^3 measured from the channel in sub-reach 13. The absence of any notable sediment gains in sub-reach 13 suggests that the material eroded from sub-reach 12 has not been re-deposited in the sub-reach immediately downstream. However, the sediment gains in sub-reaches 14 and 15 were of an equivalent magnitude to the losses in sub-reaches 12 and 13. These gains occurred in the form of vertical accretion on the right-bank lateral bar unit in both sub-reaches 14 and 15, although the head of this unit (in sub-reach 14) was dissected, which, together with loss from the channel in this sub-reach, account for the sediment losses here. The channel unit in sub-reach 15 also lost sediment. This suggests complex behaviour, whereby bar accretion could have been fed by erosion upstream (given the similarity of the magnitude lost in sub-reaches 12 and 13 and gained in sub-reaches 14 and 15, although sediment throughputs could also conceivably account for this bar accretion), while the channel unit actually exported sediment (Fig. 6a).

The patterns of sediment transfer described between 1997 and 1998 suggest that much of the sediment eroded in the Holystone reach during this period was re-deposited within the reach, and was probably not exported downstream, especially if a limited step length (*c.* 20 m) is assumed. Rather, Holystone during this period appeared to gain sediment from upstream, which was then stored in the reach. However, due to the uncertainties associated with unmeasured sediment throughputs and step lengths, these inferences must be tentative. There appears to be a pattern of alternating net gain/loss operating within the reach: sub-reaches 1–3 represent gain, 4–7 loss, 8–11 gain, 12 and 13 loss, and 14 and 15 gain. This highlights the need to study reaches of instability as a whole, rather than concentrating on single bar complexes, which in this instance would fail to accurately represent reach behaviour.

1998–1999

The patterns of sediment transfer between 1998 and 1999 are shown in Fig. 6b and c. In sub-reaches 1 and 2, sediment loss was measured from both the channels and bars, minor gain of sediment (<8 m^3) was recorded in the channel in sub-reach 3 (Fig. 6b). The first appreciable sediment gain in the reach during this period was measured in sub-reach 4, and resulted from channel aggradation (176 m^3) and bank erosion contributing sediment to the channel (99 m^3) (Fig. 6b), with losses from a right-bank lateral bar in the sub-reach due to surface lowering of the unit. Most sediment gain (554 m^3) in sub-reaches 5 and 6 was due to contribution of sediment from bank erosion, and small-scale channel aggradation (10 m^3) in sub-reach 5 and bar aggradation (45 m^3) in sub-reach 6 (Fig. 6b). Minor losses were measured from the right-bank lateral bar in sub-reach 5 (Fig. 4) and from the channel in sub-reach 6. In sub-reaches 7 and 8 sediment gains to the channel were derived again from bank erosion (322 m^3), but the left-bank lateral bar also gained >300 m^3 of sediment (Fig. 6b) due to lateral extension and sediment draping across the bar surface. Both gains result from lateral channel migration in these sub-reaches (Fig. 4). Losses were derived exclusively from the channel unit in this sub-reach. Erosion continued in the channel in sub-reach 9, accounting for most of the sediment losses here (290 m^3). The remainder was largely lost to dissection of the left-bank lateral bar tail (Fig. 4). However, the small mid-channel bar in the sub-reach extended upstream and vertically accreted (Fig. 4). Sediment was also deposited within a chute channel on the left side of the lateral bar complex, and there was overall vertical accretion of this bar complex, accounting for sediment gain in the sub-reach (Fig. 6b). In sub-reach 10, small-scale vertical lowering of a relatively stable bar surface accounted for most of the sediment lost from this sub-reach (100 m^3), the remainder was lost from the channel (70 m^3) (Fig. 6b). The more active right-bank lateral bar, vertically accreted during the period, gained 70 m^3. Sediment losses in sub-reach 11 occurred primarily due to dissection of the right-bank lateral bar and dissection of the left-bank lateral bar head in sub-reach 11 (Fig. 4). The channel also represented an area of small loss (12 m^3). Small-scale sediment gain resulted from vertical accretion of the right-bank bar surface, extension of this unit downstream and minor bank erosion (Fig. 6b).

Major dissection of a left-bank lateral bar (present in 1998 forming a mid-channel bar in 1999) (Fig. 4) contributed all sediment eroded from sub-reach 12, and most of that from sub-reach 13 (Fig. 6b). Bank erosion (40 m^3) and small-scale channel aggradation (6 m^3) resulted in the minor sediment gains measured in sub-reach 12 (Fig. 6b). However, in sub-reach 13, sediment gains (87 m^3) occurred due to sediment infilling the channel in association with an avulsion, which was measured in the subsequent sub-reach (14). This in-channel deposition is identified in Fig. 6b as bar gain, as the sediment accumulation resulted in the extension of a bar complex (Fig. 4). Most of this in-channel sediment accumulation (1258 m^3) occurred in sub-reach 14 as the channel abandoned its 1998 course and re-occupied a chute channel, resulting in the erosion of over 1000 m^3 of sediment from this sub-reach (shown as bar loss in Fig. 6b). Sediment was also removed from the 1998 right-bank lateral bar (which became a left-bank lateral bar in 1999 following the avulsion), due to surface lowering. In sub-reach 15, the avulsed channel eroded 565 m^3 from the bar across which it cut (Fig. 6b). The remaining volumetric loss resulted from surface lowering of parts of the bar unit (those in closest proximity to the avulsed channel) and loss from the 1998 channel (presumably representing erosion prior to the blocking of this feature upstream). Some parts of the lateral bar unit, further away from the avulsed channel, however measured accretion which, together with bank erosion and accretion on the bar head of the next left-bank lateral bar unit downstream, contribute the sediment measured as a gain in sub-reach 15 (Fig. 6b).

Erosion from the avulsed channel continued to produce sediment losses from sub-reach 16 (146 m^3), primarily from bar units (Fig. 6b). However, more sediment (234 m^3) was eroded from the channel in this sub-reach (Fig. 6b). The remainder was eroded from chute channels on the left-bank-lateral bar (Fig. 4), together with the area located between these channels. The areas of this bar adjacent to the channel, however, gained large quantities of sediment (1096 m^3) as this unit extended into the channel, the bulk of remaining sediment gain in this reach resulted from the deposition of a major mid-channel bar, accumulating 846 m^3 (Fig. 6b). These areas of deposition also account for much of the sediment gained in sub-reach 17, which incorporates the tails of the mid-channel and lateral bar complexes. Sediment loss from sub-reach 17 resulted almost entirely from erosion in the channel (Fig. 6b), which suggests this material (318 m^3) was exported from the reach.

The patterns of sediment transfer described between 1998 and 1999 again suggest that much of the sediment eroded from the Holystone instability zone was redeposited in the reach, especially if a limited step length (*c.* 20 m) is assumed. However, an assumption of within-reach deposition is tempered by the absence of throughput and direct step-length information. Nevertheless, the large-scale aggradation in sub-reach 14 associated with blocking of the main channel and subsequent avulsion was equivalent to the volume of sediment loss upstream from active channel and bar units, plus sediment delivered to the channel from bank erosion. This sediment could have contributed to the blocking of the channel, which occurs upstream of the main erosion associated with the avulsion. Alternatively, blocking of the main channel could have been achieved by deposition of sediment transported through the reach, but without morphological expression. Such unmeasured throughputs would not require all of the eroded material upstream of the blocked channel to have been transported to the avulsion site. Most of the sediment eroded by the avulsion was probably subsequently deposited in the mid-channel and left-bank lateral bar units in sub-reaches 16 and 17. However, the total sediment accumulation in these sub-reaches (2571 m^3) falls short of the total sediment eroded (2967 m^3) from sub-reaches 14 and 15. This suggests, together with channel losses from sub-reach 17, that the Holystone reach exported c. 700 m^3 of sediment downstream between 1998 and 1999. This export is not evident from the cumulative net curve (Fig. 6c) due to the net balance of sediment in sub-reach 14, which obscures the fact that sediment eroded from that sub-reach is exported *downstream*. The existence of alternating sedimentation zones identified in 1997–1998 is less clear during the 1998–1999 period, especially as the sediment generated in the upper sub-reaches, or throughput, is required to account for the deposition of sediment in sub-reach 14. It is unlikely, given the dynamics of bed-load transport, that upstream inputs to the reach alone could possibly explain the scale of deposition measured in that sub-reach, unless the plugging of the channel was associated with a substantial accumulation of sediment throughputs that were not measured due to the absence of morphological expression (or rapid scour and fill), but the scale of deposition would suggest this to be unlikely. This indicates that sediment within the reach is travelling greater distances in this period than occurred in the period 1997–1998.

Comparison of annual budgets

It is immediately clear from Fig. 6 that the volumes of sediment transferred in the period 1998–1999 exceed the period 1997–1998. Although the scale of activity differs, in both periods the greatest activity was measured in the lower half of the reach, notably downstream of sub-reach 12. Figure 6c also suggests that in the top half of the reach, zones of net sediment gain between 1997–1998 become zones of loss between 1998–1999, and vice versa. This suggests the downstream progression of a sediment wave through the first 10 sub-reaches. This is likely to be propagated by transfer of instability associated with local transfers of sediment between active and inactive stores, probably representing local sediment transfers rather than sediment throughput (although identification of this is beyond the resolution of the technique), as there is also a progression downstream of the zone of maximum erosion, from sub-reaches 12 and 13 (1997–1998) to sub-reaches 14 and 15 (1998–1999) (Fig. 6c).

Unit sediment budgets

Table 1 summarizes the relative contributions to the annual sediment budget at Holystone from the morphological units in the reach. These reach-summed volumes of sediment transfer are derived from the morphological changes in the specific units identified. This demonstrates:

- the contrasting scales of activity between 1997–1998 and 1998–1999;
- that bar units are the focus of activity within the reach, rather than the channel unit;
- that net changes are negligible (i.e. erosion is balanced by deposition of material), although this obscures the export of 700 m^3 discussed for 1998–1999.

Processes

The question remains as to what dominant process is driving the patterns of sediment transfer identified at Holystone. Martin & Church (1995) identify flood magnitude as the primary controlling factor conditioning sediment transport in the Vedder River. Goff & Ashmore (1994) link phases of high bed-load transport rates with periods of higher discharge. Ham & Church (2000) also link significant channel aggradation and degradation with floods on the Chilliwack River that are sufficient to erode and entrain large volumes of bed material

Table 1. *Reach-summed morphological unit sediment budgets*

Unit budget	1997–1998 (m^3)	1998–1999 (m^3)
Bank erosion	516	1234
Bar loss	1548	3115
Bar gain	1347	4180
Bar formation	–	846
Channel loss	723	1361
Channel gain	483	201
Channel avulsion	–	1756
Total gain	2346	6461
Total loss	2271	6232
Net	**+75**	**+229**

stored within the floodplain. However, Paige & Hickin (2000) found no simple relationship between changes in bed elevation and discharge in the alluvial channel of Squamish River. In their study, the bed elevation regime was considered to reflect supply-driven fluctuations in bed-load transport, as sediment moves downstream as a 'coherent bedwave' (Paige & Hickin 2000, p. 1008).

Threshold discharge

Entrainment of bed load is usually defined in terms of shear stress, however Ferguson (1994) suggests that the identification of a critical discharge may be more appropriate. The stages in the derivation of a critical or threshold discharge for sediment movement in the River Coquet at Holystone are outlined below, the method represents a first approximation based on uniform flow theory (see Milan *et al.* 2001). The shear stress threshold (τ) for sediment motion in a gravel-bed channel may be estimated from the Shields entrainment function, modified for hydraulically rough conditions, if uniform flow conditions are assumed:

$$\tau = 0.045\ (\rho_s - \rho_w) g D_g \tag{1}$$

where D_g = the sediment size that represents the median grain size present in the river channel, ρ_s = the density of the bed sediment (2650 kg m^{-3}), ρ_w = the density of the water (1000 kg m^{-3}) and g = gravitational acceleration (9.81 m s^{-2}).

The Shields entrainment function of 0.045 is derived from Komar (1988) as the best approximation for hydrodynamically rough beds, although there is some variability in the figures used to define this function (Buffington & Montgomery 1997; Milan *et al.* 2001). Samples of sediment analysed from the Coquet revealed the median grain diameter (D_{50}) to be 0.051 m and the D_{84} to be 0.083 m (Fig. 7), producing a critical shear stress for the initiation of motion of 38.8 N m^{-2}. This value may be substituted into the DuBoys equation rearranged with respect to hydraulic radius to provide an estimate of the flow average depth required to initiate motion (Milan *et al.* 2001):

$$\tau = \rho_w g R S \tag{2}$$

where R = the hydraulic radius (approximately equivalent to the average flow depth) and S = the energy slope.

This produced a critical average depth estimate of 0.43 m. This value may be used at any of the cross-sections given a local measure of water surface slope and bed-sediment size, in order to estimate the discharge required to achieve the flow depth necessary to initiate motion of the median grain size using the slope-friction method. First, the frictional characteristics of the channel may be estimated using the Colebrook White equation:

$$\frac{1}{\sqrt{f}} = 2.03\log\left(\frac{aR}{3.5D_{84}}\right) \tag{3}$$

where R = the hydraulic radius as defined in Equation (2) for the study cross-section, D_{84} = the bed-material size for which 84% of the material is finer and

$$a = 11.1\left(\frac{R}{d_{max}}\right)^{-0.314} \tag{4}$$

where d_{max} = the maximum flow depth as defined by the cross-section.

Given this value, the channel flow velocity (V) may be determined from the Darcy Weisbach equation:

$$V = \sqrt{\frac{8gRS}{f}}\ . \tag{5}$$

Using the continuity equation the bankfull discharge may be estimated:

$$Q_{crit} = VA \tag{6}$$

where A = the channel flow area as defined by the critical flow depth at the cross-section.

The threshold discharge and bankfull discharges and hydraulic characteristics have been calculated for cross-sections 1 and 2 located in the single-thread entrance reach and across the upstream instability zone (Table 2). The threshold conditions

Table 2. *Hydraulic character of section 1, located in the single-thread entrance reach to the study section at Holystone and 2, located at the upstream end of the study reach*

Section	Flow condition	Discharge (m^3s^{-1})	Velocity (m s^{-1})	Average depth (m)	Water slope	D_{50} (m)	D_{84} (m)
1	Bankfull	27.40	0.82	1.01	0.0093	0.051	0.083
	Critical motion	13.43	0.71	0.43	0.0093	0.051	0.083
2	Bankfull	38.56	0.97	0.62	0.0093	0.051	0.083
	Critical motion	22.14	0.72	0.43	0.0093	0.051	0.083

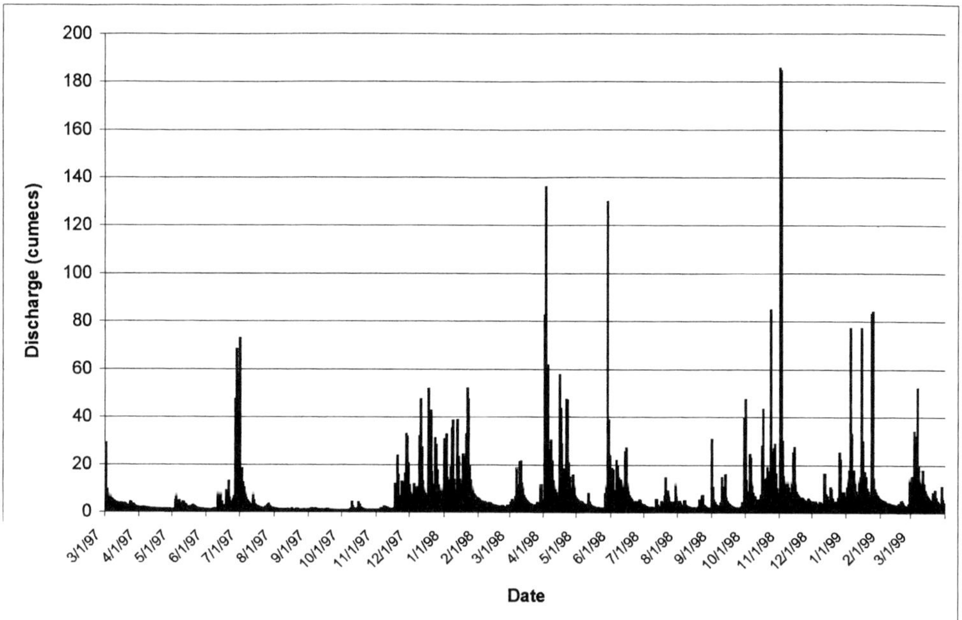

Fig. 8. Daily flow maxima, Rothbury, River Coquet: March 1997–March 1999. Copyright © Environment Agency.

represent a flow of around 49% of the bankfull condition at section 1 and 57% of the bankfull condition at section 2. This is slightly less than the 60% bankfull flow required for initiation of gravel movement for upland, single-thread channels recorded by Carling (1988). Discrepancy between these results and Carling (1988) may be attributed to having been limited to a reach-based calculation, which fails to take into account local factors affecting sediment entrainment (e.g. channel armouring, hiding/protrusion and packing). Threshold discharges may therefore be higher than this estimation.

Although no direct flow records are available from Holystone, flow data from Rothbury (gauged by the UK Environment Agency, National Grid Reference NU 067016), *c.* 12 km downstream from Holystone, are available for the 3-year survey period (Fig. 8). Using a threshold discharge of *c.* 25 $m^3 s^{-1}$ (Table 2) for sediment movement, as gauged at Rothbury, it is possible to identify the maximum (and almost certainly an overestimate) number of flow events that may have resulted in sediment transport at Holystone (Fig. 8). Between 1997–1998 20 such events were recorded, whilst 32 flows exceeded this threshold in the 1998–1999 period. In reality, discharges of 25 $m^3 s^{-1}$ gauged at Rothbury (being 12 km downstream) are likely to be considerably smaller at Holystone. A gauged discharge of 40 $m^3 s^{-1}$ (equivalent to bankfull at Holystone, Table 2) may be more realistic, in which case the number of sediment-mobilizing events is nine and 18 for 1997–1998 and 1998–1999, respectively, although the relationship between discharge at Holystone and Rothbury is at present unknown. During the 1997–1998 and 1998–1999 periods, the maximum daily discharges recorded at Rothbury were 73 and 185.8 $m^3 s^{-1}$, respectively. Peak flows during November 1998 represent the second highest discharge since daily flow maxima were monitored (1983) and the highest since 1992, and most probably explain the higher rates of fluvial activity in the Holystone reach during 1998–1999 as compared with 1997–1998. This flood is also believed to account for the movement of sediment sufficient to block the 1998 channel and cause the major avulsion at sub-reach 14. Furthermore, the channel gradient (reach-averaged) in 1999 increased from 0.0073 in 1998 to 0.0093, raising stream power in the reach during the period 1998–1999. This increase in gradient was, in part, due to the avulsion that occurred, but was also a consequence of sinuosity reduction of the channel between 1998–1999 (Fig. 4). It is possible to link observations of the timing of a significant channel change at Sharperton (*c.* 500 m upstream from Holystone) with the flow record at Rothbury. In early January 1999, an avulsion reported at Sharperton coincided with a maximum flow of 77 $m^3 s^{-1}$ on 5 January 1999 (Fig. 8). Which might

Table 3. *Comparison of sediment fluxes between Holystone, Lambley and Llandinam*

Flux	Holystone, River Coquet 1997–1998	1998–1999	Lambley, River South Tyne 1990–1991	1991–1992	1992–1993	Llandinam, River Severn August 1990–February 1991	February 1991–April 1991	April 1991–June 1992	June 1992–March 1993
Gain	2346	6461	7160	14,620	4710	336	456	852	234
Loss	2271	6232	10,320	15,630	8880	415	403	1,085	474
Net	**+75**	**+229***	**–3160**	**–1010**	**–4170**	**–79**	**+53**	**–233**	**–240**

*Loss from reach 700 m.

suggest a tentative threshold discharge at Rothbury of 77 $m^3 s^{-1}$ for *major* channel change to take place at Holystone, assuming that thresholds in the Holystone and Sharperton reaches are equivalent and behaviour is similar. However, no correlation between flow at Rothbury and Holystone is as yet available.

When flood magnitude and frequency are high, as was the case between the 1998–1999 surveys, sediment appears to be transported much greater distances, either resulting from downstream movements associated with the transfer of instability, or the unidentified throughput of sediment without morphological expression. It is likely that the series of floods prior to the avulsion (Fig. 8) made the reach unstable and 'primed' the reach for instability, which was subsequently 'triggered' by the large flood (Brewer & Lewin 1998). This sequence of priming and triggering has been proposed by Brewer & Lewin to explain longer-term planform cyclicity at Llandinam on the River Severn, and appears to typify not only upper Severn channel behaviour, but also the channel activity observed on the Coquet at Holystone. As such, this behaviour could be considered as typical of the activity style of UK piedmont rivers during the late twentieth century, given that this sequence is not confined to any specific system or region, although local factors operating in reaches (e.g. bank composition, vegetation, channel slope and sediment characteristics) are also significant. Furthermore, these results suggest that there is a strong link between the mobilization of large volumes of sediment by large flow events, to which the channel system responds when 'primed'. The passage of a 'coherent bedwave' (*sensu* Paige & Hickin 2000) appears to account for some of the sediment transfer patterns at Holystone during the survey period, however this is dependent on being mobilized by more frequent and higher magnitude flows.

Comparison with UK sediment budgeting: Severn and South Tyne

Table 3 makes a comparison between sediment budgets measured in the Coquet at Holystone with those measured using the same methodology at Lambley, on the River South Tyne (Macklin *et al.* 1998) and Llandinam on the River Severn (Brewer & Passmore 2002 and unpublished data). The River South Tyne at Lambley has a similar drainage area (276 km^2) to that of the Coquet at Holystone, but has steeper reach gradients (between 0.01 and 0.006 $m\ m^{-1}$) and an active channel zone that is narrowly confined within entrenched Holocene alluvial fills (Passmore & Macklin 2000). Here, analyses of sediment budgets over the period 1990–1993 indicate a net annual export of up to 4170 m^3 of sediment from this reach. This reflects a tendency towards valley floor incision that is also evident over Late Holocene timescales (Passmore & Macklin 2000). At Holystone, by contrast, the Coquet shows little change in channel-bed elevation over the recent historic period. However, marked lateral instability and a broad valley floor in this reach allows ample accommodation space for fluvial sediment storage and this may, in combination with lower stream powers and potential bed shear stresses, account for the comparatively small values of sediment export (up to 700 m^3) recorded at Holystone.

Comparison with sediment budget data from the River Severn at Llandinam indicates sediment fluxes in the order of magnitude equivalent to Holystone. As is the case at Holystone, there is also significantly more activity at Llandinam than the overall budget suggests due to erosion and deposition within the reach. As sediment volumes and patterns of morphological change are similar in these reaches, it is reasonable to suggest a common process of 'priming' and 'triggering' (Brewer & Lewin 1998) of channel instability at both Llandinam and Holystone, further supporting the proposal that this behaviour appears to typify late twentieth century activity styles in UK piedmont rivers that share similar structural characteristics.

Conclusions

Annual sediment budgets for an unstable, wandering gravel-bed reach of the River Coquet at Holystone provide a means of estimating sediment erosion, transfer and deposition within a laterally dynamic channel. The surveys allow a reasonably high spatial resolution of sediment transfers to be established within this 1-km reach. This is sufficient to identify the passage of a wave of sediment downstream, related to the within-reach migration of instability associated with local transfers of sediment. These transfers appear to have blocked the channel between 1998–1999, contributing to an avulsion. The reach has been characterized by re-deposition of locally eroded sediment and sediment inputs from upstream, where lateral erosion creates space for the storage of eroded sediment. Holystone appears to act as a sediment store within the Coquet system; this contrasts with the operation of the Lambley (South Tyne) reach, which was found to be a net contributor of sediment to the downstream alluvial environment over a 3-year period (Table 3). However, the magnitude of sediment flux at Holystone is equivalent to that at Llandinam

(Severn), and the reaches appear to behave in a similar way, where activity is primed and then triggered by geomorphologically effective flows. The results emphasize the need to study reaches of instability as a whole.

Further refinement to the resolution of the budgeting is desirable in order to fully appreciate specific sediment sinks and sources operating in the reach, given that bar units at Holystone display complex patterns of sediment gains/losses. With the availability of more sophisticated survey equipment, it should also be feasible to gather sufficient data to construct reliable DEMs of the reach, providing a refinement to future budgets derived from Holystone, which will also provide a more rigorous test of this budgeting methodology. In addition, planned flow gauging at Holystone as part of ongoing research will provide a clearer understanding of the frequency of sediment-mobilizing flood events.

We are very grateful to G. Renwick and B. Little for granting access to the reach. We thank A. Rooke (Newcastle) for drafting Figs 1–4. Discharge data were supplied by the Environment Agency. I. C. Fuller acknowledges financial support from the University of Northumbria Small Grants for Research Scheme. We would like to thank Dr J. Warburton (Durham) and the editors of this volume for their helpful reviews of this paper.

References

ASHMORE, P. E. & CHURCH, M. A. 1998. Sediment transport and river morphology: a paradigm for study. *In*: KLINGERMAN, P. C., BESCHTA, R. L., KOMAR, P. D. & BRADLEY, J. B. (eds) *Gravel-Bed Rivers in the Environment*, Water Resources Publications, Highlands Ranch, Colorado, 115–148.

BREWER, P. A. & LEWIN, J. 1998. Planform cyclicity in an unstable reach: complex fluvial response to environmental change. *Earth Surface Processes and Landforms*, **23**, 989–1008.

BREWER, P. A. & PASSMORE, D. G. 2002. Sediment budgeting techniques in gravel-bed rivers. *In*: JONES, S. J. & FROSTICK, L. E. (eds) *Sediment Flux to Basins: causes, controls and consequences*. Geological Society, London, Special Publications, **191**, 97–113.

BUFFIN-BÉLANGER, T., ROY, A. G. & KIRKBRIDE, A. D. 2000. On large-scale flow structures in a gravel-bed river. *Geomorphology*, **32**, 417–435.

BUFFINGTON, J. M. & MONTGOMERY, D. R. 1997. A systematic analysis of eight decades of incipient motion studies, with special reference to gravel-bedded rivers. *Water Resources Research*, **33**, 1993–2029.

CARLING, P. A. 1988. The concept of dominant discharge applied to two gravel-bed streams in relation to channel stability thresholds. *Earth Surface Processes and Landforms*, **13**, 355–367.

DAVIES, T. H. R. 1987. Problems of bedload transport in braided gravel-bed rivers. *In*: THORNE, C. R., BATHURST J. C. & HEY, R. D. (eds) *Sediment Transport in Gravel-bed Rivers*. Wiley, Chichester, 793–828.

EINSTEIN, H. A. 1937. *Bedload transport as a probability problem*. PhD Thesis, Zurich. (English translation by SAYRE, W. W. *In:* SHEN, H. W. (ed.) *Sedimentation*. Fort Collins, Colorado, 1972.)

FERGUSON, R. I. 1994. Critical discharge for entrainment of poorly sorted gravel. *Earth Surface Processes and Landforms,* **19**, 179–186.

FERGUSON, R. I. & WERRITTY, A. 1983. Bar development and channel changes in the gravelly River Feshie. *In:* COLLINSON, J. D. & LEWIN, J. (eds) *Modern and Ancient Fluvial Systems*. International Association of Sedimentologists, Special Publications, **6**, 133–143.

GARCIA, C., LARONNE, J. B. & SALA, M. 2000. Continuous monitoring of bedload flux in a mountain gravel-bed river. *Geomorphology*, **34**, 23–31.

GOFF, J. R. & ASHMORE, P. E. 1994. Gravel transport and morphological change in braided Sunwapta river, Alberta, Canada. *Earth Surface Processes and Landforms*, **19**, 195–213.

GOMEZ, B. & CHURCH, M. 1989. An assessment of bed load sediment transport formulae for gravel bed rivers. *Water Resources Research*, **25**, 1161–1186.

GOMEZ, B., NAFF, R. L. & HUBBELL, D. W. 1989. Temporal variations in bedload transport rates associated with migration of bedforms. *Earth Surface Processes and Landforms*, **14**, 135–156.

HAM, D. G. & CHURCH, M. 2000. Bed-material transport estimated from channel morphodynamics: Chilliwack River, British Columbia. *Earth Surface Processes & Landforms*, **25**, 1123–1142.

HASSAN, M. A. & CHURCH, M. 1992. The movement of individual grains on the streambed. *In*: BILLI, P., HEY, R. D., THORNE, C. R. & TACCONI, P. (eds) *Dynamics of Gravel-bed Rivers*. Wiley, Chichester, 159–175.

HEY, R. D. & THORNE, C. R. 1983. Accuracy of surface samples from gravel bed material. *American Society of Civil Engineers, Journal of Hydraulic Engineering*, **109**, 842–851.

HOEY, T. B. 1992. Temporal variations in bedload transport rates and sediment storage in gravel-bed rivers. *Progress in Physical Geography*, **16**, 319–338.

HUBBELL, D. 1987. Bed load sampling and analysis. *In*: THORNE, C. R., BATHURST, J. C. & HEY, R. D. (eds) *Sediment Transport in Gravel-bed Rivers*. Wiley, Chichester, 89–118.

KOMAR, P. D. 1988. Flow-competence evaluations of the hydraulic parameters of floods: an assessment of the technique. *In*: BEVEN, K. J. & CARLING, P. A. (eds) *Floods: Hydrological, Sedimentological and Geomorphologic Implications*. Wiley, Chichester, 107–134.

LANE, S. N., CHANDLER, J. H. & RICHARDS, K. S. 1994. Developments in monitoring and terrain modelling small-scale river-bed topography. *Earth Surface Processes and Landforms*, **19**, 349–368.

LEOPOLD, L. B. 1992. Sediment size that determines

channel morphology. *In*: BILLI, P., HEY, R. D., THORNE, C. R. & TACCONI, P. (eds) *Dynamics of Gravel-bed Rivers*. Wiley, Chichester, 297–311.

LEWIN, J. 1976. Initiation of bedforms and meanders in coarse-grained sediments. *Geological Society of America Bulletin*, **87**, 281–285.

MACKLIN, M. G. 1999. Holocene river environments in prehistoric Britain: human interaction and impact. *Quaternary Proceedings*, **7**, 521–530

MACKLIN, M. G., PASSMORE, D. G. & NEWSON, M. D. 1998. Controls of short term and long term instability: processes and patterns in gravel-bed rivers, the Tyne basin, North England. *In*: KLINGERMAN, P. C., BESCHTA, R. L., KOMAR, P. D. & BRADLEY, J. B. (eds) *Gravel-Bed Rivers in the Environment*. Water Resources Publications, Highlands Ranch, Colorado, 257–278.

MARTIN, Y. & CHURCH, M. 1995. Bed-material transport estimated from channel surveys: Vedder River, British Columbia. *Earth Surface Processes and Landforms*, **20**, 347–361.

MCCLEAN, D. G. & CHURCH, M. 1999. Sediment transport along the lower Fraser River 2. Estimates based on the long-term gravel budget. *Water Resources Research*, **35**, 2549–2559.

MILAN, D. J., HERITAGE, G. L., LARGE, A. R. G. & BRUNSDON, C. F. 1999. Influence of particle shape and sorting upon sample size estimates for a coarse-grained upland stream. *Sedimentary Geology*, **129**, 85–100.

MILAN, D. J., HERITAGE, G. L., LARGE, A. R. G. & CHARLTON, M. E. 2001. Stage-dependant variability in shear stress distribution through a riffle–pool sequence. *Catena* **44**, 85–109.

NANSON, G. C. & CROKE, J. C. 1992. A genetic classification of floodplains. *Geomorphology*, **4**, 459–486.

NEILL, C. R. 1987. Sediment balance considerations linking long-term transport and channel processes. *In*: THORNE, C. R., BATHURST, J. C. & HEY, R. D. (eds) *Sediment Transport in Gravel-bed Rivers*. Wiley, Chichester, 225–240.

NICHOLAS, A. P., ASHWORTH, P. J., KIRKBY, M. J., MACKLIN, M. G. & MURRAY, T. 1995. Sediment slugs: large scale fluctuations in fluvial sediment transport rates and storage volumes. *Progress in Physical Geography*, **19**, 500–519.

PAIGE, A. D. & HICKIN, E. J. 2000. Annual bed-elevation regime in the alluvial channel of Squamish River, southwestern British Columbia, Canada. *Earth Surface Processes & Landforms*, **25**, 991–1009.

PASSMORE, D. G. & MACKLIN, M. G. 2000. Late Holocene channel and floodplain development in a wandering gravel-bed river: the River South Tyne at Lambley, northern England. *Earth Surface Processes and Landforms*, **25**, 1237–1256.

PASSMORE, D. G., MACKLIN, M. G., BREWER, P. A., NEWSON, M. & LEWIN, J. 1993. Variability of late Holocene braiding in Britain. *In*: BEST, J. L. & BRISTOW, C. S. (eds) *Braided Rivers*. Geological Society, London, Special Publications, **75**, 205–229.

REID, I., LARONNE, J. B. & POWELL, D. M. 1999. Impact of major climate change on coarse-grained river sedimentation: a speculative assessment based on measured flux. *In*: BROWN, A. G. & QUINE, T. A. (eds) *Fluvial Processes and Environmental Change*. Wiley, Chichester, 105–115.

SCHUMM, S. A. 1977. *The Fluvial System*, Wiley, New York.

WATHEN, S. J., HOEY, T. B. & WERRITTY, A. 1997. Quantitative determination of the activity of within-reach sediment dstorage in a small gravel-bed river using transit time and response time. *Geomorphology*, **20**, 113–134.

WERRITTY, A. 1997. Short-term changes in channel stability. *In*: THORNE, C. R., HEY, R. D. & NEWSON, M. D. (eds) *Applied Fluvial Geomorphology for River Engineering and Management*. Wiley, Chichester, 47–65.

WERRITTY, A. & FERGUSON, R. I. 1980. Pattern changes in a Scottish braided river over 1, 30 and 200 years. *In*: CULLINGFORD, R. A., DAVIDSON, D. A. & LEWIN, J. (eds) *Timescales in Geomorphology*. Wiley, Chichester, 53–68.

WOLMAN, M. G. 1954. A method of sampling coarse river-bed material. *Transactions of the American Geophysical Union*, **35**, 951–956.

Tracer pebble entrainment and deposition loci: influence of flow character and implications for riffle-pool maintenance

D. J. MILAN[1], G. L. HERITAGE[2] & A. R. G. LARGE[3]

[1]*School of Environment, University of Gloucestershire, Cheltenham GL50 4AZ, UK (e-mail: dmilan@glos.ac.uk)*

[2]*Division of Geography, School of Environmental and Life Sciences, University of Salford, Manchester, M5 4WT, UK*

[3]*Department of Geography, University of Newcastle upon Tyne NE1 7RU, UK*

Abstract: This paper considers the influence of flow character upon scour and deposition loci of tracer clasts in a gravel-bed river, and discusses implications for riffle–pool maintenance. Overall, bars were found to be the dominant depositional zones where over 54% of the tracer clasts accumulated during a 13-month period, followed by riffles (31%) and, finally, pools (<15%). Variability in the location of scour and deposition zones were apparent and could be broadly linked to four flow categories: (i) low-magnitude, high-frequency flows below 29% bankfull appeared responsible for intra-unit re-distribution of sediment particles; (ii) medium-magnitude and -frequency flows (up to 70% bankfull) appeared capable of inter-unit transfer, with pool scour and immediate deposition on riffle heads downstream, and some movement from riffles to bar edges and heads; (iii) high-magnitude, low-frequency flows (70–90% bankfull) appeared capable of riffle–riffle transport, with routing around bar edges; and (iv) very high-magnitude, very low-frequency flows (bankfull and over) capable of bar to bar transport and clast transport distances exceeding the length of a single riffle–pool unit. Tracers originating from riffles do not appear to be fed into pools on the outside of meander bends, instead they appear to be routed over shallower bar surfaces. High competence and low sediment supply explains the coarse nature of the pools (D_{50} = 110 mm) in comparison to the riffles (D_{50} = 85 mm). An improved understanding of the sediment transport mechanisms operating during different flood types is needed to better predict morphological response to changes in hydrological regime and sediment supply.

Moderate slope single-thread gravel-bed rivers displaying pool–riffle–bar morphology play an important role in the storage and transfer of the sediment load within fluvial systems. The formation of riffles and pools is a mechanism of self-adjustment that minimizes the rate of potential energy expenditure per unit mass of water within the river channel (Leopold & Langbein 1962). The riffle–pool sequence has significance for the attainment and maintenance of quasi-equilibrium, and is often considered to be the primary feature in the development of a meandering planform (Tinkler 1970; Keller 1972; Knighton 1998). Further information regarding the dynamics of sediment transport through riffle–pool morphology may improve understanding of how these meso-scale bedforms are controlled by sediment supply, which is known to be a first-order control on the pattern and distribution of sedimentary facies in depositional basins (Houvis & Leeder 1998).

Riffle–pool maintenance

Tractive force reversal

Riffle–pool maintenance is traditionally based around the hypothesis of tractive force reversal (Keller 1971), whereby at low discharge and for most of the flow range the riffles are more competent relative to the pools. During these periods, sediment transport across riffles is supply limited, and restricted to sand and fine gravel (Jackson & Beschta 1982; Lisle & Hilton 1992). It is generally agreed that, as discharge rises, the pool tractive force increases at a faster rate than the riffles, and a number of workers (e.g. Andrews 1979; Lisle 1979; Teisseyre 1984; Ashworth 1987; Petit 1987; Sear 1996; Milan *et al.* 2001) have demonstrated that it can eventually exceed that of the riffles. If tractive force reversal occurs gravel mobilized on the riffle and bar is transferred through the pool before being deposited on a riffle

From: JONES, S. J. & FROSTICK, L. E. (eds) 2002. *Sediment Flux to Basins: Causes, Controls and Consequences.* Geological Society, London, Special Publications, **191**, 133–148. 0305-8719/02/$15.00

or bar further downstream, provided the tractive force is below the critical value for entrainment (Jackson & Beschta 1982). On the falling limb of the hydrograph, coarse material is stranded on the riffles and fines are winnowed into the pools, where they will reside so long as the tractive force of the pool remains below that of the riffle upstream (Lisle & Hilton 1992).

The reversal hypothesis has generated considerable debate regarding the maintenance of pools and riffles. The main argument is centred on the continuity of mass principle, which states the velocity of the pool cannot exceed that of the riffle unless the flow area of the pool is less than the riffle. A number of researchers (e.g. Richards 1978; Carling & Wood 1994; Thompson *et al.* 1996) have shown that, during most discharges contained within the channel boundary, pools have larger cross-sectional areas and should therefore have lower cross-section average velocities in comparison to riffles. Consequently, tractive force reversal has not gained complete acceptance and alternative theories of riffle–pool maintenance have been proposed.

Particle packing and grain velocity

Sear (1992) and Clifford (1993*a*) have provided an alternative theory based on particle packing differences between riffles and pools. Their ideas support Langbein & Leopold's (1968) kinematic wave theory, where grains on riffles act as particle queues and move more slowly than those in the pools. For the regulated River North Tyne in England, Sear (1992) demonstrated that virtual velocities of tracer sediments transported through pools were 4.4 times faster than riffles. Although particle velocities may reflect the high-flow tractive force pattern, they may also be influenced by contrasts in size and packing of the channel boundary sediments. Riffle sediments are generally reported as being coarser (e.g. Leopold *et al.* 1964; Keller 1971; Yang 1971; Cherkauer 1973; Bhowmik & Demissie 1982; Teisseyre 1984) and better sorted than pools (Hirsch & Abrahams 1981; Carling 1991). Riffle sediments also have a tendency to be more tightly packed (Sear 1992, 1996; Clifford 1993*a*) and exhibit a variety of structures involving grain interlock, imbrication or clustering (e.g. Brayshaw *et al.* 1983; Kirchener *et al.* 1990; Church *et al.* 1998). Structuring has been shown to influence transport distances of individual grains, with a tendency for locked and buried clasts to move shorter distances than free surface clasts (Hassan & Church 1992). Furthermore, Sear (1996) has demonstrated that well-structured riffle sediments have higher critical thresholds for motion compared with equivalent sized material in pools. Consequently, the riffles are maintained as topographic highs as sediments are more difficult to mobilize, whereas pools are maintained as topographic lows as they are more susceptible to scour.

Role of channel obstructions

Thompson *et al.* (1998, 1999) suggest that physical constrictions within the channel (e.g. obstacle clasts, bank protrusions and bedrock outcrops) play an important role in maintaining riffle–pool morphology as they encourage tractive force reversal and generate large re-circulating eddies. Ponding upstream of the obstacle creates locally high water-surface gradients that increase boundary shear stress. The obstacle also has a tendency to reduce the cross-sectional wetted area, which results in an increase in mean velocity for a given discharge. Clifford (1993*b*) has also highlighted the significance of large obstacles in both riffle–pool formation and maintenance. Flow against the upstream face of an obstruction produces high-energy horseshoe eddies that scour the bed upstream and downstream. The obstacle is thought to persist for long enough to fix these flow patterns and induce significant modification of the channel boundary. Deposition of scoured material produces embryonic riffles both upstream and downstream of the obstacle, and eventually the obstacle itself is scoured leaving behind a pool. Once established, the riffles act as larger-scale obstacles initiating further scour through the production of macro-scale eddies. Spatial contrasts in near-bed turbulence induced by incipient riffle–pool topography then create differences in entrainment (Clifford 1993*a*; Sear 1996) that enhance and maintain the sequence through a form–process feedback.

Sediment continuity

Wilkinson *et al.* (2000) propose an alternative mechanism for riffle–pool maintenance based upon sediment continuity (sediment mass balance). Where more sediment is removed from a reach than is supplied from upstream a positive sediment transport gradient exists, resulting in a decrease in bed elevation as the bed supplies sediment to make up the difference. Scour (and pool maintenance) occurs in regions of the channel with a positive shear stress gradient, whereas deposition (and riffle maintenance) occurs in regions of the channel where there is a negative shear stress gradient. In Wilkinson *et al.*'s (2000) model, shear stress maxima occur on riffle heads/pool tails rather than pool troughs (mid-point), whilst shear stress minima occur at riffle tails/pool heads. This contrasts with Keller's (1971) reversal hypothesis, which requires a tractive force minimum over the riffle crest for it to aggrade and a tractive force maximum in the trough of the pool for it to scour.

Influence of flood character upon sediment flux through the riffle–pool sequence

It is clear that riffle–pool maintenance is strongly linked to stage-dependent variation in hydraulics and sediment transport. Although change in hydrograph character is known to influence sediment transport through riffle–pool morphology, present knowledge concerning the precise effects are sparse. Sear (1992) provides one of the few assessments of the effects of a change in hydrograph character on sediment transport processes through riffle–pool morphology. By examining the influence of flow regulation on the River North Tyne, Sear forecast that riffles and pool heads would experience enhanced durations of bed mobility, whilst pool tails would be less frequently mobilized. This was due to reduced flood peaks and to extended peak and recession durations. Such a change in the spatial distribution of scour and deposition has implications for the stability of riffle–pool morphology and sediment flux, as it would encourage aggradation of pool tails and scouring of riffles and pool heads. With current emphasis on the possible impacts of climate and land-use change (e.g. Archer 2000) on hydrological and sedimentological regimes, a fuller understanding of sediment transport mechanisms operating during floods with different hydrograph characteristics should aid efforts to predict morphological response of riffle–pool sequences. This paper addresses this by examining the influence of a range of flood hydrographs and competence durations on the spatial pattern of scour and deposition loci on a gravel-bed upland stream, and considers the implications for riffle–pool maintenance and sediment flux.

Methodology and study site

The field study was performed on the River Rede, a 58-km long tributary of the River North Tyne, rising at 490 m above ordnance datum in the Cheviot Hills in Northumberland, UK (Fig. 1a). The selected reach is 4.5 km from the headwaters (National Grid Reference NT 721 043), has a Strahler stream order of 4, and comprises a single-thread, gravel-bed channel with four pools and three riffles along a 175-m long study reach (Fig. 1b). The reach is situated approximately 0.5 km downstream from that surveyed by Milne (1982). The Rede drains a catchment of 18 km^2 in area, with an impermeable geology of Carboniferous sandstones and shales overlain by peat and boulder clay. The mean annual runoff is 1026 mm and the flow regime is flashy. Flow data obtained from a gauging station situated 30 m above the study reach indicate bankfull discharge to be 8.52 m^3 s^{-1}, which flows through a channel with a bankfull width of between 9 and 18 m.

Grain-size assessment

Grain-size distributions for surface sediments were obtained by measuring the intermediate axis of 2050 particles throughout the study reach. A cell-based sampling strategy was used similar to the patch count of Wohl *et al.* (1996), in order to provide information on spatial grain-size variability. Riffles were split into nine cells, bars into three cells and pools were treated as single cells, approximately 3 × 2 m^2 in area. A total of 450 particles were sampled per riffle, 150 particles per bar and 100 particles per pool, thus satisfying the criteria of Wolman (1954).

The use of tracers to assess sediment flux through the riffle–pool sequence

Tracing techniques have been widely employed within gravel-bed rivers (see Sear *et al.* 2000), however only a limited number have provided detailed information on riffle–pool maintenance. While recent studies have employed 'active' magnetic (e.g. Sear 1992; Ferguson & Wathen, 1998) or radio tracers (McEwan *et al.* 2000), painted tracers can still provide useful data (e.g. Thompson *et al.* 1996), and have the added advantage of being significantly cheaper. In addition, the limited morphological change recorded, between 1995 and 1996 (Milan *et al.* 2001), suggested that loss through burial may not be problematic if painted tracers were to be used on the Rede study reach.

Determination of tracer scour and deposition loci

To provide information on scour and deposition loci 288 painted tracer clasts, with intermediate diameters ranging in size from 10 to 240 mm, were emplaced at random in riffles and pools situated throughout the study reach. This range of grain sizes used in the tracer study permitted an assessment of size selectivity and sediment sorting through the Rede riffle–pool sequence. After a period of competent flow to allow tracer clasts to be incorporated into the bed fabric (Hassan & Church

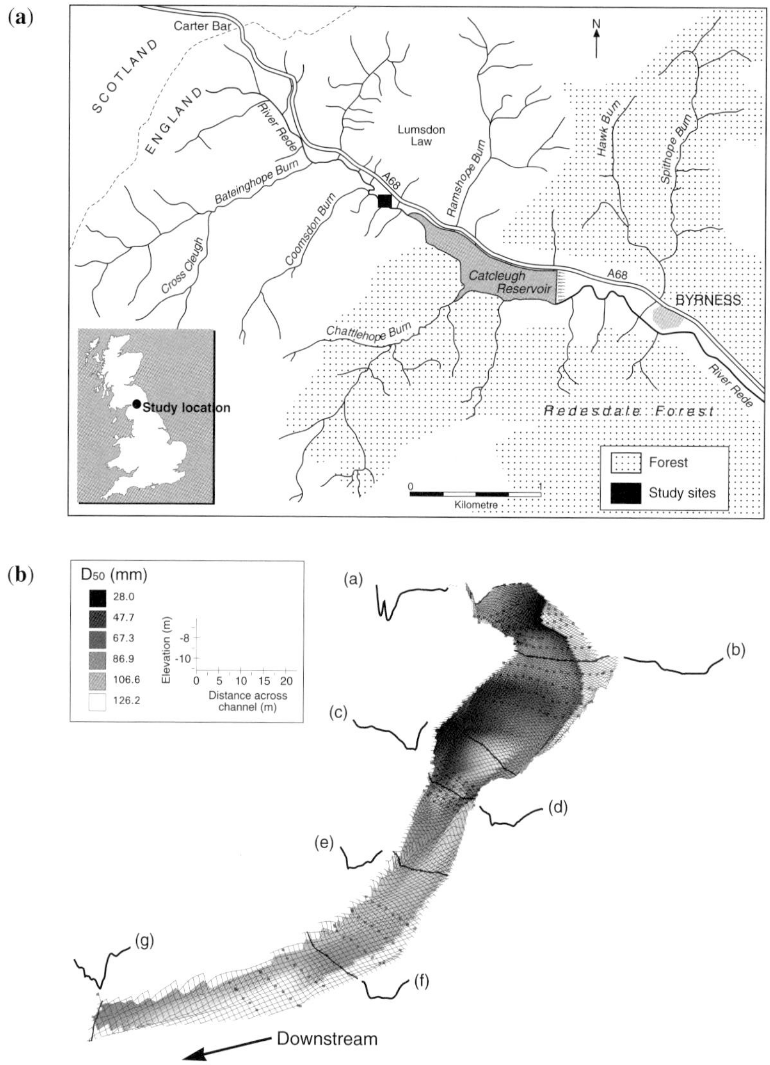

Fig. 1. The River Rede, Northumberland: (**a**) site location; (**b**) perspective plot indicating spatial variability of median grain size for surface sediments draped over channel morphology. Cross-sectional survey positions are indicated, as is the morphology of seven representative cross-sections (a–g). (**a**) Pool 1, (**b**) Riffle 1, (**c**) Pool 2, (**d**) Riffle 2, (**e**) Pool 3, (**f**) Riffle 3 and (**g**) Pool 4. Grain-size data were obtained from a cell-based Wolman (1954) grid survey. The plot is viewed looking upstream at an azimuth of 210° from N, an altitude of 30°, and a distance of 119 m (from centre). Mesh size is 1 m.

1992), all tracer pebbles were surveyed using a Sokkia 5F Total Station (Fig. 2). Recovery of tracers was restricted to those found on the bed surface. A glass-bottomed bucket was used to identify submerged clasts and, where found, each tracer was re-surveyed after five flow events with hydrograph peaks ranging from 2.43 to 9.92 $m^3\ s^{-1}$ (over bankfull). These five events occurred between tracer emplacement on 21 January 1998 to the final re-survey on 28 February 1999. The study thus covered the full hydrological year.

Determination of threshold for mobilization

As well as determining scour and deposition loci, it was necessary to estimate the threshold for gravel motion so that: (i) approximations could be made of competence durations for riffle and pool sediments;

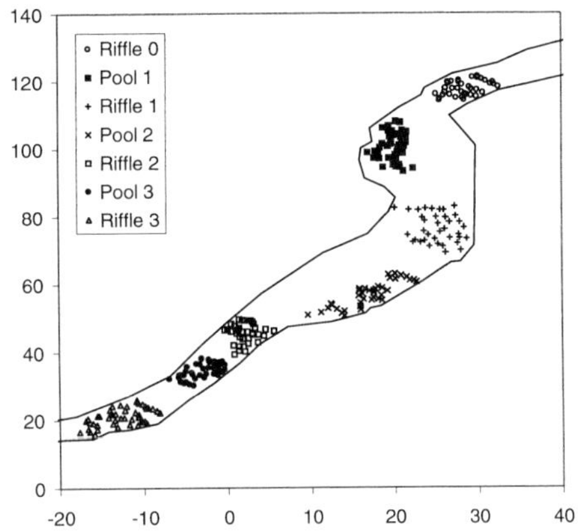

Fig. 2. Tracer seeding loci (excluding bar surfaces).

and (ii) differences between the threshold for motion for pool and riffle sediments could be distinguished. The threshold for motion was estimated for pools and riffles separately by fitting a log–log regression through a plot of mean tracer travel distance (inclusive of all size fractions) against total stream power (Ω). Total stream power was calculated using $\rho g Q s$ for each flow event, where ρ is the density of water, g is gravitational acceleration, Q is peak discharge and s is the mean gradient (here 0.0062) of the bed throughout the study reach. For the overbank flow that occurred, bankfull discharge (8.52 m^3 s^{-1}) was used to calculate Ω rather than 9.92 m^3 s^{-1} because the overbank excess discharge did not contribute to the in-channel stream power. The potential duration of mobilization above the threshold Ω value for riffles and pools was then determined from hydrograph data.

Results

Grain-size characteristics of the Rede riffle–pool sequence and tracers

Grain-size data indicate that pools are coarser (D_{50} = 110 mm) than the riffles (D_{50} = 85 mm) (Table 1 and Fig. 1b). The grain size of tracers should ideally be representative of the sediment body of interest moving within the active layer (Hubbell & Sayre 1964). Parker & Klingeman (1982) indicate that the size of the mobile sediment in gravel-bed channels corresponds with the distribution of the sub-surface rather than of the surface material. Tracers used in this study tended to be finer than the surface layer (Table 1 and Fig. 3), but resembled the D_{50} of the sub-surface gravel truncated at 6.3 mm (Fig. 3). However, tracers were somewhat coarser in the fine tail of the grain-size distribution and finer in the coarse tail, in comparison to sub-surface sediments.

Threshold for mobilization

Overall, the finer tracer sizes tended to plot higher than the coarser tracers, indicating smaller mean transport distances for coarser material and selective transport of finer grained material for both the riffle and pool data (Fig. 4). Extrapolation of the fitted curves through mean travel distance vs Ω data for riffle and pool tracers allowed an estimate of Ω required for initial mobilization of gravel which, assuming a minimum movement of 0.2 m, equated to an Ω of 105 W m^{-2} for riffles and 112 W m^{-2} for pools (Fig. 4). Outlying data taken from the lowest flow (2.43 m^3 s^{-1}) are included here as this reflects the lowest discharge where any tracer movement was recorded. Critical Ω values are equivalent to dimensionless Shields function values of 0.049 for riffles and 0.055 for pools (Milan 2000), suggesting that riffle sediments require a lower shear stress for mobilization and that they may be mobile for longer periods during floods. This is significant as both Clifford (1993*a*) and Sear (1996) suggest that, due to poorer packing, it is the pool sediments that should have greater mobility.

Tracer scour and deposition loci

A summary of tracer movements between morphological units and sub-units is presented in Fig. 5a–e. The mobilization index, ranging from 0 to 10, highlights the relative dominance of scour and deposition loci, and is defined by:

$$\frac{N_{\text{mobilized}}}{N_{\text{recovered}}} \times 100 \qquad (1)$$

Table 1. *Grain-size characteristics of tracers and bed surface sediments for the River Rede riffle–pool sequence*

Unit	No. of tracers	D_{50} (mm) Tracer	D_{50} (mm) Bed
Riffle 0	38	46	70
Pool 1	46	46	120
Riffle 1	38	46	75
Pool 2	40	66	120
Riffle 2	38	51	84
Pool 3	41	60	115
Riffle 3	47	60	96
Bar 1	–	–	39
Bar 2	–	–	64
Total	**288**		
Mean		**60**	**82**

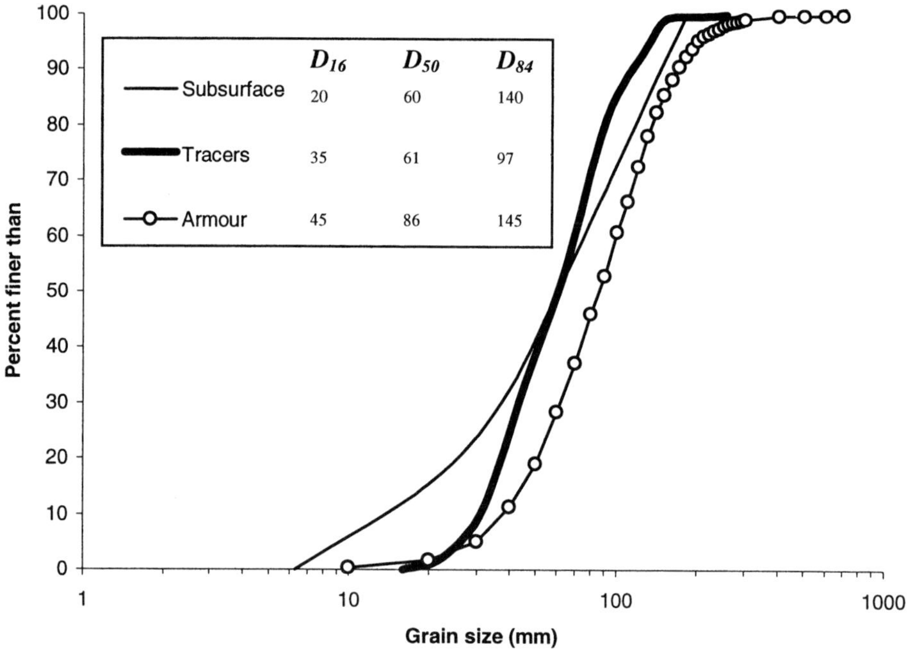

Fig. 3. A comparison of the grain-size characteristics of the tracers with the surface and sub-surface bed sediments. Sub-surface sediments were sampled using a freeze-coring method (Milan *et al.* 1999), and have been truncated at 6.3 mm. D_{16}, D_{50} and D_{84} values are given in the inset in mm.

where N is the number of clasts. The hydrographs for the intervening periods between tracer surveys, where Q has been converted to Ω, are plotted as insets. Figure 6 demonstrates the numbers and size distribution of tracer clasts retrieved after each event, whilst Fig. 7 demonstrates the number of tracers that were mobilized further than 0.2 m.

Event 1: 21 January 1998 – 22 March 1998, peak discharge 2.43 m^3 s^{-1}. The first re-survey of tracers was conducted following three small flood peaks up to 2.43 m^3 s^{-1} (29% bankfull). Figure 5a shows that most transport distances were <5 m, and limited to a small amount of redistribution mainly in the pools; from heads to troughs and from troughs to tails. Bar edges also appeared to receive some tracer particles from riffle tails and pools. Although some grains were sheltered by coarser clasts, the majority were in unlocked positions on the surface of the bed and, therefore, were more likely to be transported further in comparison to grains that were either locked into the bed fabric or buried (Hassan & Church 1992). Tracer retrieval was high with a total of 267 clasts, reflecting limited transport of the surface layer (Fig. 6), however a much smaller proportion (109 clasts) were mobilized (Fig. 7). The maximum Ω for the study reach during this period was calculated at 148 W m^{-2}, and the total duration of competent flow, defined using hydrograph data as the time above the threshold Ω for mobilization, was 23 h for riffles and 19 h for pools.

Event 2: 22 March 1998 – 8 April 1998, peak discharge 5.95 m^3 s^{-1}. The second re-survey followed a period with two significant floods that peaked at 5.95 m^3 s^{-1} (70% bankfull). The maximum Ω during this period was 360 W m^{-2}, and the duration of competent flow was 27 h for riffles and 23 h for pools. A total of 224 clasts were retrieved equating to a retrieval of 78% (Fig. 6), and a total of 138 tracers were mobilized during this event (Fig. 7). This higher magnitude flow event appears to be responsible for greater redistribution of tracers (Fig. 5b). Clasts were transferred between units; from pool troughs to riffle heads and from riffle crests and tails to bar edges, with typical transport distances in the order of 5–15 m. The most mobile size class appeared to be the 40–70-mm range on the riffles and the 20–40-mm range in the pools (Fig. 4). Scour of pool troughs and tails was shown with deposition evident on the downstream riffle heads (Fig. 5b). Movement of tracers situated on the crest and tail of

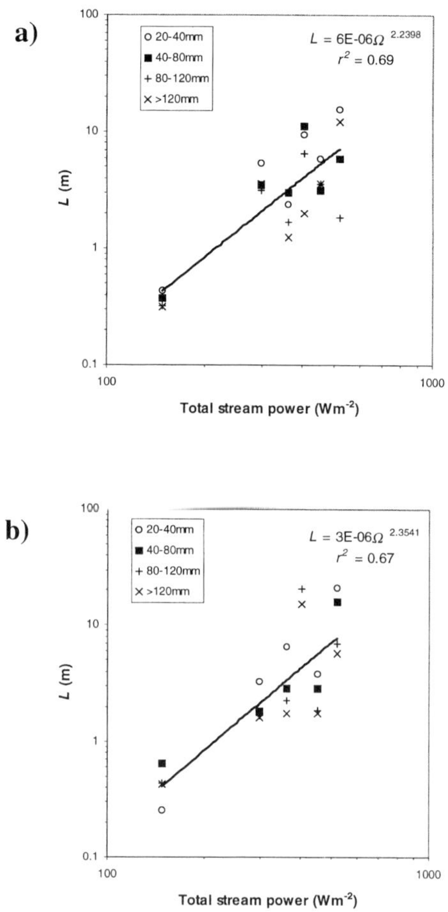

Fig. 4. Stream power (Ω) vs mean travel distance (*L*) for different grain-size fractions of tracer: (**a**) riffles; (**b**) pools. Mean travel distance was calculated using total number of recorded tracers for each flow event (see Fig. 6).

riffles was also notable, with subsequent deposition tending to occur along the edges of the bar heads. However, most movement of clasts on riffles was restricted to intra-unit redistribution (<5 m).

Event 3: 8 April 1998 – 8 June 1998, peak discharge 7.44 m^3 s^{-1}. The third re-survey was conducted after a single larger event that peaked at 7.44 m^3 s^{-1} (87% bankfull), which gave a maximum Ω of 450 W m^{-2}. Although this was one of the larger floods responsible for tracer movement, competence duration was short in comparison to the previous two events; 16 h for riffles and 15 h for pools. A total of 214 tracers were retrieved after this flow (Fig. 6), which equated to a retrieval of 74%. Fewer tracers moved (only 85) and transport distances were lower compared with Event 2 (Fig. 7), which possibly reflected the integration of tracer clasts into the bed fabric (Sear 1996) and/or the shorter competence duration. More tracer movement was recorded on riffles in comparison to the pools (Fig. 5c), with some deposition on bar edges and riffle tails, although there was relatively little inter-unit movement from pools to riffles (Fig. 5c). Intra-unit tracer movement (<5 m) was most notable at riffle crests, tails and pool troughs, although some minor redistribution was also evident on bar surfaces. Some tracers jumped morphological units following a similar pattern as the previous flow, and there was evidence of tracers being routed from bar edge to bar edge.

Event 4: 8 June 1998 – 5 November 1998, peak discharge 9.92 m^3 s^{-1}. The fourth tracer re-survey followed a series of hydrographs which peaked at 9.92 m^3 s^{-1} (over bankfull 116%). Maximum Ω, using bankfull discharge, was 518 W m^{-2} and riffles were competent to transport tracers for 53 h and pools for 43 h. This flow was responsible for mobilizing 91 tracers and for moving tracers the furthest (Fig. 4). Unfortunately, by this stage (10 months into the investigation) most of the finer tracer particles (<40 mm) had been lost (Fig. 6), with a retrieval of only 112 tracers (39% retrieval). Loss was probably due to infiltration into interstitial spaces between coarse surface clasts and burial. This suggested that vertical bed sediment redistribution (infiltration/scour and fill) may have been greater than anticipated. In addition, some tracers were retrieved which were so badly chipped or broken, due to the effects of abrasion, that they could not be identified even after re-sizing and weighing. During this event tracers were transported from most zones throughout the reach, however deposition of sediments on pool tails/riffle heads (which originated from pools) was not as dominant at this high flow in comparison to smaller floods. Least tracer movement was evident in pool troughs, even though tractive force data suggested this to be a zone of maximum tractive force during bankfull flows (Milan *et al.* 2001). This reflected exhaustion of smaller tracer clasts rather than lack of scour potential. Bars were the favoured deposition zone for clasts originating from both riffles and pools, with the main area of deposition situated along the edges of bar heads. Furthermore, it appears that some clasts often skipped over intervening riffle–pool units instead of being transported sequentially from one unit to the next.

Event 5: 5 November 1998 – 28 February 1999, peak discharge 6.66 m^3 s^{-1}. The fifth and final re-survey was conducted after a series of floods that peaked at 6.66 $m^3 s^{-1}$ (78% bankfull), with a maximum Ω of 405 W m^{-2}, and competence duration of 13 h for riffles and 10 h for pools (the shortest throughout the survey). Retrieval and transport of tracers tended to be restricted to material >35 mm as most of the finer tracers had been lost (Fig. 6). The total number of clasts retrieved after Event 5 had reduced to 60, equating to a retrieval of 21%. Of these, 51 clasts were mobilized (Fig. 7). Limited movement from the pool tails onto riffles was evident, however the main scour zones were located at the riffle crests and bar head edges, which also appeared to be the most important deposition sites (Fig. 5e). Tracer clasts tended to avoid pools, and instead appeared to leap-frog from bar to bar, or riffle head to bar, and were routed over the bar or around bar edges.

Discussion

Sediment flux patterns

The measured tracer movement broadly supports the findings of some past studies that suggest that coarse clasts are transported from riffle to riffle (e.g. Jackson & Beschta 1982; Petit 1987). Scour of tracers from the pool head and trough with deposition at the pool tail/riffle head has been a feature previously noted by Petit (1987), Sear (1996) and Thompson *et al.* (1996), and can be attributed to the upstream-sloping portion of the channel that controls the size and quantity of the bed load that leaves the pool. The tail merges into the riffle head and finishes at the riffle crest, creating strong flow divergence (Schmidt *et al.* 1993; Thompson *et al.* 1996, 1998) and down-welling (Vaux 1968) that increases the likelihood of sediment deposition.

(5a)

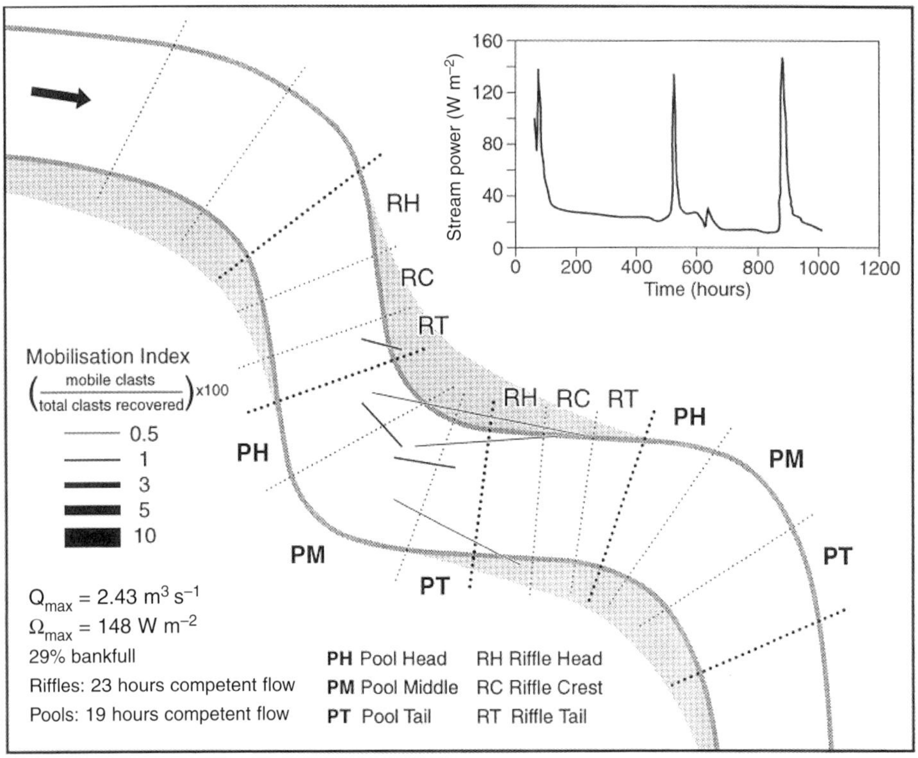

Fig. 5. A planform morphological representation of a pool–bar–riffle sequence on the River Rede, demonstrating a summary of tracer paths for: (**a**) Event 1 (2.43 $m^3 s^{-1}$); (**b**) Event 2 (5.95 $m^3 s^{-1}$); (**c**) Event 3 (7.44 $m^3 s^{-1}$); (**d**) Event 4 (9.92 $m^3 s^{-1}$); (**e**) Event 5 (6.66 $m^3 s^{-1}$). The thickness of the line describes the percentage of tracers mobilized that follow a particular pathway. Hashed lines lying orthogonal to the channel separate the following morphological sub-units: RH, riffle head; RC, riffle crest; RT, riffle tail; PH, pool head; PM, pool mid-point (referred to as trough in the text); PT, pool tail. Mobilization index is the number of mobile clasts (all grain sizes considered) divided by the total number of clasts recovered, multiplied by 100. Solid lines connect scour and deposition loci, however they do not indicate transport distance to scale.

(5b)

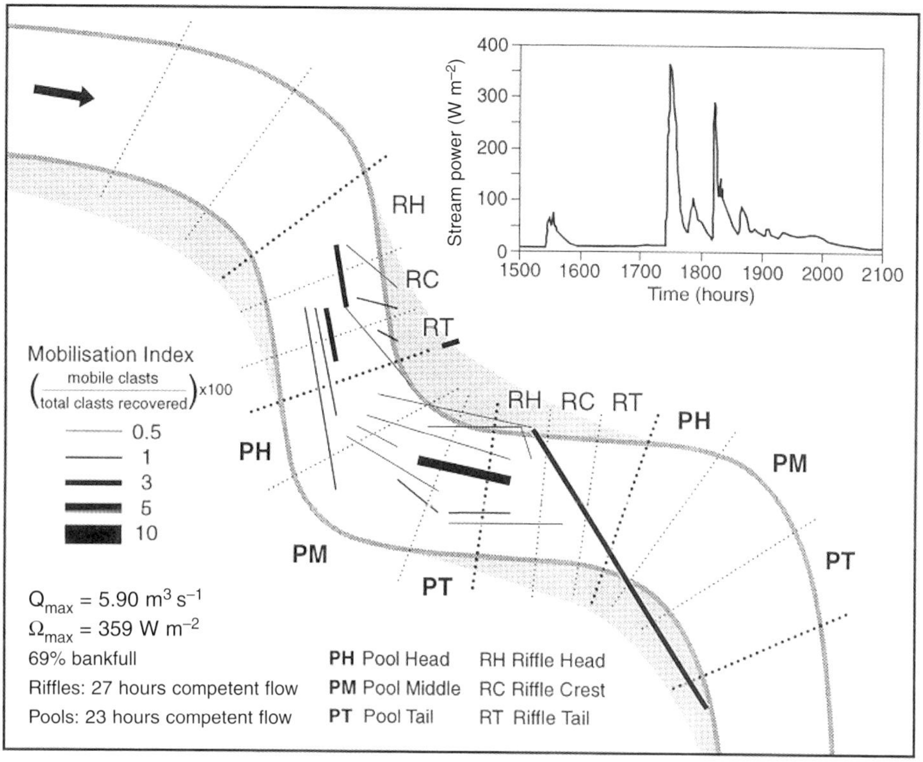
Stream power (W m⁻²)
Time (hours)
RH
RC
RT
RH RC RT
PH
PM
PT
Mobilisation Index
(mobile clasts / total clasts recovered) x100
0.5
1
3
5
10
Qmax = 5.90 m3 s-1
Ωmax = 359 W m-2
69% bankfull
Riffles: 27 hours competent flow
Pools: 23 hours competent flow
PH Pool Head
PM Pool Middle
PT Pool Tail
RH Riffle Head
RC Riffle Crest
RT Riffle Tail

(5c)

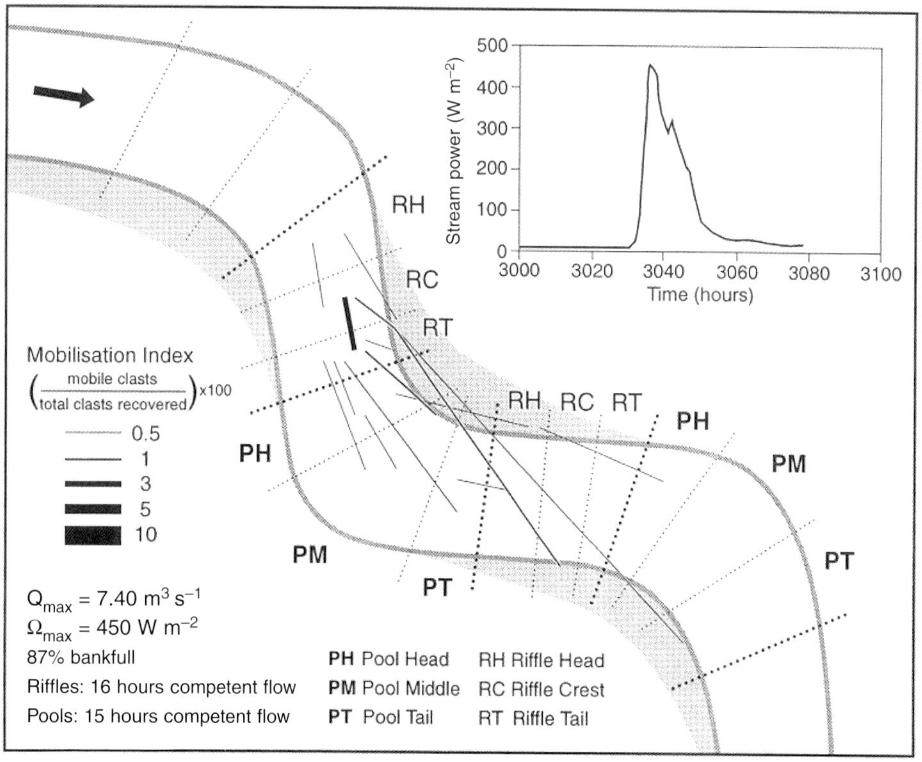
Stream power (W m⁻²)
Time (hours)
RH
RC
RT
RH RC RT
PH
PM
PT
Mobilisation Index
(mobile clasts / total clasts recovered) x100
0.5
1
3
5
10
Qmax = 7.40 m3 s-1
Ωmax = 450 W m-2
87% bankfull
Riffles: 16 hours competent flow
Pools: 15 hours competent flow
PH Pool Head
PM Pool Middle
PT Pool Tail
RH Riffle Head
RC Riffle Crest
RT Riffle Tail

(5d)

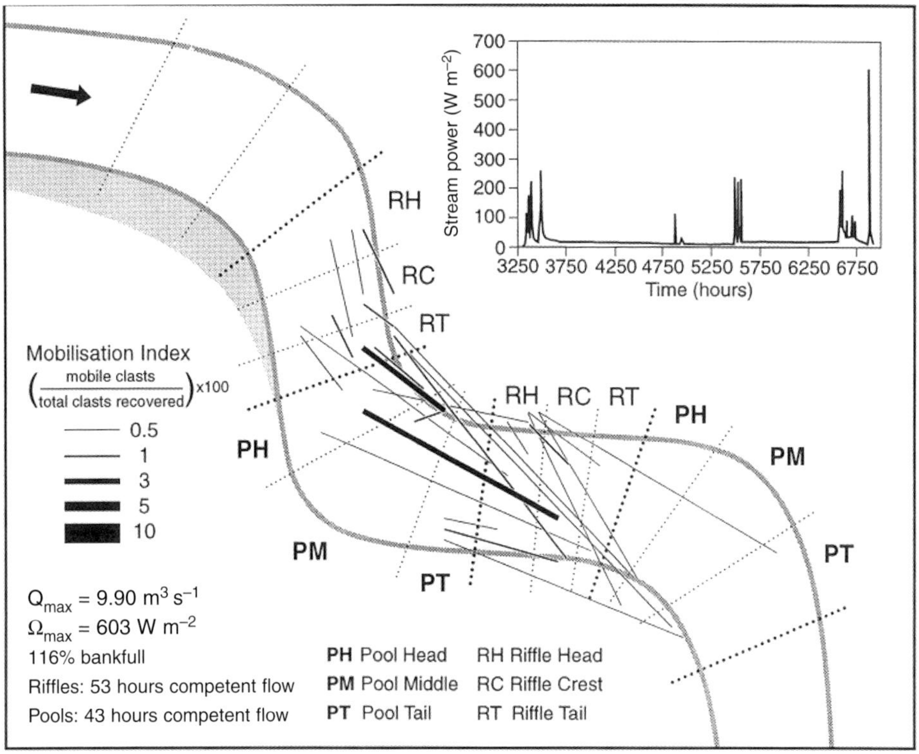

Stream power (W m⁻²)
700
600
500
400
300
200
100
0
3250 3750 4250 4750 5250 5750 6250 6750
Time (hours)
RH
RC
RT
RH RC RT
PH
PM
PT
PH
PM
PT
Mobilisation Index
(mobile clasts / total clasts recovered) x100
0.5
1
3
5
10
Qmax = 9.90 m3 s-1
Ωmax = 603 W m-2
116% bankfull
Riffles: 53 hours competent flow
Pools: 43 hours competent flow
PH Pool Head
PM Pool Middle
PT Pool Tail
RH Riffle Head
RC Riffle Crest
RT Riffle Tail

(5e)

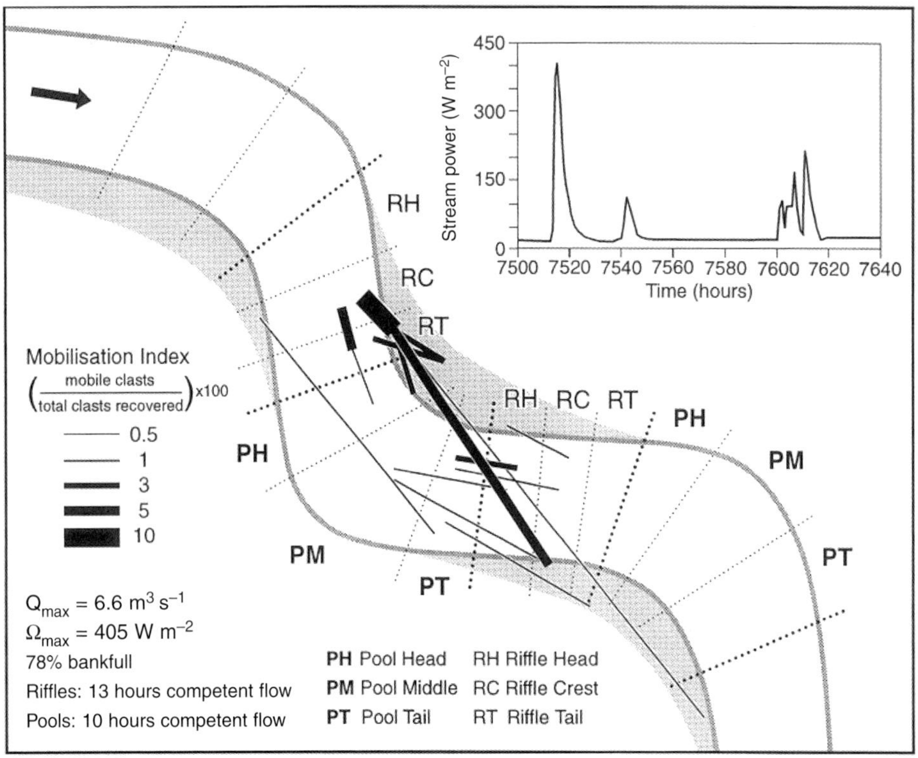

Stream power (W m⁻²)
450
300
150
0
7500 7520 7540 7560 7580 7600 7620 7640
Time (hours)
RH
RC
RT
RH RC RT
PH
PM
PT
PH
PM
PT
Mobilisation Index
(mobile clasts / total clasts recovered) x100
0.5
1
3
5
10
Qmax = 6.6 m3 s-1
Ωmax = 405 W m-2
78% bankfull
Riffles: 13 hours competent flow
Pools: 10 hours competent flow
PH Pool Head
PM Pool Middle
PT Pool Tail
RH Riffle Head
RC Riffle Crest
RT Riffle Tail

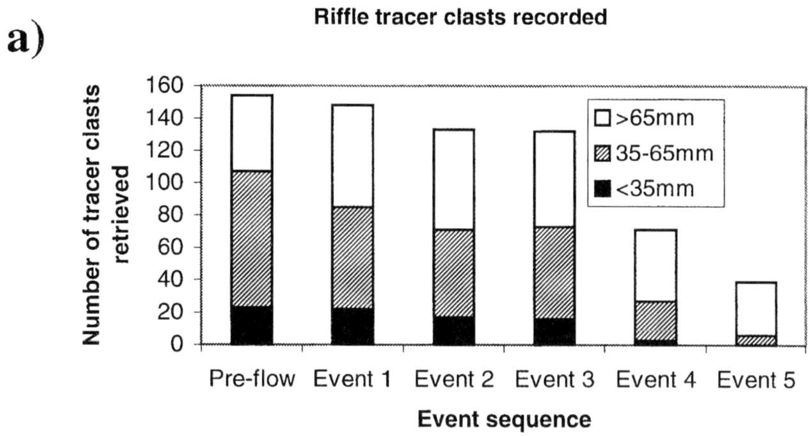

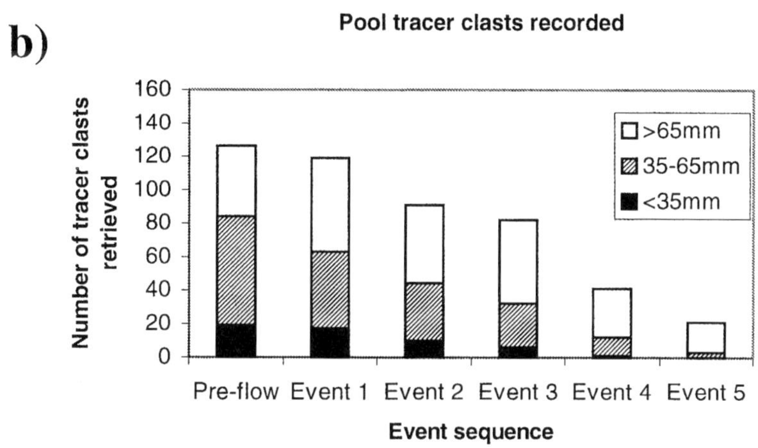

Fig. 6. Event-based retrieval of tracer clasts, demonstrating a gradual overall depletion due to burial, infiltration and abrasion, particularly evident for the finer clasts.

Bars were the dominant depositional zones, where over 54% of the tracer clasts were found to accumulate (taking into consideration the final recorded positions of all 288 tracers), followed by riffles with over 31% of the tracers, and pools (15%). This result contrasts with Haschenburger & Church's (1998) tracer investigation on Carnation Creek, British Columbia, Canada. In their study there was a tendency for tracers to travel towards, and be deposited in, pools regardless of the morphology of their starting positions. Around 80% of the tracers recovered were found in pools or adjacent bars, and the majority of each size (>60%) was routed from pool to pool. These contrasts in sediment routing have significance for ideas of riffle–pool maintenance and warrant further investigation.

Flow character and tracer movements

A morphologically-based model of tracer movement in relation to flow magnitude and frequency is presented in Fig. 8. For low-magnitude, high-frequency flows below 29% bankfull, intra-unit movement is seen to predominate. Medium magnitude and frequency flows (up to 70%

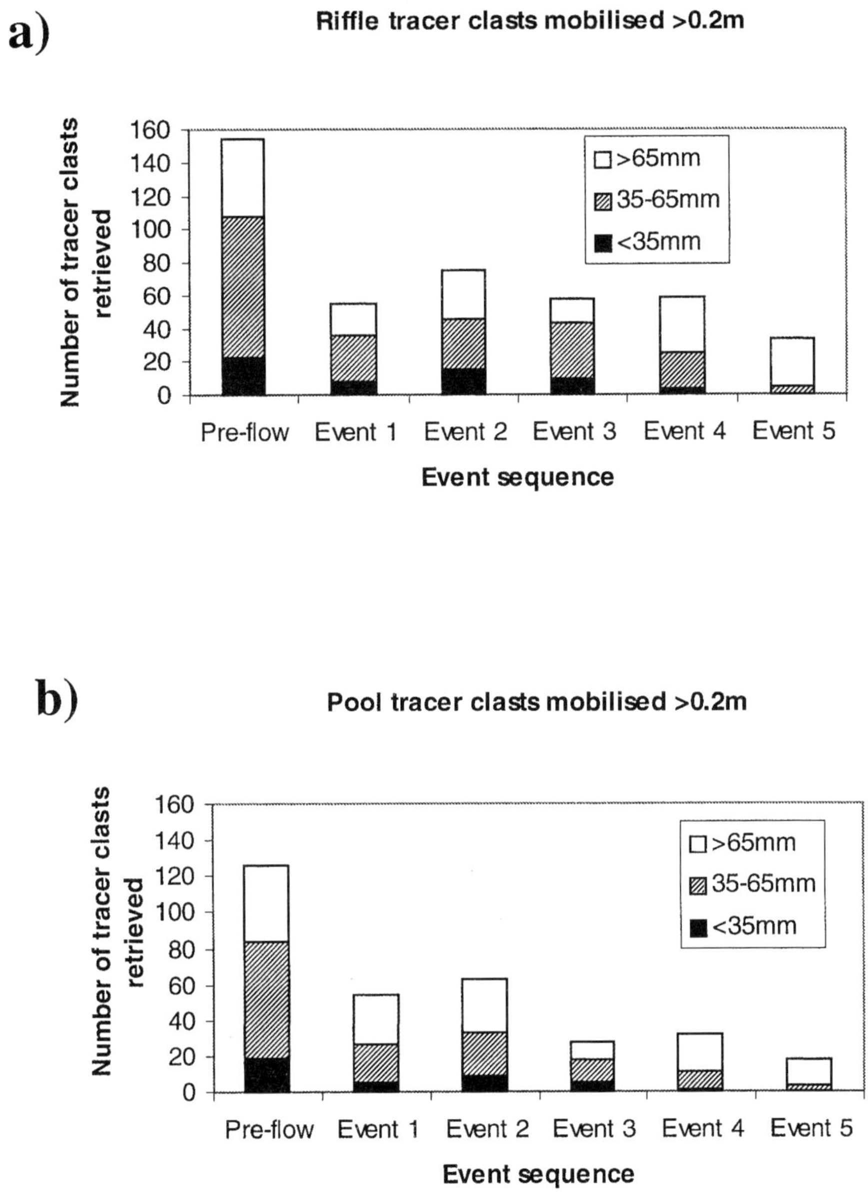

Fig. 7. Numbers of tracer mobilized >0.2 m for each flood event.

bankfull) appear capable of inter-unit transport, scouring tracers from pool troughs and tails, and depositing material on pool tails/riffle heads, and moving material from riffles to bar edges and from bar to bar. Transfer of gravel from riffle to riffle, with routing around bar edges, occurs at 70–90% bankfull and less scour of tracers from pool troughs is evident. At flows of bankfull and above (very high magnitude, very low frequency) bar to bar transport predominates, and transport distances may exceed the length of a single riffle–pool unit.

Size selectivity

A summary of the tracer grain sizes deposited on different morphological units after all five events

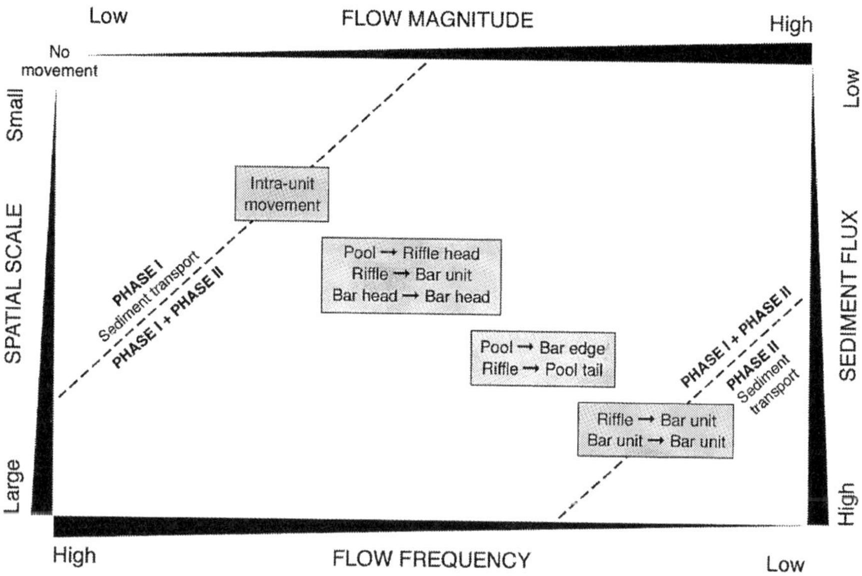

Fig. 8. Model demonstrating the influence of flow magnitude and frequency upon sediment transport within the Rede riffle–pool sequence.

(Fig. 9) implies that riffles and bars have a tendency to accumulate finer gravel in comparison to pools, particularly for particle sizes between 32 and 64 mm. The coarse grain size of tracers (128–256 mm) recorded in the pools reflect the presence of static 'lag' tracers that have not moved significantly from their original seeding location, rather than coarse tracers transported into the pool from upstream. Fine gravel is flushed out of pools without being replaced, which implies that competence exceeds supply. The grain size of the tracers left in the pool at the end of the study closely resembled the coarser size of the naturally occurring pool bed sediments. The overall sorting patterns shown by tracers appear to reflect the high flow tractive force pattern (Milan *et al.* 2001); with the coarsest areas of the bed indicative of greatest high-flow competence (pools), and the finer areas indicative of least competence (bars).

Riffle–pool maintenance and sediment sorting

Tracer evidence for the river Rede provides some support for Keller's (1971) tractive force reversal hypothesis as a mechanism for maintaining riffle–pool morphology. However, Milan *et al.* (2001) have shown that tractive force reversal can occur below bankfull, and does not necessarily happen at the same time on the flood hydrograph for all the riffle–pool units throughout the Rede reach. This is supported by the scour of pool tracers and deposition on pool tails and riffle heads, which occurs at 70% bankfull; for tracers to be deposited on the downstream riffle head implies this to be a zone of comparatively lower tractive force in comparison to the upstream pool trough.

However, the observed sediment sorting pattern on the Rede riffle–pool sequence contrasts with many previous studies which support the reversal hypothesis (e.g. Leopold *et al.* 1964; Keller 1971; Yang 1971; Cherkauer 1973; Bhowmik & Demissie 1982; Teisseyre 1984). The findings of this investigation concur with those studies which contest that the coarsest material should be found in the pool if this is the zone of maximum competence (e.g. Bhowmik & Demissie 1982; Thompson *et al.* 1999). This contention is supported by scour to bedrock in some pools (Keller & Melhorn 1973; Keller 1983), and the observation of coarse lag gravels in others (e.g. Gilbert 1914; Hack 1957; Ashworth 1987; Thompson *et al.* 1999). Sediment supply and routing strongly influence sediment sorting on the River Rede. Tracer data appear to demonstrate a supply limited system where every competent grain is flushed out of the pools onto the riffles without being replaced. Any grains fed into the pool from upstream are either evacuated or deflected onto bar surfaces. It is questionable as to

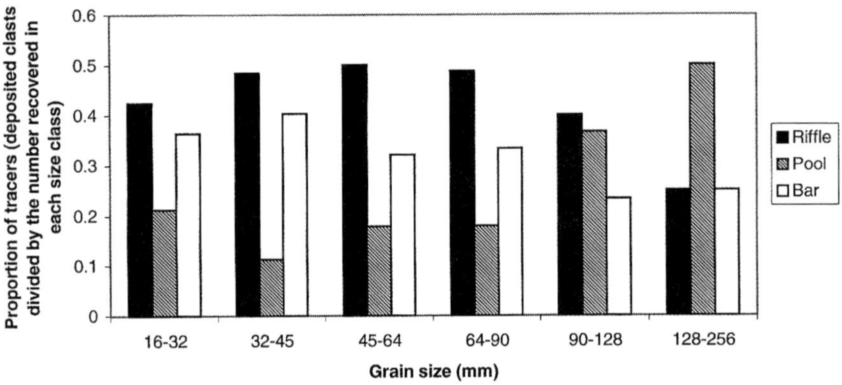

Fig. 9. Depositional zones for tracers, taking into consideration the final recorded positions of all 288 tracer clasts. The bar chart demonstrates the proportion of different grain sizes deposited on each morphological unit.

whether Keller's (1971) reversal hypothesis provides a complete explanation for riffle–pool maintenance, and the alternative models suggested earlier in this paper, as well as the role of sediment supply highlighted in the present investigation, require further verification.

Riffle–pool stability

In the Rede study reach, sediments are transferred between storage units (bars), and new sediment is introduced through bank scour and hillslope–channel coupling. Tracer data suggests that any new sediment introduced to the channel gets sorted in such a manner that grains are eventually deposited and stored on bar surfaces or edges. This is likely to lead to an increase in bar volume and bar apex accretion, which in turn will encourage bank scour due to the increased deflection of flow towards the outer bank. Point bar accretion, bank scour at pools and slight accretion of pool tails/riffle heads has been recorded between 1996 and 1998 within the Rede study reach (Milan 2000), which may reflect lateral channel migration around a quasi-equilibrium condition. However, there is little evidence to suggest a reduction in the elevation of pool beds, which may be due to the development of a coarse 'lag' gravel that protects the bed from further scour, similar to that reported by Petit (1987).

Conclusions

In summary there appears to be four types of tracer movement that are linked to flow magnitude:

- intra sub-unit movement that occurs at flows around 29% bankfull;
- inter-unit movement that appears to occur at flows of up to 70% bankfull, which is dominated by pool scour and pool tail/riffle head deposition;
- transfer of gravel from riffle to riffle, with routing around bar edges that occurs at 70–90% bankfull;
- bar to bar transport of gravel and inter-unit skipping, which occurs at flows close to bankfull and above.

This paper has demonstrated the effect of a range of flood magnitudes and durations upon sediment flux patterns within a riffle–pool sequence. Different flood magnitudes and competence durations result in characteristic spatial patterns of sediment transport and deposition through riffle–pool morphology. This is significant as any changes to the flood hydrograph induced by factors such as climate, land-use change, or flow regulation, could potentially lead to an alteration in the dominant spatial zonation of scour and deposition. This in turn could disrupt riffle–pool stability and thus upset quasi-equilibrium within the channel.

We would like to acknowledge the assistance of Dr I. Fuller in the field. M. Charlton is acknowledgement for his ARC INFO plot (Fig. 1b), whilst A. Rooke kindly drew Fig. 1a and G. Dobrzynski drew Figs 5 and 8. We would also like to acknowledge the helpful comments of Professor I. Reid and Dr D. Sear.

References

Andrews, E. D. 1979. Scour and Fill in a Stream Channel: East Fork River, Western Wyoming. US Geological Survey Professional Paper, **1117**.

Archer, D. 2000. Indices of flow variability and their use in identifying the impact of land use changes. *In*: *BHS 7th National Hydrology Symposium, Newcastle, 2000.*

ASHWORTH, P. J. 1987. *Bedload Transport and Channel Change in Gravel-bed Rivers*. PhD Thesis, University of Stirling.

BHOWMIK, N. G. & DEMISSIE, M. 1982. Bed material sorting in pools and riffles. *American Society of Civil Engineers, Journal of Hydraulic Engineering*, **108**, 1227–1231.

BRAYSHAW, A. C., FROSTICK, L. E. & REID, I. 1983. The hydrodynamics of particle clusters and sediment entrainment in coarse alluvial channels. *Sedimentology*, **30**, 137–143.

CARLING, P. A. 1991. An appraisal of the velocity-reversal hypothesis for stable pool–riffle sequences in the river Severn, England. *Earth Surface Processes and Landforms*, **16**, 19–31.

CARLING, P. A. & WOOD, N. 1994. Simulation of flow over pool–riffle topography: a consideration of the velocity reversal hypothesis. *Earth Surface Processes and Landforms*, **19**, 319–332.

CHERKAUER, D. S. 1973. Minimisation of power expenditure in a riffle–pool alluvial channel. *Water Resources Research*, **9**, 1613–1628.

CHURCH, M. A., HASSAN, M. A. & WOLCOTT, J. F. 1998. Stabilising, self organised structures in gravel-bed stream channels: field and experimental observations. *Water Resources Research*, **34**, 3169–3179.

CLIFFORD, N. J. 1993*a*. Differential bed sedimentology and the maintenance of riffle-pool sequences. *Catena*, **20**, 447–468.

CLIFFORD, N. J. 1993*b*. Formation of riffle-pool sequences: field evidence for an autogenetic process. *Sedimentary Geology*, **85**, 39–51.

FERGUSON, R. I. & WATHEN, S. J. 1998. Tracer-pebble movement along a concave river profile: virtual velocity in relation to grain-size and shear stress. *Water Resources Research*, **34**, 2031–2038.

GILBERT, G. K. 1914. *Transportation of Debris by Running Water*. US Geological Survey Professional Paper, **86**.

HACK, J. T. 1957. *Studies in Longitudinal Stream Profiles in Virginia and Maryland*. US Geological Survey Professional Paper, **294-B**.

HASCHENBURGER, J. K. & CHURCH, M. 1998. Bed material transport estimated from the virtual velocity of sediment. *Earth Surface Processes and Landforms*, **23**, 791–808.

HASSAN, M. A. & CHURCH, M. 1992. The movement of individual grains on the streambed. *In*: BILLI, P., THORNE, C. R. & TACCONI, P. *Dynamics of Gravel-bed Rivers*, Wiley, Chichester, 159–173.

HIRSCH, P. J. & ABRAHAMS, A. D. 1984. The properties of bed sediments in pools and riffles. *Journal of Sedimentary Petrology*, **51**, 757–760.

HOUVIS, N. & LEEDER, M. 1998. Clastic sediment supply to basins. *Basin Research*, **10**, 1–5.

HUBBELL, D. W. & SAYRE, W. W. 1964. Sand transport studies with radioactive tracers. *American Society of Civil Engineers, Journal of the Hydraulics Division*, **90**, 39–68.

JACKSON, W. L. & BESCHTA, R. L. 1982 A model of two-phase bedload transport in an Oregon Coast Range stream. *Earth Surface Processes and Landforms*, **7**, 517–527.

KELLER, E. A. 1971. Areal sorting of bed-load material: the hypothesis of velocity reversal. *Geological Society of America Bulletin*, **82**, 753–756.

KELLER, E. A. 1972. Development of alluvial stream channels: a five stage model. *Bulletin of the Geological Society of America*, **83**, 1531–1536.

KELLER, E. A. 1983. Bed material sorting in pools and riffles. Discussion. *American Society of Civil Engineers, Journal of the Hydraulics Division*, **109**, 1243–1245.

KELLER, E. A. & MELHORN, W. N. 1973. Rhythmic spacing and origin of pools and riffles. *Geological Society of America Bulletin*, **89**, 723–730.

KIRCHENER, J. W., DIETRICH, W. E., ISEYA, F. & IKEDA, H. 1990. The variability of critical stress, friction angle and grain protrusion in water-worked sediments. *Sedimentology*, **37**, 647–672.

KNIGHTON, D. 1998. *Fluvial Forms and Processes: A New Perspective*. Arnold, London.

LANGBEIN, W. B. & LEOPOLD, L. B. 1968. *River Channel Bars and Dunes – Theory of Kinematic Waves*. US Geological Survey Professional Paper, **422-L**.

LEOPOLD, L .B. & LANGBEIN, W. B. 1962. *The Concept of Entropy in Landscape Evolution*. US Geological Survey Professional Paper, **282-B**, 56.

LEOPOLD, L. B., WOLMAN, M. G. & MILLER, J. P. 1964. *Fluvial Processes in Geomorphology*. Freeman & Co., San Francisco, CA.

LISLE, T. E. 1979. A sorting mechanism for a riffle–pool sequence. *Geological Society of America Bulletin*, Part II, **90**, 1142–1157.

LISLE, T. E. & HILTON, S. 1992. The volume of fine sediment in pools: an index of sediment supply in gravel-bed streams. *Water Resources Bulletin*, **28**, 371–383.

MCEWAN, I. K., HABERSACK, H. M. & HEALD, J. G. 2001. Discrete Particle Modelling and Active Tracers: New Techniques for Studying Sediment Transport as a Lagrangrian Phenomenon. *In*: MOSLEY, M. P. (ed.) *Gravel-bed Rivers V*, New Zealand Hydrological Society Inc., 339–360.

MILAN, D. J. 2000. *Sand and Gravel Transport Through a Riffle–pool Sequence*. Unpublished PhD thesis, Department of Geography, University of Newcastle upon Tyne.

MILAN, D. J., HERITAGE, G. L., LARGE, A. R. G. & BRUNSDEN, C. F. 1999. Influence of particle shape and sorting upon sample size estimates for a coarse-grained upland stream. *Sedimentary Geology*, **129**, 85–100.

MILAN, D. J., HERITAGE, G. L., LARGE, A. R. G. & CHARLTON, M. E. 2001. Stage-dependent variability in shear stress distribution through a riffle-pool sequences. *Catena*, **44**, 85–109.

MILNE, J. A. 1982. Bed forms and bend-arc spacings of some coarse-bedload channels in upland Britain. *Earth Surface Processes and Landforms*, **7**, 227–240.

PARKER, R. T. G. & KLINGEMAN, P. C. 1982. On why gravel bed streams are paved. *Water Resources Research*, **18**, 1409–1423.

PETIT, F. 1987. The relationship between shear stress and the shaping of the bed of a pebble-loaded river, La Rulles – Ardenne. *Catena*, **14**, 453–468.

RICHARDS, K. S. 1978. Simulation of flow geometry in a riffle-pool stream. *Earth Surface Processes and Landforms*, **3**, 345–354.

SCHMIDT, J. C., RUBIN, M. & IKEDA, H. 1993. Flume simulation of recirculating flow and sedimentation. *Water Resources Research*, **29**, 2925–2939.

SEAR, D. A. 1992. *Sediment Transport Processes in Riffle–pool Sequences and the Effects of River Regulation for Hydroelectric Power in the River North Tyne*. Unpublished PhD Thesis, University of Newcastle upon Tyne.

SEAR, D. A. 1996. Sediment transport processes in pool–riffle sequences. *Earth Surface Processes and Landforms*, **21**, 241–262.

SEAR, D. A. LEE, M. W. E., OAKEY, R. J., CARLING, P. A. & COLLINS, M. B. 2000. Coarse sediment tracing technology in littoral and fluvial environments: a review. *In*: FOSTER, I. D. L. (ed.) *Tracers in Geomorphology*. Wiley, Chichester, 21–55.

SIDLE, R. C. 1988. Bedload transport regime of a small forest stream. *Water Resources Research*, **24**, 207–218.

TEISSEYRE, A. K. 1984. The River Bobr in the Blazkowa study reach (central Sudetes): a case study in fluvial processes and fluvial sedimentology. *Geological Sudetica*, **19**, 7–71.

THOMPSON, D. M., NELSON, J. N. & WOHL, E. E. 1998. Interactions between pool geometry and hydraulics. *Water Resources Research*, **12**, 3673–3681.

THOMPSON, D. M, WOHL, E. E. & JARRETT, R. D. 1996. A revised velocity-reversal and sediment-sorting model for a high gradient, pool–riffle stream. *Physical Geography*, **17**, 142–156.

THOMPSON, D. M., WOHL, E. E. & JARRETT, R. D. 1999. Velocity reversals and sediment sorting in pools and riffles controlled by channel constrictions. *Geomorphology*, **27**, 229–241.

TINKLER, K. J. 1970. Pools, riffles, and meanders. *Geological Society of America Bulletin*, **81**, 547–552.

VAUX, W. G. 1968. Intragravel flow and interchange of water in a streambed. *Fishery Bulletin*, **66**, 499–489.

WILKINSON, S. N., KELLER, R. J. & RUTHERFORD, I. D. 2000. Predicting the behaviour of riffle-pool sequences in natural rivers: a conceptual approach. *In*: MAIONE, U., MAJONE LEHTO, B. & MONTI, R. (eds) *New Trends in Water and Environmental Engineering and Life*. Balkema, Rotterdam.

WOHL, E. E., ANTHONY, D. J., MADSEN, S. W. & THOMPSON, D. M. 1996. A comparison of surface sampling methods for coarse fluvial sediments. *Water Resources Research*, **32**, 3219–3226.

WOLMAN, M. G. 1954. A method of sampling coarse river gravels. *Transactions of the American Geophysical Union*, **35,** 951–956.

YANG, C. T. 1971. Formation of riffles and pools. *Water Resources Research*, **7**, 1567–1574.

Drainage basin structure, sediment delivery and the response to environmental change

KEITH RICHARDS

Department of Geography, University of Cambridge, Cambridge CB2 3EN, UK

Abstract: The purpose of this paper is to explore the role that drainage basins and networks play in filtering and modulating the effects of environmental change. The record of this change is often inferred from the properties of alluvial stratigraphy in river valleys. However, the variable time lags and smoothing effects caused during the flux of sediment through a drainage basin are very poorly understood, and need to be clarified before the depositional record can be linked to the causes of change. Interactions amongst drainage network elements – main valleys and tributaries – result in complex patterns of storage and delivery in time and space, and these interactions are reviewed in this paper. While the combination of cosmogenic isotopic dating of erosional records, and the use of shallow seismic and ground penetrating radar methods may provide data on the empirical linkages between sediment sources and sinks, a significant research need is for a modelling framework to enable cut-and-fill records to be simulated, and to provide a rigorous connection between the valley-fill depositional record and the environmentally-driven variations in sediment production. This will require a return to the long-neglected analysis of drainage network structure, and an improved understanding of the within-catchment variations in the sediment delivery ratio. Until this is achieved, it will be premature to assume direct connections between environmental change and the record of alluviation in many valleys.

Causes of change in the sediment yield from drainage basins are numerous, but may be broadly categorized into those driven by human impacts (usually associated with land-use change) and those that reflect climatic changes. Analysis of the varying sediment fluxes arising from these different causes involves different assumptions and languages. Human impacts are thought to be largely effective on shorter timescales, and are referred to in relation to phenomena such as sedimentation or scour, which are practical problems in the economics of land and river management. Climatic changes occur over longer timescales, and generate the sedimentary stratigraphy of valley-fill and river-terrace forms via processes referred to as aggradation and incision. The history of variation in sediment yield is, in both cases, reconstructed from properties of material contained in the depositional record. However, such reconstruction often fails to recognize the effects of the drainage basin and its drainage network in smoothing and delaying variation in the yield of sediment resulting from particular environmental changes. These effects, which considerably obscure the correlation between cause and effect, are the subject of this paper, which focuses particularly on: (i) the need for a modelling framework for the interpretation and correlation of alluvial depositional records; (ii) the utility, in the interpretation of longer-term aggradation, of methods developed for the study of modern sedimentation; (iii) the drainage basin and network structure as controls of depositional patterns; and (iv) the question of sediment delivery variation within drainage systems.

In fact, assumptions and languages employed in the study of sediment flux and storage are not as distinctive as the above summary implies. Agricultural and land-use change have been quite capable of causing valley-fill aggradation on human timescales in, for example, California (Gilbert 1917) and the mid-west USA (Trimble 1983). Subtle climatic changes have also been shown to be perfectly capable of causing significant cut-and-fill cycles in arroyos in semi-arid valleys, even on the timescales of human occupance (Cooke & Reeves 1976). For example, Fig. 1 shows a typical valley-fill cross-section of inset and overlapping aggradations such as might be generated by a history of changing sediment supply and flood frequency (Brakenridge 1984, 1985). Modelling and simulation of this kind of complex cut-and-fill behaviour is still in its infancy, but there is increasing need to develop models that can predict such behaviour, and the associated changing patterns of sediment storage and flux within drainage basins. This development is an essential

From: JONES, S. J. & FROSTICK, L. E. (eds) 2002. *Sediment Flux to Basins: Causes, Controls and Consequences.* Geological Society, London, Special Publications, **191**, 149–160. 0305-8719/02/$15.00

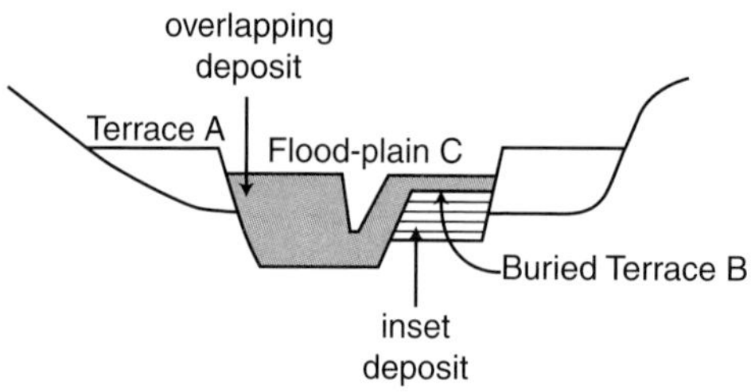

Fig. 1. A typical cut-and-fill sequence in a river valley, showing inset and over-lapping sedimentary units and a buried terrace.

prerequisite for the reliable interpretation and correlation of dated evidence of cut-and-fill history, and for successful connection to be made between the record of alluviation and the ultimately causal environmental changes. The empirical evidence of valley-fill stratigraphy, obtained from exposures, sediment cores, shallow seismography and ground-penetrating radar, is often spatially intermittent, and value is added to these data if they can be assimilated with models which predict the effects of changing sediment fluxes on valley-fill storage. Furthermore, such models would be invaluable aids to the assessment of long-term pollution problems, as a result of simulating release of sediment-attached pollutants, or of providing evolving stratigraphic and topographic boundary conditions for modelling groundwater pollution dispersal. Considerable research and development is required before such modelling is operationally straight-forward. However, there may be more potential than has yet been realized for the application of relatively simple modelling approaches developed in one area (human impact studies) to be applied in the other (climate change studies).

The need for simple strategies arises because there remain some limitations to existing sediment modelling approaches that inhibit their successful application to the problem of catchment-scale sediment yield, storage and delivery. The more rigorous physically-based models, in which sediment transport process representation is driven using hydrological sub-models, tend to suffer from parameter uncertainties when applied at the catchment scale: lack reliable input data on relevant timescales; accumulate error in their process cascade and, therefore, generally have at present relatively wide prediction uncertainties; continue to have relatively poor representation of within-catchment sediment routing and storage processes; and contain sediment loss functions whose capacity to describe the transfer into storage of eroded material remains approximate. These models are often acceptable at the scale of plots or hillslopes, but are not readily scaled-up to catchments. Furthermore, the strategy for scaling-up sediment transport process laws has often been to use morphological variables as surrogate control variables, in forms such as:

$$q_s \propto x^m y^n$$

where q_s is sediment transport rate, x is distance downslope (or area drained) and y is the local slope angle (Kirkby 1971; Ahnert 1976). This results in an element of circularity in this kind of model, since topographic variables determine topographic evolution directly without the mediation of a process relation, and simulations tend towards characteristic forms that are dependent on the exponent values. These catchment- or landscape-scale models are also applicable to timescales that are long relative to those of storage and release of sediment, and have thus tended to ignore valley-fill evolution, particularly because they pay limited attention to the physical representation of the deposition process. New approaches to modelling that do capture the sediment storage history are thus accordingly desirable. One possibility is to adopt the pragmatic methods used to assess reservoir sedimentation in the study of catchment sediment yield, and to cross the divide between the two scales, and languages, of the research questions summarized above.

Sedimentation in reservoirs and catchments

Engineering assessments of the sediment storage characteristics of reservoirs have used a variety of empirical methods, particularly involving the concept of the 'trap efficiency' (TE, %). This is a measure of the proportion of the sediment inflow into the reservoir that enters storage within it. The Brune curves (Brune 1953) represent this as an increasing function either of the ratio of the reservoir capacity to the basin area, or of the ratio of the reservoir capacity to the average annual streamwater inflow (Fig. 2A). The trap efficiency is inevitably dependent on a range of other variables and, for example, tends to be higher in catchments with coarser sediments. An alternative approach, which is nevertheless convertible into Brune-type curves, is provided by the Churchill curves (Churchill 1948; Verstraeten & Poesen 2000), which use the alternative index (100 –TE; %). This is a measure of the percentage of supplied sediment which passes through the reservoir. These curves relate (100 –TE) to the reservoir retention period, or the mean flow velocity through the reservoir (Fig. 2B). In this case, the reservoir sediment storage index tends to be higher in catchments (reservoirs)

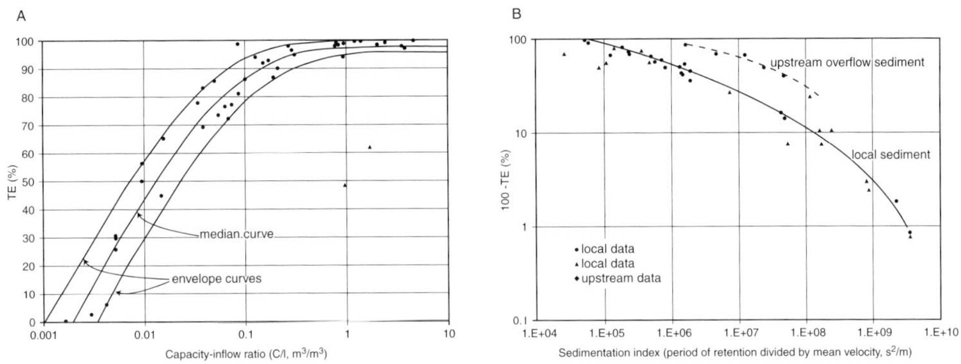

Fig. 2. (**A**) The Brune curve, defining reservoir trap efficiency (TE) as a function of the reservoir capacity : inflow ratio (after Brune 1953). (**B**) The Churchill curve, in which the reservoir sediment throughput is a function of reservoir retention period (after Churchill 1948).

where the transported sediments are finer, since the curves are essentially the inverse of the Brune curves.

Similar, simple empirical models exist to describe the sediment delivery characteristics of catchment-scale sediment yield. The models that exist to describe and predict sediment yield at plot, field or hillslope scales (variants of the Universal Soil Loss Equation, and other more physically-based models such as ANSWERS or EUROSEM; Beasley *et al.* 1980; Morgan, 1994) require a loss function to reduce the yield to reflect the lower yield per unit area that occurs at the larger scale of the whole catchment. This loss function is often taken to be the 'delivery ratio', which is the sediment yield as a percentage total erosion, and which decreases as a function of basin area with an exponent of the order of –0.125 (Fig. 3). The delivery ratio is, in fact, the same parameter as Churchill's (100 –TE) index, as used in the case of reservoir sediment storage, and varies as an inverse function of the basin area because area is a measure of both capacity and inflow. Unfortunately, there is still no reliable basis for estimating catchment delivery ratios, and this continues to be a major research lacuna in studies of catchment erosion and sediment yield. A recent investigation for a large UK catchment was forced to rely on an empirical regression model in which the percentages of the basin area under crops and urban land represented independent variables (Naden & Cooper 1999); clearly this is of limited value in longer-term modelling of sediment fluxes. A more physically-realistic approach to sediment delivery is that described by Dickinson *et al.* (1986), who pay explicit attention to the fact that field-scale sediment production only becomes catchment sediment yield if the source area is connected to the drainage network; this emphasizes the fact that delivery is a function of connectivity, and that the drainage basin structure modulates the delivery of sediment. This is further discussed below.

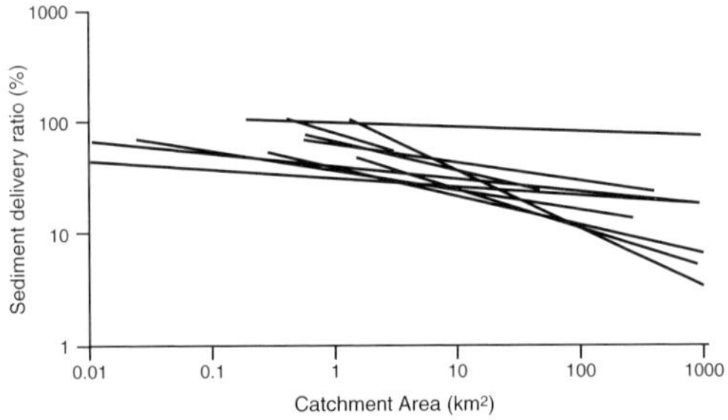

Fig. 3. Catchment delivery ratio as a function of drainage basin area in a number of cases.

One basis for developing a simple conceptual model of valley-fill evolution might be that suggested in Fig. 4. Here, the valley-fill surface elevation is suggested to vary with the ratio of water to sediment discharge from the catchment, and the time series of this ratio then translates into a history of aggradation and incision that results in a cut-and-fill history; inset and overlapping terrace sediments are generated as the runoff : erosion ratio varies through time. A question that needs to be considered is whether this conceptual model can be quantified using ideas based on the Brune–Churchill–delivery ratio parameters. A simple, quantifiable model might consider the aggradation–incision history of a particular reach within a drainage basin/drainage network. The inflow volume of sediment at the head of this reach is defined by the sediment supply from the upstream

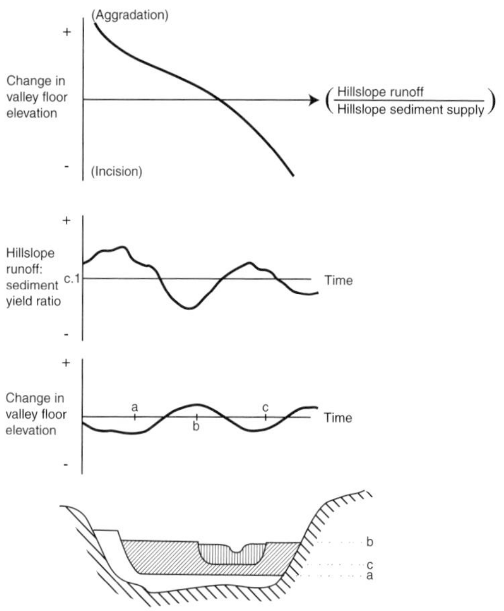

Fig. 4. A hypothetical model of valley-fill evolution as a function of variation in the ratio of runoff to sediment production. In the top diagram, the abscissa is a measure of the ratio of hillslope runoff to sediment supply, both averaged over an appropriate period. A low ratio (inadequate runoff to transport the sediment supply) results in aggradation of the valley fill. The second and third diagrams suggest time series of the runoff : sediment supply ratio, and the associated change in valley floor elevation. These histories generate the hypothetical valley-fill evolution in the bottom diagram; the timing of the limits of the incision and aggradation phases are noted on the time axes of the lower time series.

basin slopes, which may be represented as the upstream basin area multiplied by the average erosion depth (to a first approximation changes in sediment density between soil erosion and subsequent aggradation might be ignored). The outflow volume from the reach could then be represented as the inflow volume multiplied by the delivery ratio for the reach, plus the additional slope erosion generated by the slopes draining directly into the reach. The delivery ratio would have to be determined as a function of basin area, basin relief or slope, and drainage network structure, to provide a delivery ratio that can be estimated as a variable within a catchment, rather than being determined as a variable between catchments. Given the paucity of reliable data on delivery ratios, constructing a sediment routing model of this kind at the catchment scale poses a considerable challenge.

However, a reach of river valley might in this model be regarded as a trough (rectangular or trapezoidal) with a specified width, initial fill thickness and a specified ratio of lateral to vertical river erosion. The net storage volume in a time step within the river reach would be defined by continuity as the inflow minus the outflow. Thus, if the net storage volume is >0, aggradation would occur and would be distributed across the valley width (both the channel bed and the floodplain). If, however, the net storage volume is <0, incision would occur at a rate controlled by the ratio of lateral to vertical river erosion. A succession of linked valley reaches would then allow a solution of the continuity equation for particular patterns of valley sediment conveyance (implying that differential storage capacities in successive reaches could reflect changes in gross valley topography; and that network geometry and topology would control the delivery process at the catchment scale). Such a simple model requires parameters that are presently difficult to determine in individual cases, but would allow both backward and forward modelling for particular assumed parameter values, and sensitivity analyses of the parameter significance. The former approach would allow reconstruction of the time series of sediment supply needed to replicate geological cut-and-fill behaviour, so that this time series could be assessed against the climate record to determine whether the necessary series is plausible. The latter approach could begin with an independently generated time series of sediment production, a plausible set of parameter value and could seek to simulate a cut-and-fill history to compare with the stratigraphic record in particular valley fills.

Effects of drainage basin structure

The foregoing discussion of a modelling framework is predicated on the assumption that it is desirable to explore the role of the drainage network in modulating the effect of environmental change on sediment flux; and that this can be explored using simulation methods. There are several pointers to the nature of this effect in existing research.

First, there are the now well-known complex response ideas of Schumm (1977). These are heavily dependent on the interaction of patterns of aggradation and incision between main river valleys and their tributaries, and on the assumption that threshold valley gradients trigger incision, being intrinsic controls of changes in sediment storage and release behaviour. This is illustrated in Fig. 5, in which a damped oscillation of sediment yield is superimposed on a negative exponential decline following a base-level perturbation of a

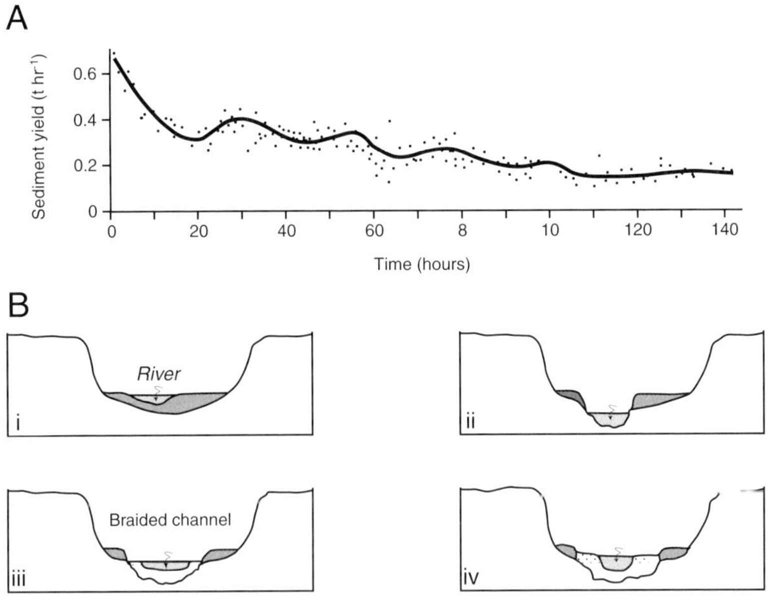

Fig. 5. (**A**) Complex response of basin sediment yield response to base-level change, as a superimposition of an exponential decrease and a damped oscillation as a result of tributary : mainstream interaction (after Schumm 1977). (**B**) The series of cut-and-fill episodes resulting from the damped oscillation of sediment yield in (A).

model drainage basin. The successive waves of sediment delivery from the basin are triggered when aggradation in the trunk valley, caused by erosion in tributaries, attains a threshold slope, and the sediment supply : transport capacity ratio in the trunk stream changes to favour incision. This model clearly emphasizes the role of the drainage network structure, and the interaction of drainage network components, in modulating the sediment yield from a catchment following an environmental change. However, details of the model may require some reconsideration in the light of evidence of tributary–mainstream interactions in some real catchments (Small 1973; Rice & Church 1998; see below). Furthermore, the complex response model was developed for the relatively unusual case of abrupt base-level lowering; a case of downstream control of catchment change. When the environmental change involves altered runoff and sediment supply regimes from the catchment surfaces (the hillslopes) the control is upstream, and the patterns of tributary–mainstream interaction are likely to differ significantly.

Thus, the second example of the effect of network structure on sediment storage and flux arises when the initial drainage basin response to accelerated erosion occurs in tributary catchments, and results in the accumulation of sediment in alluvial fans at the tributary junctions. In this case, the tributary may create a constriction in the main valley that causes sedimentation and a lessening of the main valley-fill gradient immediately upstream. This occurs until the main valley sedimentation creates a gradient across the fan toe that is steep enough to scour the base of the fan, trigger incision of the fill in the main valley upstream from the fan and generate terrace sequences in this fill. This interaction was described by Small (1973) in an example of fan–terrace systems in the Swiss Valais. It is illustrated in Fig. 6, where the Allt Lorgaidh can be seen to have built a fan that has constricted the valley of the River Feshie in the Cairngorms, Scotland, and to have encouraged mainstream aggradation phases upstream from the fan. This is a localized form of complex response, in which the tributary and mainstream sediment supply :transport capacity ratios vary relative to one another over time, and result in changing locations of aggradation and incision. The role of tributary–mainstream interaction in relation to sediment transfer has also been demonstrated by Rice & Church (1998), in terms of resetting the downstream decline in bed material size commonly observed along rivers. Thus, the drainage network structure strongly influences spatial patterns of sediment storage and flux within the drainage basin, as a result of the interaction of tributary sediment supply, tributary fan growth and main

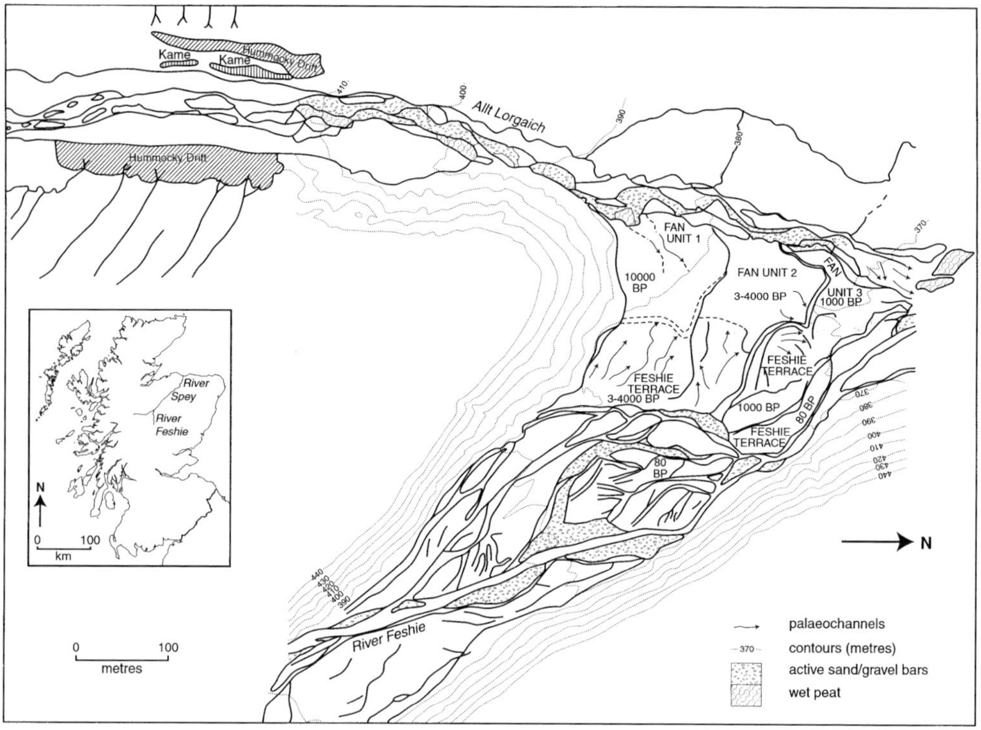

Fig. 6. Aggradation patterns at the junction of the Allt Lorgaidh tributary with the River Feshie in the Scottish Cairngorms (based on data from Robertson-Rintoul 1986).

valley aggradation, and these spatial patterns then determine the temporal variation of sediment yield at whichever point is taken as the catchment outlet. A critical implication is that the relationship between the environmental changes driving the time variation of catchment sediment output and the observed record of that output (for example, in a lake basin at the catchment outlet) can only be understood through definition of a transfer function defining those internal properties of the drainage basin that modulate its response characteristics.

The third piece of evidence of the role of the drainage network structure in modulating the sediment delivery from a catchment is provided by Boyce's (1975) analysis of the contrast between the 'upland' and 'lowland' sediment yield, reflecting the variation in sediment delivery ratio within a drainage basin. Figure 7a shows a simple, hypothetical example of a gridded drainage basin, in which the delivery ratios of sub-areas 5, 7, 8 and 9 must be significantly lower than those of sub-areas 1, 3, 4 and 6 in order that the inverse relationship of sediment delivery ratio to catchment area (Fig. 3) can occur. Assume that this hypothetical catchment consists of 10 sub-areas, each of 15 km^2. The delivery ratio for the whole basin might be 10% and, on the basin of area alone, a single sub-area might have a delivery ratio of 20%. If the average soil erosion from a sub-area is 100 t, the total sediment production is 1000 t; the overall catchment delivery ratio then suggests an output of only 100 t. However, the output from a single sub-area is implied to be 20 t, on the basis of the area alone and, when cumulated across the 10 sub-areas, this is twice the yield from the whole catchment. The implication is that the delivery ratios of those sub-areas that are associated with 'interior links' of the drainage network (5, 7, 8 and 9) must be less than the 20% implied by their area alone. In fact, for the cumulated sub-area sediment yield to equal the basin sediment yield requires interior link delivery ratios of the order of 3%. Thus, the flux of sediment within a basin, which determines how the basin temporarily stores the products of environmentally-driven changes in sediment production, is dependent on the network organization. However, any drainage basin with n stream sources and associated exterior links has n–1 interior links, which implies that the balance between the numbers of links may be less relevant in controlling

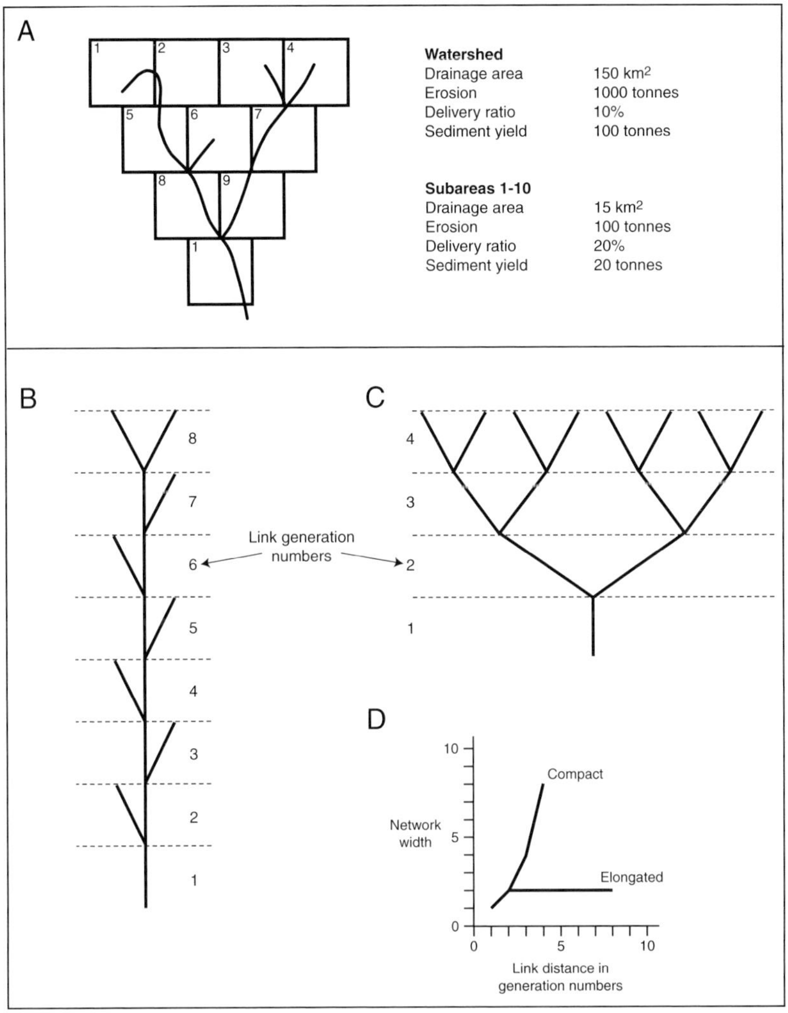

Fig. 7. (**A**) A hypothetical drainage basin showing how the internal sub-areas must have a low delivery ratio to compensate for the higher delivery ratio of external sub-areas (even if all have the same surface area) (after Boyce 1975). (**B**) & (**C**) Different drainage network topologies for networks of similar size (measured by the number of exterior links), and different numbers of link generations (B is an elongated, C is a compact network; both are of magnitude $n = 8$, where n is the number of exterior links). (**D**) The plot of network width (measured as the number of links across the network) as a function of link distance headward of the network mouth, showing the difference between an elongated and a compact network.

sediment flux and storage than the structure of the basin. For example, a network with n exterior links could be elongated, with n link 'generations' from the outlet to the headwaters, or it could be compact with as few as $n/2$ link generations (Fig. 7b). The pattern of sediment storage and release in these two networks may be very different, even for a comparable history of sediment production. This structural characteristic of drainage networks – known as the network diameter – has been shown to have geomorphological significance by Jarvis (1976). An additional indicator of network structure that has hydrological significance, and therefore might be expected to relate to sediment flux, is the network width (Kirkby 1976) that is based on a graph of the number of links measured across the network at successively increasing link distances from the network outlet. The descriptive

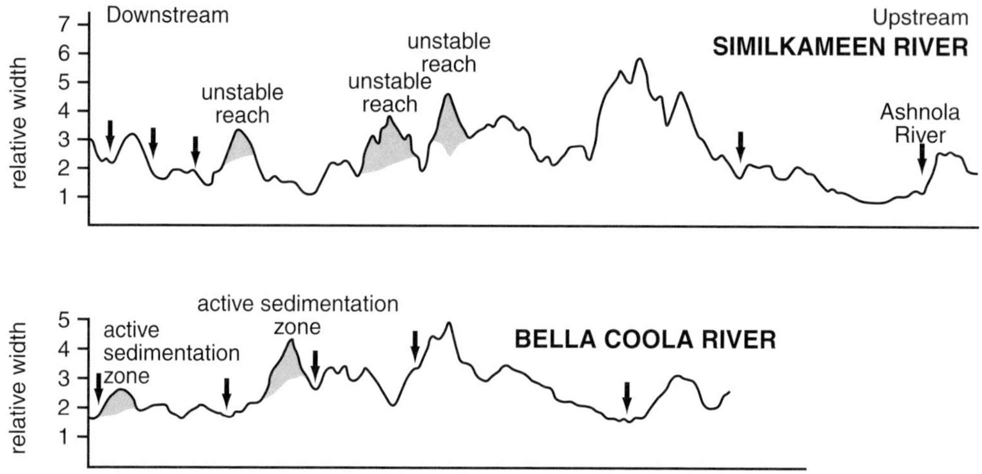

Fig. 8. Sedimentation zones separated by transport reaches in river valleys (after Church & Jones 1982). The ordinate is the 'relative width' of valley floor deposit (in multiples of the local channel width). Arrows show tributary junctions with alluvial fans. Sedimentation zones have laterally extensive valley fill, and these reaches are separated by relatively narrow transport reaches.

and predictive value of these measures of network topology suggests that they could form an important element of a model of within-basin sediment flux and storage.

A fourth issue, however, concerns less the topology of the drainage network structure than the topography of the valley system. Sediment storage patterns along river valleys, defined as sedimentation zones by Church & Jones (1982), may in some cases represent the temporary location of waves of sediment moving down-valley (Fig. 8). However, they also appear to be related to, and anchored by, the form of the valley system as determined by its geological structure, tectonic history and the topographic evolution of its drainage basin. The fan constrictions described above constitute one of the topographic constraints that determine the locations of sedimentation zones. However, there are several other factors that create alternations between broad, gently-sloping valley reaches and narrow, deeply-incised and steep reaches (gorges). The former develop as sediment storage zones and the latter as transport reaches. For example, Himalayan river valleys such as that of the Alaknanda River in the Garhwal Himal (Fig. 9) appear to have broad reaches upstream from the locations where the river crosses the major thrusts and steep gorges downstream. This reflects the gradient changes caused as uplift occurs on the thrust surface, the associated river response being to aggrade upstream where the thrusting reduces the river gradient, and to incise downstream. Rivers in many mountain regions display this pattern of behaviour, and exhibit sequences of broad, low-gradient, sediment-filled valley reaches separated by steep gorges. Unravelling the history of aggradation and incision, and terrace formation, in such valleys is a real challenge since the timing of sediment flux from one storage reach to another is not necessarily common to all storages in a given catchment. Rather, one storage reach may be accumulating sediment while another is releasing it, dependent in particular on the climatic and tectonic history of the basin. As different thrusts are active at different times, some storage reaches may be accumulating while others are delivering, and

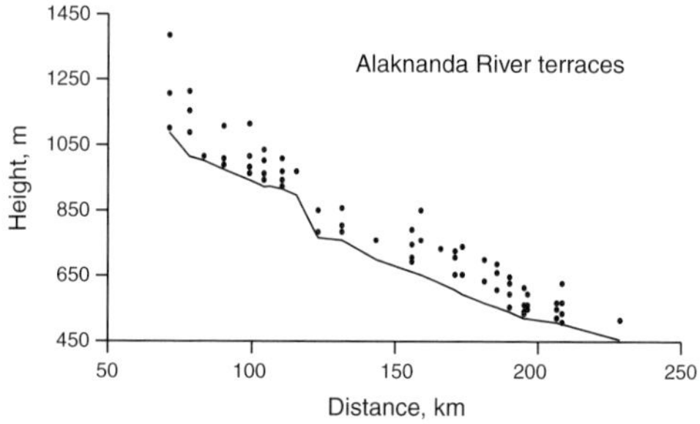

Fig. 9. The pattern of terrace fragments along the Alaknanda River in the Garhwal Himal, showing a tendency for convergence of terrace levels through wider sedimentation reaches to the downstream transport reaches where the river has incised as it cuts across thrust zones.

this means that the terraces evident in one reach are uncorrelated with those in adjacent reaches. A modelling framework would assist in unravelling these complex spatial relationships, which again emphasise that the drainage basin transfer function that links erosion to sediment yield is liable to confuse the relationships amongst environmental change, sediment yield, and aggradation and incision.

The conclusion to draw from the foregoing discussion is that drainage basins act as filters of environmental change; they induce variable time lags and alter the amplitude of the change of the output response relative to that of the input variation, through their role in storing and releasing sediments. The kind of modulation effected by a basin will depend on whether the extrinsic changes are imposed upstream or downstream, on how the intrinsic relationships operate and on the scale of the system relative to the amounts of sediment moving through the basin. The filtering role is dependent on the size of basin: large basins are insensitive to high-frequency external variations, while small basins are more reactive and may respond to high-frequency and short-lived change. This is evident from the preserved evidence of aggradation phases in river systems in the UK; the valley-fill deposits of a small river such as Jugger Howe Beck in the North York Moors may all be Holocene in age (Richards *et al.* 1987), while a large river such as the Thames has terraces that relate to the climate history of the whole Late Pleistocene (Gibbard 1985).

Interpretation of environmental history

If the drainage basin acts as a filter, and one whose properties are, *a priori*, unknown, it follows that the nature of the environmental signal that drives the sediment yield record cannot be read directly from that record. The reason that a spatially-structured model of storage within, and release of sediments from, drainage basins is so vital is that it alone can deliver for terrestrial sedimentary environments the reliability of interpretation that the marine sedimentary record already provides. It is worth recalling that the reason that the marine record of climate change can be read so successfully is because there is a reliable transfer function – that which translates foraminiferal oxygen isotope ratios into temperature. No comparable transfer function for the storage–sediment flux functions of drainage basins exists to underpin the interpretation of the sparse dating evidence that exists.

There have been attempts to read, and to generalize, the record of environmental change preserved in the fluvial stratigraphy by accumulating dates from alluvial sediments and inferring the changes of environmental conditions from the resulting histogram (Knox 1975; Macklin & Lewin 1993; Macklin 1999). Figure 10 illustrates an example of this approach; and in some cases, this may be an acceptable practice. However, in general, it cannot be acceptable if the drainage basins from which the dates have been obtained are highly varied in size, topological structure and topography (shape, gradient, relief, etc.), since the backwards inference of the timing of environmental change from that of the change in accumulation of alluvium will necessarily involve different time lags and amplitude changes in each drainage basin from which dated evidence has been derived.

This is not to imply that some level of generalization cannot be achieved; it is possible in cases where some level of matching of basin types is feasible. Figure 11, for example, illustrates the

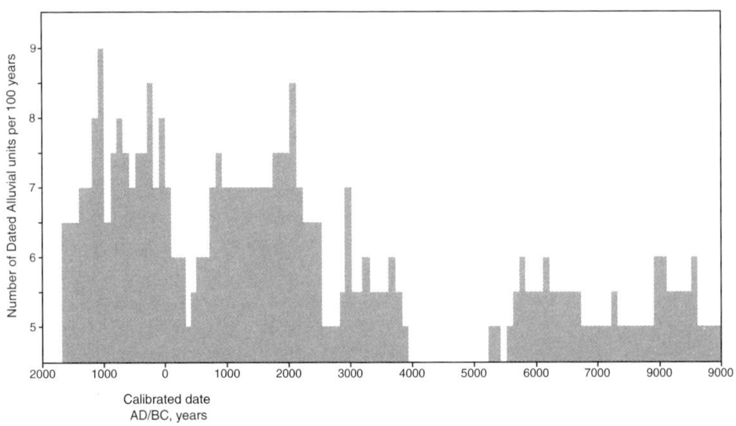

Fig. 10. The frequency distribution of radiocarbon dates in UK river basins (after Macklin 1999).

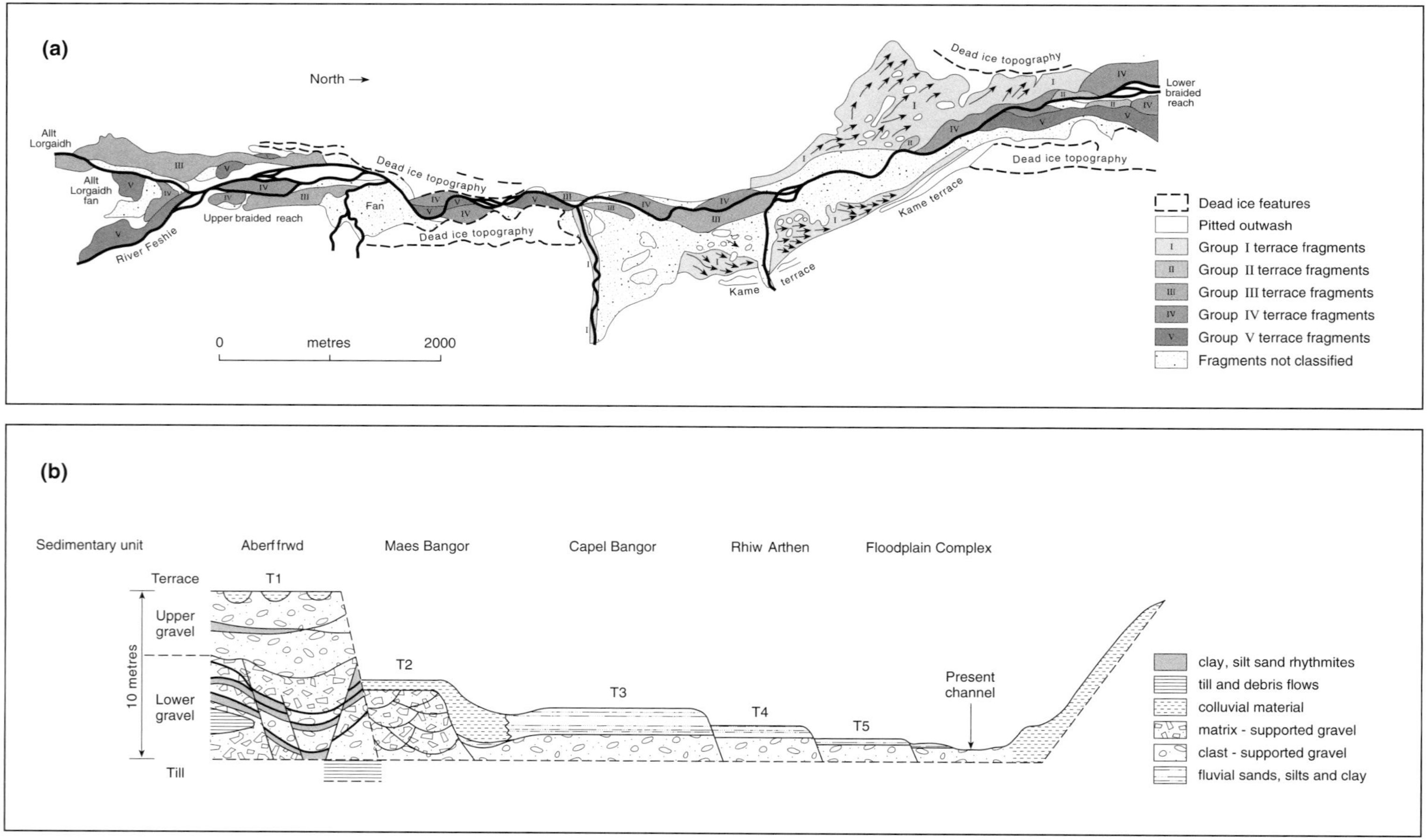

Fig. 11. A comparison between the post-glacial alluvial records of (**a**) the River Feshie (after Robertson-Rintoul 1986) and (**b**) the River Rheidol (after Macklin & Lewin 1986). Approximate ages of the terraces in the two cases are given in Table 1.

Table 1. *Hydrological, sedimentological and geomorphological similarities between the rivers Rheidol and Feshie*

	River Rheidol	River Feshie
Basin area (km^2)	182	240
Basin relief (m)	700-25	900-450
Annual precipitation (mm)	1500-2000	1000-1500
Bedrock	Lower Palaeozoic shales, siltstones, grits	Acid Moinian schists
Floodplain width (m)	800	300-500
Channel width (m)	30	30
Channel gradient	0.004	0.01
Bed material	Gravel–cobble	Gravel–cobble
Mean annual flood ($m^3 s^{-1}$)	71	85
Dates: Terrace 1	Late glacial	*c.* 13 000 BP
Terrace 2	Late glacial?	*c.* 10 000 BP
Terrace 3	<2775 BP	<3600 BP
Terrace 4	<650 BP	*c.* 1000 BP
Terrace 5	*c.* 100 BP	*c.* 100 BP

basic nature of the record of alluviation in two catchments of broadly similar physical character, the Feshie in Scotland (Robertson-Rintoul 1986) and the Rheidol in Wales (Macklin & Lewin 1986). As Table 1 shows, these two basins have comparable physical properties. They are both about 200 km^2 in area, with a similar relief and gradient. They are both coarse gravel-bed rivers with floodplains about 5–10 times the channel width, and similar mean annual floods. Their physical similarity implies that their filtering function of environmental change is broadly comparable, and as a result it is interesting that both have a terrace record in which there are late-glacial surfaces that probably reflect climatically controlled alluviation (*c.* 13 000 and *c.* 10 000 BP), and three Late Holocene terraces (at about 3000, 1000 and 100 BP) that in all probability reflect the effects of human influences on the runoff : sediment supply ratio.

Conclusions

The conclusions to be drawn from this discussion of the effects of drainage basins on the spatial and temporal variations of sediment storage and flux are primarily couched as a set of requirements. First, new approaches are required to the study of valley-fill development, and these might follow from combining the methods used in studies of catchment sediment yield and reservoir trap efficiency. However, secondly, there is a need to understand the valley-scale ‘trap efficiency’ and ‘delivery ratio’ as distributed variables *within* basins, not just as *between*-basin variables. Thirdly, there is a need to develop new, simple models of valley-fill evolution to assist in the interpretation of the often fragmentary evidence preserved in terrace sediments; to permit back-calculation of the sediment yield required to generate a particular cut-and-fill history, so that this reconstructed record may be compared with climate and land-use histories; and to allow simulation of cut-and-fill stratigraphies for comparison with the observed record of valley-fill development. Finally, there is a considerable need to return to the catchment scale in fluvial geomorphology in order to understand how drainage basin structure modulates the effects of environmental change when generating a record of sediment yield response.

References

AHNERT, F. 1976. Brief description of a comprehensive 3-dimensional process–response model of landform development. *Zeitschrift für Geomorphologie Supplement-Band*, **25**, 29–49.

BEASLEY, D. B., HUGGINS, L. F. & MONKE, E. J. 1980. ANSWERS: a model for watershed planning. *Transactions of the American Society of Agricultural Engineers*, **23**, 938–944.

BOYCE, R. C. 1975. Sediment routing with sediment delivery ratios. *In*: *Present and Prospective Technology for Predicting Sediment Yields and Sources*. US Department of Agriculture Publications, **ARS-S-40**, 61–65.

BRAKENRIDGE, G. R. 1984. Alluvial stratigraphy and radiocarbon dating along the Duck River, Tennessee – implications regarding flood-plain origin. *Geological Society of America Bulletin*, **95**, 9-25.

BRAKENRIDGE, G. R. 1985. Gradational thresholds and landform singularity – significance for Quaternary studies: Discussion. *Quaternary Research*, **23**, 417–419.

BRUNE, G. M. 1953. The trap efficiency of reservoirs. *Transactions of the American Geophysical Union*, **34**, 407–418.

CHURCH, M. & JONES, D. 1982. Channel bars in gravel-bed rivers. *In*: HEY, R. D., BATHURST, J. C. &

THORNE, C. R. (eds) *Gravel-bed Rivers*. Wiley, Chichester, 291–324.

CHURCHILL, M. A. 1948. Discussion of Analyses and use of reservoir sedimentation data by L.C. Gottschalk. *In*: *Proceedings of the Federal Inter-Agency Sedimentation Conference, Denver, Colorado*. US Geological Survey, Washington, DC, 139–140.

COOKE, R. U. & REEVES, R. W. 1976. *Arroyos and Environmental Change in the American Southwest*. Clarendon Press, Oxford.

DICKINSON, W. T., RUDRA, R. P. & WALL, G. J. 1986. Identification of soil erosion and fluvial sediment problems. *Hydrological Processes*, **1**, 111–124.

GIBBARD, P. L. 1985. *The Pleistocene History of the Middle Thames Valley*. Cambridge University Press, Cambridge.

GILBERT, G. K. 1917. *Hydraulic Mining Debris in the Sierra Nevada*. US Geological Survey Professional Paper, **105**.

JARVIS, R. J. 1976. Link length organisation and network scale dependencies in the network diameter model. *Water Resources Research*, **12**, 1215–1225.

KIRKBY, M. J. 1971. Hillslope process-response models based on the continuity equation. *In*: BRUNSDEN, D. (ed.) *Slopes: Form and Process*. IBG Special Publicatio, **3**, 15–30.

KIRKBY, M. J. 1976. Tests of the random network model, and its application to basin hydrology. *Earth Surface Processes*, **1**, 197–212.

KNOX, J. C. 1975. Concept of the graded stream. *In*: MELHORN, W. N. & FLEMAL, R. C. (eds) *Theories of Landform Development*. SUNY, Binghamton. Publications in Geomorphology, 169–198.

MACKLIN, M. G. & LEWIN, J. 1986. Terraced fills of Pleistocene and Holocene age in the Rheidol valley, Wales. *Journal of Quaternary Science*, **1**, 21–34.

MACKLIN, M. G. & LEWIN, J. 1993. Holocene river alluviation in Britain. *Zeitschrift für Geomorphologie, Suppl.-Bd*, **88**, 109–122.

MACKLIN, M. G. 1999. Holocene river environments in prehistoric Britain: human interaction and impact. *Journal of Quaternary Science*, **14**, 521–530.

MORGAN, R. P. C. 1994. The European Soil Loss Model: an update on its structure and research base. *In*: RICKSON, R. J. (ed.) *Conserving Soil Resources: European Perspectives*. CAB International, Cambridge, 286–299.

NADEN, P. S. & COOPER, D. M. 1999. Development of a sediment delivery model for application in large river basins. *Hydrological Processes*, **13**, 1011–1034.

RICE, S. & CHURCH, M. 1998. Grain size along two gravel-bed rivers: statistical variation, spatial pattern and sedimentary links. *Earth Surface Processes and Landforms*, **23**, 345–363.

RICHARDS, K. S., PETERS, N. R., ROBERTSON-RINTOUL, M. S. E. & SWITSUR, V. S. 1987. Recent valley sediments in the North York Moors: evidence and interpretation. *In*: GARDINER, V. (ed.) *International Geomorphology 1986 Part I*. Wiley, Chichester, 869–883.

ROBERTSON-RINTOUL, M. S. E. 1986. A quantitative soil-stratigraphic approach to the correlation and dating of post-glacial river terraces in Glen Feshie, Southwest Cairngorms. *Earth Surface Processes and Landforms*, **11**, 605–617.

SCHUMM, S. A. 1977. *The Fluvial System*. Wiley, Chichester.

SMALL, R. J. 1973. Braiding terraces in the Val d'Herens, Switzerland. *Geography*, **58**, 129–135.

TRIMBLE, S. W. 1983. A sediment budget for Coon Creek basin in the Driftless Area, Wisconsin, 1853–1977. *American Journal of Science*, **283**, 454–474.

VERSTRAETEN, G. & POESEN, J. 2000. Estimating trap efficiency of small reservoirs and ponds: methods and implications for the assessment of sediment yield. *Progress in Physical Geography*, **24**, 219–251.

A fractionation model for sediment delivery

JOHN C. TIPPER

Geologisches Institut, Albert-Ludwigs-Universität, Albertstrasse 23B, D-79104 Freiburg im Breisgau, Germany (e-mail: tipper@uni-freiburg.de)

Abstract: The delivery of sediment into basins can readily be studied using process-independent models. The simplest such model is the discrete fractionation model in two dimensions, without erosion. It represents the path down which sediment is delivered as a simple set of discrete steps, and assumes that the sediment is delivered down those steps in a series of discrete events. This model has two parameters – the feed volume and the fractionation coefficient. The feed volume is the volume of sediment fed to the first step in the delivery path, and the fractionation coefficient is the proportion of the sediment reaching a step that is then moved on. Several variants of this model are described here, each involving restrictions on one or both of these parameters. The model has obvious and important applications, but these will be difficult to realize until the problem of parameter estimation has been solved. Fortunately, it seems possible to solve this problem by removing some of the model's restrictions; these were built into it initially for analytical reasons. Removing these restrictions also makes the model applicable to a wider range of sedimentation systems.

The delivery of sediment into basins is a conceptually simple phenomenon: sediment is picked up in predominantly erosional source areas, it is transported down delivery paths and it is laid down finally in predominantly depositional sinks. In detail, of course, there is considerable complexity involved, for even in the simplest delivery system there are numerous ways in which erosion, transport and deposition can take place. Models of sediment delivery need always to be capable of recognizing this complexity, and most are made so by being process-oriented (Slingerland *et al.* 1994; Syvitski *et al.* 1998; Weltje *et al.* 1998). They are built by coupling together numerous quasi-independent sub-models, each of which represents some specific delivery process, for example bed-load transport in rivers, wave action in shallow seas, or hill creep. The greater the number of separate processes a process-oriented delivery model takes into account, the more realistic that model is usually deemed to be.

There are certainly many situations in which process-oriented delivery modelling is sensible. There are others, however, in which it is not. Consider the following situation, for instance: a sediment delivery model is to be constructed for some area of an ancient continental shelf, perhaps as part of a larger-scale modelling exercise in which sediment delivery into an entire marine basin is being investigated. The construction of a suitable process-oriented model in this situation demands an extraordinarily large amount of constraining data. Full details of the shelf bottom topography are required, together with information on sea level, tidal regime, sea-surface winds, and flows across all open boundaries. If these data are not available – *and for most ancient shelves it is difficult to see how they ever could be available* – then the near-bottom velocity field on the shelf cannot be calculated. It is then not possible to go any further (cf. Smith 1977, p. 540). In this type of situation it may be better to take a different approach from the start, and to work instead with process-independent delivery models. These are generalized models that represent directly the changes in sediment flux along defined source-to-sink delivery paths.

The simplest process-independent delivery model is the discrete fractionation model in two dimensions, without erosion (Fig. 1). The model treats sediment delivery as occurring in a series of discrete events; each event results in sediment being moved along a defined delivery path, which is topologically equivalent to a straight line and made up of a set of discrete steps. The model operates in the following way for each delivery event. First, some volume of sediment is fed to the first (most upstream) step; then a proportion of this sediment is deposited on that step; then the remaining sediment is moved to the second (next downstream) step; then a proportion of that sediment is deposited on that step; and so on until the last step is reached. The next event then begins

From: JONES, S. J. & FROSTICK, L. E. (eds) 2002. *Sediment Flux to Basins: Causes, Controls and Consequences.* Geological Society, London, Special Publications, **191**, 161–170. 0305-8719/02/$15.00

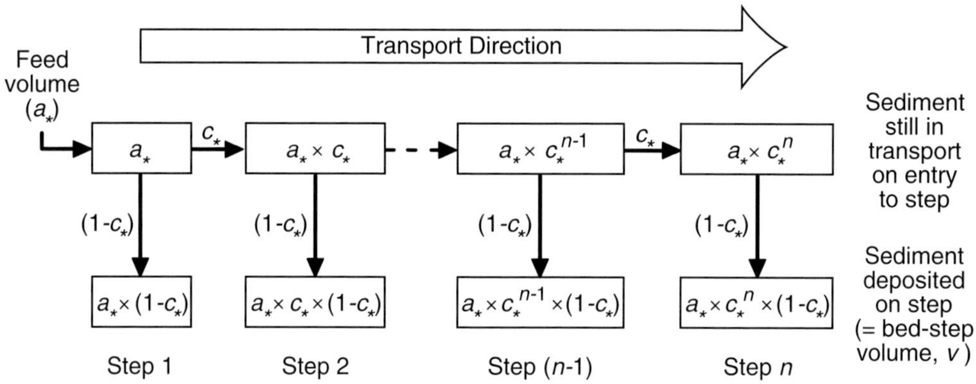

Fig. 1. The discrete fractionation model in two dimensions, without erosion (variant DF1). A fixed proportion (c_*) of the sediment still in transport on entry to any step is moved downstream to the next step. The remaining proportion ($1-c_*$) of the sediment is deposited.

with another volume of sediment being fed to the first step, etc. Each delivery event results in the deposition of a single bed of sediment.

The aim of this paper is to describe the theory behind some of the variants of this model, and to point to contexts in which these variants might find application. The model is also used in an analysis of patterns of bed-thickness variation found along simple proximal-to-distal transects through basins. Some important practical questions relating to the model's application to real sedimentation systems are addressed at the end of the paper. These questions concern: (1) the problem of parameter estimation; (2) the application of the model to sedimentation systems that are not non-erosional; (3) the implementation of the model in three dimensions; and (4) the application of the model to systems in which there is systematic spatial variation in fractionation coefficients.

The discrete fractionation model in two dimensions, without erosion

Variables and notation

There are four basic variables in the model: a is the feed volume, c is the fractionation coefficient, n is the step number and v is the bed-step volume. The feed volume is the volume of sediment fed to the first step in a delivery path; it is a measure of external sediment supply. The fractionation coefficient is the proportion of the sediment reaching a step that is then moved on; it is a measure of internal sediment transport efficiency. Step number is an integer measure of the distance down a delivery path. Bed-step volume is the volume of sediment deposited on a step during a single delivery event, ie. the change in sediment flux across that step during that event. Delivery paths are of unit width and steps are of unit length; therefore bed-step volume is numerically identical to bed thickness.

A subscripted variable name generally denotes the value of that variable for a particular bed and/or a particular step: thus a_j denotes the value of a for bed j, $c_{j,i}$ denotes the value of c for bed j and step i, and $v_{j,n}$ denotes the value of v for bed j and step n. The subscript * denotes a value that is the same for all beds and/or for all steps: thus a_* denotes a value of a that is the same for all beds, c_{j*} denotes a value of c that is the same for all steps for bed j, and v_{*n} denotes a value of v that is the same for all beds for step n. c_* denotes a value of c that is the same for all beds *and* all steps. The subscript cr is used to mark particular critical values of c and n; these values are defined later in the paper. Pr{ } denotes a probability, $F()$ denotes a distribution function, and $f()$ denotes a frequency function.

The model's equations, and its variants

The structure of the model is expressed in equations relating bed-step volume to step number. In their most general form, these equations are:

$$v_{j,n} = a_j \times \left(\prod_{i=1}^{n-1} c_{j,i}\right) \times (1-c_{j,n}), \text{ for } n > 1 \quad (1)$$

$$v_{j,n} = a_j \times (1-c_{j,n}), \text{ for } n = 1. \quad (2)$$

Clearly, $a_j > 0$ and $0 < c_{j,i} < 1$.

The model has several variants, each of which involves some restriction on the feed volume and/or the fractionation coefficient. In the first variant ('DF1'), the feed volume is kept the same for all beds, and the fractionation coefficient is kept the same for all beds *and* for all steps. Then:

$$v_{*n} = a_* \times c_*^{n-1} \times (1-c_*). \quad (3)$$

In the second variant ('DF2'), the fractionation coefficient is kept the same for all steps and all beds, and the feed volume is allowed to vary from bed to bed. Then:

$$v_{j,n} = a_j \times c_*^{n-1} \times (1-c_*). \quad (4)$$

In the third variant ('DF3'), the feed volume is kept the same for all beds, and the fractionation coefficient is allowed to vary from bed to bed. Then:

$$v_{j,n} = a_* \times c_{j*}^{n-1} \times (1-c_{j*}) \quad (5)$$

In the fourth variant ('DF4'), the feed volume and the fractionation coefficient are both allowed to vary from bed to bed. Then:

$$v_{j,n} = a_j \times c_{j*}^{n-1} \times (1-c_{j*}) \quad (6)$$

In the final variant ('DF5'), the feed volume is kept the same for all beds, and the fractionation coefficient is allowed to vary both from bed to bed *and* from step to step. Then:

$$v_{j,n} = a_* \times \left(\prod_{i=1}^{n-1} c_{j,i}\right) \times (1-c_{j,n}), \text{ for } n > 1 \quad (7)$$

$$v_{j,n} = a_* \times (1-c_{j,n}), \text{ for } n = 1 \quad (8)$$

Variant DF1 (Equation 3). Variant DF1 applies only to sedimentation systems that are spatially and temporally uniform. It is therefore unlikely ever to be of any practical value. Its principal uses are in introducing the discrete fractionation model (e.g. Fig. 1), and in demonstrating that model's properties. Two properties are particularly important: (1) the linkage between the discrete fractionation model and the continuous fractionation model – this is a fractionation model with which geologists are already familiar; and (2) the effect that changes in fractionation coefficient necessarily have on values of bed-step volume.

The basic assumption behind the continuous fractionation model is that the instantaneous rate of change of the modelled variable is directly proportional to that variable's current value. Radioactive decay is one such process for which this assumption is justified, and the continuous fractionation model is therefore used in the calculation of radiometric ages. The assumption that the instantaneous rate of deposition in a sedimentation system is directly proportional to the volume of sediment that that system is transporting is probably also justified for some systems, especially for those in which sediment is transported mostly as suspended load.

The nature of the linkage between the continuous and the discrete fractionation models is most easily demonstrated by comparing expressions for the volume of sediment still in transport during any given delivery event, at a given distance down a delivery path. The expression for this volume given by the continuous fractionation model is $d \times e^{-bx/u}$, where x is the path distance, d is the volume at $x = 0$, b (> 0) is the coefficient of proportionality between the rate of deposition and the volume of sediment in transport, and u is the transport velocity. The corresponding expression given by variant DF1 of the discrete fractionation model is $a_* \times c_*^n$ (Fig. 1), where n is the number of steps moved, and a_* and c_* are the feed volume and fractionation coefficient, respectively. Clearly the expressions are equivalent, with the term $e^{-b/u}$ in the continuous model corresponding to c_* in the discrete model.

The effects that changes in the value of the fractionation coefficient necessarily have on the value of the bed-step volume are best seen by first rewriting Equation (3) as $V_{*n} = c_*^{n-1} \times (1-c_*)$, where $V_{*n} = v_{*n}/a_*$. (The variable V is simply the bed-step volume for a bed whose feed volume is equal to 1.) Provided that n is greater than 1, a value of V_{*n} can generally be obtained using two different values of c_*. These two values are denoted here by cl and cu, where $0 < cl \leq cu < 1$; cl can readily be calculated when cu is given, or vice versa, because $cl^{n-1} \times (1-cl) = cu^{n-1} \times (1-cu)$. Only when $cl = cu$ is a single value of V_{*n} ever obtained for a given n. This single value of V_{*n} – the maximum possible for that value of n – is given by $(n-1)^{n-1}/n^n$. The value of the fractionation coefficient that gives this maximum value of V_{*n} is referred to here as the critical fractionation coefficient for that value of n, and is denoted by $(c_{cr})_n$. The value of n corresponding to $(c_{cr})_n$ is referred to here as the critical path length, and is denoted by n_{cr}. It is easily shown that $(c_{cr})_n = (n_{cr}-1)/n_{cr}$, and hence that $n_{cr} = 1/(1-(c_{cr})_n)$. The relationships between these various variables are illustrated in Fig. 2.

Variant DF2 (Equation 4). Variant DF2 differs from variant DF1 only in that the feed volume is allowed to vary randomly in time according to some given distribution. The fractionation coefficient is held constant. Variant DF2 can therefore be seen as potentially applicable to sedimentation systems that are supply-limited, and whose transport efficiency is essentially unaffected by the amount of sediment they are carrying.

The straightforward effects that changes in the value of the feed volume have on the value of the bed-step volume are clear from Equation (4); for

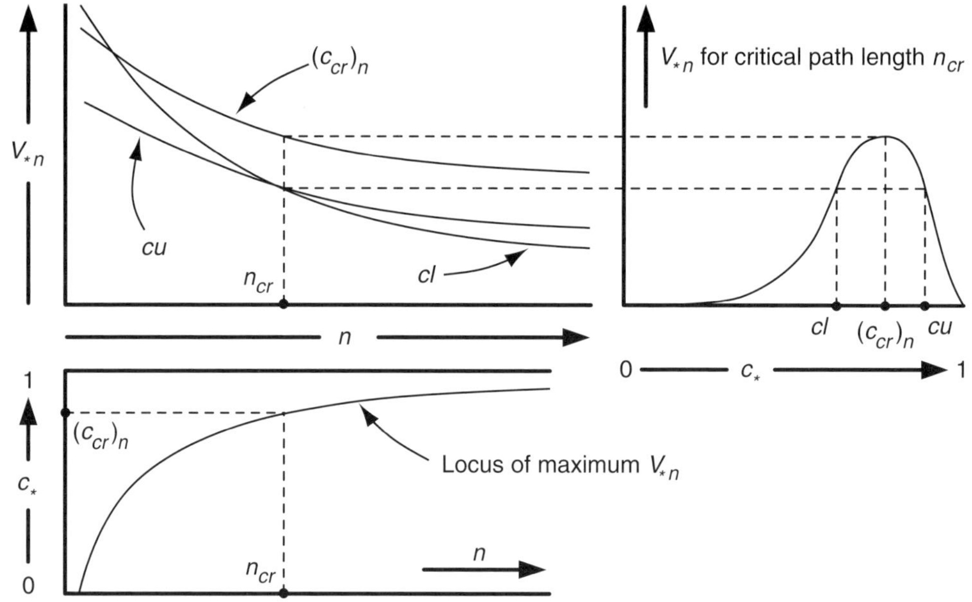

Fig. 2. The relationships between fractionation coefficient (c_*), bed-step volume (V_{*n}), and step number (n). Top-left: V_{*n} plotted against n, for fractionation coefficient values cl, cu and $(c_{cr})_n$; $(c_{cr})_n$ is the critical fractionation coefficient for marked critical path length n_{cr}. Top-right: V_{*n} for $n = n_{cr}$ plotted against c_*; maximum value of V_{*n} occurs for $c_* = (c_{cr})_n$. Bottom-left: c_{cr} plotted against n, showing locus of maximum V_{*n} and correspondence of $(c_{cr})_n$ and n_{cr}.

instance, for $c_* = 0.9$ and $n = 10$, $v_{j,n} = 0.0387 \times a_j$. The standard change-of-variable technique (Hoel 1962) can be used to calculate the frequency distribution of v from the given frequency distribution of a. For the values of c_* and n just given, $f(v) = f(a)/0.0387$.

Variant DF3 (Equation 5). Variant DF3 is complementary to variant DF2, in that it holds the feed volume constant and allows the fractionation coefficient to vary randomly in time according to some given distribution. Variant DF3 is potentially applicable to steadily supplied sedimentation systems that vary in time in their transport efficiency. An example – albeit a rather artificial one – might be a sedimentation system on some relatively shallow, wind-dominated part of the continental shelf, offshore of a stable, low-relief continent. Wave action will be the principal means of sediment transport for this system, and the system's efficiency in transporting sediment will therefore be determined largely by wind strength. This can be expected to vary both with the day-to-day weather conditions and with the time of year. In contrast, the supply of sediment into the system can be expected to be substantially constant. The shelf will be relatively starved of sediment, because of the stability and low relief of the neighbouring continent, and the amount of sediment from that continent that actually does manage to get out over the coastline and onto the shelf is unlikely to vary at all significantly in time, except over relatively long time spans.

The frequency distribution of v can readily be calculated for variant DF3, as c is assumed to be a random variable for which the frequency distribution is known. This calculation is performed in two stages. First, the frequency distribution of V is calculated, where $V_{j,n}$ $(= v_{j,n}/a_*) = c_{j*}^{n-1} \times (1-c_{j*})$. Then the change-of-variable technique is used to obtain the frequency distribution of v.

I know of no published data on values of fractionation coefficients in real sedimentation systems, and I can therefore illustrate frequency distributions of v here only by first choosing some apparently reasonable model for the frequency distribution of c. A highly peaked beta distribution is one obvious possibility, with its single mode located at some value of c slightly less than 1 (Fig. 3). Its frequency function is $f(c) = c^{p-1} \times (1-c)^{q-1}/\mathrm{B}(p,q)$, where p ($\gg 1$) and q (> 1) are two parameters, and $\mathrm{B}(p,q)$ is the standard beta function. Another possibility is a monomial distribution (Fig. 3). This is equivalent to a beta distribution with the parameter q set equal to 1. It has a J-shaped frequency function, $f(c) = p \times c^{p-1}$,

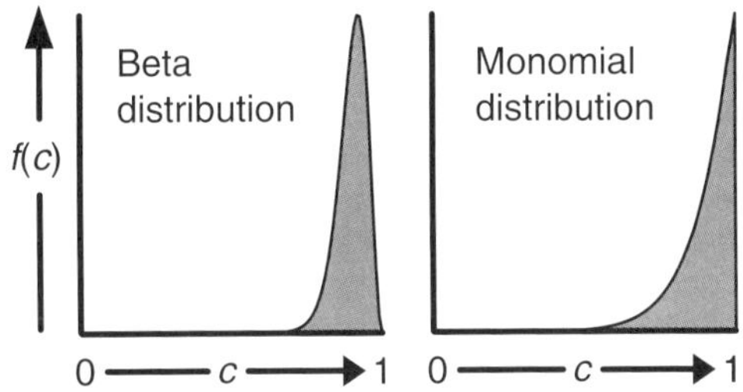

Fig. 3. Frequency functions for typical beta and monomial distributions. Mean value of c for both distributions was set to 0.9; standard deviation (SD) of beta distribution was set to 0.05. For beta distribution: $p = 31.5$, $q = 3.5$, mode is at $c = 0.9242$. For monomial distribution: $p = 9$, SD = 0.0905.

where $p \gg 1$. Differences in calculated frequency distributions of v resulting from using the monomial rather than the beta distribution will probably be small, especially when the mean value of c is large. The monomial distribution can therefore be seen as the better alternative, because it is much quicker to compute. Its only drawback is that its mean ($= p/(p+1)$) and variance ($= p/((p+2)(p+1)^2)$) are not independent; the variance of c automatically decreases as the mean of c increases.

Three distributions of v are shown in Fig. 4. These are for 50, 100 and 200 steps. They were calculated using a monomial distribution for c, with the mean value of c set equal to $(c_{cr})_{100}$, the critical fractionation coefficient value for 100 steps; $(c_{cr})_{100} = 0.99$, therefore p was set equal to 99. For simplicity, a_* was set equal to 1. The three distributions should be interpreted as follows: (a) that for step 50 is representative of sites with path lengths significantly less than the average critical path length; (b) that for step 100 is representative of sites with path lengths approximately equal to the average critical path length; and (c) that for step 200 is representative of sites with path lengths significantly greater than the average critical path length.

The three distributions differ systematically from each other in two respects: they differ in their overall form; and they differ in their balance of fractionation coefficients. The differences in overall form reflect the basic relationships between V_{*n}, $(c_{cr})_n$, n_{cr} and n (Fig. 2), and are therefore relatively easy to predict. Thus: (1) the maximum value of v decreases with increasing path length; and (2) the kurtosis (peakedness) of the distribution of v increases with increasing path length until the average critical path length is reached, after which it decreases. The differences in the balance of fractionation coefficients are less easy to predict. In general, however, the following result holds (Fig. 4): if nl is the number of beds in a distribution that have bed-step volumes less than x and fractionation coefficient values less than cl, and if nu is the number of beds in that same distribution that have bed-step volumes less than x and fractionation coefficient values greater than cu, then the ratio $nl : nu$ increases with increasing path length. The values of cu and cl – these variables were defined earlier for variant DF1 – are, of course, those that apply to volume x at the site in question.

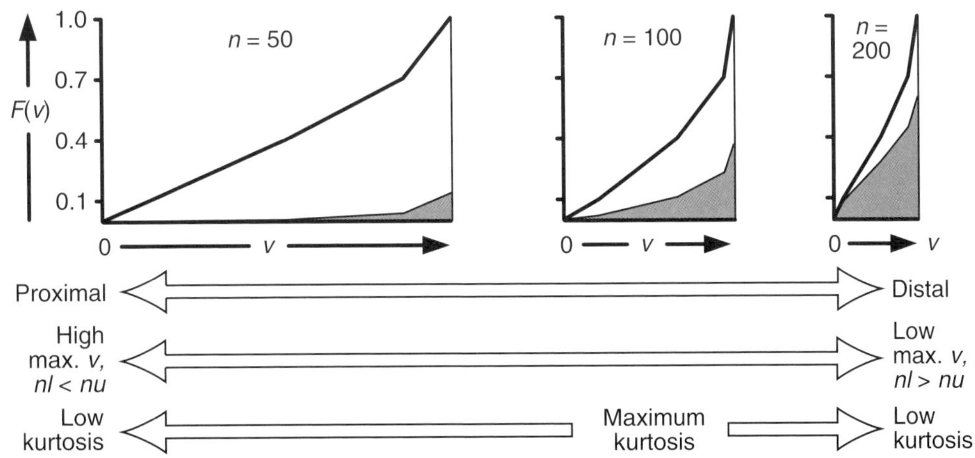

Fig. 4. Distribution functions of v for variant DF4, for delivery paths of 50, 100 and 200 steps; see text for details of calculation and interpretation. The shaded area under each distribution function corresponds to beds with values of c less than cl; the unshaded area under each distribution function corresponds to beds with values of c greater than cu.

Variant DF4 (Equation 6). Variant DF4 allows both the fractionation coefficient and the feed volume to vary randomly in time according to given distributions, and is the most general of the variants so far considered. This generality would seem theoretically to be a considerable advantage, for it should apparently make variant DF4 applicable to numerous types of sedimentation system. In practice, however, there is no advantage, for variant DF4 can actually be applied to a system only under extremely restrictive conditions: *either* the fractionation coefficient and the feed volume in the system must be known to be mutually independent random variables, *or* the exact form of the interdependency betwen fractionation coefficient and feed volume must be known. Neither of these conditions is likely to be fulfilled for natural sedimentation systems, and variant DF4 will therefore probably find use only as a basis for theoretical delivery simulations. In that case the frequency distribution of v will always be capable of being calculated directly from Equation (6), using Monte Carlo methods.

Variant DF5 (Equations 7 and 8). Variant DF5 should be seen as an elaboration of variant DF3. It holds the feed volume constant, but allows the fractionation coefficient to vary randomly in space as well as in time, according to given distributions. The frequency distribution of v can be calculated for variant DF5 in much the same way as for variant DF3, albeit with more difficulty. Again, the calculation is carried out in two stages. The frequency distribution of V is calculated first, and the distribution of v is then obtained via the change-of-variable technique. The calculation of the distribution of V will probably always best be performed using Monte Carlo methods, simply because of the potentially rather complex form of the expression for $V_{j,n}$ in Equation (7). There are two distinct terms in this expression, and it is difficult to predict which one will be dominant for any particular distribution of c. The first of these terms, $\left(\prod_{i=1}^{n-1} c_{j,i}\right)$, is the volume of sediment still in transport after step n–1; $(\prod c)$ can be expected to be log-normally distributed when n is large, irrespective of the form of the distribution of c. The second term, $(1-c_{j,n})$, is the proportion of the sediment deposited on step n; the form of the distribution of $(1-c)$ is obviously going to be intimately dependent on the form of the distribution of c. Two examples of distributions of v for variant DF5 are given in Fig. 5.

The general form of the model (Equations 1 and 2). The general form of the discrete fractionation model has no restrictions on feed volume and no restrictions on fractionation coefficient. It should therefore be applicable to almost any non-erosional sedimentation system, provided, of course, that the conditions stipulated earlier for variant DF4 are fulfilled. There is unlikely ever to be much sense in trying to obtain the frequency distribution of v other than by Monte Carlo methods, and an example of a distribution of v calculated in this way is given in Fig. 6. The following assumptions are made in this example: (1) a and c are mutually independent random variables; (2) c has a monomial distribution; and (3) a has a strongly bimodal distribution. This latter distribution of a might suggest that floods have influenced the way in which this sedimentation system is fed.

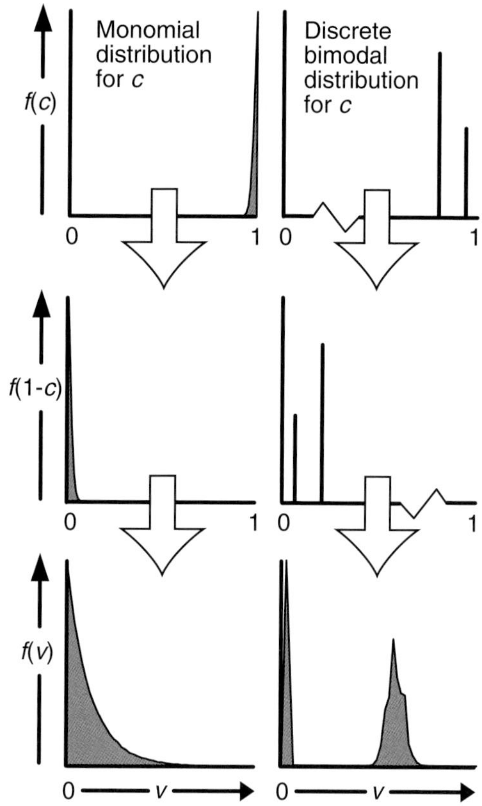

Fig. 5. Two distributions of v for variant DF5. The top line shows given frequency functions of c; the centre line shows corresponding frequency functions of $(1-c)$; the bottom line shows resulting frequency functions of v, for delivery path of 100 steps. Distribution of v in the left-hand column is based on monomial distribution for c, with mean = 0.99. Distribution of v in the right-hand column is based on a discrete bimodal distribution for c, with $\Pr\{c = 0.985\} = 0.6429$ and $\Pr\{c = 0.999\} = 0.3571$.

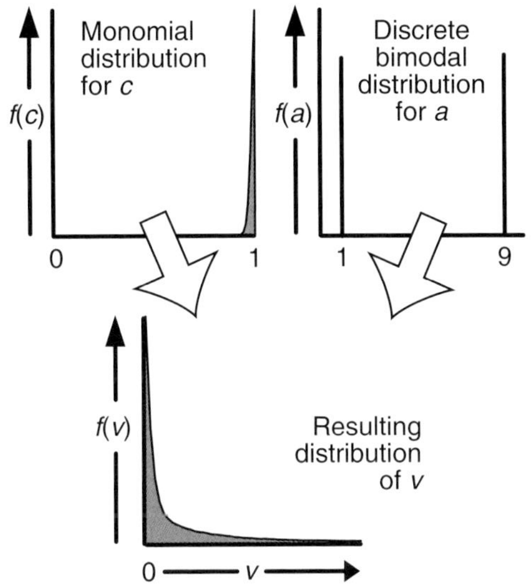

Fig. 6. A distribution of v for the discrete fractionation model in its general form. Top left: frequency function of c, with mean = 0.99. Top right: frequency function of a, with $\Pr\{a = 1\} = 0.5$ and $\Pr\{a = 9\} = 0.5$. Bottom: resulting frequency function of v, for a delivery path of 100 steps. Note that distribution of v is heavy in its high-v tail; it is not log-normal.

Proximal-to-distal patterns of bed-thickness variation

Proximal successions in clastic sedimentary basins are usually found to be more thickly bedded than distal successions. Stratigraphers therefore usually accept that individual beds of sediment normally get thinner as they are traced distally down their delivery paths (Sadler 1982; Zeng & Lowe 1997*b*, fig. 16). Instances of beds getting thicker distally are not uncommon, of course, but such instances tend usually to be seen as demanding special deterministic explanations. Among the explanations that can be found in the literature are ones involving changes in flow depth along a single delivery path (Kneller 1995), ones involving convergence and divergence of delivery paths (Kneller 1995), ones involving deposition from highly variable grain-size populations (Zeng & Lowe 1997*a*), and ones involving locally restricted developments of particular sedimentation mechanisms (Zeng & Lowe 1997*b*, fig. 3). The 'distal-thinning' and 'distal-thickening' patterns of bed-thickness variation are both reproduced by the discrete fractionation model (albeit only in its general form and as variant DF5), and they are seen there simply as end-members of a continuum of bed-thickness variation patterns. This suggests that it may now be appropriate to re-evaluate some accepted ideas on bed-thickness variation.

The discrete fractionation model reproduces the 'distal-thinning' and 'distal-thickening' patterns in a probabilistic sense, and these patterns need therefore to be defined in probabilistic terms. This is most logically done by using what can be termed the 'distal-thickening probability'. This is the probability that bed-step volume (and hence bed thickness) increases in m steps in a distal direction down a specified delivery path; it is abbreviated here as $\Pr\{\delta v_m > 0\}$, where $\delta v_m = v_{j,n+m} - v_{j,n}$. Clearly $\Pr\{\delta v_m > 0\}$ will be expected to decrease as m increases. The 'distal-thinning' and 'distal-thickening' patterns of bed-thickness variation correspond, respectively, to low and to high values of $\Pr\{\delta v_m > 0\}$.

Values of $\Pr\{\delta v_m > 0\}$ can readily be calculated when the distribution of c is known. For $m > 1$, $\Pr\{\delta v_m > 0\}$ can be shown to be equal to $\Pr\left\{\left(\prod_{i=1}^{m-1} c_i\right) \times c_k \times (1-c_l)/(1-c_k) > 1\right)$, where c_i, c_k and c_l are all values of c drawn independently from the given distribution. Some values of $\Pr\{\delta v_m > 0\}$ are presented in Table 1, for several different distributions of c. These values have all been calculated for m equal to what can be termed the 'mean path half-length' – this is the number of steps down a delivery path over which bed-step volume can be expected to halve. The mean path half-length is the natural length scale for the discrete fractionation model.

The values of $\Pr\{\delta v_m > 0\}$ presented in Table 1 are mostly rather high. This suggests that instances of beds getting thicker distally may be expected to be not uncommon in many sedimentation systems, even over distances for which the mean bed

Table 1. *Distal-thickening probabilities ($\Pr\{\delta v_m > 0\}$) calculated at the mean path half-length (m) for several distributions of* c

Distribution type	Mean	SD	m	$\Pr\{\delta v_m > 0\}$
Monomial	0.9330	0.06265	10	0.310
Monomial	0.9659	0.03296	20	0.334
Monomial	0.9862	0.01361	50	0.333
Monomial	0.9931	0.00685	100	0.331
Beta	0.9330	0.03133	10	0.172
Beta	0.9330	0.01566	10	0.030
Beta	0.9659	0.01648	20	0.164
Beta	0.9659	0.00824	20	0.020
Beta	0.9862	0.00681	50	0.162
Beta	0.9862	0.00340	50	0.026
Beta	0.9931	0.00343	100	0.162
Beta	0.9931	0.00171	100	0.020

thickness decreases by a factor of 2. These instances of distal thickening are due simply to the normal spatial variability that exists in a sedimentation system's transport efficiency. They do not demand special deterministic explanations.

This conclusion has two important corollaries, and these concern the way in which stratigraphers currently tend to interpret bed-thickness data. The first corollary is that decreasing bed thickness, *when measured for single beds*, can evidently be an extremely unreliable indicator of distal direction. Decreasing bed thickness should be taken as an indicator of distal direction only when it can be demonstrated for many beds together. The second corollary is that beds in isolated stratigraphic sections are probably best never correlated on the basis of relative thickness – this is effectively what some trace-matching algorithms do that are used in well-log correlation programs. It is more than possible that some unusually thick bed in the more distal of a pair of sections that are being correlated corresponds to a readily overlooked thin bed in the more proximal of the sections. The relative thickness of beds should be used as a criterion for lithostratigraphic correlation only when it is backed up by other independent evidence.

Discussion: putting theory into practice

The work reported here has concentrated solely on setting out the basic theory of the discrete fractionation model; the application of the model to real sedimentation systems has been left entirely for the future. There are, nevertheless, some application-related questions that need addressing now. These are: (1) the problem of parameter estimation; (2) the application of the model to sedimentation systems that are not non-erosional; (3) the implementation of the model in three dimensions; and (4) the application of the model to systems in which there is systematic spatial variation in fractionation coefficients.

The discrete fractionation model will be capable of being applied to real sedimentation systems only once the problem of estimating its parameters has been solved. This is a problem for which there seems to be no simple general solution, at least at present. The best that can be hoped for is therefore that the model can be verified for artificial sedimentation systems – ones that can be set up in the laboratory. This verification would seem to be feasible, because ways certainly exist by which feed volumes and fractionation coefficients should be capable of being estimated in some laboratory systems.

The fractionation coefficient will always be the more important of the model's parameters to estimate correctly, for it is principally by specifying different combinations of fractionation coefficients that the model can be made to reproduce different types of sediment delivery behaviour. In contrast, the feed volume will usually serve just to describe the input boundary condition of the sedimentation system being modelled. It will play a part in reproducing a system's sediment delivery behaviour only for systems in which feed volume and fractionation coefficient are non-independent.

Estimating feed volumes in laboratory sedimentation systems should be relatively straightforward, but the same will certainly not be true for fractionation coefficients. Indeed, the task of estimating the complete set of fractionation coefficients for an entire sedimentation system will be an extremely tedious one indeed. In practice, it will probably always be necessary to estimate every single value of c indirectly, with the help of the continuous fractionation model referred to earlier in this paper; c is equal to $e^{-b/u}$, where b and u are two of the parameters in that model. Estimating the complete set of fractionation coefficients for an entire sedimentation system will therefore require that continuous sets of measurements of b and u be made simultaneously at large numbers of equally-spaced sites, with the spacing of the sites equal to the step length for which c is being determined. These measurements will effectively define how the values of c for the system are distributed in space and time.

It is logical to ask if it might not be possible to estimate the distribution of fractionation coefficients in a sedimentation system by using information from sediments that that system has laid down. For instance, might it not be possible to estimate this distribution from patterns of lateral variation in bed thickness? The obvious way to do this would seem to be to make systematic measurements of mean path half-length; this was defined earlier as the number of steps on a delivery path over which bed-step volume (and hence bed thickness) can be expected to halve. Unfortunately, this is a dead end. Any measured value of mean path half-length (measured in metres, for instance) is ultimately a value of $m \times k$, where m is the mean path half-length measured in terms of numbers of steps and k is the individual step length in metres. As k is unkown, m cannot be determined, and nothing can therefore be said about the distribution of c.

The discrete fractionation model, as it has so far been analysed here, has been restricted: (a) to two dimensions; (b) to non-erosional sedimentation systems; and (c) to systems in which there is no systematic spatial variation in fractionation coefficients. With these stringent restrictions there is a substantial parameter estimation problem. It would therefore seem strange to suggest that it might be worth relaxing these restrictions, because relaxing restrictions usually makes problems more difficult to solve. For this particular model, however, relaxing restrictions makes sense. It makes sense: (a) because the model becomes applicable to a wider range of sedimentation systems; and (b) because relaxing the restriction prohibiting systematic spatial variation in fractionation coefficients appears to be the key by which the parameter estimation problem finally may be solved.

Relaxing the restriction that the model applies only to non-erosional sedimentation systems involves simply removing the upper bound on the value of c (= 1). The only restriction on c is then that it must be positive. Deposition will occur on steps for which c is less than 1; stasis (*sensu* Tipper 2000) will occur on steps for which c is equal to 1; erosion will occur on steps for which c is greater than 1. Bed-step volume will, of course, necessarily be negative for any step for which c is greater than 1, and this means that the thickness of beds deposited on that step during one or more previous delivery events will be reduced, possibly to zero. The distributions of bed-step volumes calculated using variants of the discrete fractionation model that allow for erosion will therefore *not* be distributions of finally preserved bed thicknesses. Those bed-thickness distributions will have to be calculated from their corresponding bed-step volume distributions, for instance using the approach described by Kolmogorov (1951) and Schwarzacher (1975).

Relaxing the restriction that the model be two-dimensional changes fundamentally the nature of the delivery paths with which it deals. These were required originally to be topologically equivalent to straight lines, but they have now only to lie on a surface (the lithic surface) that is single-valued in its elevation. Provided that they lie on that surface, they can have any geometry: they can diverge, they can converge, they can run parallel and they can cross.

The simplest way to allow the model to have diverging and converging delivery paths is also to remove the upper bound on the value of c (Fig. 7). Divergence in a delivery path effectively produces the same delivery result as deposition – it reduces the volume of sediment moved to the next downstream step. Divergence in a delivery path can therefore be represented by a value of c less than 1. Convergence in a delivery path effectively produces the same delivery result as erosion – it increases the volume of sediment moved to the next downstream step. Convergence in a delivery path can therefore be represented by a value of c greater than 1.

Relaxing the restriction prohibiting systematic spatial variation in fractionation coefficients is

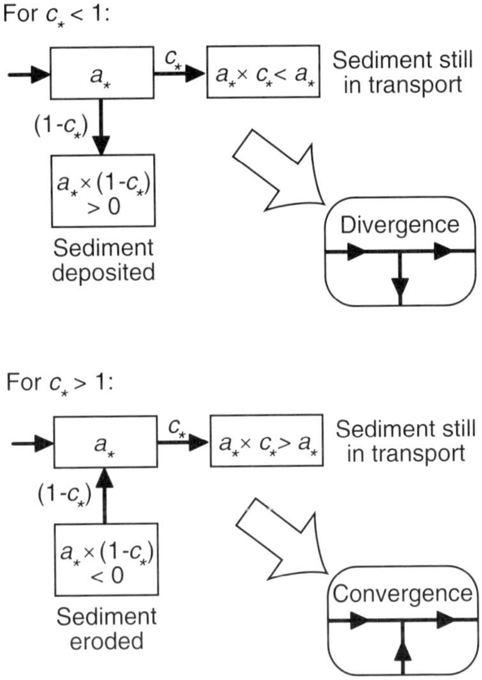

Fig. 7. Elaboration of the discrete fractionation model to allow erosion as well as deposition, and to allow diverging and converging delivery paths. Compare with Fig. 1. Upper diagram shows that deposition is equivalent to divergence (see arrowed inset). Lower diagram shows that erosion is equivalent to convergence (see arrowed inset).

equivalent to letting c depend in some way on n. This dependency can be described deterministically, using some function $c(n)$, or it can be described probabilistically, by making the parameters of the distribution of c depend on n. The latter alternative is by far the more powerful of the two, and is very strongly to be recommended. The reason for this is that c is thereby automatically converted into a regionalized variable, which makes available many standard geostatistical techniques (see, for example, Journel and Huijbregts 1978). Among these are a number of proven practical estimation techniques. It is probably by using these techniques that the parameter estimation problem for the discrete fractionation model will finally be solved.

A short final assessment

The discrete fractionation model is the simplest of all generalized sediment delivery models. It provides the possibility of looking at sediment delivery in basins in a process-independent way. The several variants of the model described here apply only to non-erosional systems with simple delivery paths, but the model can be elaborated so that it is more generally applicable. The greatest problem with the model lies in its practical application, because fractionation coefficient distributions are likely to be extremely difficult to estimate for real sedimentation systems. It will only be when this parameter estimation problem is solved – probably with the use of geostatistical estimation techniques – that the full potential of the discrete fractionation model will be realized.

Discussions with fellow participants at the Millennium Flux conference convinced me to formulate the results of this work in terms of volume rather than bed thickness. I thank P. Burgess and B. Dade for their helpfully critical reviews.

References

HOEL, P. G. 1962. *Introduction to Mathematical Statistics*, 3rd Edition. Wiley, New York.

JOURNEL, A. G. & HUIJBREGTS, C. J. 1978. *Mining Geostatistics*. Academic Press, London.

KNELLER, B. 1995. Beyond the turbidite paradigm: physical models for deposition of turbidites and their implications for reservoir prediction. *In*: HARTLEY, A. J. & PROSSER, D. J. (eds) *Characterisation of Deep Marine Clastic Systems*. Geological Society, London, Special Publications, **94**, 31–49.

KOLMOGOROV, A. N. 1951. Solution of a problem in probability theory connected with the problem of the mechanism of stratification. *American Mathematical Society Translations*, **53**, 171–177.

SADLER, P. M. 1982. Bed-thickness and grain size of turbidites. *Sedimentology*, **29**, 37–51.

SCHWARZACHER, W. 1975. *Sedimentation Models and Quantitative Stratigraphy*. Elsevier, Amsterdam.

SLINGERLAND, R. L., HARBAUGH, J. W. & FURLONG, K. P. 1994. *Simulating Clastic Sedimentary Basins*. Prentice-Hall, Englewood Cliffs, NJ.

SMITH, J. D. 1977. Modeling of sediment transport on continental shelves. *In*: GOLDBERG, E. D., MCCAVE, I. N., O'BRIEN, J. J. & STEELE, J. H. (eds) *The Sea*, Vol. 6. Wiley, New York, 539–577.

SYVITSKI, J. P., MOREHEAD, M. D. & NICHOLSON, M. 1998. HYDROTREND: a climate-driven hydrologic-transport model for predicting discharge and sediment load to lakes or oceans. *Computers & Geosciences*, **24**, 51–68.

TIPPER, J. C. 2000. Patterns of stratigraphic cyclicity. *Journal of Sedimentary Research*, **70**, 1262–1279.

WELTJE, G. J., MEIJER, X. D., & DE BOER, P. L. 1998. Stratigraphic inversion of siliciclastic basin fills: a note on the distinction between supply signals resulting from tectonic and climatic forcing. *Basin Research*, **10**, 129–153.

ZENG, J. & LOWE, D. R. 1997*a*. Numerical simulation of turbidity current flow and sedimentation: I. Theory. *Sedimentology*, **44**, 67–84.

ZENG, J. & LOWE, D. R. 1997*b*. Numerical simulation of turbidity current flow and sedimentation: II. Results and geological applications. *Sedimentology*, **44**, 85–104.

Transverse rivers draining the Spanish Pyrenees: large scale patterns of sediment erosion and deposition

STUART J. JONES

Department of Geological Sciences, University of Durham, South Road, Durham DH1 3LG, UK (stuart.jones@durham.ac.uk)

Abstract: Data collected from six transverse rivers draining the Spanish Pyrenees highlight how the rate of change in river gradients downstream controls long-term deposition and erosion patterns across the mountain belt. The rivers draining the Spanish Pyrenees provide evidence for an important relationship between gradient (S) and distance from the drainage divide (x), such that S is proportional to x^{Ω}. This can be expressed quantitatively as a power function relation $S \propto L^{-R}$. In areas dominated by erosion Ω has a value > –5, while in areas of deposition Ω has a value < –0.4. The contrast between areas of erosion and deposition is also reflected in downstream increases and decreases, respectively, in stream power and bed shear stress along the whole of the mountain front.

It is proposed, based upon previous empirical studies of loose to semi-armoured channels, that the downstream variations of stream power and bed shear stress result in variations of bed-load transport rates, which in turn create the erosional and depositional patterns associated with variations of channel gradient.

These results highlight the complex response of the fluvial system, emphasizing the need for regional approaches for examining long-term changes in rivers draining orogenic belts and the importance for incorporation of hydraulic data into geomorphological models to assist interpretations of the mass distribution of sediment.

The boundary between areas of fluvial erosion and deposition is often abrupt, especially in regions of active tectonics. The rapid change from an erosional topographic landscape to a smoother depositional landscape is well exhibited in many mountain belts. The surface gradients in depositional areas such as the Ebro Basin (Fig. 1) or in other depositional basins, e.g. the Gangetic Plain in India, are so low (less than 80 cm km^{-1}) that it is difficult to see the direction of slope in the field. However, within a few kilometres it is possible to find deeply incised valleys that cut into the adjacent mountain belt. Erosion and deposition in these adjacent areas has been sufficiently prolonged to strip away or deposit several hundreds of metres of sediment (e.g. Hirst & Nichols 1986; Burbank, 1992; Vincent & Elliott 1997; Jones 1997; Friend *et al.* 1999; Jones *et al.* 1999).

Understanding the process of why and where rivers erode and deposit sediment is fundamental to explaining landscape evolution. The patterns of erosion and deposition play a vital and critical role in shaping the Earth's surface (Fig. 1) (Milliman & Meade 1983). Geologists have long used the record of fluvial sediments within depositional basins to provide valuable information on the neighbouring active mountain belts and river systems. (e.g. Burbank 1992; Gawthorpe & Leeder 2000). In many tectonically active areas stream long profiles provide a critical relationship between relief, elevation and denudation rates (Hack 1957; Howard *et al.* 1994; Seidl *et al.* 1994; Whipple *et al.* 1999). Recent studies have concentrated on deriving quantitative estimates of the rates of tectonic uplift from topography and stream profile analysis (Snyder *et al.* 2000; Kirby & Whipple 2001). However, a quantitative explanation is needed on how uplift and subsidence controls erosion and deposition, respectively, for large-scale sediment transportation by river systems.

This research aims to identify the important factors that control large-scale basin-wide erosion and deposition patterns, and illustrates that the change of gradient downstream is quantifiably different in areas of erosion and deposition. Other hydraulic parameters are used (e.g. velocity, channel width, channel depth and drainage area) to determine the variations in stream power and bed shear stress for six main transverse rivers draining the Spanish Pyrenees (Figs 1 and 2). It is shown

From: JONES, S. J. & FROSTICK, L. E. (eds) 2002. *Sediment Flux to Basins: Causes, Controls and Consequences.* Geological Society, London, Special Publications, **191**, 171–185. 0305-8719/02/$15.00

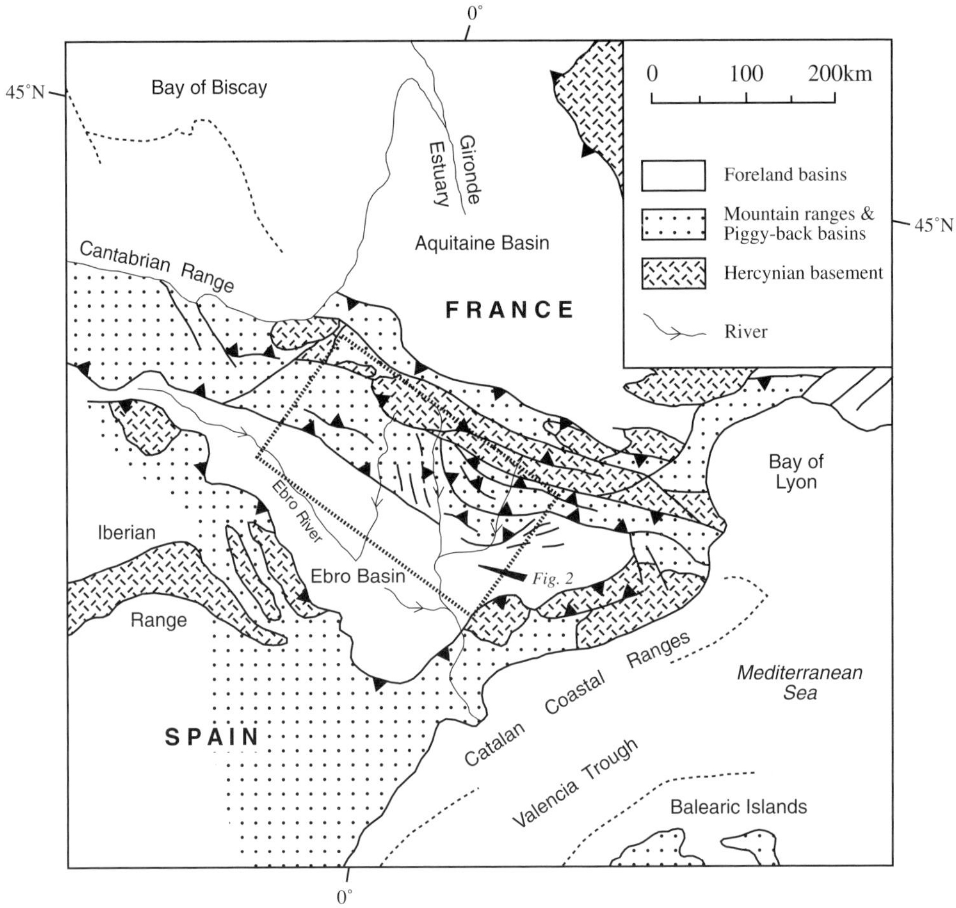

Fig. 1. Regional tectonic setting of the Pyrenean mountain belt between the Aquitaine (France) and Ebro (Spain) foreland basins. The map also indicates the location of the study area as referred to in the text and shown in more detail in Fig. 2.

that these two parameters change downstream according to the spatial distribution of the Plio-Pleistocene to Recent sediments and their associated erosion and deposition. The importance of these two parameters has been clearly shown in previous studies for the determination of bed-load transport rates (Laronne & Reid 1993; Jones *et al.* 1999; Garcia *et al.* 2000). This paper also suggests an important link between long-term downstream variations in erosion and deposition patterns and associated bed shear stress and stream power for the gravel-bed rivers of the Spanish Pyrenees.

Methodology

Channel profiles were derived by measuring downstream distance and elevation along the river channels on 1:25 000 Instituto Geografico Nacional topographic maps (Fig. 2). The channel elevations for each of the major rivers draining the Pyrenees were measured at intersection points with contour lines and from multiple spot heights in the near vicinity or adjacent to the channel. Gradients for the river long profiles were calculated using the drop in elevation and the downstream distance between two adjacent points on the profile. This procedure allows for some smoothing of data and positioning of gradient changes precisely along the profile. The channel long profiles were used to create plots of channel gradient and downstream distance (Fig. 3). For straight sections of the log–log plots (Figs 3 and 4) linear regression equations were derived with the following form:

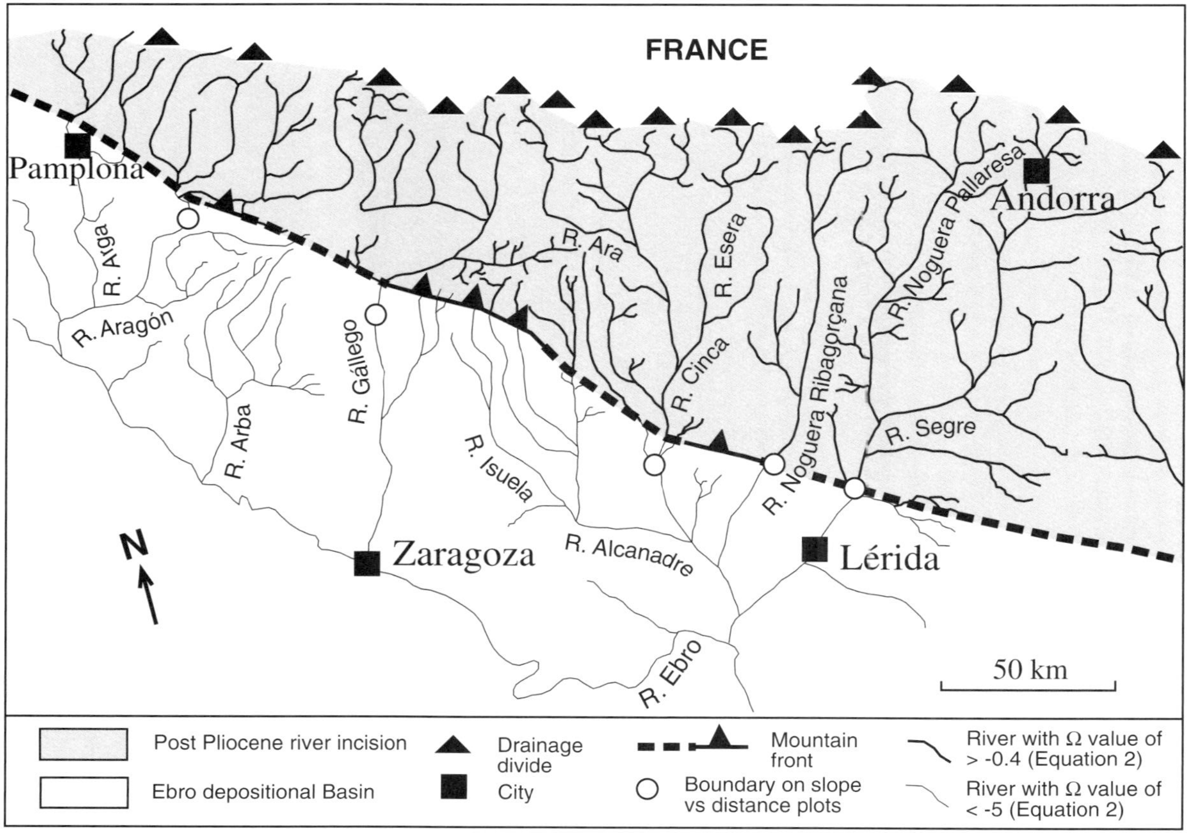

Fig. 2. Map of the southern flank of the Spanish Pyrenees and northern margin of the Ebro foreland basin. The map shows the position of the main transverse rivers draining the mountain belt and the divide between post-Pliocene incision (erosion) and deposition. The river long profiles and plots of slope vs distance downstream (Figs 6 and 7) were used to define the portion of each river with variable rates of slope and areas of long-term erosion and deposition. Sections of rivers with $\Omega > -0.4$ and $\Omega < -5$ (see Equation 2) are indicated, and the boundaries between such areas are shown with open circles. The criteria for choosing the boundary positions are discussed in the text.

$$\log_{10} S = \Omega \log_{10} x + \Psi. \quad (1)$$

This transforms to

$$S = 10^{\Psi} x^{\Omega} \quad (2)$$

where S is the channel long profile gradient (m km $^{-1}$) and x is the distance downstream (km) from the topographic ridge pole to the confluence with the axially draining Ebro River. The regression best fit defines the intercept and gradient (slope at $x = 1$) with the values of Ω and Ψ (Fig. 4). This approach for relating gradient and downstream distance allows bed shear stress and stream power to be calculated when Equation (2) is combined with other hydraulic equations.

Longitudinal river profiles

River long profiles are an important element of drainage basin evolution in that, together with channel networks, they fix the boundary conditions for slope processes. At any given point along a

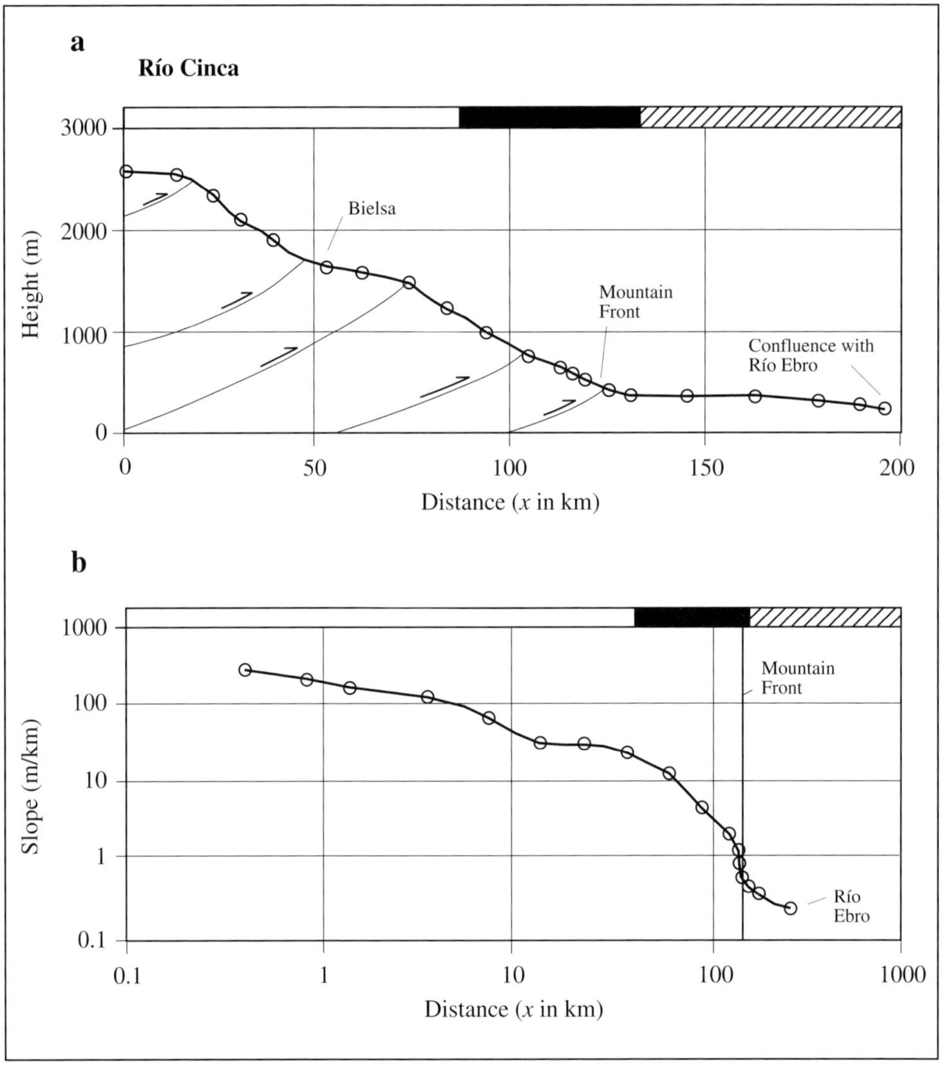

Fig. 3. (**a**) River long profile for the Río Cinca. Position of the mountain front, main thrusts along the river course and confluence with the Río Ebro are shown (adapted from Jones *et al.* 1999). (**b**) Logarithmic plot of downstream changes in slope in relation to distance from drainage divide. The bar at the top of each graph indicates areas of erosion (clear), areas of incision and aggradation (shaded) and areas of deposition (diagonal lines).

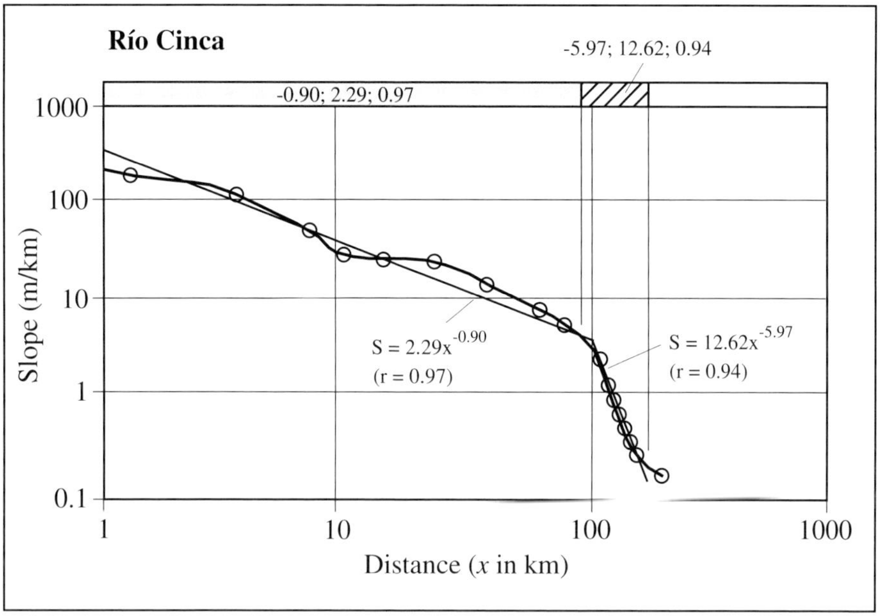

Fig. 4. Example of a logarithmic plot of channel slope (S) and downstream distance (x) for the Río Cinca that has been divided into two segments. Best-fit lines based on a power-law equation produces a good fit for each segment (this equally applies to the other rivers in this study, see Fig. 7). The bar at the top of the graph indicates the range of points used for each segment for a best-fit line.

stream profile, the altitude of the river bed will change at a rate equal to the rate of bedrock uplift minus the rate of incision. The shape of the channel profile is a direct result of the spatial distribution of erosion and deposition along a river course. There has been an extensive history of theoretical and experimental research about the form and development of stream profiles (e.g. Mackin 1948; Hack 1957; Langbein & Leopold 1964; Begin *et al.* 1981; Knighton 1984, 1987; Ohmori 1991; Hovius 1999). Long profiles have been used to decipher many extrinsic controls upon rivers, such as tectonic uplift (e.g. Seeber & Gornitz 1983; Frostick & Reid 1989; Jones *et al.* 1999; Snyder *et al.* 2000), changes in base-level (e.g. Koss *et al.* 1994; Seidl *et al.* 1994) and changes in bedrock lithology (e.g. Seidl & Dietrich 1992; Wohl 1992; Howard *et al.* 1994).

A river that is in quasi-equilibrium, such that it is neither depositing nor eroding, is described as having a graded river profile. However, many parameters other than gradient need to be taken into account before a river will attain a nearly 'ideal' river profile (Knighton 1987). In this study the river profiles are not in quasi-equilibrium, but are ungraded and characterized by long-term erosion and deposition during the Quaternary. The long-term erosion of the rivers is inferred from the surrounding mountainous topography with rugged relief, and from the geological record with transverse rivers draining the southern flank of the Pyrenees since Oligocene times (Jones *et al.* 1999, 2001). A depositional history over a similar timescale is inferred from thick Quaternary deposits and terrace sequences. Many of the modern river channels are currently depositing sediments, but this may represent a phase of aggradation amongst periods of incision (Jones 1997).

The overall geometry of long profiles varies significantly between mountain belts, and differences in profile shape have been attributed to the balance between bedrock uplift and fluvial incision (e.g. Howard *et al.* 1994; Burbank *et al.* 1996; Hovius 1999). The dominance of research investigating changes in river profiles due to long-term erosion have neglected the important effects of deposition on long profiles. This study focuses on whether there are fundamental differences in the general shape of river long profiles in areas of erosion and deposition, with profiles measured in the Spanish Pyrenees and joining the northern margin of the Ebro Basin (Figs 1 and 2).

Plio-Pleistocene–Recent patterns of sedimentation

Rivers draining the southern flank of the Spanish Pyrenees, some of which rise at or near the drainage divide in the Axial Zone, take an incised, southerly flowing transverse route to the main thrust fault structures of the mountain belt. This simple pattern of drainage, transverse to young actively uplifting mountain belts' main structural trend (Fig. 1), is a common occurrence (Hovius 1996). The present-day rivers (Fig. 2) are very well spaced with a regular distance of 20–30 km between each of them. This equates to a spacing ratio of 2.0, and is defined as the ratio of the half-width of the mountain range to the distance between the outlets of adjacent catchments with headwaters immediately below the main divide, measured at the mountain front (Hovius 1996). This spacing ratio is similar to many other juvenile mountain ranges (e.g. Southern Alps of New Zealand and the Taiwanese Central Range). The regularity of transverse drainage networks in the Pyrenees illustrates that at the present day, even after tectonic uplift has ceased, a regularity is still maintained that may represent an optimal catchment shape and area along the southern flank of the mountain belt (Langbein & Leopold, 1962; Jones *et al.* 1999, 2001).

The Pyrenean rivers are incised along the whole length of the mountain chain, but the distance of incision varies in a downstream distance from the mountain front (Figs 1 and 2) (Coney *et al.* 1996; Jones *et al.* 1999). This incision was initiated in the latest Pliocene–Early Pleistocene, but the boundary between longer-term erosion and deposition varies along the Pyrenees.

Several rivers draining the Pyrenees are diverted from their transverse course with bends up to 90° to the east or west, taking a temporary axial direction often due to thrust faulting and associated growth anticlines. The Río Gállego (Fig. 2) is diverted to the west, north of the External Sierras in the Jaca piggyback basin, and subsequently takes an axial drainage course for about 35 km. Along this route it gathers the drainage of smaller transverse rivers becoming incised along its upper to middle reaches. The river regains a transverse drainage pattern before entering the Ebro foreland basin. Other transverse rivers that are diverted to an axial drainage pattern at some point along their drainage pathways include the Río Segre, which turns sharply to the west to eventually continue southwards, merging with other transverse rivers on route before joining the Río Ebro (Fig. 2). This diversion seems likely to have caused the Río Segre to become fixed in position and to incise deeply into the bedrock.

The depositional character of the Pyrenean rivers is preserved in river terraces, valley fill and valley floor deposits, with some evidence of historical avulsions from the spatial distribution of such deposits (Fig. 5a) (Jones 1997; Jones *et al.* 1999). They are all characterized by cobble–boulder conglomerates, are usually clast-supported and occur in erosionally based channelized units (over 50–300 m wide; Fig. 5b). Cobble conglomerates with diffuse stratification and imbricated clasts are common (Fig. 5c). Cross-stratified conglomerates and coarse-grained sandstones occur as variable components of the river deposits and are usually found associated with channel-fill complexes.

Changes in channel slope

Figure 6 shows longitudinal profiles and their systematic trends for five rivers draining the southern flank of the Pyrenees. In Fig. 7 logarithmic plots of channel gradient (slope) are shown as a function of downstream distance. For all the rivers, an abrupt change in gradient occurs downstream of the mountain front or in the very near vicinity. Considering the approximations and uncertainities involved with the data measurement and plotting lines on the logarithmic plots, the changes occur at the downstream termination of predominant incision and the location of the thrust front or an associated fold (Figs 2–4).

However, the Río Alcanadre and Aragón do not show an abrupt change in gradient at the mountain front and the inflection point is found further to the south (Figs 6 and 7). The inflection points in the plots correspond with the position of sub-surface folds along the northern margin of the Ebro Basin.

All of the plots from along the southern flank of the Pyrenees show a striking coincidence between areas of Plio-Pleistocene–Recent incision and deposition, with the gradient of the river profile (Fig. 7). Depositional areas have a relationship between channel gradient and downstream distance, such that the value of Ψ given in Equations (1) and (2) is found to be in the range of 3–11 and the value of Ω is in the range of –1.4 to –6. This contrasts to areas of erosion where the value of Ψ is far smaller at between 1.5 to 3.4, and the value of Ω is typically between –2.2 and –0.15.

This coincidence between the pattern of erosion and deposition during Plio-Pleistocene–Recent and the gradient of the logarithmic plots of channel gradient and downstream distance plots is probably not that surprising considering the importance of gradient in controlling river behaviour and development (e.g. Richards 1982; Knighton 1984, 1987; Hovius 1999). However, by placing values on areas of deposition and erosion allows for better comparison between rivers and the predominant controls. It

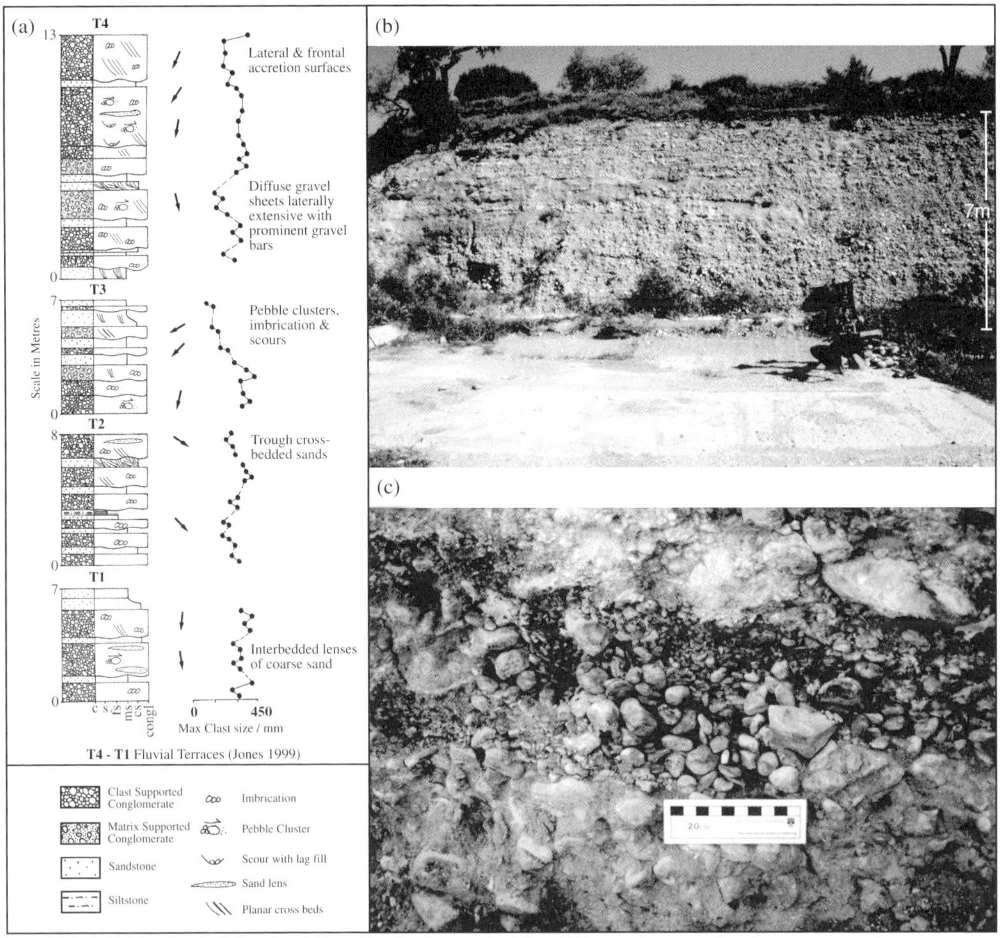

Fig. 5. (**a**) Detailed graphic logs of four terrace levels from the Río Cinca (see Jones *et al.* 1999 for further details of the terrace relationships). (**b**) Photograph of a T3 with cobble–boulder conglomerates typically found in channel fills and associated gravel bars. Supra-bar sandstones are also common. (**c**) Photograph of diffuse texture that is common amongst many of the sections for all the rivers. Allan & Frostick (1999) commented that diffuse open framework textures are formed from winnowing of sediment once the gravel is in place and stabilized, mostly during flood recession. This texture tends to be restricted to the terrace deposits found in the depositional areas of the river profiles (see text for discussion).

also provides support for the rate of adjustment at which a river gradient will decrease downstream and the long-term patterns of deposition and erosion for a whole mountain belt.

Stream power and bed shear stress

Rates of sediment transfer from the mountain belt to the foreland basin are unlikely to be solely controlled by changes in river gradient. Semi-theoretical equations and many experimental studies have adopted many approaches to the problem of predicting bed-load transport. These approaches illustrate that sediment flux is related to the rate of energy expenditure by the river flow over the bed (stream power; Bagnold 1966) and the balance between the driving and resisting forces (bed shear stress; Allen 1985, 1993). The link between these parameters and bed-load transport rates is supported by field measurements using sediment monitoring stations (e.g. Laronne & Reid 1993), modelling of fluvial processes in bedrock channels (e.g. Seidel & Deitrich 1992; Howard *et al.* 1994) and studies investigating bed-load behaviour and entrainment in small streams (e.g.

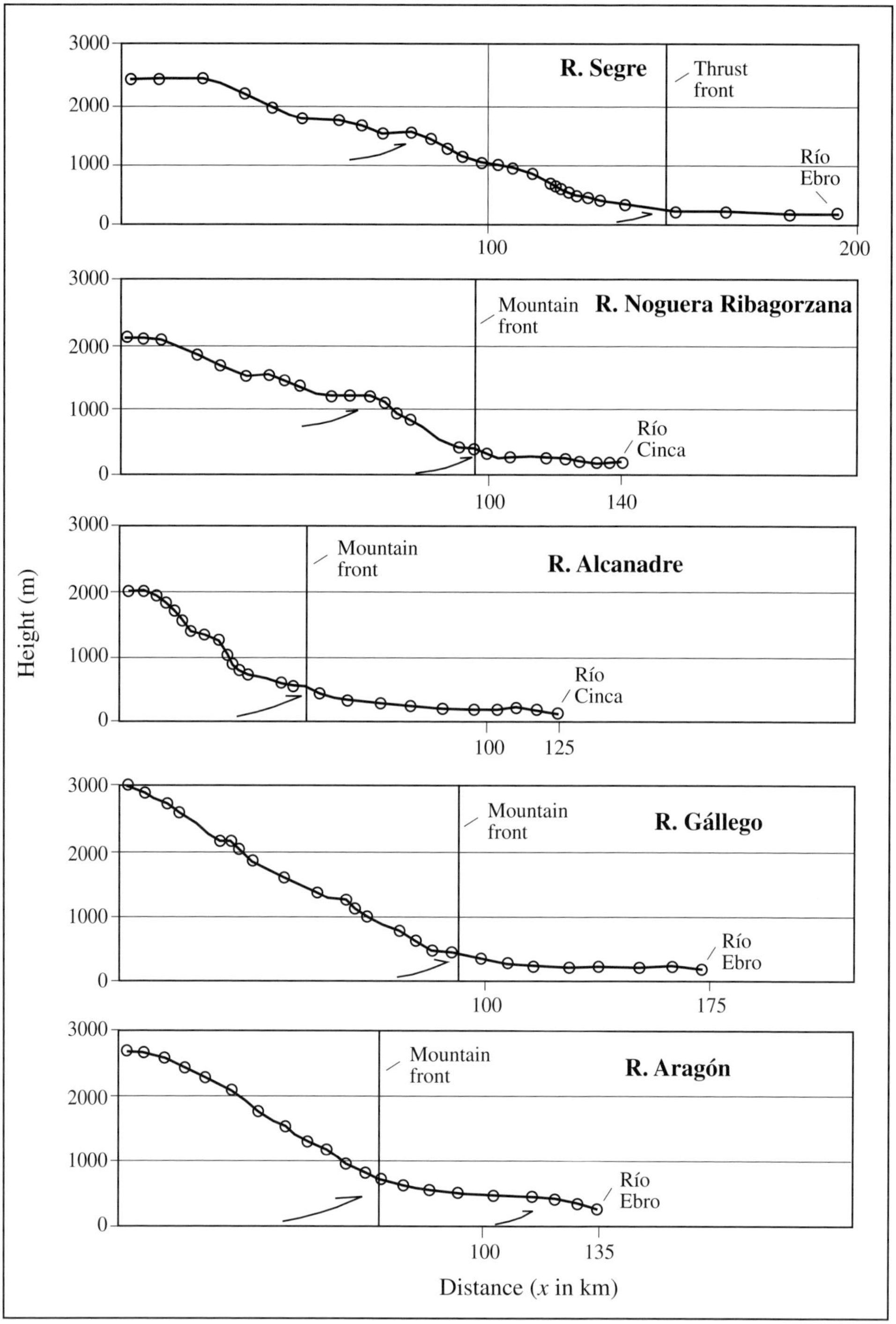

Fig. 6. Longitudinal profiles of five transverse rivers draining the Spanish Pyrenees. The location of the mountain front, important thrusts that cause perturbations of the profiles and the name of the confluence rivers are shown and discussed in the text.

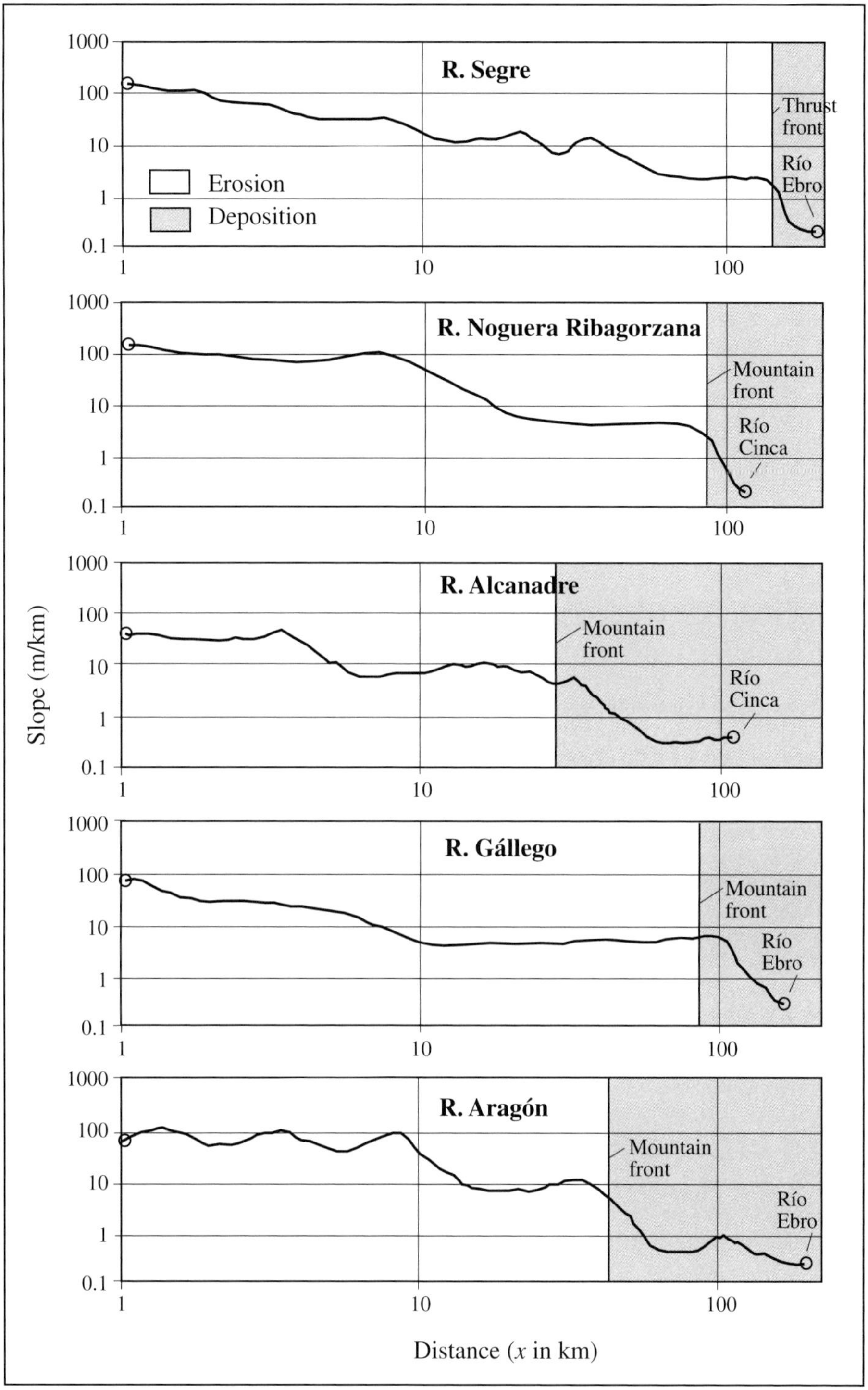

Fig. 7. Logarithmic profiles of slope (S) vs distance downstream (x) for the same five transverse rivers as in Fig. 6. Clear areas indicate erosion, whilst shaded areas refer to areas of long-term deposition. Refer to Fig. 2 for spatial relationship between inflection points for each of the rivers.

Carling 1983; Reid & Frostick 1984; Hoey 1992; Reid & Laronne 1995).

Stream power (ω in W m^{-2}) represents the work done by a flow on a unit area of a channel bed (Bagnold 1966), such that:

$$W = \rho gdSu = \tau_o u \quad (3)$$

where ρ is fluid density of water (1000 kg m^{-3}), g is the acceleration of gravity (9.81 m s^{-2}), d is flow depth, S is the channel gradient, u is the mean flow velocity and τ_o (in kg m^{-1} s^{-2}) is shear stress exerted by the fluid at the channel bed. Bed shear stress is expressed as:

$$\tau_o = \rho gRS \quad (4)$$

where R is the hydraulic radius (m) and closely equates to channel depth. Trends in these parameters and relationships can be compared with those of gradient and distance downstream (Figs 6 and 7) to investigate downstream trends in stream power and bed shear stress.

Trends in stream power and bed shear stress

As bed shear stress is closely related to stream power any trends closely resemble those displayed by stream power. Thus, stream power provides a valuable guide to the variability of the parameters and relationship to areas of erosion and deposition. The downstream variations of stream power have been calculated using Equations (2) and (3). It is shown in Fig. 8 that stream power does not attain a mid-basin maximum in accordance with the work of Graf (1983) and Lawler (1992, 1995), but lies further downstream. This is probably due to the strong structural control by thrust faults diverting rivers and capturing tributaries, providing a large increase in discharge approaching the mountain front (Figs 2 and 8). Upstream locations for stream power maxima are probably largely the reflection of steep headwater slopes, although subsidiary peaks can also be produced by localized tectonic uplift through thrust tip migration.

Assessment of the downstream variation in stream power reveals that when values of Ω greater than –0.4 and less than –5 occur they coincide with areas of erosion and deposition, respectively. As a result, channel reaches that experienced erosion during the Plio-Pleistocene underwent increases in stream power, whereas channel reaches experiencing deposition underwent decreases in stream power (Fig. 8). A similar correlation occurs with increases and decreases of bed shear stress and areas of erosion and deposition. This relationship is an important result of the study, but both parameters are strongly dependent upon the nature of the stream bed for thresholds of initiation and cessation of sediment movement, and therefore consideration is needed as to whether the river bed is loose gravel or armoured. The implications for the form of a river bed of the results presented in this paper are discussed below.

Discussion

Given that stream power affects sediment transport capacity and has a controlling influence on several aspects of channel form (Lecce 1987; Knighton 1999), the way in which it varies downstream has important implications for the distribution of process activity and channel adjustment within the fluvial system. There is now a need to discuss whether there is an important link between stream power and bed shear stress, and if large-scale patterns of erosion and deposition are related to these hydraulic parameters.

Bed-load transport rate

Bed-load transport formulae are founded upon the premise that a specific relationship exists between hydraulic variables, sedimentological parameters and the rate at which the bed-load is transported. Many bed-load equations are available for generating an estimate of sediment transport rates. However, although equations arise from different Einstein–stochastic approaches, the fact that the problem they are trying to solve is reasonably well defined inevitably means that there is considerable genetic similarity. The review of predictive power of bed-load transport equations by Gomez & Church (1989) concluded that many of the formulae used to estimate bed-load transport in gravel-bed rivers were of limited use and none universally applicable. Despite this conclusion, Bagnold's (1966) equation based upon stream power and bed shear stress provides the best fit for both flume and field measurements (Reid & Frostick 1993; Reid *et al.* 1997; Garcia *et al.* 2000; Martin & Church 2000). The Bagnold equation is:

$$i_b = w\,(e_b/\tan\alpha) \quad (5)$$

where i_b is the sediment flux by immersed weight (in kg m^{-1} s^{-1}), e_b is a bed-load transport efficiency factor and tan α is a friction coefficient for the bed material. Downstream changes in bed-load transport rate across the entire channel, at bankfull discharge, will change the channel-bed elevation. There are several important assumptions in such a proposal (i.e. that bed-load sediment mass is conserved in a downstream direction), but, importantly, downstream increases in bed-load transport rate will result in erosion and any decrease will cause deposition. It is proposed that the downstream

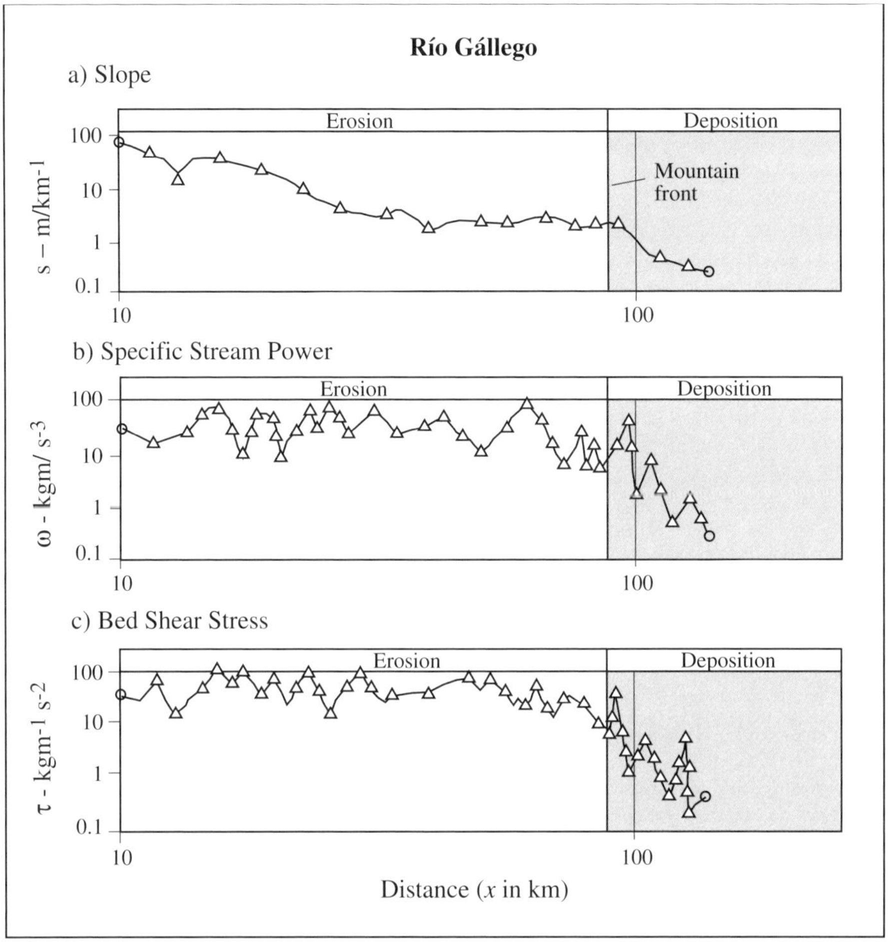

Fig. 8. Graphs showing the relationships between slope (**a**), stream power (**b**) and bed shear stress (**c**) for their downstream trends. The Río Gállego is used as an example, as this river most characteristically illustrates the change from areas of erosion to areas of deposition.

variations of stream power and bed shear stress result in variations of bed-load transport rates and which in turn create the erosional and depositional patterns associated with variations of channel gradient.

Bedrock, armoured and loose bed channel form

Along the southern flank of the Spanish Pyrenees the change from erosion to deposition occurs in semi-loose bed channels (Figs 2 and 5). The processes involved in erosion of bedrock, armoured and loose bed channels are completely different (Reid & Frostick 1993; Howard *et al.* 1994; Seidl *et al.* 1994), and any bed-load transport equations need to take such differences into account. Field observations of the modern rivers draining the Spanish Pyrenees and ancient terrace deposits (Fig. 5) shows that the erosion–deposition transition occurs at points in the rivers where the river bed changes from armoured to semi-loose. This transition is supported by the physical character of the beds with pebble clusters and diffuse texture abundant in areas of deposition where the beds are stabilized (Fig. 5c) (Reid *et al.* 1992; Allan & Frostick 1999). Field studies have shown that the flow required to initiate bed-load transport is consequently higher than would be expected with a uniform size distribution along a channel reach. Measurements by Reid & Frostick (1984) show that, during the rising stage of the first flood after a long period of inactivity, relatively little transport may occur. Eventually, though, the bed structure is broken, and on the falling stage of the flood, transport occurs to a much greater degree. Measurements show that transport may continue to a value of the bed shear stress, typically one-third of the value characterizing initiation of motion. On the other hand, when floods follow each other closely, the bed material remains comparatively loose and larger amounts of bed-load can be transported. The critical transport conditions are, therefore, to some extent dependent on the former high flow history.

Along the upper to middle reaches of the river sections studied, bedrock incision must occur and is an important driving force for areas of long-term erosion and changes in slope. Channel incision into bedrock occurs when the supply of sediment to the channel cannot keep it continuously mantled with an alluvial cover, usually due to either steeper gradients or a meager sediment supply. Because of scouring and plucking that occurs during high flood stages, channels with a thin alluvial cover can slowly erode the underlying bedrock while maintaining an alluvial cover during low-flow conditions (Howard & Kerby 1983); but the bedrock erosional capacity of alluvial channels is limited, so that if downstream erosion rates exceed this capacity local gradients steepen and bedrock becomes exposed (Merritts & Vincent 1989).

The Pyrenean rivers illustrate that river long profiles reflect the different processes operating in bedrock and/or armoured to semi-loose bed channels with a prominent change in river behaviour from areas of erosion to deposition. Further work is needed to determine whether bedrock incision occurs during high-stage events and what controls the transition between bed character along the channel reach. However, such an observation allows inferences to be made about long profiles from other orogenic belts, and how the rivers draining mountain belts change in character.

Spatial and temporal variability

Research using a bed-load sampler along the Tordera River, NE Spain, has shown that there is a low correlation between bed-load flux and shear stress mainly due to the different thresholds of initiation of motion for separate flows and the variable character of the bed between events (Garcia *et al.* 2000). However, the stream power and bed shear stress results in this study appear to indicate that channel beds will tend to experience higher values during short-term high-magnitude flood events. This has been shown by Laronne & Reid (1993), whereby such flows produce disproportionally high bed-load transport rates. This study also indicates that the disproportionally high values need not only occur in ephemeral streams (Laronne & Reid 1993), but in perennial rivers such as those draining the Pyrenees with a changing river bed character (Jones *et al.* 1999). Further research using short-term measurements of discharge, river width and flow velocity needs to be undertaken to characterize the changes with more accuracy and to show how such parameters change the river's long profile during high-magnitude events.

Extrinsic or intrinsic controls?

Sediment budgets in catchments draining the frontal sides of orogens are among the highest in the world (Milliman & Syvitsky 1992). As fluvial processes are responsible for the transport of most erosion material out of the mountain belts (Milliman & Meade 1983; Hovius 1998), any observed trends in measurable hydraulic parameters have important consequences for our understanding of not only the erosion of the orogen, but also the depositional history in the adjacent sedimentary basin.

The observed river long profiles for the Pyrenees and many other mountain belts (e.g. Hovius 1999) correlate well with the trends in stream power and bed shear stress. As a result it is important to distinguish between the extrinsic and intrinsic controls on the rivers, with the long-term evolution of channel form reflecting both controls. The extrinsic controls include: (a) the effects of tectonic uplift on channel gradient, leading to increased sediment supply whose rate of mass transfer to the fluvial system must relate directly to rates of incision and sediment transport; (b) climate change through its control on sediment supply; and (c) base-level changes. In comparison, the varied intrinsic controls include channel width, channel depth, channel pattern, grain-size bed armouring and drainage network patterns. Both extrinsic and intrinsic controls can influence the observed trends in stream power and bed shear stress.

The consistent relationship between stream power, bed shear stress and the river long profiles supports the link between bed-load flux, erosion and deposition, and the influence of tectonic deformation on the river profiles (Fig. 3) (Jones *et al.* 1999). Rivers can only attain a smooth, concave, graded profile when they are capable of removing bedrock at a rate equal to, or greater than, the rate of rock uplift. Such river profiles would not erode or deposit sediment over extended periods of time. The river long profiles in this study are not graded and have probably existed in a state of persistent imbalance continually affected by the extrinsic controls and influencing the long-term erosion and deposition patterns along the mountain front. Such an occurrence in an orogen may represent an optimal river profile and catchment geometry (transverse rivers) that embodies a 'most probable state' in an uplifting–erosional–depositional system along the mountain belt front.

Conclusion

The results from the Pyrenees clearly indicate that downstream changes in gradient are different in areas of fluvial erosion compared to deposition across the entire southern flank of the mountain belt–foreland basin system. Rivers draining the Spanish Pyrenees provide evidence for a relationship between river gradient and distance from drainage divide such that gradient (S) is inversely proportional to distance (x). Areas of deposition have $\Omega < -0.4$, while in areas of erosion $\Omega > -5$. The observed relationship correlates well with trends in stream power and bed shear stress. The relationship between stream power, bed shear stress and the river long profiles supports an important link between bed-load flux, patterns, and distribution of erosion and deposition influenced through intrinsic and extrinsic controls on the river profiles.

This study demonstrates the complexities in drainage evolution and associated sedimentation histories along mountain belts that can occur in the long-term record, principally driven by changes in slope through tectonics. Several attempts have been made to integrate drainage evolution and mountain building into models of landscape development (e.g. Beaumont *et al.* 1992; Chase 1992). The incorporation of hydraulic data into the interpretation of tectonically active orogens is of crucial importance to the quality of these geomorphological models. The ability to understand the long-term erosional and depositional pattern of drainage systems, allowing for comparison between rivers and their hydraulic behaviour, has important consequences for appreciating the mass distribution of sediment, the erosion of uplifting orogens and the infill of adjacent sedimentary basins.

This paper greatly benefited from discussions with B. Dade and G. Evans at the Millennium Flux Conference held in Southampton. I thank K. Richards and A. Arche for their very thoughtful and comprehensive reviews, and editor L. Frostick for the final revision of the text. K. Davis is thanked for drawing the figures.

References

ALLAN, A. F. & FROSTICK, L. E. 1999. Framework dilation, winnowing and matrix particle size: the behaviour of some sand–gravel mixtures in a laboratory flume. *Journal of Sedimentary Research*, **69**, 21–26.

ALLEN, J. R. L. 1985. *Principles of Physical Sedimentology*. Allen & Unwin, London.

ALLEN, J. R. L. 1993. Fundamental properties of fluids and their relation to sediment transport processes. *In*: PYE, K. (ed.) *Sediment Transport and Depositional Processes*. Blackwell, Oxford, 25–60.

BAGNOLD, R. A. 1966. *An approach to the Sediment Transport Problem From General Physics*. US Geological Survey Professional Paper, **422I**.

BAGNOLD, R. A. 1980. An empirical correlation of bedload transport rates in flumes and natural rivers. *Proceeding of the Royal Society*, **372A**, 453–473.

BEAUMONT, C., FULLSACK, P. & HAMILTON, J. 1992. Erosional control of active compressional orogens. *In*: MCCLAY, K. R. (ed.) *Thrust Tectonics*. Chapman & Hall, London, 1–18.

BEGIN, Z. B., MEYER, D. F. & SCHUMM, S. 1981. Development of longitudinal profiles of alluvial channels in response to base-level lowering. *Earth Surface Processes and Landforms*, **6**, 49–68.

BURBANK, D. W. 1992. Causes of recent Himalyan uplift deduced from depositional patterns in the Ganges basin. *Nature*, **357**, 680–683.

BURBANK, D. W., LELAND, J., FIELDING, E., ANDERSON, R. S., BROZOVIC, N., REID, M. R. & DUNCAN, C. 1996. Bedrock incision, rock uplift and threshold hillslopes in the north western Himalayas. *Nature*, **379**, 505–510.

CARLING, P. 1983. Threshold of coarse sediment transport in broad and narrow rivers. *Earth Surface Processes and Landforms*, **8**, 1–18.

CHASE, C. G. 1992. Fluvial landsculpting and the fractual dimension of topography. *Geomorphology*, **5**, 39–57.

CONEY, P. J., MUNOZ, J. A., MCCLAY, K. R. & EVENCHICK, C. A. 1996. Syntectonic burial and post-tectonic exhumation of the southern Pyrenees foreland fold–thrust belt. *Journal of the Geological Society, London*, **153**, 9–16.

FRIEND, P. F., JONES, N. E. & VINCENT, S. J. 1999. Drainage evolution in active mountain belts: extrapolation backwards from present-day Himalayan river patterns. *In*: SMITH, N. D. & ROGERS, J. (eds) *Fluvial Sedimentology VI*. International Association of Sedimentologists Special Publications, **28**, 305–313.

FROSTICK, L. E. & REID, I. 1989. Climatic versus tectonic controls of fan sequences; lessons from the Dead Sea, Israel. *Journal of the Geological Society*, **146**, 527–538.

GARCIA, C., LARONNE, J. B. & SALA, M. 2000. Continuous monitoring of bedload flux in a mountain gravel-bed river. *Geomorphology*, **34**, 23–31.

GAWTHORPE, R. L. & LEEDER, M. R. 2000. Tectono-sedimentary evolution of active extensional basins. *Basin Research*, **12**, 195–218.

GOMEZ, B. & CHURCH, M. 1989. An assessment of bed load transport formulae for gravel bed rivers. *Water Resources Research*, **25**, 1161–1181.

GRAF, W. L. 1983. Downstream changes in stream power in the Henry Mountains, Utah. *Annals of the Association of American Geographers*, **73**, 373–387.

HACK, J. T. 1957. *Studies of Longitudinal Stream Profiles*

in Virgina Maryland. US Geological Survey Professional Paper, **504B**.

HIRST, J. P. P. & NICHOLS, G. 1986. Thrust tectonic controls on alluvial sedimentation patterns, southern Pyrenees. *In*: ALLEN, P. A. & HOMEWOOD, P. (eds) *Foreland Basins*. International Association of Sedimentologists Special Publications, **8**, 247–258.

HOEY, T. 1992. Temporal variations in bedload transport rates and sediment storage in gravel-bed rivers. *Progress in Physical Geography*, **16**, 319–338.

HOVIUS, N. 1996. Regular spacing of drainage outlets from linear mountain belts. *Basin Research*, **8**, 29–44.

HOVIUS, N. 1998. Controls on sediment supply by large rivers. *In*: SHANLEY, K. W. & MCCABE, P. J. (eds) *Relative Role of Eustacy, Climate and Tectonics in Continental Rocks*. SEPM Special Publications, **59**, 3–16.

HOVIUS, N. 1999. Macroscale process systems of mountain belt erosion. *In*: SUMMERFIELD, M. A. (ed.) *Geomorphology and Global Tectonics*. Wiley, New York, 77–105.

HOWARD, A. D. & KERBY, G. 1983. Channel changes in the badlands. *Geological Society of America Bulletin*, **94**, 739–752.

HOWARD, A. D., DIETRICH, W. E. & SEIDL, M. A. 1994. Modeling fluvial erosion on regional to continental scales. *Journal of Geophysical Research*, **99**, 13 971–13 986.

JONES, S. J. 1997. *The Evolution of Alluvial Systems in the South Central Pyrenees, Spain*. PhD Thesis, University of Reading.

JONES, S. J., FROSTICK, L. E. & ASTIN, T. R. 1999. Climatic and tectonic controls on fluvial incision and aggradation in the Spanish Pyrenees. *Journal of the Geological Society, London*, **156**, 761–769.

JONES, S. J., FROSTICK, L. E. & ASTIN, T. R. 2001. Braided stream and flood plain architecture: The Río Vero Formation, Spanish Pyrenees. *Sedimentary Geology*, **139**, 229–260.

KIRBY, E. & WHIPPLE, K. 2001. Quantifying differential rock-uplift rates via stream profile analysis. *Geology*, **29**, 415–418.

KNIGHTON, A. D. 1984. *Fluvial Forms and Processes*. Arnold, London.

KNIGHTON, A. D. 1987. River channel adjustment: the downstream dimension. *In*: RICHARDS, K. (ed.) *River Channels: Environment and Process*. Institute of British Geographers Special Publications, **18**, 95–128.

KNIGHTON, A. D. 1999. Downstream variation in stream power. *Geomorphology*, **29**, 293–306.

KOSS, J. E., ETHERIDGE, F. G. & SCHUMM, S. A. 1994. An experimental study of the effects of base level change on fluvial, coastal plain and shelf systems. *Journal of Sedimentary Research*, **B64**, 90–98.

LANGBEIN, W. B. & LEOPOLD, L. B. 1964. Quasi-equilibrium states in channel morphology. *American Journal of Science*, **262**, 782–794.

LARONNE, J. B. & REID, I. 1993. Very high rates of bedload transport by ephemeral desert rivers. *Nature*, **306**, 148–150.

LAWLER, D. M. 1992. Proceess dominance in bank erosion systems. *In*: CARLING, P. A. & PETTS, G. E. (eds) *Lowland Floodplain Rivers: Geomporphological Perspectives*. Wiley, Chichester, 117–143.

LAWLER, D. M. 1995. The impact of scale on the processes of channel-side sediment supply: a conceptual model. *In*: *Effects of Scale on Interpretation and Management of Sediment and Water Quality. Proceedings of a Boulder Symposium, July 1995*. International Association of Hydrological Sciences Publications, **226**, 175–184.

LECCE, S. A. 1997. Nonlinear downstream changes in stream power on Wisconsin's Bluw River. *Annals of the Association of American Geographers*, **87**, 471–486.

MACKIN, J. H. 1948. Concept of a graded river. *Geological Society of America Bulletin*, **59**, 463–512.

MARTIN, Y. & CHURCH, M. 2000. Re-examination of Bagnold's empirical bedload formulae. *Earth Surface Processes and Landforms*, **25**, 1011–1024.

MERRITTS, D. J. & VINCENT, K. R. 1989. Geomorphic response of coastal streams to low, intermediate and high rates of uplift. Mendocino triple junction region, northern California. *Geological Society of America Bulletin*, **101**, 1373–1388.

MILLIMAN, J. D. & MEADE, R. H. 1983. World-wide sediment delivery of river sediment to the Oceans. *Journal of Geology*, **91**, 1–21.

MILLIMAN, J. D. & SYVITSKI, J. P. M. 1992. Geomorphic/tectonic controlof sediment discharge to the ocean: the importance of small mountainous rivers. *Journal of Geology*, **100**, 525–544.

OHMORI, H. 1991. Changes in mathematical function type describing the longitudinal profile of a river through an evolutionary process. *Journal of Geology*, **99**, 97–110.

REID, I. & FROSTICK, L. E. 1984. Particle interaction and its effect on the thresholds of initial and final bedload motion in coarse alluvial channels. *In*: KOSTER, E. H. & STEEL, R. J. (eds) *Sedimentology of Gravels and Conglomerates*. Canadian Society of Petroleum Geologists Memoirs, **10**, 61–68.

REID, I. & FROSTICK, L.E. 1993. Fluvial sediment transport and deposition. *In*: PYE, K. (ed.) *Sediment Transport and Depositional Processes*. Blackwell, Oxford, 89–155.

REID, I. & LARONNE, J. B. 1995. Bedload sediment transport in an ephemeral stream and a comparison with seasonal and perennial counterparts. *Water Resources Research*, **31**, 773–781.

REID, I., BATHURST, J. C., CARLING, P. A., WALLING, D. E. & WEBB, B. W. 1997. Sediment erosion, transport and deposition. *In*: THORNE, C. R., HEY, R. D. & NEWSON, M. D. (eds) *Applied Fluvial Geomorphology for River Engineering and Management*. Wiley, Chichester, 95–135.

REID, I., FROSTICK, L. E. & BRAYSHAW, A. C. 1992. Microform roughness elements and the selective entrainment and entrapment of particles in gravel-bed rivers. *In*: BILLI, P., THORNE, C. R., HEY, R. D. & TACCONI, P. (eds) *Dynamics of Gravel-bed Rivers*. Wiley, Chichester, 251–272.

RICHARDS, K. 1982. *Rivers: Form and Process in Alluvial Channels*. Methuen, London.

SEEBER, L. & GORNITZ, V. 1983. River profiles along the

Himalayan arc as indicators of active tectonics. *Tectonophysics*, **92**, 335–367.

SEIDL, M. A. & DIETRICH, W. E. 1992. The problem of channel erosion into bedrock. *Catena Supplement*, **23**, 101–124.

SEIDL, M. A., DIETRICH, W. E. & KIRCHNER, J. W. 1994. Longitudinal profile development into bedrock: an analysis of Hawaiian channels. *Journal of Geology*, **102**, 457–474.

SNYDER, N. P., WHIPPLE, K. X., TUCKER, G. E. & MERRITTS, D. J. 2000. Landscape response to tectonic forcing: digital elevation model analysis of stream profiles in the Mendocino triple junction region, northern California. *Geological Society of America Bulletin*, **112**, 1250–1263.

VINCENT, S. J. & ELLIOTT, T. 1997. Long-lived transfer-zone palaeovalleys in mountain belts: an example from the Tertiary of the Spanish Pyrenees. *Journal of Sedimentary Research*, **B67**, 303–310.

WHIPPLE, K. X., KIRBY, E. & BROCKLEHURST, S. H. 1999. Geomorphic limits to climate-induced increases in topographic relief. *Nature*, **401**, 39–43.

WOHL, E. E. 1992. Bedrock channel incision along Piccaninny Creek, Australia. *Journal of Geology*, **101**, 749–761.

Process-based modelling of the climatic forcing of fluvial sediment flux: some examples and a discussion of optimal model complexity

P. W. BOGAART, R. T. VAN BALEN, J. VANDENBERGHE & C. KASSE

Netherlands Centre for Geo-ecological Research, Faculty of Earth and Life Sciences, Vrije Universiteit Amsterdam, De Boelelaan 1085–1087, 1081 HV Amsterdam, The Netherlands (e-mail bogw@geo.vu.nl)

Abstract: During the complex and highly dynamic climate of the Late Quaternary, precipitation and temperatures were highly variable and generally did not change synchronously. Reconstructed fluvial sedimentological response to this climatic forcing was previously shown to be complex. This response can be understood only by accounting for all the relevant catchment processes. This paper reviews a number of simple numerical models that aim to enhance our understanding of the role of different processes in this climate–sedimentology system. The applicability of these models depends on the scale of the models involved, the required input data and the way that parameters are estimated. Simple process-based models perform best on Quaternary timescales because their data needs are low and parameters can be interpreted in physical terms, which enables an *a priori* estimation. Process-based models of fluvial dynamics may provide a methodology for the understanding of the mechanisms responsible for the dynamics of sediment flux to basins.

Sediment supplied to basins is generally delivered by fluvial systems such as rivers. Any understanding of the causal coupling between external forcing factors such as climate and tectonics on sediment supply is therefore dependent on the understanding of the coupling between climate and fluvial systems, both from the prediction point of view and from the interpretation of ancient and recent records point of view. Of these external forcing mechanisms climate is the most complex. Not only because it is multivariate (temperature, precipitation, seasonality), but also because it drives earth surface processes along a number of causal pathways. Climate has changed constantly throughout Earth's history, but the most dynamic changes are generally associated with the Quaternary. The response of the (fluvial) surface processes to these climatic forcing, are now considered to non-linear, and, as will be reviewed below, are most intense and complex during rapid climatic changes. The unravelling of the relative role of different climate-related changes in surface processes and – ultimately – the sedimentological record is thus dependent on high-quality reconstruction and time control of these processes, both of which suffer from degradation with time. The last glacial–interglacial transition (Last Termination or Late Glacial, *c*. 18.0–11.5 cal. ka BP) (Walker *et al.* 1999) is the best investigated major climatic change in Earth's history. Terrestrial records for this period are relatively undisturbed and well dated, and, as a consequence, climate and environment can be reconstructed with a high degree of accuracy (e.g. Hoek 1997; Hoek *et al.* 1999; Isarin & Renssen 1999). This makes the Late Glacial of the most relevant periods for learning about the coupling between climate and sedimentology.

Previous research has shown that the response of alluvial rivers to these climate changes was of a complex nature (Kasse 1995; Vandenberghe 1995; Huisink 1997). Phases of river incision and aggradation alternated, and channel patterns varied between meandering and braiding. Figure 1 shows a summary of Late Glacial climate and environment, as reconstructed from various proxy data sources.

Mol *et al.* (2000) compared fluvial evolution, during the last glacial cycle, for nine European rivers spanning a W–E gradient from Britain to Poland. They found that all rivers experienced similar behaviour: incision during major climate change (cold to warm, as well as warm to cold), multiple-thread planform (braided or anastomosing

From: JONES, S. J. & FROSTICK, L. E. (eds) 2002. *Sediment Flux to Basins: Causes, Controls and Consequences*. Geological Society, London, Special Publications, **191**, 187–198. 0305-8719/02/$15.00

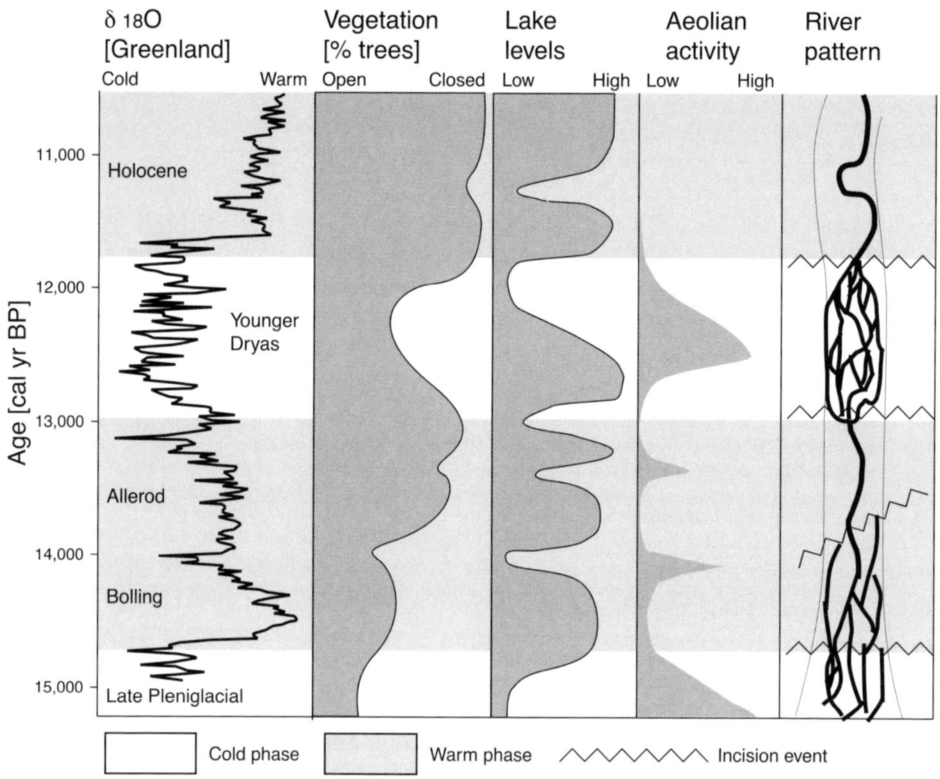

Fig. 1. Reconstruction for the Late Glacial environment in The Netherlands, modified after Hoek (1997) and Huisink (1997).

during warm phases) and single-thread (meandering) channel planform during warm phases. Incision events take place during sharp transitions from warm to cold, as well from cold to warm.

The aim of this study is to present an overview of various modelling approaches that deal with processes that are relevant for climate–sediment supply questions, to make an assessment of these models for geological applications and to give an assessment of the value of process-based modelling to geological applications. We will explore how process-based numerical modelling might extend our understanding of the causal coupling between climate change and (alluvial) river response to these changes. We will do so by critically evaluating a number of modelling approaches, give some examples and discuss the benefits, if any, of these approaches. We will limit ourselves to alluvial rivers within the NW European lowlands during the last few thousand years of the Quaternary, since their evolution is well investigated and therefore provides the control required for any theoretical modelling.

Geomorphological modelling

The simplest model describing the spatio-temporal evolution of river channels is a one-dimensional model in which water and sediment input drive fluvial incision, aggradation (Fig. 2), as described in detail elsewhere (Bogaart & van Balen 2000).

This model (or any comparable one) can provide a better insight in the coupling between rivers as a transport system and rivers as a morphological feature. Attributes such as channel pattern style (meandering, braided) can be calculated on the basis of model state variables such as grain size, stream power and channel gradients (Fig. 3). Channel pattern index is defined as the ratio of actual potential unit stream power to a threshold value (see Bogaart & van Balen 2000 for details)

Given a sedimentological reconstruction of past phases of incision and aggradation, this kind of model can be used to infer the rates of water and sediment discharge that could have produced this response. However, water discharge or sediment supply are not independent in the sense that either

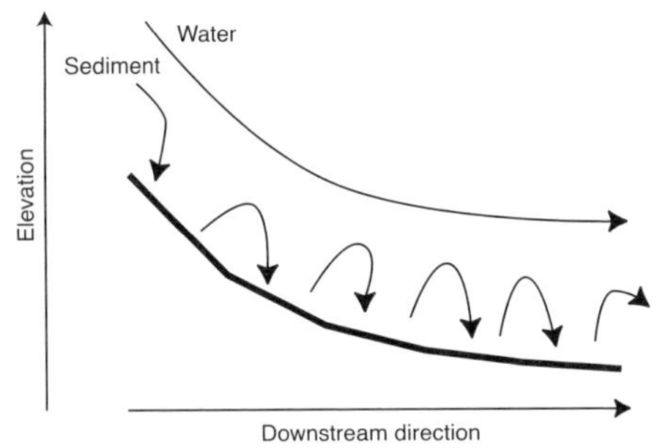

Fig. 2. Cartoon showing the general structure of a one-dimensional fluvial geomorphological model. The arrows indicate the movement of water and sediment through the model's segments.

one is able to control independently the river's fate. Instead, it is the specific combination of water and sediment supply that determines fluvial evolution. The number of resulting degrees of freedom could, in principle, be minimized by using palaeo-hydrological techniques that retrodict (bankfull) channel discharges (e.g. Rotnicki 1983, 1991). Apart from prediction errors that arise from uncertainty in palaeo-roughness reconstruction (Dury 1985), the application of these techniques is, probably, in many cases limited because former channels are often not preserved or are unsuitable for proper analysis, such as non-prototypic channels in-between the braided and meandering domains. For example, in a reconstruction of the Late Glacial evolution of the Maas River in The Netherlands, only a few palaeo-channels of the braided Younger Dryas system were found in terrace remnants (Huisink 1997). The dominant channel(s) from this system are most probably replaced by the subsequent Holocene single-channel system.

Simple numerical experiments using this model also reveal the hidden complexity of a fluvial system. For instance, it can be assumed that, in general, an increase in river discharge is associated with an increase in sediment load, because the same

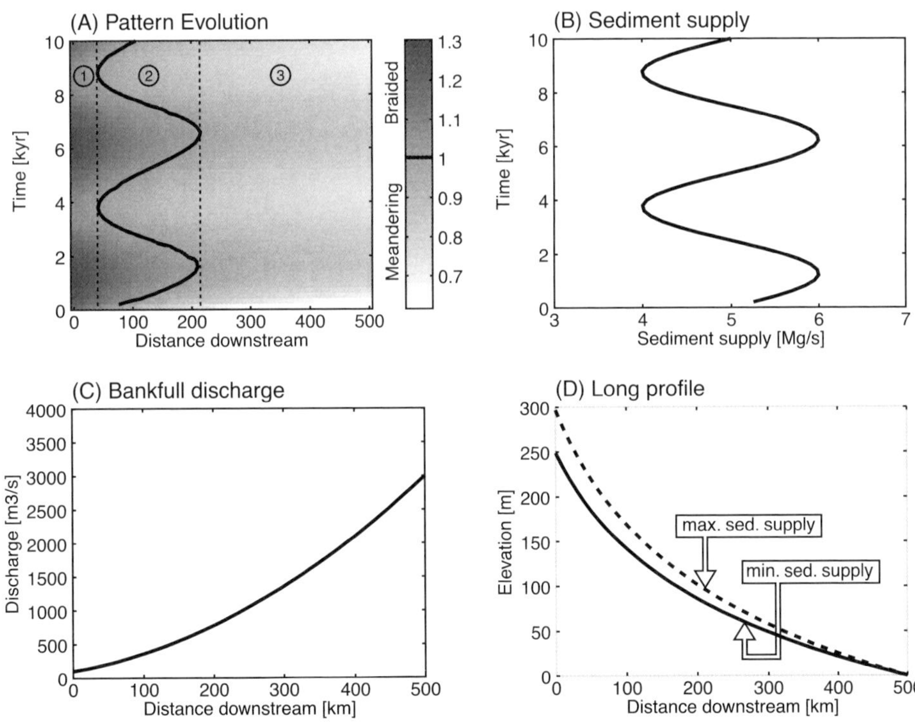

Fig. 3. The response of a channel pattern index (**A**, indicated in grey-scale) to temporal changes in sediment load (**B**), and spatial changes in bankfull discharge (**C**) and sediment load (B), (**D**) shows the longitudinal river profiles under conditions of minimum and maximum sediment supply. From Bogaart & van Balen (2000).

water source (precipitation) is responsible for both, due to runoff generation and soil erosion processes. However, a time lag between increase in discharge and increase in sediment supply might exist due to differences in travel velocity for water and sediment within the channel network. The numerical experiment, shown in Fig. 4, gives two scenarios of changes in sediment and water supply. The fluvial response of the scenario where discharge and sediment supply are in phase is minimal. In this experiment, the sediment load, as well as the water, just flows faster through the channel network. Channel width, however, is more likely to adjust as it has a close relationship to bankfull discharge. Phases of channel widening and narrowing are therefore likely to occur, but – as argued above – do not necessarily leave traces within the sedimentological record.

It should be noted, that this response is an 'extreme' example, as the probability that a change in transporting capacity is exactly compensated for by a change in sediment supply is unlikely to occur. If the increase in sediment supply is smaller than the equivalent increase in discharge, the river starts to incise; if it is larger, the river starts to aggrade. The other extreme scenario, where discharge is in anti-phase with sediment supply, gives the most drastic response, as one might expect. Two scenarios, which are in-between these extremes, are also provided. They show a relatively large response, and thus indicate that an incision–aggradation response is non-linearly related to the time lag between sediment supply and water discharge.

It should be further noted that this example does not consider the valley cross-sectional dimension, that is aggradation occurs by vertical stacking of sediment instead of deposition across the river valley floor. Vertical dimensions are therefore exaggerated with respect to a real-world case.

Given the above considerations, the one-dimensional approach outlined above might be too simplistic for some research problems, especially when one realizes that a direct causal coupling between climate and fluvial processes does not exist at all. For instance, rivers do not 'know' how much precipitation falls on the catchment, only how much water enters the channel network. Figure

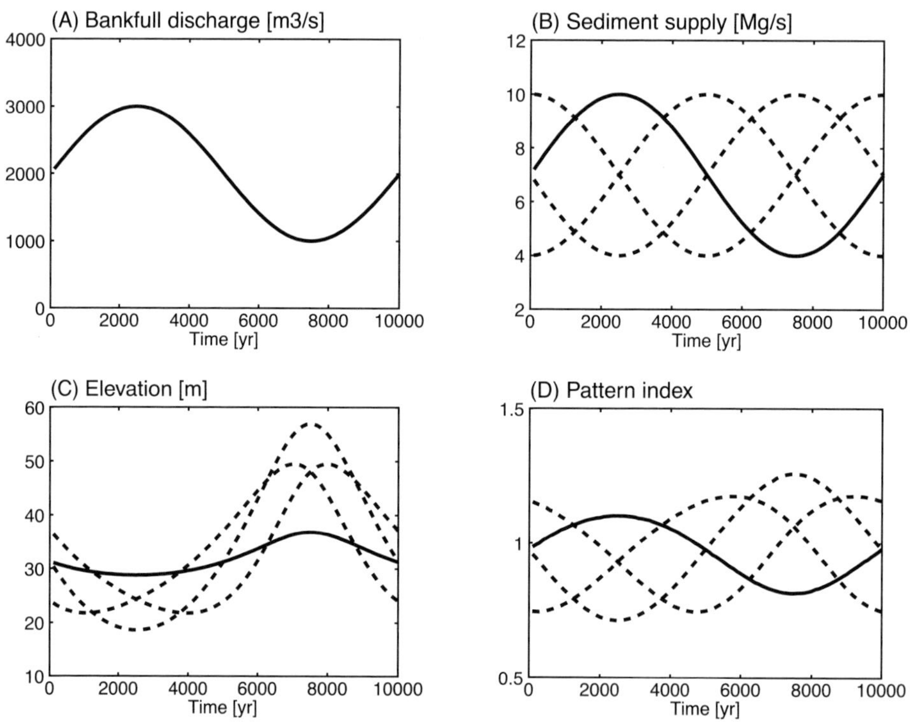

Fig. 4. The response of channel elevation (**C**) and channel pattern index (**D**) to changes in discharge (**A**) and sediment supply (**B**). The solid lines indicate a scenario where discharge and sediment supply are in phase, and the dotted lines where they are out of phase. From Bogaart & van Balen (2000).

5 outlines the proper causal chain, which couples climate with fluvial response and, ultimately, the (fluvial) sedimentological record. In general, this is a two-step process. By means of precipitation and temperature, climate operates on the catchment (hillslope-) morphology, hydrological processes, vegetation and development of a soil profile. These attributes, in turn, regulate the amounts of runoff production and sediment that enter the fluvial system through processes such as infiltration, overland flow, soil erosion, etc.

It is therefore clear that the coupling between fluvial processes or sedimentology and climate change can only be understood fully by explicit consideration of the near-surface catchment dynamics that provide the link between them. Applied to the problem of this paper, this means that it is necessary to introduce process knowledge of catchment hydrology and sediment production under temperate, as well as periglacial, climates in the modelling framework, as these were the dominant climatic modes during the Late Quaternary within NW Europe (Aalbersberg & Litt 1998; Coope *et al.* 1998, Huijzer & Vandenberghe 1998; Isarin & Renssen 1999). The following sections will give outline how these processes can be tackled on millennial, and longer, timescales. Brief reviews of existing approaches will be given, along with some new numerical experiments.

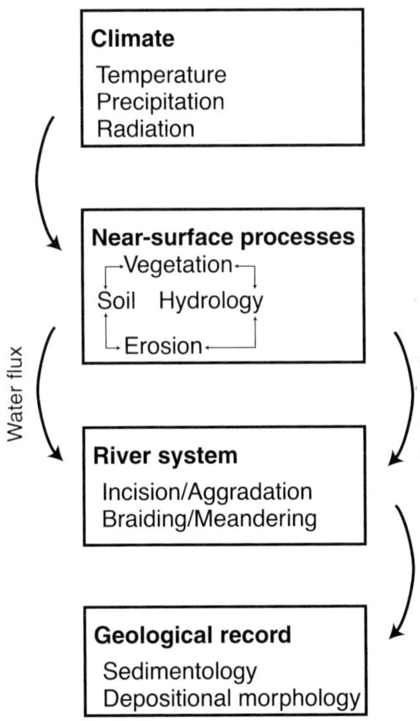

Fig. 5. The causal pathway leading from climate (change) to the fluvial sedimentological record.

Runoff modelling

As discussed in the previous section, the characteristics of the fluvial sedimentological record are determined by two main controls: water and sediment. In this section, we review a number of published modelling approaches that deal with catchment hydrology in general, and river runoff prediction in particular. The main criterion that will be used to assess these models will be their ability to give predictions for palaeo-applications.

In the last few decades a number of highly sophisticated catchment hydrological models have been developed, such as SHE – the European Hydrologic System model (Abbott 1987). The applicability of these models to the larger spatial and temporal scales, which are of particular interest to the earth scientist, is, however, limited, as the amount of data needed to run and calibrate these models is much larger than we are ever able to provide. For instance, high-resolution discharge measurements for the last 20 000 years are not readily available.

On the other extreme of the model spectrum are the purely empirical models. To give just one example, Jones & Fahl (1994) present a set of regionalized flood frequency prediction equations for Alaskan rivers. Stream discharge for a variety of return intervals are calculated from precipitation, basin area, January temperature, etc., using a non-linear multiple regression analysis. However, these specific predictors are not applicable outside the regions for which they were derived due to the lack of physical meaning of the empirical model parameters. It should be noted, however, that empirical models do remain useful, as long as their limits are known and respected. For instance, a general non-linear regression of mean annual flood Q_{af} against mean discharge Q_m yields the predictor $Q_{af} = 27.1\ Q_m^{0.74}$ ($R^2 = 0.78$, based on data collected in van den Berg 1995), which can be used to predict channel forming discharges when no further information about flow regime is available. In reverse, this equation may be used to predict mean discharge from channel forming discharge, as inferred from palaeo-hydrological reconstructions.

The approach that we will discuss here in more detail is simple process-based water balance modelling, where the dynamics of the main storages of the hydrological cycle are simulated using well-established empirical equations. Note that the empirical nature of this kind of model has not disappeared completely. Process-based models can be seen to display a hierarchy, where the top level consists of process descriptions (e.g. soil water infiltration, sediment transport capacity) and lower levels consists of physical based or empirical implementation of these concepts (e.g. a non-linear regression sediment transport model, or the law of gravitation). Empirism cannot be eliminated from models, only pushed towards a lower level.

An example of such a model is the hydrological sub-model of the BIOME3 global vegetation model (Haxeltine & Prentice 1996). This model consists of a feedback between soil hydrology and plant physiology. The soil hydrological model uses daily precipitation data (e.g. linearly interpolated from monthly estimates) and divides this between soil moisture storage and overland flow, depending on the soil texture and current moisture state. Soil moisture can be transported to a lower layer (and eventually discharged as groundwater recharge) and taken up by the vegetation. As vegetation dynamics is the primary concern of the BIOME3 model, the latter is extremely well modelled by a sophisticated plant physiological module, which, although the internal sub-models are fairly complex, requires no other input data than leaf area index, plant functional type, temperature, precipitation and carbon dioxide (CO_2) levels. All of which can be estimated fairly well from palaeo-environmental records such as pollen, fossil leaf stomata counts and oxygen isotopes.

One of the design philosophies that underlies this model is the general applicability that it must have: the model is designed to predict vegetation response to global change. Vegetation structure of the future cannot be measured, and calibration of the model as a whole is thus not possible. All the internal sub-models, however, are extensively calibrated. It is exactly these specifications, detailed process descriptions and low data needs that make the BIOME3 model suitable for application to the past, for example vegetation dynamics and water balance throughout the Quaternary, both of which have a profound effect on sediment production and transport.

An example of how BIOME3 could be used as a hydrological model is shown in Fig. 6, where the Late Glacial warming is simulated in five time slices. Input data for precipitation, temperature and CO_2 levels are taken from General Circulation model (GCM) experiments for these periods (Isarin & Renssen 1999; Renssen pers. comm.), or interpolated from these. Vegetation type and cover are estimated based on palaeo-botanical evidence (Hoek *et al.* 1999). The results show a change from the spring snowmelt-dominated discharge regime during the relatively colder Late Pleniglacial and

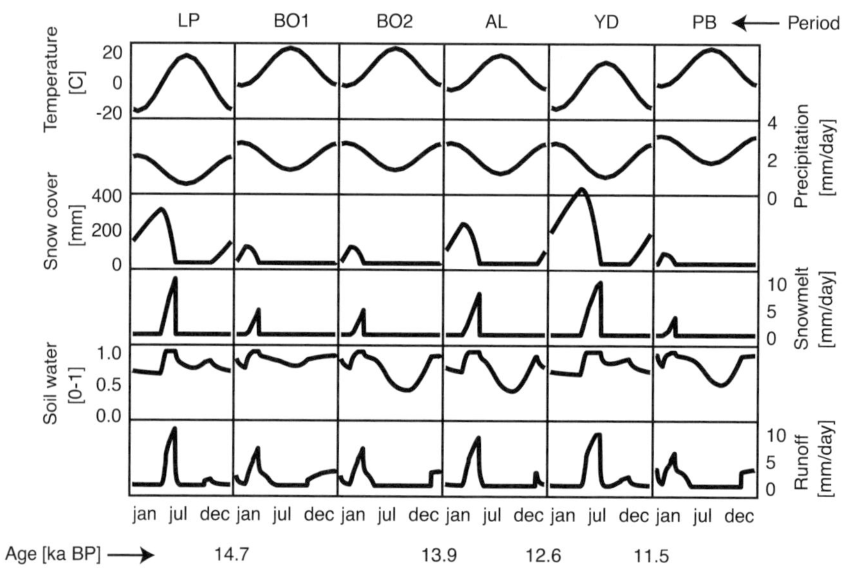

Fig. 6. Diagram showing the results of a generalized hydrological model. Model input is temperature, precipitation, vegetation and soil type. Periods indicated are: LP, Late Pleniglacial; BO, Bølling; AL, Allerød; YD, Younger Dryas; PB, Preboreal. Approximate ages for major boundaries are given according to the $\delta^{18}O$ chronology based on the GRIP ice core (*Walker et al.* 1999).

Younger Dryas towards an increasing contribution of autumn saturation excess overland flow in the warmer Bølling–Allerød and Preboreal periods. The general shapes of these annual hydrographs are seen to correspond with modern periglacial river regimes (Woo 1993).

Another interesting feature is the difference between the two time slices for the Bølling period. All environmental settings are identical except for vegetation cover, which is low in the first time slice and high during the second corresponding with a reconstructed out-of-equilibrium vegetation that slowly increases due to succession limited by nutrient (nitrogen) shortage (van Geel 1996). The resulting annual hydrographs (Fig. 6, bottom row) clearly shows the effect of an increasing vegetation cover on generated runoff, on two scales. On the annual scale, the total amount of runoff decreases due to increasing evapotranspiration during the summer. On the daily to monthly scale, we see that runoff production *magnitude* does not decrease, but runoff production *frequency* does. This also has an effect on long-term average sediment transport capacity, which is defined as the integral of the distribution of flow or transport events (Wolman & Miller 1960).

Care must be taken, however, when scaling this essentially zero-dimensional approach up to a whole catchment, as heterogeneities and channel routing are not taken into account. Still, this runoff-generation model approach, in combination with the fluvial model outlined above, enhances our understanding of the coupling between climate and river hydrology on long (graded and longer) timescales.

Macroscale soil erosion modelling

Models for soil erosion or sediment production on large spatial and temporal scales also typically belong to one of the extremes of the model complexity spectrum. On one extreme of the model spectrum are the event-based, high-resolution, spatially distributed soil erosion models, such as WEPP – the Water Erosion Prediction Project (Flanagan & Nearing 1995). This model provides a high level of detail and requires a massive amount of data and parameters, which have to be calibrated. In a recent comparison of a number of these models it was shown that even when these data are available, the model prediction uncertainty is very large (Jetten *et al.* 1999).

On the other side of the model spectrum we find empirical sediment yield models, which are based on regression analysis of sediment yield and catchment parameters as mean or maximum elevation, basin area and total precipitation (Hovius 1998; Harrison 2000). The main problem with this approach is the lack of sensitivity for possible relevant attributes such as vegetation and/or catchment structure parameters such as elongation ratio, heterogeneity and hypsometric integral. A much more serious limitation might be that these regression equations do not readily allow for the analysis of out-of-equilibrium scenarios where spatial or temporal lags in system adaptation might directly determine the fate of an alluvial river. Their application is therefore limited to longer time spans (i.e. longer than the temporal scale of interest for the issues discussed in this paper).

There are only a few simple process-based models that fill the gap between these two extremes, and that combine applicability on larger scales with low data demands. One of these is the Cumulative Seasonal Erosion Potential model (CSEP) (Kirkby & Cox 1995), which explicitly models the amount of soil erosion based on a distribution of rainfall events and a coupled soil–vegetation model that caters for the strong dependence of soil erodibility on soil organic matter. Leeder *et al.* (1998) describe how the CSEP model can be used to compare contemporary soil erosion in the Mediterranean with that during the Last Glacial Maximum. As with the BIOME model, the numbers generated by this model need to be treated with care and its qualitative value could be more significant than its quantitative value.

Another promising approach is the use of landscape evolution models, which simulate the morphological evolution of catchments over long time spans. (e.g. Howard 1994; Kooi & Beaumont 1994; Tucker & Slingerland 1997). In essence, these spatially distributed models calculate the amount of denudation or deposition on a per-cell basis, as a function of local sediment entrainment, transport capacity and input from upstream. Current model applications are, however, limited to very small catchments (Howard 1997; Tucker & Slingerland 1997) or very large scales (Kooi and Beaumont 1994; Braun & Sambridge 1997), due to numerical complexity.

An even more simple model (Zhang *et al.* 1998) reduces the problem to a simple equation where the amount of erosion is a function of only soil erodibility, overland flow depth, slope gradients and vegetation cover. These parameters are probably the maximum that can be estimated on the timescale of 100–10 000 year. Slope and discharge can be calculated from catchment topography and hydrology, as outlined above; vegetation cover can be estimated on the basis of pollen analysis, and soil erodibility can be estimated on the basis of texture and estimated soil organic matter.

An example of how this model might be used to assess sediment production during a Glacial–Interglacial transition is given in Fig. 7. This is a simple experiment in which the model is applied to a conceptualized representation of the Late Pleniglacial – Bølling (15–13 ka) warming in the Meuse catchment (The Netherlands). We used the same data sources as in the hydrological experiment described above.

Figure 7A–C shows the data input. For each variable, an estimated scenario for the temporal evolution of that variable is used. For vegetation, the most probable scenario is that in which vegetation cover slowly increases from very low during the cold phase, representing a semi-arid tundra-like environment, towards an almost closed (forest) cover during the warm phase, and a considerable period during which vegetation adapts. Soil erodibility is assumed to evolve in correspondence with vegetation cover, since this parameter is to a large extent a function of soil organic matter (Morgan 1995), especially in non-cohesive unconsolidated Quaternary sediment which makes up a large part of the NW European landscape.

Figure 7D–G shows the output of the experiment. Figure 7d shows the calculated amounts of soil erosion over time. The onset of the warm period is characterized by a temporary increase in erosion, followed by a persistent decrease. This is caused by the time lag introduced by the slow response of vegetation to climate. First, overland flow discharges are high, but vegetation cover is too low to protect the soil fully from eroding, while later on vegetation cover increases and overland flow decreases. The response of soil erosion to climate has therefore a clear non-linear character.

The analysis can be further extended to the fluvial network by assuming that (apart from independent factors like base level and tectonics) the sediment supply–discharge ratio is the most prominent control on fluvial sedimentation or incision. If this number is high, rivers will tend to deposit because not all sediment entering the system can be transported out of it. If the number is low, channels will tend to incise as the water flow entering the system carries sediment below its transport capacity and will entrain sediment before leaving the stream segment. Figure 7E shows the temporal evolution of this ratio. This figure has a character equal to Fig. 7D, except that the erosion peak during the start of the warm period is lacking. The reason for this is that the large amount of sediment supply (from soil erosion) is balanced by the higher amounts of river discharge (from overland flow), leaving the ratio unchanged. However, it is unlikely that this model will apply to the drainage network as a whole. If we assume that water and sediment both originate in the upland parts of the catchment, and that sediment and water travel trough the network with different rates, we can investigate the response in more downstream channel segments by introducing a temporal shift between the time series of discharge and sediment supply. Figure 7F shows a scenario where sediment supply significantly lags behind discharge (lag = 250 years). The fluvial response is now much more complex. First we have incision, then aggradation, then again incision and, finally, a new equilibrium. Reconstructed river responses (e.g. Vandenberghe *et al.* 1986; Vandenberghe 1995; Huisink 1997; Tebbens *et al.* 1999; Mol *et al.* 2000) do indeed indicate complex incision–aggradation successions during the climate changes of the Late Quaternary. For example, Vandenberghe (1995), in an integration of previous work, presents a general model in which climate changes of all kinds are characterized by a temporarily period of instability, consisting of an incision–aggradation succession. Our current model offers a simple explanation for this phenomenon, as the result both of slow adjustment of relevant parameters (e.g. vegetation related) to climate change, and the time lags caused by the differences in travel velocity of water and sediment trough the channel network.

Figure 7G shows the results when the time lag between sediment and water movement through the channel network is further increased to 1000 years. It clearly shows that the overall response is characterized by a general incision trend, superimposed on which is a deposition–incision phase, the timing of which is determined by the location within the channel network. The interpretation of these graphs must be treated with caution, however. In these simulations the different travel behaviour of sediment and water is implemented by a simple temporal shift, resulting in a downstream travelling wave of sediment. In reality, however, the diffusional character of sediment transport and fluvial hydraulics will attenuate this wave. The full impact of climate-induced hillslope erosion on alluvial channel behaviour can therefore only be modelled in a quantitative sense by explicitly catering for the channel sediment routing processes, such as travel velocity and temporary sediment storage. The impact of the latter factor to fluvial morphodynamics or sediment throughput rates cannot easily be assessed as this depends, to a large extent, on local conditions such as valley morphology.

Concluding remarks

The models outlined above only describe a few of the relevant catchment processes, such as channel incision–aggradation and runoff generation. In order to model the complete causal pathways that couple the climate and sedimentological record, it is necessary to integrate these models. One of the complications, however, is the mismatch of the

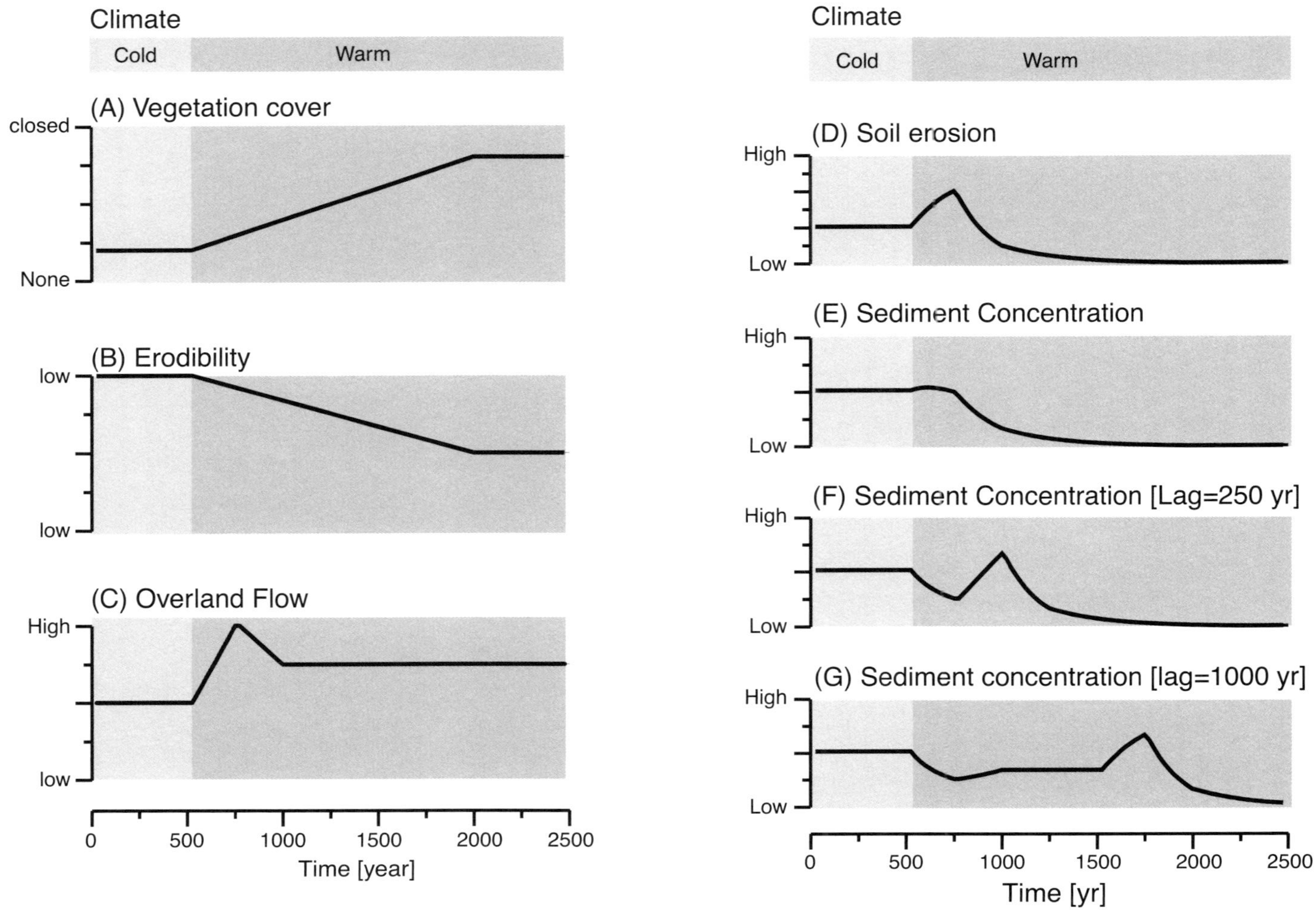

Fig. 7. Diagram showing the effect of climate-related changes in soil erodibility and vegetation cover on soil erosion and stream channel sediment concentration. The timescale serves as a rough indication only.

different model scales. Long-term sediment fluxes through an alluvial channel network is, for example, not a simple function of *a priori* known effective discharge, but of the integration of sediment fluxes over the complete frequency distribution of discharges within a channel that is adapted to mean annual discharge. A potential technique to solve this problem is, for instance, the approach based on Dimensionless Geomorphological Unit Hydrographs (Hebson & Wood 1986). The inputs of this model are catchment morphometrics and a simple distribution of excess rainfall events. The former can be derived from available digital elevation models, and the latter can be parameterized on the basis of the General Circulation model (GCM) output. The model's output is the discharge distribution that we are looking for.

The examples given in this paper were constrained to temperate and periglacial climates. A similar approach can be followed for other climatic settings, such as tropical and (semi)-arid climates. For example, palaeo-estimates of runoff, based on reconstructed monsoon strength, can also be used to drive hydrological and sedimentological models.

Process modelling is a valuable approach in order to understand landscape evolution processes on longer timescales. These larger temporal and areal scales also put severe constraints on feasible modelling techniques. Data availability and model parameter estimation are particular problems. Simple, process-based models form an intermediary between highly sophisticated physically based models (which are data greedy) and regression-based empirical models (which lack physical meaning), and are probably the best class of models for earth science problems.

One of the main concerns within modelling in general is the question of how much detail must be taken into account. For example, river discharge within the channel network is increasing downstream. Within some of the models described above, this increase is continuous, while in reality it is not. Each tributary adds water to the main channel, and thus a step-like model might seem to be a better approximation. Whether or not this is, indeed, the case depends on our scale of interest, and the effect of simplifications on our results. A good reference in this case is Hack's law (e.g. Rigon *et al.* 1996), which states that the length of the main channel L and upstream catchment area A are related by a power law $L = A^{0.59}$. Reversing Hack's law such that L is the independent yields $A = L^{1.69}$. Close inspection of the actual A–L relationship, as derived from a 75-m resolution Digital Elevation model (Fig. 8) reveals that Hack's law is, indeed, a good description of the A–L relation on the scale of a whole basin. Individual

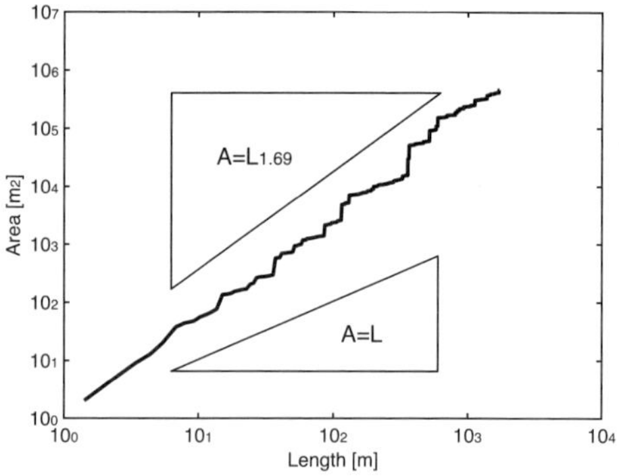

Fig. 8. Relationship between upstream catchment area A and main channel length L, as derived from a 75-m resolution DEM of an area in Luxembourg.

reaches, however, are separated by discrete jumps in A as an effect of tributaries, and are themselves better represented by a more linear relation $A = L$. This implies that a reach-scale model of fluvial dynamics or sediment supply should employ the linear relation $A = L$, while a large (catchment) scale model should employ $A = L^{1.69}$.

A similar issue is the assumption of sediment availability. Here, we made the simplifying assumption that we are dealing solely with alluvial rivers, that sediment is available in all cases and that sediment transport rates consequently are transport-limited rather than supply-limited. This assumption is clearly not applicable to all catchments in all cases. Trivial examples are rapid incising rivers and headwaters in bedrock areas. A less trivial example is that of a river in which the transport capacity changes with climate, and which during periods of low transport capacity (e.g. interglacials without high snowmelt peak discharges) is not able to transport gravels that were deposited during period of higher discharges (e.g. glacials). Such a river will be an alluvial river in the strict sense, but will behave more like a bedrock river. We have not explored these issues within the current paper, but existing concepts and model experiments (e.g. Gasparini *et al.* 1999) may be combined with the concepts discussed here.

The coupling of models for a (potentially large) number of sub-processes, as advocated in this paper, introduces the danger of uncertainty growth beyond acceptable limits. However, it is our opinion that the advantages of using a process-oriented model outweigh the disadvantages. Two important points should be noted here. First, the stacking process might increase uncertainty with respect to the quantitative interpretation of model results. therefore the focus should be on the qualitative interpretation of results. Second, a number of available techniques such as error propagation analysis (Heuvelink & Burrough 1993) might be employed to analyse the growth of uncertainty within the modelling process.

It has been mentioned a few times in this paper that real fluvial systems do not behave as smoothly as numerical models assume. For instance, temporary storage of sediment within a valley might greatly alter the temporal dynamics of sediment fluxes. Although not explored in detail here, some general remarks can be made on how this issue can be dealt with in terms of numerical models. First, it is paramount to realize that processes such as sediment storage and release within valley reaches can be highly contingent in nature. For instance, a wide river valley with steep tributaries offers ample opportunities for sediment storage. In such a geometry, sediment delivered by the tributaries might be deposited in alluvial fans on the valley floodplain. Each of these fans is a potential sediment point source with respect to the main river in that valley. If this river is meandering, then the timing of sediment removal from these fans is a random process regulated by the meander dynamics of the main river. It is clear that sediment input from these fans to the main river system is inherently unpredictable on short timescales (i.e. the timescales on which the meandering process operates). On longer timescales, individual fan-cutting events become less relevant, and the process as a whole might well be described using a 'leaking bucket' model (output is a linear function of storage), after a analogy widely used in hydrological modelling.

In conclusion, we have shown that a process-based numerical modelling approach to sediment production, transport and deposition contributes to the understanding of how climate impacts on sediment supply to basins. The control of climate on this flux is non-linear and depends on numerous sub-processes such as hillslope erosion, catchment hydrology, soil formation and vegetation dynamics. A number of existing models for these processes have been reviewed and discussed. It has been emphasized that the application to millennial timescales and the uncertainty in model parameters on this scale puts constraints on the model choice. It has been argued that simple process-based models form the ideal intermediate between simpler statistical models, that lack general applicability, and pure physical-based models, that require too many parameters.

We recognize that the current trend within the 'Global Change' modelling community is towards these simple models, with physically interpretable parameters and relatively low input data needs. It is these properties that also makes them attractive to Quaternary geologists or earth scientists in general, as they face the same problems. It is therefore likely that geological applications of these models can contribute to their development, given the abundance of geological records against which comparison is possible. We therefore also foresee a closer cooperation between those who study the Earth's past with those who study the Earth's future. Not only is the present the key to the past, but might also the future be the key to the past, and vice versa.

S. Jones, D. Maddy and an anonymous reviewer are gratefully thanked for valuable comments on an earlier version of this paper.

This is a contribution to The Netherlands Environmental Earth System Dynamics Initiative (NEESDI) research program.

References

AALBERSBERG, G. & LITT, T. 1998. Multiproxy climate reconstructions for the Eemian and Early Weichselian. *Journal of Quaternary Science*, **13**, 367–390.

ABBOTT, M. B. 1987. An introduction to the European Hydrological System – Système Hydrologique Européen, 'SHE' – 2: Structure of a physically-based, distributed modelling system. *Journal of Hydrology*, **87**, 61–77.

BOGAART, P. W. & VAN BALEN, R. T. 2000. Numerical modeling of the response of alluvial rivers to Quaternary climate change. *Global and Planetary Change*, **27**, 147–163.

BRAUN, J. & SAMBRIDGE, M. 1997. Modelling landscape evolution on geologic time scales: A new method based on irregular spatial discretization. *Basin Research*, **9**, 27–52.

COOPE, G. R., LEMDAHL, G., LOWE, J. J. & WALKLING, A. 1998. Temperature gradients in northern Europe during the last glacial–Holocene transition (14–9 ^{14}C kyr BP) interpreted from coleopteran assemblages. *Journal of Quaternary Science*, **13**, 419–433.

DURY, G. H. 1985. Attainable standards of accuracy in the retrodiction of palaeodischarge from channel dimensions. *Earth Surface Processes and Landforms*, **10**, 205–213.

FLANAGAN, D. C. & NEARING, M. A. 1995. *USDA – Water Erosion Prediction Project: Hillslope Profile and Watershed Model Documentation*. NSERL Report 10. USDA-ARS-NSERL, West Lafayette, IN.

GASPARINI, N. M., TUCKER, G. E. & BRAS, R. L. 1999. Downstream fining through selective particle sorting in an equilibrium drainage network. *Geology*, **27**, 1079–1082.

HARRISON, C. G. A. 2000. What factors control mechanical erosion rates? *International Journal of Earth Sciences*, **88**, 752–763.

HAXELTINE, A. & PRENTICE, I. C. 1996. BIOME3: An equilibrium terrestrial biosphere model based on ecophysiological constraints, resource availability, and competition among plant functional types. *Global Biogeochemical Cycles*, **10,** 693–709.

HEBSON, C. S. & WOOD, E. F. 1986. A study of scale effects in flood frequency curves. *In:* GUPTA, V. K., RODRÍGUEZ-ITURBE, I. & WOOD, E. F. (eds) *Scale Problems in Hydrology*. Reidel, Dordrecht, 133–158.

HEUVELINK, G. B. M. & BURROUGH, P. A. 1993. Error propagation in cartographic modeling using boolean logic and continuous classification. *International Journal of Geographical Information Systems*, **7**, 231–246.

HOEK, W. Z. 1997. *Palaeogeography of Lateglacial Vegetations; Aspects of Lateglacial and Early Holocene Vegetation, Abiotic Landscape, and Climate in The Netherlands*. KNAG, Utrecht.

HOEK, W. Z., BOHNCKE, S. J. P., GANSSEN, G. M. & MEIJER, T. 1999. Lateglacial environmental changes recorded in calcareous gyttja deposits at Gulickshof, southern Netherlands. *Boreas*, **28**, 416–432.

HOVIUS, N. 1998. *Controls on Sediment Supply by Large Rivers*. SEPM Special Publications, **59**, 3–16.

HOWARD, A. D. 1994. A detachment-limited model of drainage basin evolution. *Water Resources Research*, **30**, 2261–2285.

HOWARD, A. D. 1997. Badland morphology and evolution: interpretation using a simulation model. *Earth Surface Processes and Landforms*, **22**, 211–227.

HUIJZER, B. & VANDENBERGHE, J. 1998. Climatic reconstruction of the Weichselian Pleniglacial in northwestern and central Europe. *Journal of Quaternary Science*, **13**, 391–417.

HUISINK, M. 1997. Late-glacial sedimentological and morphological changes in a lowland river in response to climatic change: The Maas, southern Netherlands. *Journal of Quaternary Science*, **12**, 209–223.

ISARIN, R. F. B. & RENSSEN, H. 1999, Reconstructing and modelling Late Weichselian climates: the Younger Dryas in Europe as a case study. *Earth Science Reviews*, **48**, 1–38.

JETTEN, V., DE ROO, A. & FAVIS-MORTLOCK, D. 1999. Evaluation of field-scale and catchment-scale soil erosion models. *Catena*, **37**, 521–541.

JONES, S. H. & FAHL, C. B. 1994. *Magnitude and Frequency of Floods in Alaska and Conterminous Basins of Canada*. US Geological Survey Water-Resources Investigations Report, **93-4179**.

KASSE, C. 1995. Younger Dryas cooling and fluvial response (Maas River, the Netherlands). *Geologie en Mijnbouw*, **74**, 251–256.

KIRKBY, M. J. & COX, N. J. 1995. A climatic index for soil erosion potential (CSEP) including seasonal and vegetation factors. *Catena*, **25**, 333–352.

KOOI, H. & BEAUMONT, C. 1994. Escarpment evolution on high-elevation rifted margins: Insights derived from a surface process model that combines diffusion, advection and reaction. *Journal of Geophysical Research*, **B99**, 12 191–12 209.

LEEDER, M. R., HARRIS, T. & KIRKBY, M. J. 1998. Sediment supply and climate change: implications for basin stratigraphy. *Basin Research*, **10**, 7–18.

MOL, J., VANDENBERGHE, J. & KASSE, C. 2000. River response to variations of periglacial climate in mid-latitude Europe. *Geomorphology*, **33**, 131–148.

MORGAN, R. P. C. 1995. *Soil Erosion and Conservation*. Second Edition. Longman, Harlow.

RENSSEN, H. & ISARIN, R. F. B. 2001. The two major warming phases of the last deglaciation at ~14.7 and ~11.5 kyr cal BP in Europe: climate reconstructions and AGCM experiments. *Global and Planetary Change*, **30**, 117–153.

RIGON, R., RODRIGUEZ-ITURBE, I., MARITAN, A., GIACOMETTI, A., TARBOTON, D. G. & RINALDO, A. 1996. On Hack's law. *Water Resources Research*, **32**, 3367–3374.

ROTNICKI, K. 1983. Modelling past discharges of meandering rivers. *In:* GREGORY, K. J. (ed.) *Background to Palaeohydrology*. Wiley, Chichester, 321–354.

ROTNICKI, K. 1991. Retrodiction of palaeodischarges of meandering and sinuous alluvial rivers and its palaeohydroclimatic implications. *In:* STARKEL, L., GREGORY, K. J. & THORNES, J. B. (eds) *Temperate Palaeohydrology*, Wiley, Chichester, 431–471.

TEBBENS, L. A., VELDKAMP, A., WESTERHOF, W. & KROONENBERG, S. B. 1999, Fluvial incision and channel downcutting as a response to Late-glacial and Early Holocene climate change: the lower reach of the River Meuse (Maas), The Netherlands. *Journal of Quaternary Science*, **14**, 59–75.

TUCKER, G. E. & SLINGERLAND, R. 1997. Drainage basin response to climate change, *Water Resources Research*, **33**, 2031–2047.

VAN DEN BERG, J. H. 1995. Prediction of alluvial channel pattern of perennial rivers. *Geomorphology*, **12**, 259–279.

VANDENBERGHE, J. 1995. Timescales, climate and river development, *Quaternary Science Reviews*, **14**, 631–638.

VANDENBERGHE, J., BOHNCKE, S., LAMMERS, W. & ZILVERBERG, L. 1986. Geomorphology and palaeoecology of the Mark valley (southern Netherlands): geomorphological valley development during the Weichselian and Holocene. *Boreas*, **19**, 55–67.

VAN GEEL, B. 1996. Factors influencing changing AP/NAP ratios in NW-Europe during the Late-Glacial period. *Il Quaternario*, **9**, 599–604.

WALKER, M. J. C., BJÖRCK, S., LOWE, J. J., CWYNAR, L. C., JOHNSEN, S., KNUDSEN, K.-L., WOLHFARTH, B. & INTIMATE GROUP 1999. Isotopic 'events' in the GRIP ice core: a stratotype for the Late Pleistocene. *Quaternary Science Reviews*, **18**, 1143–1150.

WOLMAN, M. G. & MILLER, J. P. 1960. Magnitude and frequency of forces in geomorphic processes. *Journal of Geology*, **68**, 54–74.

WOO, M. K. 1993. Northern Hydrology. *In:* FRENCH, H. M. & SLAYMAKER, O. (eds) *Canada's Cold Environments*. McGill–Queen's University Press, Montreal.

ZHANG, X., DRAKE, N. & WAINWRIGHT, J. 1998. *Downscaling Land Surface Parameters for Global Soil Erosion Estimation Using no Ancillary Data. Proceedings of the 3rd International Conference on GeoComputation, 17–19 September 1998, University of Bristol, UK*, Geocomputation CD-ROM, ISBN: 0-9533477-0-2.

The flux of siliciclastic sediment from the Iberian Peninsula, with particular reference to the Ebro

GRAHAM EVANS[1*] & ALFREDO ARCHE[2]

[1]*School of Ocean and Earth Science, Southampton Oceanography Centre, European Way, Southampton SO14 3ZH, UK*

**Present address: Cranford, La Route de la Haule, St. Brelade, Jersey JE3 8BA Channel Islands, UK*

[2]*Consejo Superior de Investigaciones Cientificas, Instituto de Geologia Económica, Facultad de Ciencias Geológicas, Universidad Complutense, 28040 Madrid, Spain*

Abstract: Today, the main flux of sediment from the Iberian Peninsula is carried to the Atlantic Ocean by westerly or northwesterly flowing rivers. The only major contribution of sediment from Iberia to the Mediterranean is that carried by the Ebro River, which flows along a foreland basin before cutting through the Catalan Ranges to form a delta in the adjacent Valencian Trough of the Mediterranean. During the Cretaceous and Early Eocene the sediment flux was to the northwest in this part of Iberia. During Late Eocene–Tortonian times it was mainly centripetal, as in other Tertiary Basins in Iberia. Beginning in the Seravallian–Tortonian the main flux of sediment has been to the adjacent Valencian Trough of the Mediterranean. This change in direction, it is proposed, was initiated in Pre-Messinian times by a small river cutting back into the Ebro Basin by headward erosion and was possibly aided, or even induced, by changes in palaeo-slope during contemporaneous rifting along the Mediterranean coast.

The direction of dispersal of siliciclastic sediment on the Iberian Peninsula throughout the latter part of the Mesozoic and Cenozoic and the deposition on the modern coastlines has received considerable attention (Hirst & Nichols 1986; Puigdefábregas & Souquet 1986; Maldonado & Nelson 1988; Friend 1989; Nelson & Maldonado 1990; Jones *et al.* 2001). It has been shown that many of the Tertiary basins of Iberia filled with sediment deposited in mainly fluvial–lacustrine environments during Late Eocene–Miocene times (Fig. 1). In the Ebro Basin, the infilling was succeeded by the development of a new drainage system that flowed southeastwards, breached the Catalan Ranges and developed a delta on the northern margins of the Valencia Trough of the adjacent Mediterranean. The causes and history of the development of the Ebro drainage network are examined and discussed. The Messinian origin of the Ebro drainage system is questioned and an alternative time of its conception, as well as the reason for its direction of flow into the Mediterranean, is proposed.

Drainage pattern of Iberia

Most of the major modern rivers of the Iberian peninsula (the Miño, Duero, Mondego, Tagus, Guadiana and Guadalquivir) flow westwards and carry their sediment to the Atlantic west coast, whilst a series of smaller rivers flow northwards to the Bay of Biscay. A drainage divide between easterly and southeasterly draining rivers and those that flow to the west and north lies close to the east coast of the peninsula; consequently, the rivers that flow to the Mediterranean, with the exception of the Ebro, have short courses (Fig. 2). These small rivers are, however, affected by flash flood events of high magnitude but low frequency, this combined with the small storage areas of their catchments mean that they may be very important sources of coastal sediment despite their size (see Schumm & Hadley 1961 for a general discussion of this point).

The importance of such small streams with violent episodic flows is well illustrated by the records from the Santa Clara River (southern California, USA), which during a single flood in 1969 transported 50×10^6 tonnes (t) of sediment to the sea, and of this 22×10^6 t were transported in 1 day (Curtis *et al.* 1973). Similarly, the Oued Medjerah in Tunisia carried 25×10^6 t of sediment to the sea in a 6-day flood in 1973 (Claude & Loyer 1977). Milliman & Meade (1983) gave some interesting examples of this pattern of behaviour,

From: JONES, S. J. & FROSTICK, L. E. (eds) 2002. *Sediment Flux to Basins: Causes, Controls and Consequences.* Geological Society, London, Special Publications, **191**, 199–208. 0305-8719/02/$15.00

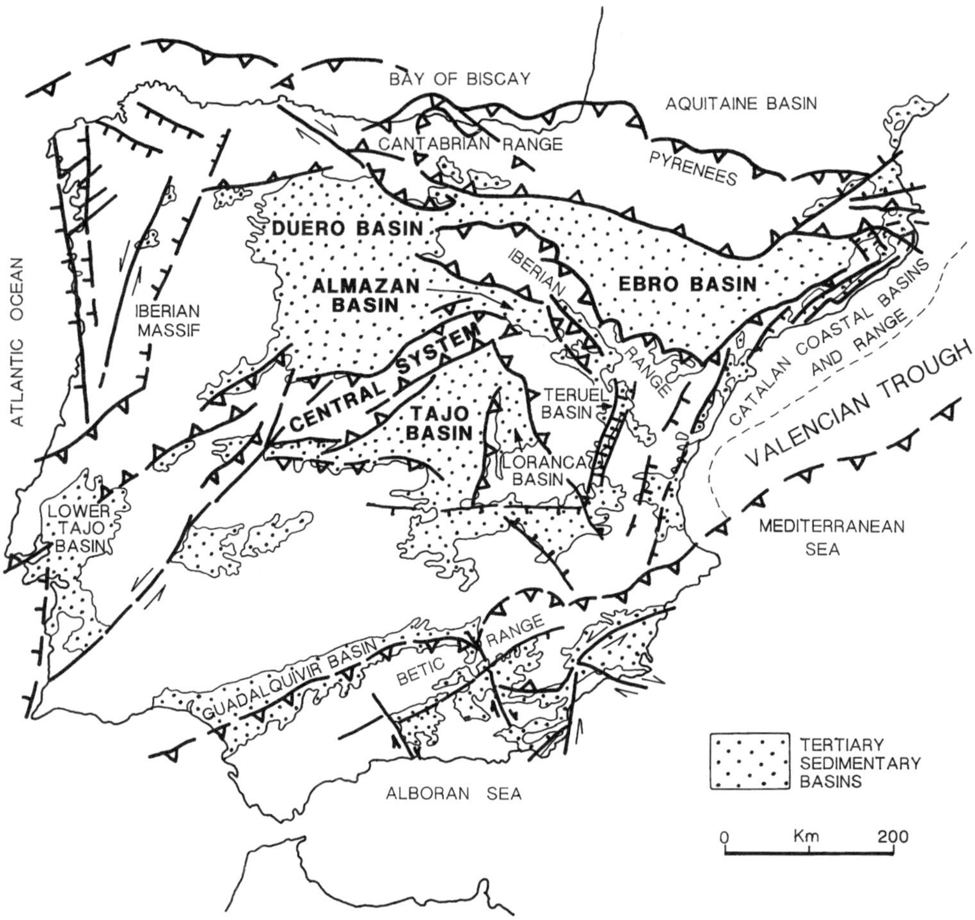

Fig. 1. The tectonic framework of Iberia showing the Tertiary basins and the setting of the Ebro.

and Milliman & Syvitski (1992) suggested that the importance of small mountainous rivers have been underestimated as suppliers of sediment to the coast.

Most of the westerly flowing rivers pass into large estuarine systems where they reach the coast. In contrast, the Guadalquivir has developed as a fluvio-deltaic system in a well-developed foreland basin between the orogenic belt of the Betic Ranges to the south and the relatively stable Iberian massif to the north (Maldonado & Nelson 1988). Similarly, the Ebro flows in a foreland basin between the orogenic belt of the Pyrenees to the north and the Iberian massif to the south in a similar fashion to the Guadalquivir, and has developed a superb example of an A-zone delta (nomenclature of Audley-Charles *et al.* 1977, 1979). This foreland basin setting is one in which many of the major modern deltas, as well as those of the past, have developed (Audley-Charles *et al.* 1977, 1979; Dickinson 1988).

The Ebro is the exception to the general westward flow of the major rivers in Iberia as it is the only major Iberian river that flows southeastwards into the Mediterranean (Fig. 2). It has a drainage area of 85 820 km^2 and its catchment extends westwards from the Mediterranean coast into the foothills of the Cantabrian Mountains. Also, it is unusual in that it dissects deeply into the line of the normal watershed that separates the Atlantic-flowing from the Mediterranean-flowing Iberian rivers.

A curious feature of the Ebro is that before reaching the coast to form its delta, it cuts directly across the Catalan Ranges of northeastern Spain (Fig. 4). The crossing of a mountain range by a major river, that is transverse drainage, is by no means a unique phenomenon, for example in Spain

Fig. 2. The drainage basins of the Iberian Peninsula, from Atlas Nacional de España (Anon 1998).

the Rio Cinca cuts across the whole of the Pyrenees (Jones *et al.* 1999). However, although there are major modern rivers that cut transversely across tectonic belts, for example the Snake–Columbia and Fraser rivers of North America, and the Brahmaputra of India, few major rivers that funnel sediment to the coast to form major deltas behave in this way (Inman & Nordstrom 1971; Audley-Charles *et al.* 1977, 1979; Potter 1978; Dickinson 1988). Where transverse drainage does occur antecedence, superimposition and river capture are usually invoked (see Oberlander 1985 for an interesting discussion of this point).

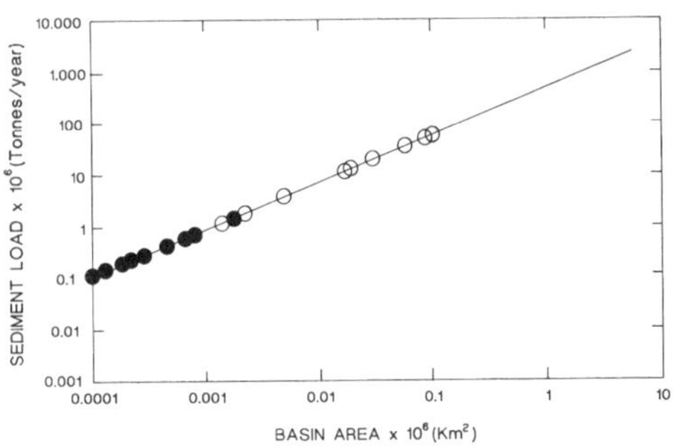

Fig. 3. A plot of catchment area and suspended sediment concentration of Iberian rivers. The solid circles are measured values from Cantabrian rivers (A. Uriarte pers. comm.). The hollow circles (○) have been calculated from the catchment area using the formula of Poulos *et al.* (1996). The distribution occupies the central part of the plot of Milliman & Syvitski (1992).

Catchment area, water and sediment discharge

Today, 4.18% of the total catchment area of the peninsula drains to the north coast (between the Galician–Asturian border and frontier with France), 72.08% to the west coast and 23.74% to the Mediterranean (of the latter, 18.89% is the catchment of the Ebro). On the other hand, 14.93% of the total discharge of the peninsula flows to the north coast, 64.82% to the west coast and 20.25% to the Mediterranean coast, of which 17.28% is carried by the Ebro. Obviously, if it were not for the Ebro there would be very little water reaching the

Table 1. *Percentages of total catchment area, water discharge and potential suspended sediment discharge reaching the coasts of the Iberian Peninsula*

	Catchment area ($\times 10^6$ km^2)	Water discharge ($\times 10^6$ km^3)	Sediment discharge ($\times 10^6$ t)
North coast	4.18	14.93	4.84
West coast	72.08	64.82	74.52
Mediterranean coast	23.74	2.25	20.64
	(of which 18.89 is Ebro)	(of which 17.28 is Ebro)	(of which 17.6 is Ebro)

Mediterranean (2.37% of the total Iberian discharge) (Table. 1).

Unfortunately, there are only limited data available on the sediment load of Iberian rivers. Furthermore, any data are only of limited use as many of the rivers are dammed and, consequently, the recorded sediment discharges are not representative of natural conditions. It has been estimated that the Ebro carried approximately 25 × 10^6 t of sediment per year in 1880; but by 1990 this had reduced to 0.12 × 10^6 t, i.e. 2.5% of its former load (Guillen & Palanques 1992).

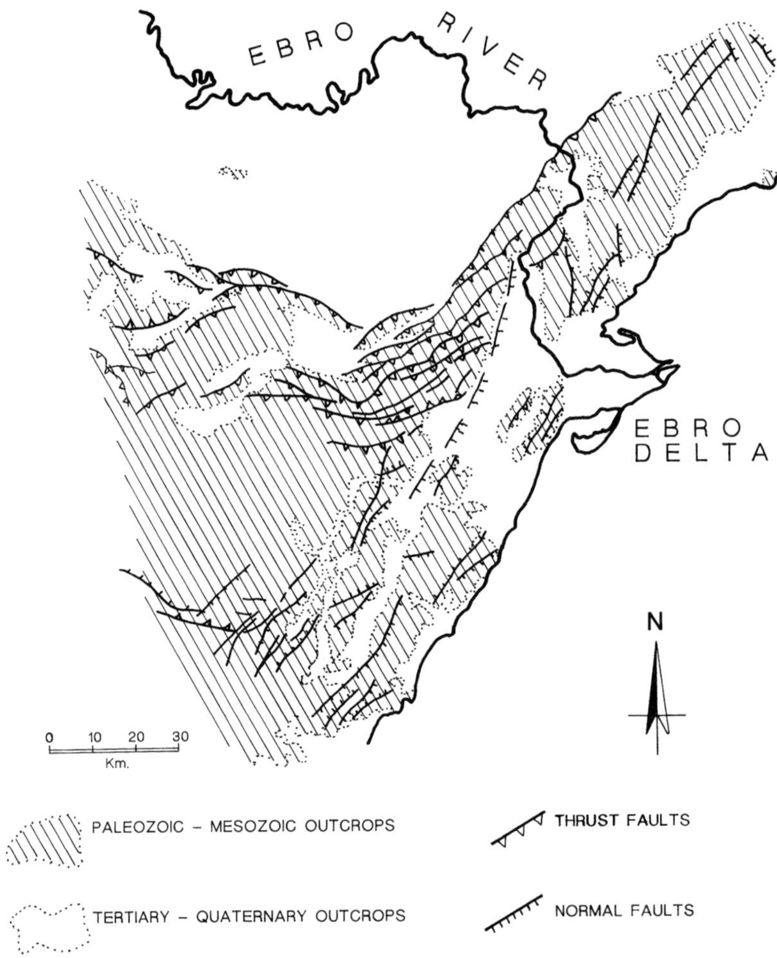

Fig. 4. The course of the Ebro across the Catalan Ranges.

However, there are some interesting compilations of data on the suspended sediment load of world rivers; in particular, small mountain rivers and Mediterranean rivers that drain similar terrains as regards topography, geology and climate as those of the Iberian peninsula (Milliman & Meade 1983; Milliman & Syvitski 1992; Poulos *et al.* 1996; Poulos & Collins 2002). Using the relationship between catchment area and suspended sediment discharge, it is possible to give a first approximation of the relative importance of the various catchments as potential suppliers of the flux of sediment to the seas around the Iberian peninsula (Fig 3). Using the data of these authors it appears, not surprisingly, that the potential sediment discharge to the Cantabrian coast is 4.84%, to the west coast 74.52% and to the Mediterranean coast 20.64%, with 17.6% of the total sediment yield of Iberia being carried by the Ebro (see Table. 2 for the actual values).

Temporal variations of sediment flux from NE Iberia

The polarity of the drainage pattern and sediment flux of the Ebro, as in many other large drainage systems, is a relatively recent geological phenomenon (Friend 1989) but contrasts with other major rivers that appear to have courses of considerable antiquity, such as the Mississippi and Amazon (Potter 1978). Stratigraphic studies by Puigdefábregas & Souquet (1986) revealed that during the Late Cretaceous–Eocene the flux of sediment in the northeast of Spain was to the west and northwest, together with subsidiary flows from the north and south in the south Pyrenean Basin (which lay to the north of the present-day Ebro Basin). This pattern of sediment flux terminated in a series of fluvio-marine deltas and their associated distal offshore deposits in the Jaca and Cantabrian Basins in the northwest.

During the succeeding Oligocene, the depocentre moved to the south on to the site of the present-day Ebro Basin, at this time the main flux of sediment was south from the Pyrenees and westwards from the Catalan Ranges and terminated in continental lacustrine settings (Allen & Mange-Rajetzky 1982). Throughout the Early Miocene alluvial fans with radial drainage developed along the Ebro Basin margins (Hirst & Nichols 1986; Nichols 1987*a*, *b*; Friend 1989; Jones *et al.* 2001) and, although the sediment flux was dominantly radial, it has been pointed out by Friend (1989) that this pattern of centripetal drainage that issued from the main tectonic highlands of the Pyrenees along the modern margin of the Ebro Basin, particularly in the case of Huesca fan system, fed other rivers that flowed to the north-northwest, that is at right angles to the direction of progradation of the small fan system (Friend 1989, fig. 3). What is clear, and of particular interest for the present discussion, is that there is no evidence of any flux of sediment to the east or southeast in the Ebro Basin. Hence, at this time, the flux of sediment differed from that of today. Again, as in the Oligocene, the system terminated in a continental lacustrine setting. The youngest Miocene continental deposits have been dated as Middle Turolian (approximately Middle Miocene); sedimentation ceased at this time and the basin became essentially an exporter of sediment rather than a depocentre (Calvo 1990; Costa *et al.* 1990).

Similar changes occurred in other Iberian Tertiary basins. In the Tajo Basin the uppermost units of evaporites were succeeded abruptly during the Turolian (Middle Tortonian) by fluvial siliciclastics and lacustrine carbonates. This change

Table 2. *Catchment area, discharge and potential suspended sediment load of some Iberian rivers and drainage areas*

	Catchment area ($\times 10^3$ km^2)	Discharge (km^3 year^{-1})	Suspended sediment load ($\times 10^6$ t year^{-1})
Cantabria	19.00	1.26	11.38
Miño	17.76	10.70	10.72
Total Galicia	30.63	24.80	17.32
Duero	98.16	20.81	48.28
Mondego	4.9	2.60	2.47
Tajo	80.95	15.76	40.75
Guadiana	55.21	2.48	29.10
Gualiquivir	57.39	5.17	30.11
Segura	1.5	0.95	1.21
Jucar	2.1	1.6	1.64
Ebro	85.82	19.09	42.90
Llobregat	4.9	0.69	3.45

is interpreted as being due to the onset of tensional tectonics that were coeval with the opening of the basin to the Atlantic (Portero & Olive 1983; Calvo *et al.* 1996). In the neighbouring Duero Basin the fill of alluvial fan, fluvial and lacustrine sediments was tapped by an ancestry Duero during the Late Turolian (Late Tortonian); a change dated using the rich vertebrate faunas found in the Duero terraces (Santiesteban *et al.* 1996).

The Catalan Range, which separates the Ebro Basin from the Mediterranean, is a major NW-directed thick-skinned thrust sheet cut by a system of NE–SW-oriented horsts-and-graben, and forms the southeastern margin of the Ebro Basin (Roca *et al.* 1999). This range developed with an initial compressive phase in the Lower–Upper Oligocene, with sediments accumulating in piggy-back basins in the adjacent offshore, as well as in the Ebro Basin. This compressive phase was followed by an extensional stage with the development of NE–SW grabens that became infilled with alluvial fan and related sediments (Cabrera & Calvet 1996), and, finally, as the tectonic activity gradually subsided in the late Langhian (post-rift), deposition occurred over the horsts (Roca & Deseglaux 1992; Roca *et al.* 1999).

Seaward, the Valencian Trough lying between the Catalan Ranges and the Balearic promontory became a major depocentre with more than 3000 m of Tertiary sediments of Miocene–Quaternary age (Fig.5) (Stoekinger 1976; Batrina *et al.* 1992; Maillard *et al.* 1992). A major unconformity, caused by a fall in the relative sea level during the Messinian, separates the two latter groups (Martinez del Olmo 1996). The basal breccias and overlying muds are succeeded by the dominantly siliciclastic deposits of the Middle–Upper Miocene Castellón Group. The Castellón Sands and Shales were produced by the introduction of a large amount of siliciclastic sediments derived from the adjacent land areas to form a basinward-prograding wedge of deltaic shallow shelf and slope sediments, with an original thickness of more than 1000 m (Fig 5) (Garcia-Sineriz *et al.* 1979; Johns *et al.* 1989; Stampli & Hocker 1989; Martinez del Olmo 1996; Melendez-Hevia & Alvarez del Buergo 1996).

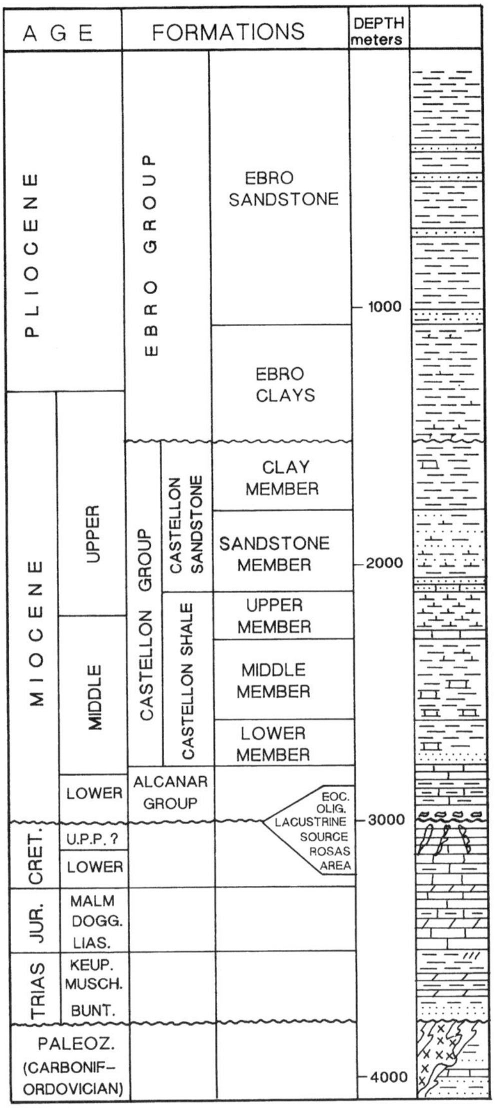

Fig. 5. A generalized stratigraphic section of the Valencian Trough (slightly modified after Melendez-Hevia & Alvarez del Buergo, 1996).

Canyons were cut into these deposits during the lowstand of the Messinian (this is now thought to have been strictly intra-Messinian; Fig. 5; see Melendez-Hevia & Alvarez del Buergo 1996). They were then filled by a succession of Pliocene–Quaternary valley-fill deposits of the Ebro Group (Stampli & Hocker, 1989; Nelson 1990; Nelson & Maldonado 1990). The latter commenced with a transgressive sequence followed by a regressive sequence. Dañobeitia *et al.* (1990), in one of the relatively recent discussions on the Ebro, followed the generally accepted view and suggested that the commencement of flow of the Ebro drainage system to the Mediterranean may have been caused by the drastic lowering of sea level during the Messinian salinity crisis and the latter event obviously must have had a very significant affect on the development of the Ebro drainage network. However, the presence of a thick succession of deposits in the adjacent Valencian Trough of the Mediterranean (>1 km Castellón

Shale and Sandstone Formations) of pre-Messinian age with deltaic, shelf and bathyal characteristics clearly indicate a considerable flux of sediment before the lower base levels of the Mediterranean during the Messinian (Fig. 6). The progradation of siliciclastic sediments, which preceded the Messinian lowering of sea level, is evident from the work of Bartrina *et al.* (1992): 'two progradational terrigenous shelf slope complexes (Latest Seravallian–Tortonian and Pliocene–Pleistocene in age) are separated by the Messinian erosive surface' (the Castellón and Ebro megasequences; Fig. 5). This considerable volume of sediment was presumably supplied by a substantial river (a proto-Ebro?), as a small mountain stream with its catchment limited to the Catalan Ranges is unlikely to have produced such a substantial accumulation. Detailed petrographic studies on these submerged deposits are required to provide conclusive confirmation of this point; but due to the economic sensitivity of the offshore area, as regards petroleum exploration, no data are available. Thus, it appears likely that a proto-Ebro river broke across the Catalan Ranges and developed a considerable drainage basin some time in the Middle Miocene, that is by pre-Messinian times. The down-cutting resulting from the lowered Messinian sea level led to further incision and, presumably, headward erosion inland into the Ebro Basin to tap the extensive volumes of sediment of the previously land-locked Oligocene–Miocene basin of deposition.

During Pliocene–Recent times, the Ebro Basin, as have the Duero and Tajo Tertiary Basins, has been an area from which sediment has been exported and where few deposits accumulated, except for fluvial terrace and floodplain deposits of the Ebro River and its tributaries (Riba *et al.* 1987). Instead, sediment has been transported to the Mediterranean coast to form the fluvio-marine deltaic complex of the modern Ebro delta (Maldonado 1972; Nelson 1990; Nelson & Maldonado 1990). However, anthropogenic changes, particularly the construction of dams, has greatly decreased sediment supply in recent times and has resulted in erosion of the delta (Palanques 1987; Palanques *et al.* 1990).

Discussion

Why did these changes in sediment flux occur? Was it, as stated or implied in earlier papers, merely the result of headward erosion during the Messinian of a small river flowing into the Mediterranean tapping the thick Cenozoic fill of the Ebro Basin? Such an event would have been encouraged by the low base levels during Messinian times. However, a point not stressed in earlier papers on the Ebro was that in pre-Messinian (Seravallian–Tortonian) times a southeastwards flux of siliciclastic sediment had already started to flow into the Valencian Trough in the northeastern Mediterranean. This southeastern flux of sediment was enhanced by, but not initiated by, the Messinian event.

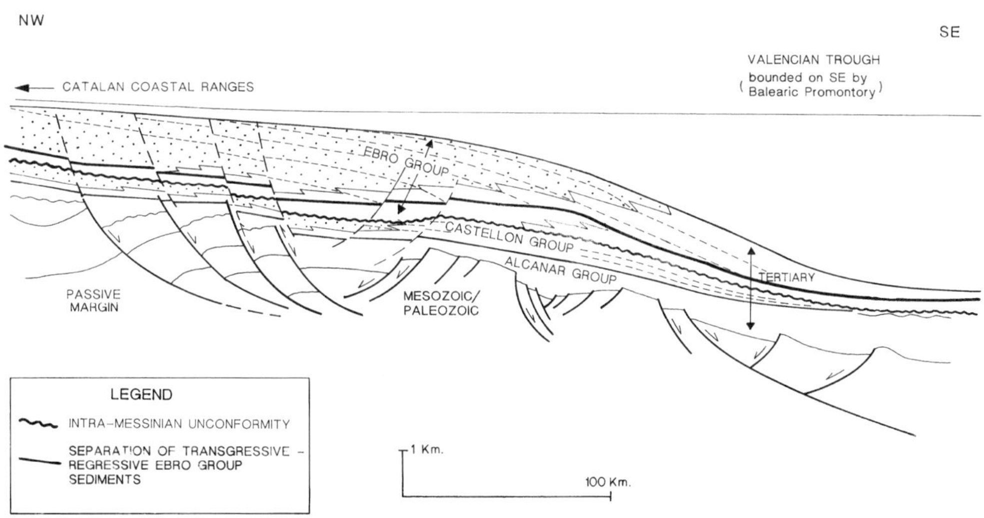

Fig. 6. A schematic cross-section of the major Tertiary sedimentary groups and their lateral and vertical extent. The pre-Messinian Castellón Sands and Shale clastic wedge forms a prograding unit in the middle part (after ESCAL UGS unpublished).

The release of the huge volume of sediment that had accumulated during Late Eocene–Tortonian times was probably initiated by a river (proto-Ebro?) cutting across and exploiting aligned depressions produced by the NE–SW faults that cut the Catalan Range. In addition, it seems possible that such an initiation of southeasterly flux of sediment may have been due to, or was aided by, tectonic events. Tectonic changes such as faulting, folding and tilting have resulted in significant changes in drainage pattern, including drainage reversals, dissection, development of lakes and river capture (Russell 1958; Harvey & Wells 1987; Bailey *et al.* 1993; Jones *et al.* 1999; Leeder 1993; Gawthorpe & Leeder 2000; Peakall *et al.* 2000). In some cases, changes of flow direction, as deduced from paleo-current studies, are assumed to have been due to such tectonic events. A change in direction of flux of sediment occurred in another foreland basin – the Swiss Molasse Basin (Homewood *et al.* 1986). There, the lower freshwater molasse was deposited on fans that led into a longitudinal drainage system, flowing from southwest to northeast, and was succeeded by the upper freshwater molasse deposited on fans that led to a longitudinal drainage system flowing from northeast to southwest. There is no clear explanation for such a reversal, but it is assumed that it was in some way of tectonic origin. In addition, it is of interest to note that successive transgressions of the sea in Cretaceous times came from opposing directions in the northeast of Iberian, indicating possible earlier reversals in gradient of the regional palaeo-slope (Alonso *et al.* 1992).

Hence, the idea seems worthy of consideration that the change of flux sediment in the Ebro Basin in Seravallian–Tortonian times was perhaps caused or perhaps aided by a change of palaeo-slope of the Ebro foreland basin. Also, of interest as regards the latter suggestion, is that during the Late Eocene and throughout the Oligocene times a change from siliciclastic-dominated to carbonate-dominated lacustrine sedimentation in the eastern Ebro Basin was accompanied by a displacement of the latter southwards towards the Catalan Ranges (Allen & Mange-Rajetzky 1982).

The change from siliciclastic-dominated lacustrine sedimentation to carbonate–evaporite-dominated lacustrine sedimentation makes it unlikely that the southeasterly flux of sediment was due to overfilling of the Ebro Basin and consequent overspill into the adjacent Valencian Trough of the Mediterranean. In any case, during Seravallian–Tortonian times the Catalan Ranges were a formidable barrier that had been created at the end of the Eocene and had estimated heights of 1200–2000 m (Anadon *et al.* 1986; Lopez-Blanco *et al.* 2000).

Conclusions

Today, the major flux of sediment from the Iberian Peninsula is carried westwards towards the Atlantic coast by a series of large westerly flowing rivers and some small rivers that flow into the Bay of Biscay. In contrast, much of the erosional products of northeastern Iberia are carried to the Mediterranean by the River Ebro.

During the Late Cretaceous–Eocene the flux of sediment in the northeastern part of the peninsula was directed towards the northwest, where it accumulated in various fluvio-marine and marine basinal environments. Tectonic movements between the Iberian and European plates led to the closure of the westerly outlet and the Ebro Basin became an area of centripetal drainage, where more than 4000 m of continental sediments (fluviatile and lacustrine) accumulated. In Seravallian–Tortonian times an outlet developed across the Catalan Ranges in the southeast, and a southeasterly flux of sediment produced a wedge of siliciclastic deposits (the Castellón Sands and Shale) into the adjacent Valencian Trough. This breach in the enclosing Catalan Ranges marked the commencement of the present-day Ebro drainage network. Low base levels during the succeeding Messinian salinity crisis led to entrenchment and extensive development of the network to produce a second large influx of siliciclastic sediments – the Pliocene–Recent Ebro Sands and Clays, as well as the present-day Ebro delta on the northern flanks of the Valencian Trough. This southeasterly flux of sediment may have been enhanced or possibly partially caused by longitudinal tilting of the Ebro Basin.

The authors wish to record their thanks to J. Lopez, P. Friend, G. Nichols and, especially, A. Mather, A. Hartley and S. Jones for helpful comments which improved the text; to B. Bartolamé and T. Curtis for typing various versions of the manuscript and C. Sánchez for drawing the figures. G. Evans wishes to record his thanks to the Departmento de Estratigrafia Universidad Complutense for their generous hospitality (1991–1992) when the paper was first conceived. Also, the authors wish to thank GESSAL and ESCAL UGS for permission to reproduce Fig. 6, C. Amos for calculating and plotting the river data, and A. Uriarte for the data on Cantabrian rivers.

References

ALLEN, P. & MANGE-RAJETZKY, M. 1982. Sediment dispersal and palaeohydraulics of Oligocene rivers in the eastern Ebro Basin. *Sedimentology,* **29**, 705–716.

ANADON, P., CABRERA, L., COLOMBO, F., MARZO, M. & RIBA, O. 1986. Syntectonic unconformities in alluvial fan deposits, eastern Ebro Basin, NE Spain. *In*: ALLEN, P. A. & HOMEWOOD, P. (eds) *Foreland*

Basins. International Association of Sedimentologists, Special Publications, **8**, 259–271.

ALONSO, A., FLOQUET, M., MAS, R. & MELÉNDEZ, A. 1992. *Late Cretaceous Carbonate Platforms: origin and evolution, Iberian Range, Spain*. American Association of Petroleum Geologists Memoirs, **56**, 297–313.

ANON. 1998. *Atlas Nacional de España,* Vol. I. Ministerio de Obras Públicas, Transportes y Medio Ambiente, España.

AUDLEY-CHARLES, M. G., CURRAY, J. R. & EVANS, G. 1977. Location of major deltas. *Geology*, **5**, 341–344.

AUDLEY-CHARLES, M. G., CURRAY, J. R. & EVANS, G. 1979. Significance and origin of big rivers: a discussion. *Geology*, **87**, 122–123.

BAILEY, G. N., KING, G. C. P. & STUDY, D. A. 1993. Active tectonics and land use strategies: a Palaeolithic example from northwest Greece. *Antiquity,* 67, 292–312.

BARTRINA, M. T., CABRERA, L., JURADO, M. J., GUIMERA, J. & ROCA, E. 1992. Evolution of the central Catalan margin of the Valencia Trough (Western Mediterranean). *Tectonophysics,* **203**, 219–247.

CABRERA, L. & CALVET, F. 1996. Onshore Neogene record in NE Spain: Vallés Penedes and El Camp half grabens. *In:* FRIEND, P. F. & DABRIO, C. J. (eds) *Tertiary Basins of Spain.* Cambridge University Press, Cambridge, 97–105.

CALVO, J. P. 1990. Up-to-date Spanish continental Neogene synthesis and palaeoclimate interpretation. *Revista de la Sociedad Geologica de España*, **6**, 29–40.

CALVO, J. P., ALONSO, A., GARCIA, A. & ORDOÑEZ, S. 1996. Sedimentary evolution of lake systems through the Miocene of the Madrid Basin. *In:* FRIEND, P. F. & DABRIO, C. J. (eds) *Tertiary Basins of Spain*. Cambridge University Press, Cambridge, 272–277.

CLAUDE, J. & LOYER, J. Y. 1977. Les alluvions deposees par l'Oued Medjerdah lors la crue exceptionelle de Mars 1973 en Tunisie: aspects quantatif et qualitatif du transport et du depot. *International Association of Hydrological Science*, **122**, 211–218.

COSTA, J. M., RAMIREZ, J. I. & SALAZAR, A. 1990. *Hoja 356, Lanaja, Mapa*. Geologico de España 1 : 50 000. IGME, Madrid.

CURTIS, W. F., CULBERTSON, J. K. & CHASE, E. B. 1973. Fluvial sediment discharge to the oceans from the coterminous United States. *US Geological Survey Circular,* **670**.

DAÑOBEITA, J. J., ALONSO, B. & MALDONADO, A. 1990. Geological framework of the Ebro continental margin and surrounding areas. *Marine Geology*, **95**, 265–287.

DICKINSON, W. R. 1988. Provenance and sediment dispersal in relation to palaeotectonics and palaeogeography of sedimentary basins. *In*: KLEINSPEHN, K. L. & PAOLA, C. (eds) *New Perspectives in Basin Analysis*. Springerg, New York, 3–25.

FRIEND, P. F. 1989. Space and time analysis of river systems, illustrated by Miocene systems of the Northern Ebro Basin in Aragon, Spain. *Revista Sociedad Geológica de España*, **2**, 55–64.

GARCIA-SINERIZ, B., QUEROL, R., CASTILLO, F. & FERNANDEZ, J. R. 1979. A new hydrocarbon province in the Western Mediterranean. *In*: *World Petroleum Congress, Bucharest*, 10th World Petroleum Congress Bucharest (Roumania) Heydon, London, 191–197.

GAWTHORPE, G. L. & LEEDER, M. R. 2000. Tectono-sedimentary evolution of active extensional basins. *Basin Research*, **12**, 195–218.

GUILLEN, J. & PALANQUES, A. 1992. Sediment dynamics and hydrodynamics of a river highly regulated by dams: The Ebro river. *Sedimentology*, **39**, 567–579.

HARVEY, A. M. & WELLS, S. G. 1987. Response of Quaternary fluvial systems to differential epeirogenic uplift: Aguas Feos river systems, southeast Spain. *Geology,* **15**, 689–693.

HIRST, J. P. P. & NICHOLS, G. J. 1986. Thrust tectonic controls on Miocene alluvial distribution patterns, southern Pyrenees. *In*: ALLEN, P. A. & HOMEWOOD, P. (eds) *Foreland Basins*. International Association of Sedimentologists Special Publications, **8**, 247–258.

HOMEWOOD, P., ALLEN, P. A. & WILLIAMS, G. D. 1986. Dynamics of the Molasse Basin of Western Switzerland. *In:* ALLEN, P. A. & HOMEWOOD, P. (eds) *Foreland Basins*. International Association of Sedimentologists Special Publications, **8**, 199–218.

INMAN, D. L. & NORDSTROM, C. E. 1971. On the tectonic and morphologic classification of coasts. *Journal of Geology*, **79**, 1–21.

JOHNS, D. R., HERBER, M. A. & SCHWANDER, M. M. 1989. Depositional sequences in the Castellón area, offshore northeast Spain. *In*: BALLY, A. W. (ed.) *Atlas of Seismic Stratigraphy.* American Association Petroleum Geology Studies, **27**, 181–184.

JONES, S. J., FROSTICK L. E. & ASTIN, T. R. 1999. Climatic and tectonic controls on fluvial incision and aggradation in the Spanish Pyrenees. *Journal of the Geological Society, London*, **155**, 761–769.

JONES, S. J., FROSTICK L. E. & ASTIN T. R. 2001. Braided stream and flood plain architecture: The River Vero Formation, Spanish Pyrenees. *Sedimentary Geology,* **139**, 229–260.

LEEDER, M. R. 1993. Tectonic controls upon drainage basin development, river channel migration and alluvial architecture: Implications for hydrocarbon resevoir development and characterization. *In*: NORTH, C. P. & PROSSER, D. J. (eds) *Characterization of Fluvial and Aeolian Resevoirs.* Geological Society, London, Special Publications, **73**, 7–22.

LOPEZ-BLANCO, M., MARZO, M., BURBANK, D., VERGES, J., ROCA, E., ANADON, P. & PIÑA, J. 2000. Tectonic and climate controls on the development of foreland fan deltas: Montserrat and Sant Llorenc del Munts. Systems, Middle Eocene, Ebro Basin, Spain. *Sedimentary Geology*, **138**, 17–39.

MAILLARD, A., MAUFFRET, A., WATTS, A. B., TORNÉ, M., PASCAL, G., BUHL, P. & PINET B. 1992. Tertiary sedimentary history and structure of the Valencia

Trough (western Mediterranean). *Tectonophysics*, **203**, 57–75.

MALDONADO, A. 1972. El delta del Ebro. Estudio sedimentológico y estratigráfico. *Boletin Estratagrafía Universidade Barcelona,* **1**, 1–486.

MALDONADO, A. & NELSON, C. M. 1988. Dos ejemplos de márgenes continentales de la Peninsula Ibérica: el margen del Ebro y el Golfo de Cadiz. *Revista Sociedad Geológica de España,* **1**, 317–325.

MARTINEZ DEL OLMO, W. 1996. Depositional sequences in the Gulf of Valencia Tertiary basin. *In*: FRIEND, P. F. & DABRIO, C. J. (eds) *Tertiary Basins of Spain.* Cambridge University Press, Cambridge, 55–67.

MELENDEZ-HEVIA, F. & ALVAREZ DEL BUERGO, E. 1996. Oil and gas resources of the Teritary basins of Spain, *In*: FRIEND, P. F. & DABRIO, C. J. (eds) *Tertiary Basins of Spain.* Cambridge University Press, Cambridge, 20–25.

MILLIMAN, J. D. & MEADE, R. H. 1983. World wide delivery of river sediment to the oceans. *Journal of Geology*, **91**, 1–21.

MILLIMAN, J. D. & SYVITSKI, J. P. M. 1992. Geomorphic/tectonic control of sediment discharge to the ocean: the importance of small mountainous rivers. *Journal of Geology,* **100**, 525–544.

NELSON, C. H. 1990. Estimated post-Messinian sediment supply and sedimentation rates on the Ebro continental margin. *Marine Geology*, **95**, 395–418.

NELSON, C. H. & MALDONADO, A. 1990. Factors controlling Late Cenozoic continental margin growth from the Ebro delta to the Western Mediterranean deep sea. *Marine Geology*, **95**, 415–440.

NICHOLS, G. J. 1987*a*. Structural controls on fluvial distributor systems – the Luna System, Northern Spain. *In:* ETHRIDGE, F. G., FLORES, R. M. & HARVEY, M. D. (eds) *Recent Developments in Fluvial Sedimentology.* SEPM Special Publications, **39**, 269–277.

NICHOLS, G. J. 1987*b*. Syntectonic alluvial fan sedimentation southern Pyrenees. *Geological Magazine*, **124**, 121–133.

OBERLANDER, T. M. 1985. Origin of drainage transverse to structures in orogens. *In*: MORISAWA, M. & HACK, J. T. (eds) *Tectonic Geomorphology.* Unwin Hyman, Boston, 155–182.

PALANQUES, A. 1987. Determinación y dinámica de los metales pesados asociados a la materia particulada en suspension y los sedimentos de fondo del margen continental del Ebro. *Geoquimica España.* **II**, 112–116.

PALANQUES, A., PLANA, F. & MALDONADO, A. 1990. Recent influence of man on Ebro margin sedimentation system (northwestern Mediterranean sea). *In:* NELSON, C. H. & MALDONADO, A. (eds) *The Ebro Margin. Marine Geology*, **95**, 247–263.

PEAKALL, J., LEEDER, M., BEST, J. & ASHWORTH, P. 2000. River response to lateral ground tilting: A synthesis and some implications for modelling of alluvial architecture in extensional basins. *Basin Research*, **12**, 413–424.

PORTERO, J. M. & OLIVE, A. 1983. El Terciario del borde meridional del Guadarrana. *In*: COMBA, J. M. (ed.) *Geologia de España.* IGME, Madrid, 527–543.

POTTER, P. E. 1978. Significance and origin of big rivers. *Journal of Geology,* **86**, 13–33.

POULOS, P. E. & COLLINS, M. B. 2002. Fluviatile sediment fluxes to the Mediterranean Sea: a quantitative approach and the influence of dams. *In*: JONES, S. J. & FROSTICK, L. E. (eds) Sediment Flux to Basins: causes, controls and consequences. The Geological Society, London, Special Publications, **191**, 227–246.

POULOS, S. E., COLLINS, M. & EVANS, G. 1996. Water sediment fluxes of Greek rivers, south-eastern Alpine Europe: Annual yields, seasonal variability, delta formation and human impact. *Zeischrift für Geomorphologie*, *NF*, **40**, 243–261.

PUIGDEFÁBREGAS, C. & SOUQUET, P. 1986. Tecto-synsedimentary cycles and depositional sequences of the Mesozoic and Tertiary from the Pyrenees. *Tectonophysics,* **129**, 173–203.

RIBA, O., REGUANT, S. & VILLENA, J. 1987. Ensayo de sintesis estratigráfica y evolution de La Cuenca Terciaria del Ebro. *In*: LIBRO JUBILAR J. M. RIOS, *Geologia de España.* Instituto Geológico y Minero de España, Madrid, **2**, 131–159.

ROCA, E. & DESEGLAUX, P. 1992. Analysis of the geological evolution and vertical movements in the Valencia Trough area, W. Mediterranean. *Marine Petroleum Geology*, **9**, 167–185.

ROCA, E., SAMS, N., CABRERA, L. & MARZO, M. 1999. Oligocene to middle Miocene evolution of the central Catalan margin (NW Mediterranean). *Tectonophysics*, **315**, 209–233.

RUSSELL, R. J. 1958. Geological geomorphology. *Geological Society of America Bulletin,* **69**, 1–22.

SANTIESTEBAN, J. I., MEDIAVILLA, R., MARTIN, A. & DABRIO, C. J. 1996. The Duero Basin: a general overview. *In*: FRIEND, P. F. & DABRIO, C. J. (eds) *Tertiary Basins of Spain.* Cambridge University Press, Cambridge, 183–187.

SCHUMM, S. A. & HADLEY, R. F. 1961. Progress in application of landform analysis to studies of semi-arid erosion. *US Geological Survey Circular*, **437**.

STAMPLI, G. M. & HOCKER, C. F. W. 1989. Messinian palaeorelief from a 3D seismic survey in the Tarraco concession area (Spanish Mediterranean Sea). *Geologie en Mijnbouw*, **8**, 201–210.

STOEKINGER, W. 1976. Valencia Gulf offer deadline nears. *Oil and Gas Journal*, **29**, 197–204.

The history of sediment flux to Atchafalaya Bay, Louisiana

JOHN McMANUS

School of Geography and Geosciences, University of St Andrews, St Andrews, Fife KY 16 9 AL, UK (e-mail: jm@st-andrews.ac.uk)

Abstract: The Atchafalaya became a free-flowing river system in the mid-nineteenth century. It progressively increased its diversion of waters from the Mississippi until, by the mid-twentieth century, it threatened to capture the total flow of the main river. The installation of control structures has, temporarily at least, prevented total diversion, limiting the flows to 30% of the Mississippi; water discharge and sediment transport into the Atchafalaya have led to deposition throughout the basin. A major system of lakes and bayous has become largely filled with deltaic sands and layered silts, so that the waters currently drain almost directly into the Atchafalaya Bay. The natural outlet and one artificially created outlet each carry waters from the system. As a result of major flooding in 1973, sediment was flushed from the Atchafalaya Basin and began the sub-aerial emergence of deltas, which had been slowly building below the bay water surface since the 1950s. The two deltas have provided, parallel growth one largely in natural form, and the other substantially modified by the dredging needed to maintain a navigable routeway to Morgan City and the interior of the basin. The sub-aerial deltas initially grew rapidly, a process apparently slowing, but the sediment transported to the depositional areas has not diminished, rather the area of deposition within the receptor basin is increasing, so that apparent growth inevitably slows. By 1994 the two deltas occupied an area of more than 153 km^2 above the –0.6 m measuring datum, created in approximately 30 years. In each case the sediments show upward passage from fine clays to laminated silts and fine sands, within a succession rarely more that 3 m in thickness, as defined by the original 1890 bathymetry of the Bay. Within the Bay itself resuspension of deposited fine sediment due to wave activity and water set-up and set-down in response to variations in the directions of coastal winds creates plumes of material that travel through the basin and may leave to form suspended sediment plumes on the inner parts of the continental shelf. A total of 20–25 such plumes annually drift initially east of the Bay outlet but later move west towards the chenier coasts of western Louisiana and Texas.

On the northern shore of the Gulf of Mexico, between the chenier coast of Texas–west Louisiana and the Mississippi Delta, is a 120-km long embayment extending from Vermilion Bay in the west to Atchafalaya and Four League Bays in the east (Fig. 1). It is partly separated from the Gulf by the presence of Marsh Island in the west, and the Pointe au Fer Island and former shell reefs in the east. The waters of the Atchafalaya have two outlets into the bay, one from the natural river passing Morgan City and, the other, the Wax Lake Outlet, artificially constructed past the town of Calumet in 1943. Recent sedimentation in the Atchafalaya Bay area is closely linked to the history of changing flows within the Atchafalaya system during the last two centuries.

The history of building and abandonment of successive lobes of the Mississippi Delta during the Holocene has been documented by many authors, following the important pioneering study of Fisk (1952). To the southwest of the present course of the Mississippi in Louisiana lies a long, broad lowland, formerly occupied by at least three early courses of the Mississippi River as they formed the Maringouin or Sale Cypremort (6000–4500 BP), the Teche (3900–2700 BP) and the Lafourche (1900–800 BP) lobes during successive stages of delta construction (Fig. 2, after Kolb & van Lopik 1958). During both Teche and Lafourche times the Red River followed a course to the west of the present Atchafalaya, with a series of outlets to the Gulf along the courses of the Vermilion and Lower Atchafalaya. When the Mississippi underwent avulsion the abandonment of the lower part of its former courses left as remnants the modern Bayous Maringouin, Teche and Lafourche as misfit streams (Fig. 3). The limits of the basins of the former Teche to the west and both the Lafourche and modern Mississippi to the east provide topographic features which constrain the basin occupied by the present Atchafalaya. This outflow route was established after about 600 BP (Roberts & Coleman

From: JONES, S. J. & FROSTICK, L. E. (eds) 2002. *Sediment Flux to Basins: Causes, Controls and Consequences*. Geological Society, London, Special Publications, **191**, 209–226. 0305-8719/02/$15.00

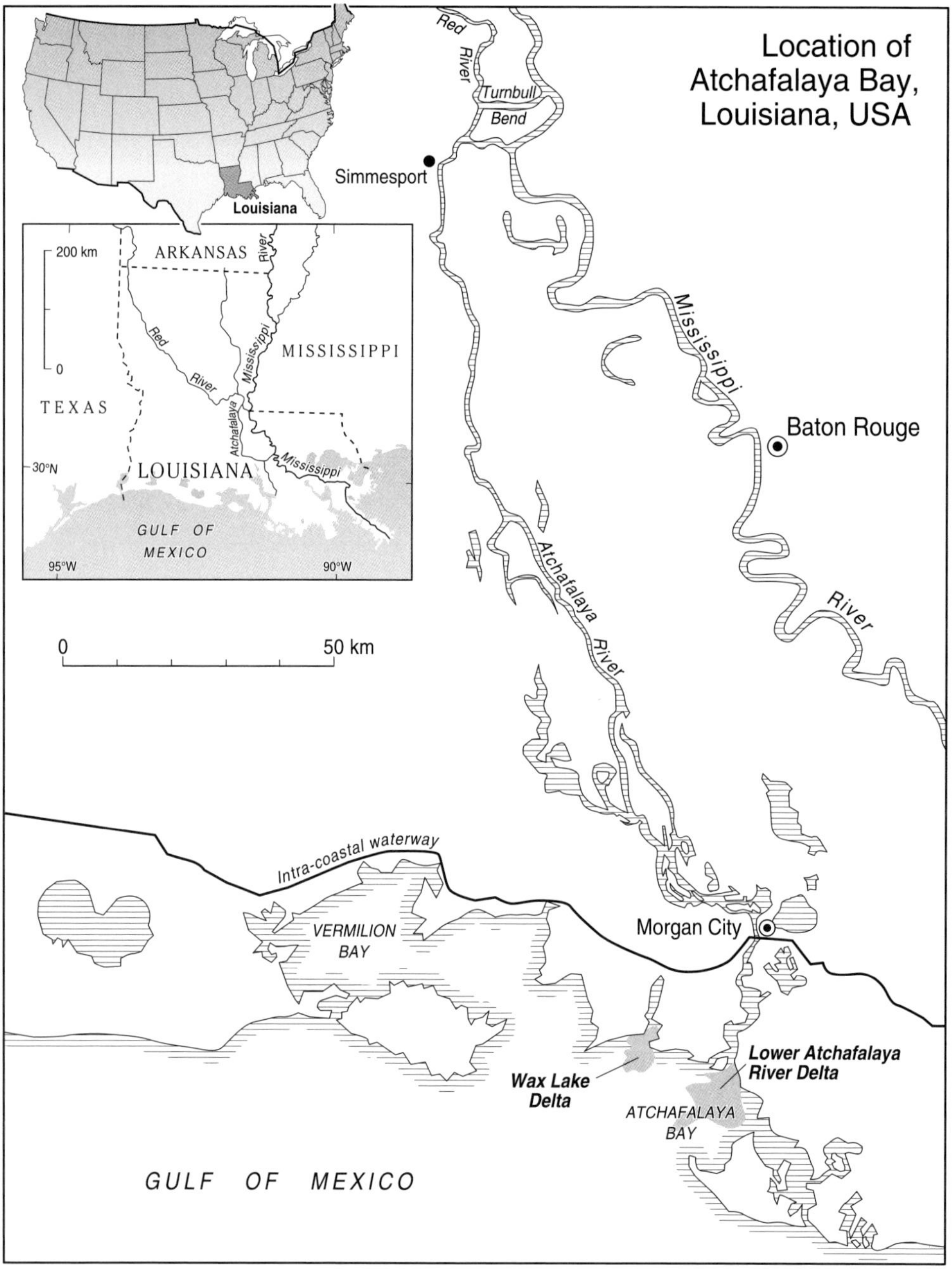

Fig. 1. Location map of the Atchafalaya in relation to the lower Mississippi and Gulf of Mexico. The positions of the Wax Lake and Lower Atchafalaya Outlet deltas are indicated.

1996), and initially carried waters of the Red River to the Gulf. Much of this land today remains as swampy wooded wetland through which the river flows in several channels, which come together in Six Mile Lake, immediately north of Morgan City.

When the great Spanish explorer Hernando de Soto turned his attention to the southern part of the interior of North America in 1539 the principal rivers provided natural routeways. Shortly before his death in 1542 he noted the existence of a small

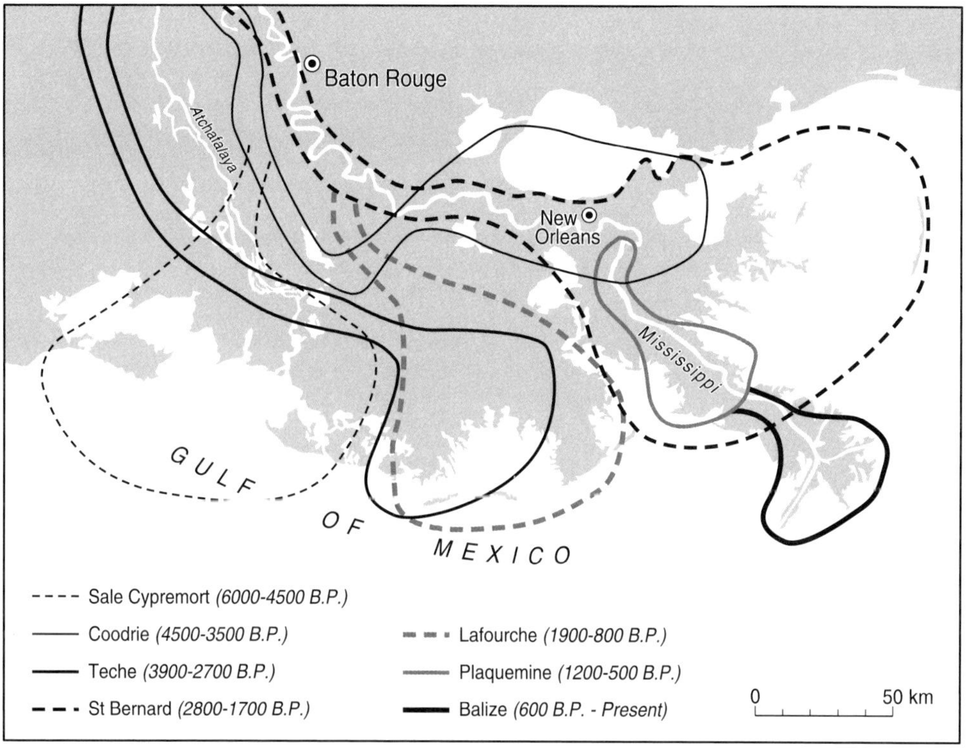

Fig. 2. Sequence of formation of lobes of the Mississippi Delta (after Kolb & van Lopik 1958).

distributary stream issuing westwards from the meandering path of the Mississippi approximately 60 km above the site of the present city of Baton Rouge (Fisk 1952). The waters flowed south-westwards towards the Gulf of Mexico. This is the earliest direct record that we have of the Atchafalaya (local Longtown Choctaw language for Long River). The area did not feature again historically until 1805, when the meander known as Turnbull Bend migrated west to capture for the Mississippi the waters of the Red River, which had hitherto flowed independently towards the Gulf (Roberts 1999). Although the pre-capture route of the Red River is not known in detail it is known that at this stage both the Red River and the Atchafalaya were heavily obstructed with fallen trees and it may be surmised there was little water or sediment moving towards the Gulf along this channel.

Captain Henry Shreve, the first major engineer of the river, is credited with designing the layer-cake shallow-draught paddle steamers of the Mississippi (McPhee 1989). He largely cleared the stranded tree trunks from the course of the lower Mississippi, to improve safety for navigation, and in 1831 he created the first artificial meander cut-off, dredging across a narrow neck of land and partly isolating the Turnbull Bend from the flows of the main stream itself. This bend, subsequently known as the Old River Channel, contained both the inflow of the captured Red River and the outflow, now known as the Atchafalaya. However, the discharge of waters flowing towards the Gulf along the Atchafalaya did not show an immediate increase, because the uppermost 45 km of its channel was blocked by a raft of trees so substantial that the main overland routeway west into Texas crossed it. Shreeve worked to remove the raft, lifting, blasting and burning the wood. The obstruction was largely cleared by 1839 (van Heerden 1983), but the waterway was not finally and fully opened until the advent of the major floods of 1863 (Barry 1998). By removal of this log-jam and a similar, but longer, one on the Red River, which was cleared by 1870, not only was navigation throughout the western part of river system established, but so, too, was freedom for water flows and associated sediment transport. From the largely unpredictable trickle of water

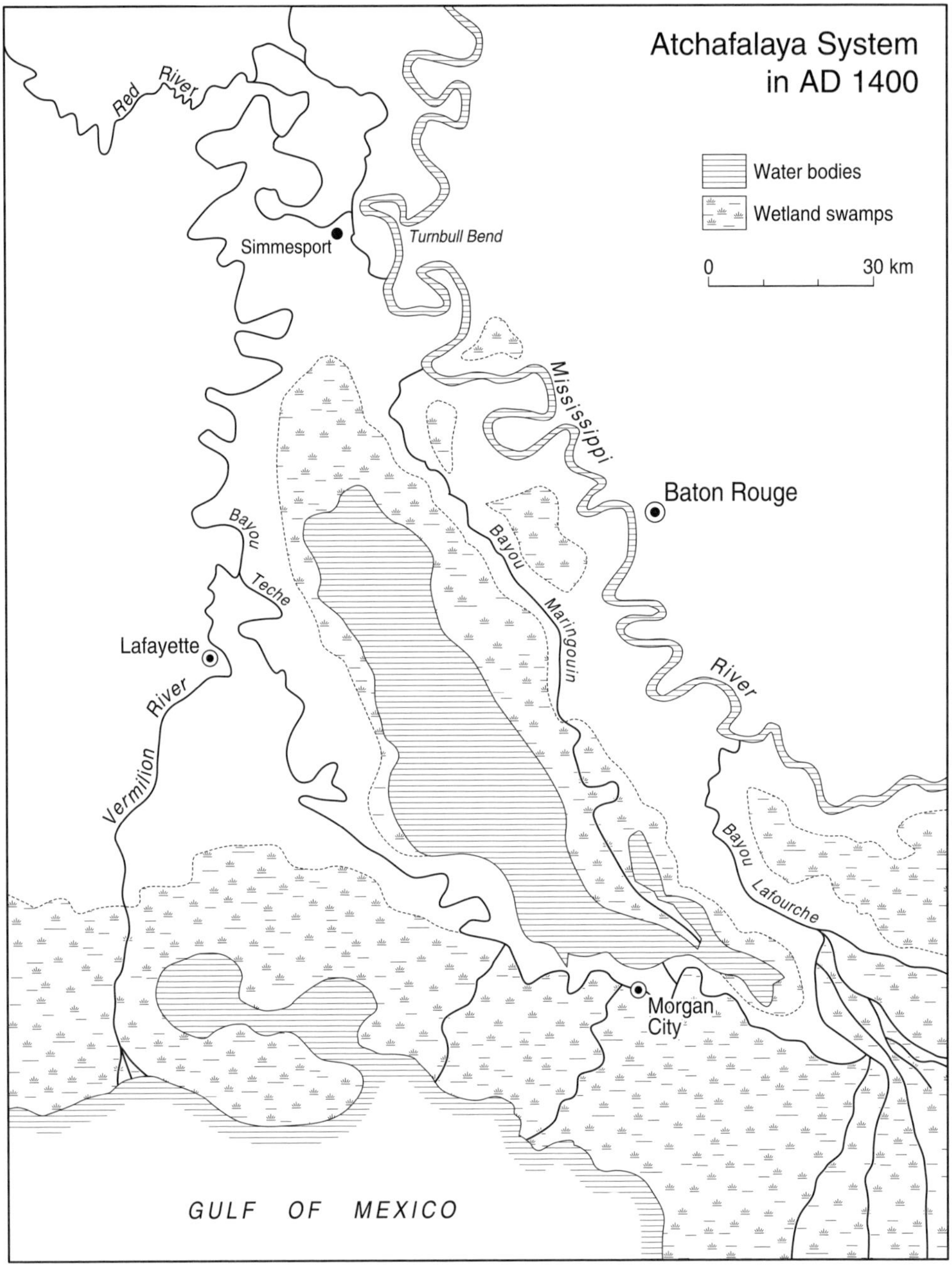

Fig. 3. Map of the Atchafalaya system in AD 1400, as suggested by Roberts & Coleman (1996).

entering the Atchafalaya in the early half of the nineteenth century (Morgan *et al.* 1953) the discharges now began to follow a regular seasonal pattern of summer and autumn low flows and winter and spring high flows, derived partly from the Mississippi and partly from the Red River. Whereas initially the Atchafalaya carried little water or sediment the flows, and presumably the sediment loads, steadily increased so that by 1900 it was taking 13% of the Mississippi waters, this had risen to 30% by 1952 (Fisk 1952). There was consequently an enormous rise in sediment transport to the Atchafalaya system.

Controlling the outflowing waters

From its diversion point on the Mississippi to the Gulf the Atchafalaya travels a distance of 220 km, whereas the Mississippi flows the much greater distance of 520 km to its outlet. In consequence, the Atchafalaya has a steeper average gradient than the Mississippi. Furthermore, the inherited basin through which it flows extends northwards to within a few kilometers of the Old River Bend, so that in the uppermost reaches there is a rapid fall in water surface level between the two rivers. This water level difference varies with river discharge, but is normally within the range of 2.4–5.2 m (Martinez & Haag 1987).

Until 1963 the steep reach was largely uncontrolled and bed erosion was considerable, leading to ever-increasing proportions of the water flows becoming diverted from the Mississippi and into the Atchafalaya. Fisk (1952) recognized that this steady increase could lead to capture of the lower Mississippi waters, which he predicted would occur some time during the decade 1965–1975 unless preventative measures were taken. This would have been to the commercial disadvantage of port cities such as New Orleans and Baton Rouge, through lack of freshwater during low discharge periods, and would cause a considerable disruption of the navigation on the major trade artery through the centre of the continent. The US Army Corps of Engineers was commissioned to prepare recommendations on how the capture could be prevented. These recommendations were linked with continued efforts to reduce flood risk in the lower Mississippi Basin, so that an initial proposal for a river control weir coupled to a navigation lock system was part of a major programme of change. This had earlier involved construction of the Morganza Spillway 20 km further south (completed in 1954), and the formation of the West Atchafalaya Floodway, a sacraficial leveed area, access to which is provided through an automatically activated fuse–plug system (low point installed in the normal levees) to be used to accommodate excess floodwaters during extreme flow conditions.

The Old River Low–Sill Control system was completed in 1960. Having been severely damaged during the major floods of 1973 (McPhee 1989) it was repaired and although it performed well during the even larger floods of 1973, an auxiliary control was created. More recently, the Sidney A. Murray Jr hydroelectricity plant has been opened to take advantage of the low head but high water discharge between the two rivers (Fig. 4). In all, 30% of the flow of the Mississippi is now permitted to enter the Atchafalaya by way of three controlled outlets, under most flow conditions, or expressed differently, the Atchafalaya now carries about half of the volume of water that is carried by the lower Mississippi. According to Mossa & Roberts (1990)

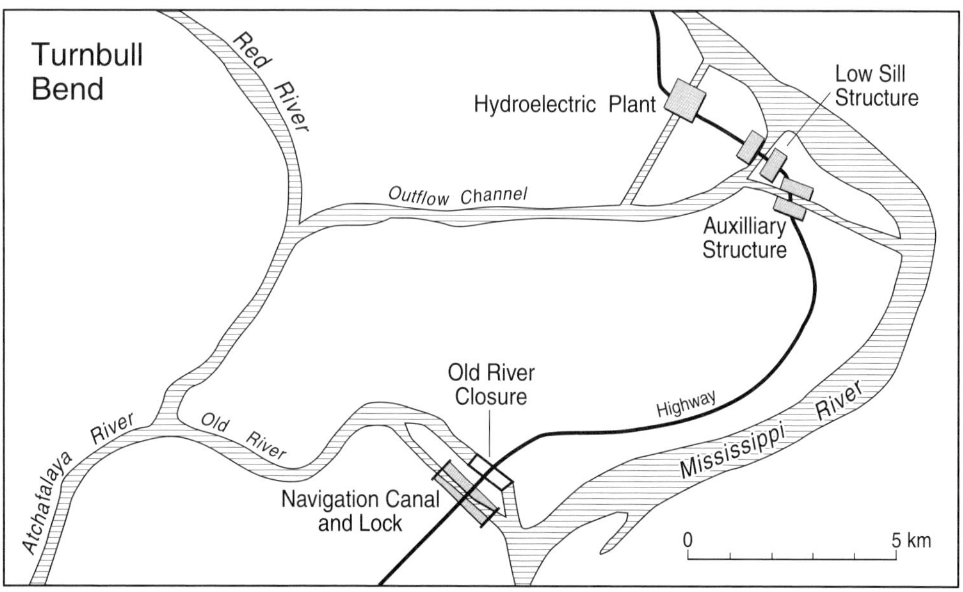

Fig. 4. Control systems at the site of the Mississippi–Aychafalaya interchange area.

on average some 60% of the present suspended sediment load of the Mississippi is now diverted into the Atchafalaya system, leaving no more than 40% of the original suspension load to flow to the Gulf past New Orleans.

Nature of the Atchafalaya basin

Prior to the opening of the river, the currents were undoubtedly slow moving. The waters passed into the basin, the upper part of which follows an unusually straight plan probably a result of known dredging activity by Shreve in the late 1830s and early 1840s, but still incorporating small lakes and bayous. The southern part of the basin was dominated by the 50-km long Grand Lake–Six Mile Lake complex. According to Fisk (1952) much of the land surface of the northern part of the Atchafalaya Basin was built of deposits from a much larger ancestral Grand Lake water body than is present today. He suggested that in 500 BP the lake was approximately 120 km in length. About 390 km^2 of the lower basin today still consists of shallow lakes. Any sediment in motion in the early years of the river's life would have become entrapped within these shallow-water bodies. As the flows increased during the late nineteenth and early twentieth centuries so the lake systems progressively filled. Fisk himself demonstrated drastic changes in the period 1930–1960 (Fig. 5), stating that in a period of about 20 years 155 km^2 of the Grand Lake system had completely filled with sediment, a pattern supported by Shlemon (1972) and Roberts *et al.* (1980). The period during which the exceptionally rapid sedimentation was recorded coincided with a period of considerable expansion of agricultural activity in the centre of the continent.

According to Keown *et al.* (1981), whereas at the start of the nineteenth century most of the centre of the continent remained as natural grassland and forest, by the mid-twentieth century more than 30% of the area had been developed for intense agriculture. Revisiting of Mississippi River survey records from the period 1884–1911 enabled Kesel *et al.* (1992) to determine that, at a stage when human impact on the system remained limited, changes in point bar volumes could be closely related to rates of bank caving. Lateral migration of the Mississippi above the Red River confluence averaged 24 m year^{-1}, and over 60% of the sediment released travelled a limited distance before entering short-term storage. The central parts of the Mississippi Basin were believed to have been aggrading during the period 1880–1920.

As accelerated settlement progressed, the changes of land use culminated in the heavy soil erosion associated with the dust-bowl conditions of the 1930s. Much of the area directly concerned is drained by the Red River–Mississippi system, so that there would have been an excess of sediment available for transportation during that period.

In any single lake the sedimentary filling begins with finely laminated alternations of silts and clays, which pass upwards into massive silts and fine sands in the upper layers. The upward coarsening of the normally highly micaceous sediments is an indication of the growth of lacustrine deltas, which Tye & Coleman (1989) also believed to have been rapidly accumulated. Within each lake fill are subfacies of deposition relating to levee, point bar, channel fill, well-drained swamp, poorly-drained swamp, open lake and lake delta conditions (Fisk *et al.* 1954; Kolb 1962; Coleman & Gagliano 1965; Roberts 1986).

Sediment loads in the Atchafalaya

A 40-year record of flows between 1951–1952 and 1988–1989 (Table 1) is available for the gauging station at Simmesport, 6 km below the control structures (and therefore including the Red River discharges). This shows long-term average flows of 6310 m^3 s^{-1}, with high flows between January and June averaging 11 910 m^3 s^{-1}, and peak recorded flows of 19 824 m^3 s^{-1} during the 1973 floods (Fig. 6). The records show that in the 10 years before 1961–1962 the flows averaged 4511 m^3 s^{-1}, while during the following 10 years they averaged 4643 m^3 s^{-1}. Thereafter, after completion of the control system, between 1972–1973 and 1988–1989 the average flows rose to 6021 m^3 s^{-1}, one third greater than previously. However, the sediment transport patterns differed considerably, with annual loads averaging 121 057 × 10^3 tonnes (t) in the decade before 1961–1962, 74 034 × 10^3 t during the following decade, before the controls were fully operational, and 75 211 × 10^3 t during the 17 succeeding years. On six occasions during the earliest decade the annual loads exceeded 117 910 × 10^3 t, but on only two subsequent years has that level been reached (Fig. 7). During the first decade the average annual suspension concentrations always exceeded 550 mg l^{-1}, a level reached only twice in the next decade, but never subsequently. The reduction in suspended sediment load is almost certainly related to changes in land management policy, including river bank protection schemes, coupled with the construction of sediment retention dams on the Arkansas and Missouri Rivers during the period 1963–1970. According to Kesel (1988), suspension loads in the Mississippi have fallen by over 70% since 1850. At Vicksburg, Mississippi, 110 km above the Old River control

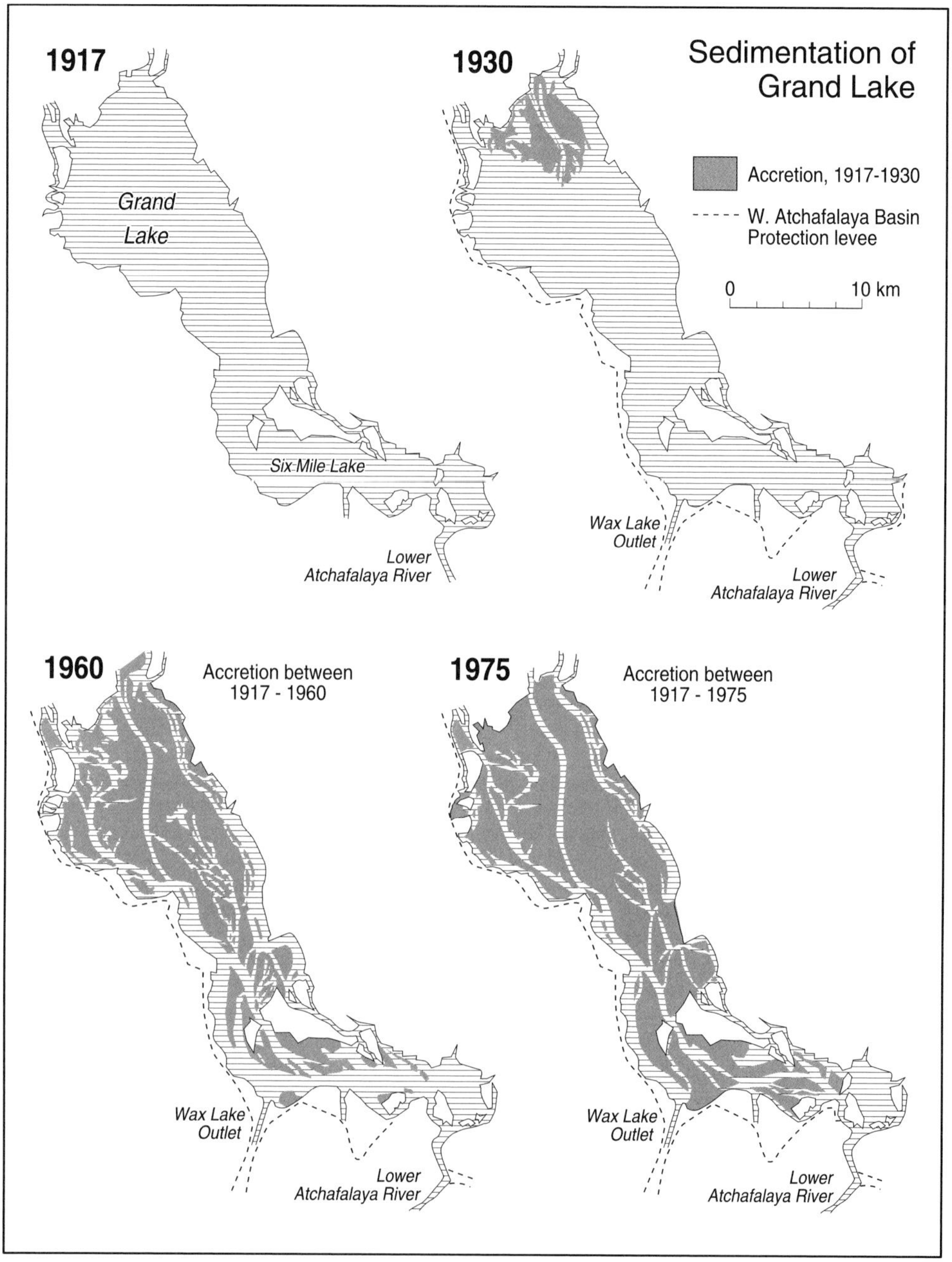

Fig. 5. History of sedimentation of Grand Lake in the Atchafalaya Basin (after Roberts *et al.* 1980).

system, annual suspended sediment loads before the improvements averaged over 400×10^6 t, whereas after completion of the structure the 10-year average fell to 225×10^6 t (Keown *et al.* 1981).

Beyond the analysis of long-term average variations it is valuable to examine some of the more extreme periods of flood discharges, especially those of the 1972–1975 period, when the annual mean flows reached 9948 $m^3 s^{-1}$, the annual

Table 1. *Summary of water discharges and sediment loads in the Atcahafalaya River at Simmesport, Louisiana, 1951–1989*

Water year	Mean discharge ($m^3 s^{-1}$)	Total sediment (× 1000 t)	Sand (%)	Silt (%)
1951–1952	5743	194 460	25	75
1952–1953	4050	135 230	21	79
1953–1954	2273	54 230	24	76
1954–1955	3584	93 360	26	74
1955–1956	3489	67 175	23	77
1956–1957	5264	225 474	25	75
1957–1958	6355	214 390	22	78
1958–1959	3961	83 230	25	75
1959–1960	4928	131 878	18	82
1960–1961	5460	133 372	30	70
1961–1962	6318	151 913	38	62
1962–1963	3345	44 876	19	81
1963–1964	2345	52 291	20	80
1964–1965	4723	108 871	25	75
1965–1966	3627	88 522	20	80
1966–1967	4074	55 710	12	88
1967–1968	5694	121 351	14	86
1968–1969	5923	115 245	24	76
1969–1970	5280	75 098	26	74
1970–1971	5098	72 441	27	73
1971–1972	5360	89 587	21	79
1972–1973	9948	124 468	36	64
1973–1974	8314	142 994	233	77
1974–1975	8326	157 938	22	78
1975–1976	4686	56 113	15	85
1976–1977	3398	57 137	11	89
1977–1978	5668	71 194	18	82
1978–1979	7451	112 343	23	77
1979–1980	5516	67 801	16	84
1980–1981	3909	51 079	10	90
1981–1982	5508	104 102	11	89
1982–1983	7718	100 894	25	75
1983–1984	6616	73 213	17	83
1984–1985	6315	116 757	15	85
1985–1986	5590	81 794	11	89
1986–1987	5707	71 855	6	94
1987–1988	4218	57 780	11	89
1988–1989	6246	52 228	5	95

sediment loads peaked at 143.2×10^6 t, annual sand transport reached 41.1×10^6 t. Although concentrations of material in suspension were not exceptionally high (499 mg l^{-1} in 1974–1975), annual silt loads reached 111.4×10^6 t. Both the sand and silt loads were about double those normally carried into the system. However, during the two succeeding years the sand transport fell to well below half the normal load, suggesting that the sediment supply had become severely depleted during the flood years. Over the period of reporting, from 1951–1952 to 1988–1989, the sediments transported in suspension at Simmesport showed a halving in the proportion of sand being carried seawards, from an average of 25% to 12.7% during the first and last decades of the record respectively. During periods of heavy flooding the sand contents rose well above normal levels.

Graphical plots of the monthly water discharge figures during the period 1956–1981 presented by van Heerden (1983) provide a clear illustration of the seasonality of the flows (Fig. 8). They also demonstrate the very substantial increase in the number of months each year in which flows in the latter half of the record (after 1968) exceeded an arbitrary figure of 5000 $m^3 s^{-1}$. This increase reflects the higher, but constant, inflow from the Mississippi through the control structures. During the high flow years of 1972–1975 discharges rarely fell below this figure.

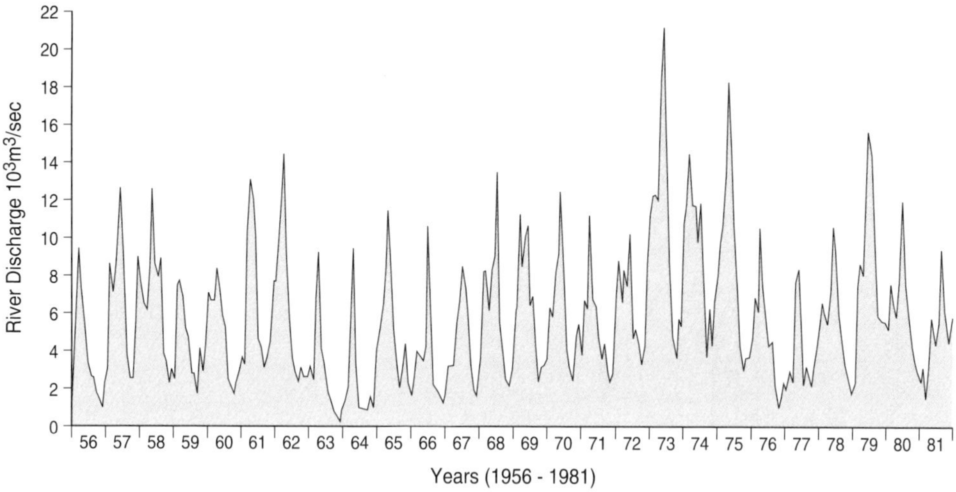

Fig. 6. Mean monthly river discharges at Simmesport, 1951–1989.

Details of the deposited bed material sediments through the Atchafalaya Basin, as provided by Keown *et al.* (1981), reveal that the grain size decreases sharply away from the diversion area. In the Old River Outflow Channel and at Simmesport, the medium and coarse sands accounted for 33 and 30% of the sediment, respectively, but by river miles 50–100 only 5% of medium sand was present, and this decreased to 1% at Morgan City and Calumet. Fine sand is more abundant, forming 67, 70, 31 and 9% at the sites concerned, respectively. The median grain sizes at the four localities were 0.3, 0.37, 0.045 and 0.015 mm.

Sediments in Atchafalaya Bay

Atchafalaya Bay is an area formerly characterized by sedimentation associated with the early delta lobes, so that the bed of the Bay is underlain by alluvial–deltaic sediments, which are known to be about 25 m in thickness (Wells *et al.* 1984). These in turn rest upon the Pleistocene Prairie Formation, which was downwarped by around 30 m during the Quaternary due to regional subsidence of the Gulf coastlands. Once avulsion took the river outflows elsewhere, little terrestrially derived sediment entered the 5-m deep Bay for many years. Indeed, Cratsley (1975) showed that between 1858 and 1952 there had been no significant sedimentation in the Bay area. Any sediments in motion were entrapped in the lake systems within the drainage basin.

Between 1889 and 1935 a layer of 2 m of clays had accumulated on the seaward side of the Pointe au Fer shell reef and a further 1 m had been deposited by 1951 (Thompson 1951). The movement of wave-induced suspensions from the Vermilion Bay area, reported by Walker & Hammack (1999), is not sufficient to have accounted for this volume of deposited material rather than derivation from the increasingly important Atchafalaya system. It has been suggested that flocculation of clays in the fresh river waters upon meeting the saline Gulf waters might account for the deposit (Roberts *et al.* 1980). The particle number concentrations and shear within the moving water bodies are certainly sufficient for the local development of orthokinetic flocculation within the Bay area, but the salinities within the bay itself are lower than normally required to permit perikinetic flocculation before the waters reach beyond the line of reefs on to the shelf.

By 1962 an area of 120 km^2 of the floor of Atchafalaya Bay had become covered with a layer of fine sediment (Shlemon 1972, 1975). During the period 1962–1972 silts and fine sands began to build up off the mouths of the two river outlets, the Wax Lake Outlet and the Lower Atchafalaya Outlet. There is no doubt that some of this was as a result of dredging the shipping channel to Morgan City, but there was also natural sedimentation during this period, and by 1972 the embryonic delta lobes off both outlets averaged 1.8 m in thickness. The outer 0.6–0.9-m thick deposits of the pro-delta slope had reached the shell reef and the organisms were showing signs of stress.

The stage was set for the major floods of 1973, whose impact in the upper parts of the basin led to the use of the Morganza Spillway, but did not

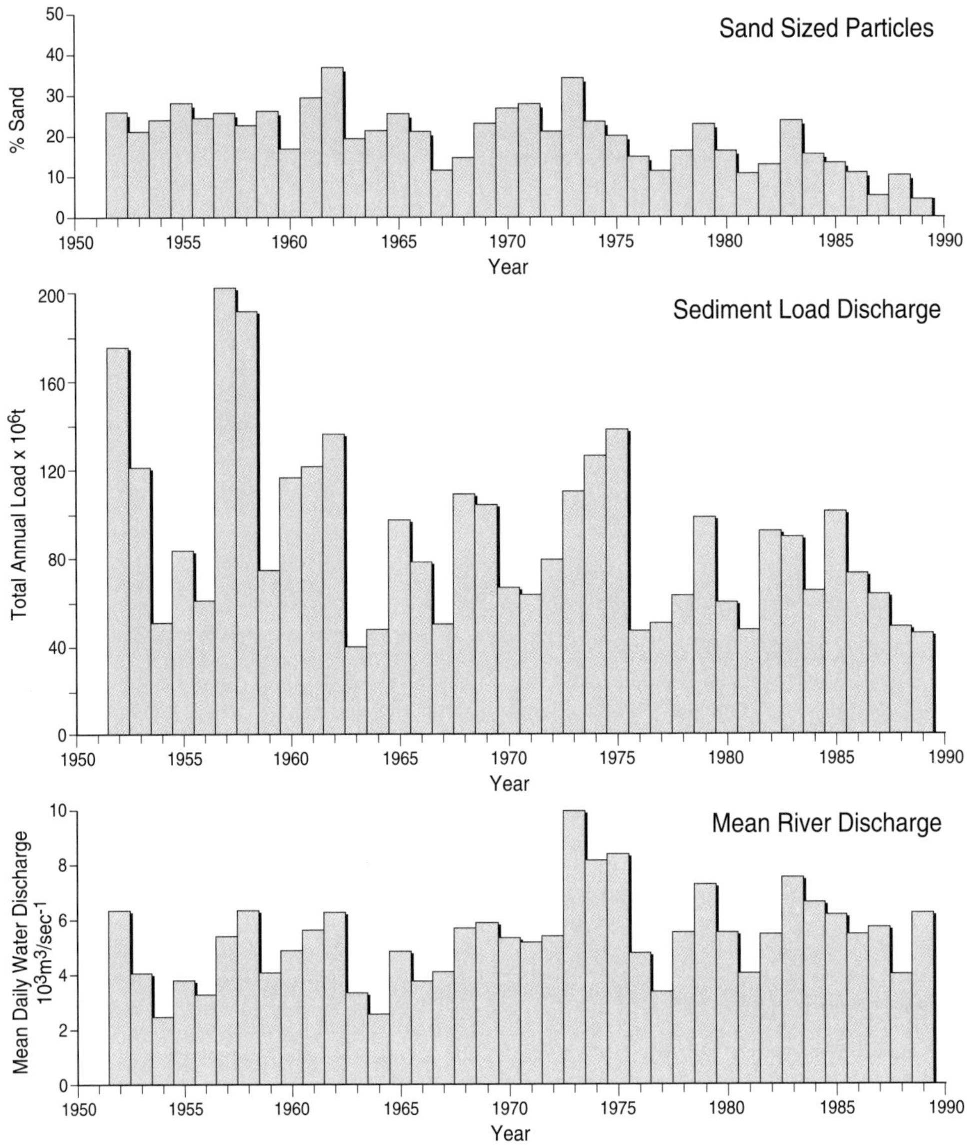

Fig. 7. Annual sediment loads discharged at Simmesport 1951–1989, with the proportion of sand-sized particles carried. Annual mean river discharges are also shown.

necessitate use of the West Atchafalaya Flood Spillway. The flood period lasted for over 6 months and at its peak the Simmesport gauge recorded mean monthly flows of more than 20×10^3 m^3 s^{-1}, the highest discharges since the control structures were erected. The swiftly flowing river carried high suspended sediment loads through the system, beyond the Six Mile Lake still-water area and out into Atchafalaya Bay, where much of the load became deposited beyond the river mouth, as also off the Wax Lake Outlet. In the period 1972–1975, although total annual loads reaching the Bay were shown to average 112×10^6 t (Wang 1985), a total of 34×10^6 t of sand-sized sediment passed Simmesport, but of that no more than 30.7×10^6 t reached Atchafalaya Bay (Roberts 1999). Whereas previously fine sands had become entrapped within the basin, they were now eroded from the point bars

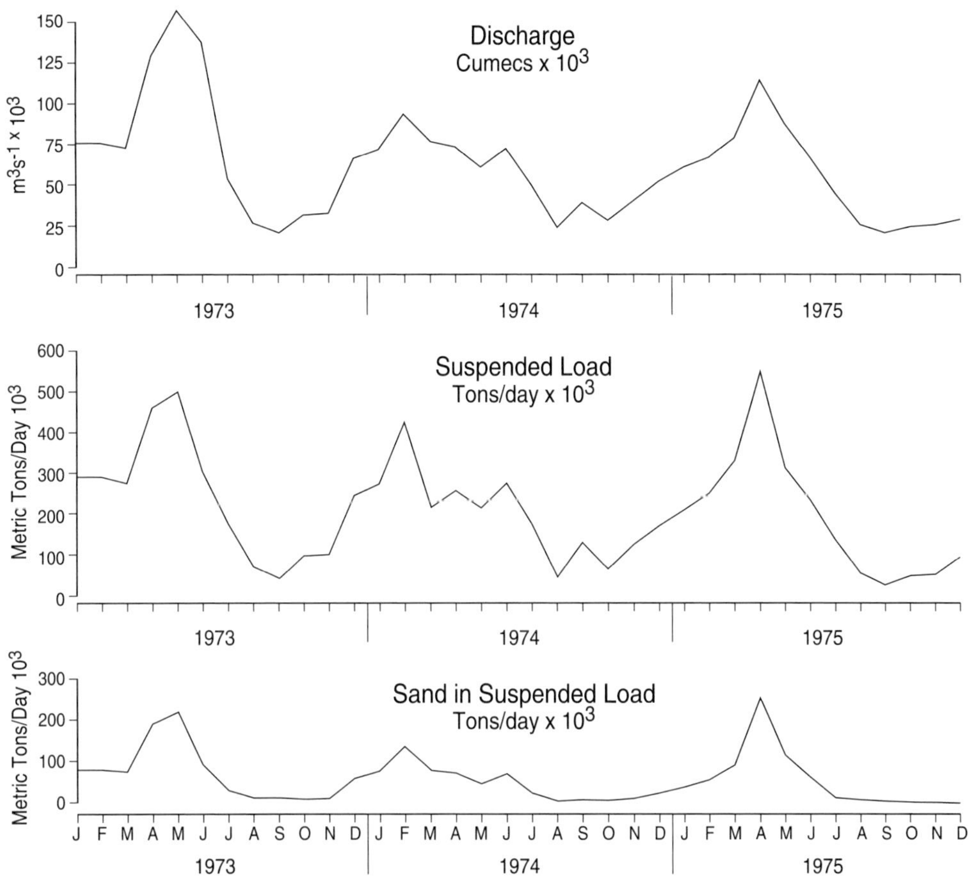

Fig. 8. Monthly variation of flow discharges, suspended sediment load and sand percentage in the load at Simmesport during the high flow period of 1973–1975 (after van Heerden 1983).

and channel floor in exchange for coarse sands and carried to the delta, so that substantial deposits of fine sand accumulated upon the pro-delta silts and clays. Off both outlets the sediments began to emerge above sea level, on either side of the exit channels, and since that time two substantial deltas have prograded into the Bay.

Delta growth

Monitoring by the US Army Corps of Engineers using air photography and satellite imagery analysis has enabled the history of sedimentation and growth of the emergent deltas to be followed during the last 30 years. Largely carried out by contractors, the analysis has produced a wide range of estimates of accretion rates, depending upon the method used and the datum applied. Thus, while accretion rates based upon analysis of satellite imagery and air photographs identify the boundary between land and sea during the overpasses, not necessarily at low water, analysis using a ground-terrain model and depending on a series of bathymetric traverses across the area are normally reported in relation to a datum at or below mean low water level (–0.3 to –0.6 m), and as a result show generally a less irregular pattern of delta growth.

By 1976 the Wax Lake Outlet had built 3.8 km^2 of emergent land (Fig. 9), whereas the Lower Atchafalaya River Delta had created 32.5 km^2 of new land. Terrain modelling using a datum at –0.6 m enabled Roberts *et al.* (1997) to show the continued patterns of growth of the two deltas, as illustrated in Table 2.

The accretion rates in the two deltas differed greatly during the early phases of development,

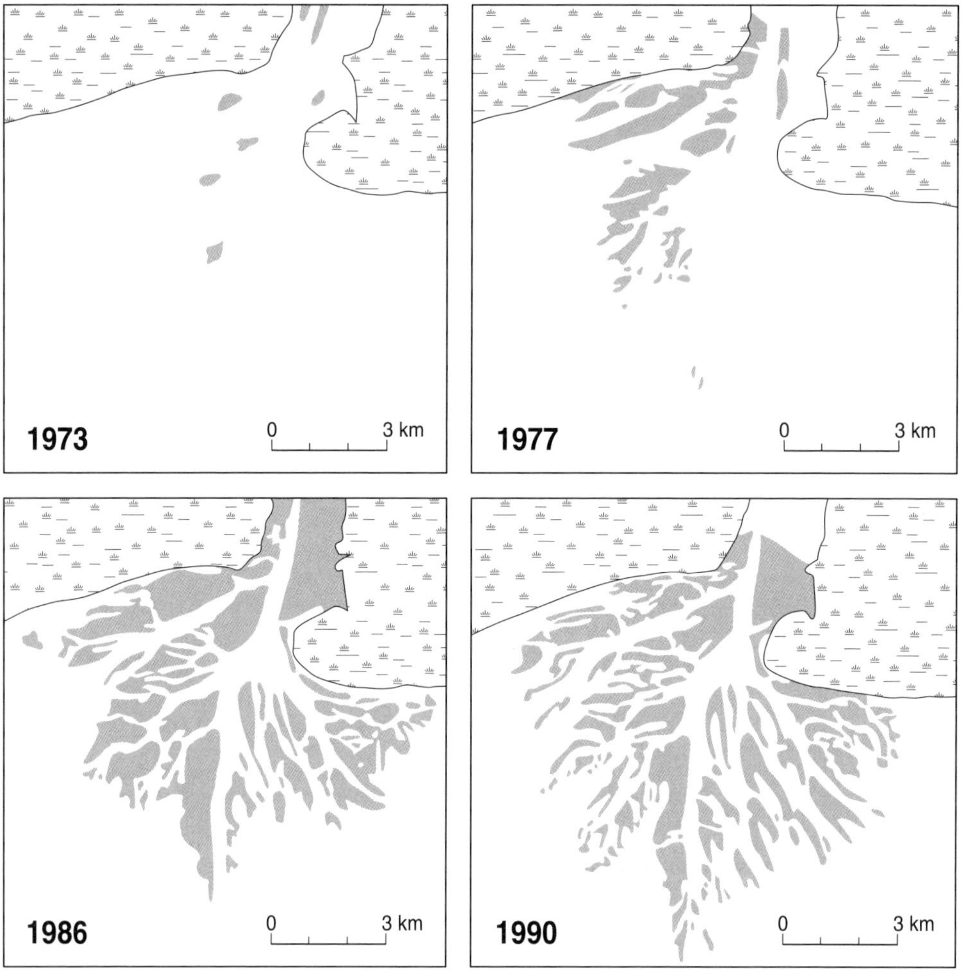

Fig. 9. Growth of the Wax Lake Outlet Delta, 1973–1990 (after Roberts 1999).

partly because coastal irregularities, lakes, bayous and embayments required to be filled above the Wax Lake Outlet, whereas the Lower Atchafalaya River Outlet was already a well-constrained feature. In addition, a weir was established in 1988 in the Six Mile Lake reach to diminish the impact of flood discharges at Morgan City. In practice this led to increased deposition and channel confinement in the Lower Atchafalaya Outlet and backing up of the waters, effectively raising floodwater levels at Morgan City and stimulating sedimentation at the entrance to Grand Lake. The restricted flows into the Wax Lake Outlet channel had reduced bed-load transport. The recommendation by Kemp *et al.* (1995) that the structure be removed was followed through and thereafter sedimentation was unhindered. Minor navigational dredging in the Wax Lake Outlet area ceased in 1980, whereas substantial dredging for navigation access to Morgan City has continually modified the Lower Atchafalaya Outlet Delta.

The two sub-deltas also differ in size and growth

Table 2. *Areas of the Atchafalaya Bay deltas (after Majersky* et al. *1997)*

Year	Wax Lake (km^2)	Lower Atchafalaya (km^2)
1976	3.8	32.5
1981	19.7	67.3
1989	47.9	85.2
1994	84.2	101.5

patterns because they receive differing quantities of sediment from their feeder rivers. Normally, the Wax Lake Outlet has carried about 40% of the sediment load entering the Bay, but during the 6 years in which the weir was in existence the sediment load passing along this route was reduced to an average of 31.7%. Removal of the weir permitted patterns to return to 'normal', but there is no doubt that its presence held back the growth of the Wax Lake Outlet Delta and may well have conditioned the development of the channel system, which Roberts (1999) noted appeared to have stopped elongating after about 1983, a short period before the weir was installed.

While the growth of the deltas is rapid during periods of high river discharge, when flows revert to normal the sediment loads carried are initially low due to supply limitation, although they rise after a few years to reach what are believed to be 'normal' levels. Thus, the delta growth takes place through discontinuous periods of rapid accretion with intervening phases of lesser build-up. The total measured sediment load passing Simmesport during each of the water years 1973, 1974, 1975, 1979, 1982, 1983, 1985, 1990 and 1991 exceeded 100×10^6 t when rapid delta growth was recognized. Long-term rates of accretion of the Wax Lake Delta are 3.0–3.5 km^2 year^{-1}, while the Lower Atchafalaya Delta shows rates of 2.2–3.2 km^2 year^{-1}, determined from terrain analysis with a –0.6 m datum horizon (Roberts 1999).

The rate of initial sub-aerial growth of the two deltas was rapid as the sediment load entering the bay was spread around the periphery of the small enlarging deposition centres. It would be expected that as the sediment supply rate was maintained and the deltas enlarged so the same volume of sediment being carried into the bay was spread over an increasingly large area, so the annual increment of sub-aerial growth should have decreased through time, but there has been no clear pattern to confirm this trend. The lack of stable continuity of load delivery between flood and low flow years may account for this, but it is also possible that anthropogenic influences on each of the sites may have played a role.

The major difference between the two deltas lies in the degree to which dredging for navigation has modified the depositional history. While the Wax Lake Outlet experienced minor dredging, which ceased in 1980, the Lower Atchafalaya Delta has, throughout its life, experienced ongoing programmes of dredging and disposal of sediment. Over the years the channel requirements have increased from 10 (3.0 m) × 100 feet (30.5 m) in 1939, to 20 (6.1 m) × 200 feet (61 m) in 1974, broadening to 400 feet (122 m) soon thereafter (Penland *et al.* 1996), which may be placing hydraulic stress on the stability of the delta deposits.

The bed form which serves as the basis for deposition within an outflow channel typically develops where flow expansion into a broader reach takes place. The typical distributary mouth bar interferes with the passage of the waters and leads to formation of a parabolic planned structure with a prominent central 'head' and two lateral levee-like ridges tailing down-current, and increasing in width. The channel divides to either side and the head of the bank continues to gather sediment and emerges, maintaining its characteristic parabolic plan. The bank tails are sub-aqueous levees which grow to a length approximately double the width of the bank. They become elevated above the level of a central protected area within which suspended sediment settles, into which fine sand is washed through crevasses and where aquatic plants become established. Eventually sedimentation in this sector fills the hollow, so that the bank assumes a trapezoidal outline. On either side of the bank the water flows along narrowed channels, but, as these waterways in turn broaden, beyond the greatest width of the bank, fresh mid-channel bars form and further division of the water body ensues as the next generation of banks is created. In this way a broad initial channel narrows in a step-like progression as a series of banks with their apices pointing towards the principal channel are formed. In the Wax Lake Outlet delta it is possible to envisage the banks distributed in such a way that each 'generation' has its apices arranged along the circumference of concentric rings centred on the principal outlet into the Bay Roberts (1999) have shown that the banks as a whole commonly migrate upstream during growth and also that narrow channels between adjacent banks may become progressively shallower until adjacent banks may merge.

The open pattern of parabolic banks with intervening channels characterizes the Wax Lake Outlet Delta today. As the channel divides around a bank the two resultant parts are unequal in width or depth. Overall, the channels broaden away from the delta head but many become shallower, so that the outflowing waters are exposed to increasing frictional resistance, slow down and permit sediment deposition. Despite the continued supply of sediment, the actual forward extension of the delta fan should slow through time as the increments of sediment are added to the ever-extending periphery of the structure. As indicated above, the known rates of accretion have been remarkably constant for this delta, partly because it is building into such a shallow bay.

The Lower Atchafalaya Outlet Delta receives a greater discharge of water and the principal outlet is maintained for navigation. As a result the opportunity for mid-channel banks to develop is limited. Nevertheless, to the east and west of the main channel (East Pass and Southeast Pass) bank systems have developed, principally of fine sands. During the early phases of growth the banks to the west were used as sites for disposal of dredged spoil. Until 1985 most of the sediment was discharged on to Big Island, Camp Island or Willow Island, but thereafter, with minor exceptions, the sediments were released on to the eastern banks, whose configuration has in consequence changed substantially (Penland *et al.* 1996).

The growth and forward progression of the individual banks demonstrated by van Heerden (1983) during the period of early dumping was often aided by disposal of the sediments directly on to the bank head and marginal levees, but additional islands were deliberately created as bird roosting and nesting sites. After 1987 the dredged material was redirected to the east side of the delta, where again bank head and marginal levees were reinforced and sediment-laden waters were encouraged to flow through the inter-bank channels into the open waters beyond, where deposition could also occur (Fig. 10) After 1991 fine sediments were pumped to low areas behind the levees which had not accumulated fresh material for some years. Further expansion and formation of bird islands along the western margin of the outer part of the navigational channel was noted, and by 1999 these had been extended to within 3 km of the Point au Fer shell reef line, where few of the original oysters survive.

The network of channels developed in the eastern part of the delta consisted of primary, secondary and tertiary channels. Roberts (1999) noted that by 1973 the primary channel of East Pass opened into a single shallow bay, with a number of smaller secondary channels, Natal Channel, Tiger Bayou and Ratcliffe Pass, leading eastwards between the banks. Further sub-divisions present by 1976 produced very shallow channels, and a network of even shallower drainage gullies formed on the banks themselves. As sedimentation occurred along the channel margins the minor and tertiary channels became closed and fine sediments settled where formerly sands were carried as bed load.

The internal structure of the deposits was discussed initially by Roberts *et al.* (1980) and van Heerden *et al.* (1981), and was later explored by means of vibracores taken along and across both of the deltas (van Heerden & Roberts 1988). They showed that the Bay bed sediments, across which aggradation has occurred, are generally fine-grained, blue–grey clays which are heavily burrowed, containing oyster shell debris. The lowermost representative of the delta sediments proper is clay, up to 0.5 m in thickness, which is overlain by interlaminated sands silts and clays, up to 2.0 m in thickness, and these are capped by sands up to 4.5 m thick in the centre of the Wax Lake Outlet Delta. Both the base and the top of the interlaminated unit rise gradually (1 m in 5 km) toward the head of the delta and the entire section is cut through by the deep channels, such as Main Pass. The Lower Atchafalaya Outlet Delta sits on similar bay floor clays, with less than 2 m of prodelta deposits, beginning with bioturbated and shell-bearing silt and clay layers overlain by interlaminated silts and clays, which may show depositional cyclicity. Above these is a complex zone of about 2 m comprising sand deposits from distributory mouth bars and levees, with intervening laminated algal muds which originated in the sheltered areas behind the levee systems of the bars. Again, the deeper channels may expose any of these sediments. Thus, both deltas provide essentially similar successions of sediments, capped by sands deposited in bars, levees and along migratory channels. van Heerden (1983) estimated that sand (principally fine sand) accounts for 67% of the Wax Lake Outlet Delta and 55% of the entire deposit in Atchafalaya Bay.

Sediments beyond the deltas

The shallow open waters of the Vermilion Bay–Atchafalaya Bay lagoon are microtidal, so moderate tidal currents are characteristic. However, and more significantly, the waters are susceptible to the generation of wind-induced currents and the formation of waves. The winds determined by the passage of cold frontal systems, especially in late winter and early spring, veer from the southeast, to southwest, to northwest and northeast. The waves have the ability to resuspend fine bottom sediments and the wind-induced currents propel the suspensions around the bay (Walker 1996). In combination with the river discharges the coastal winds exert dominant control over the release of suspended sediment plumes into the Gulf of Mexico.

Winds from the southeast cause set-up to the water surface through the introduction of saline waters into the Bay, but lead to little wave generation or resuspension. Winds from the southwest also drive saline, but suspensate-rich, waters into the Bay through the Southwest Pass. The saline waters may enhance settling of suspensions, contributing to the shoaling known to have been in progress in Vermilion Bay for at least 20 years (Cunningham & Grymes 1997). Westerly winds lead to set-down of the coastal waters, and the combination of resultant currents and wave activity cause sediment resuspension, creating plumes

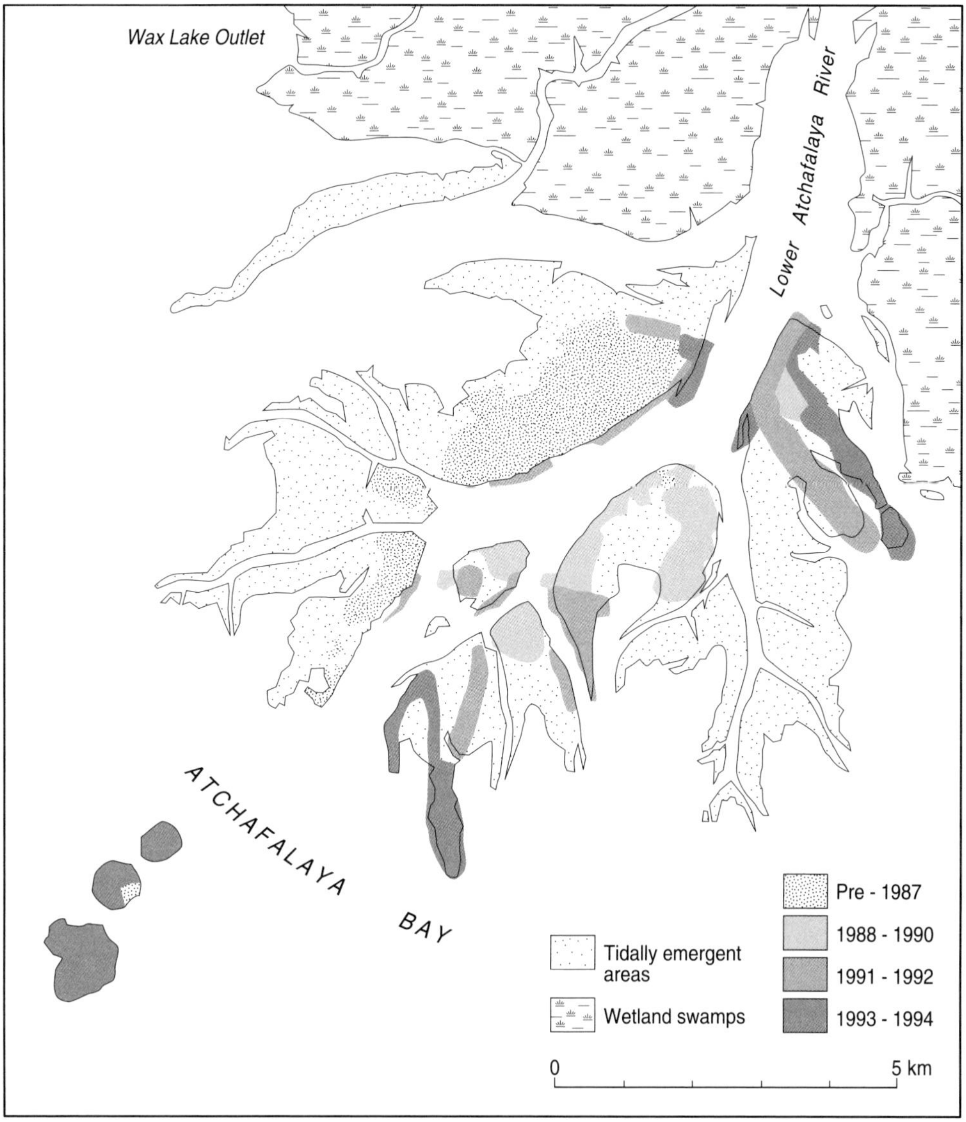

Fig. 10. Locations of deposition of dredged spoil material on the Lower Atchafalaya River Outlet Delta (after Penland *et al.* 1996).

which travel east along the lagoon and further, out onto the inner part of the continental shelf, commonly to the east of the outlet (Fig. 11). The egress of water is strengthened under northwest winds, which may serve to carry Wax Lake Outlet suspensions from the Bay. The northeasterly winds are weaker, and exert little influence within the bay, but as they become re-established over the inner shelf the sediment-rich plume becomes directed westwards with the prevailing current. As a result suspended matter is made available for deposition across a wide area of the inner shelf, but only a limited proportion of the material actually reaches the bed due to the turbulence of the waters, and the sediment is carried towards the chenier coast of western Louisiana and Texas, where settling of the muds is known to occur. In a remarkable series of satellite-based studies Walker *et al.* (1997) documented the effects of the passage of several frontal systems through the area. During one typical event

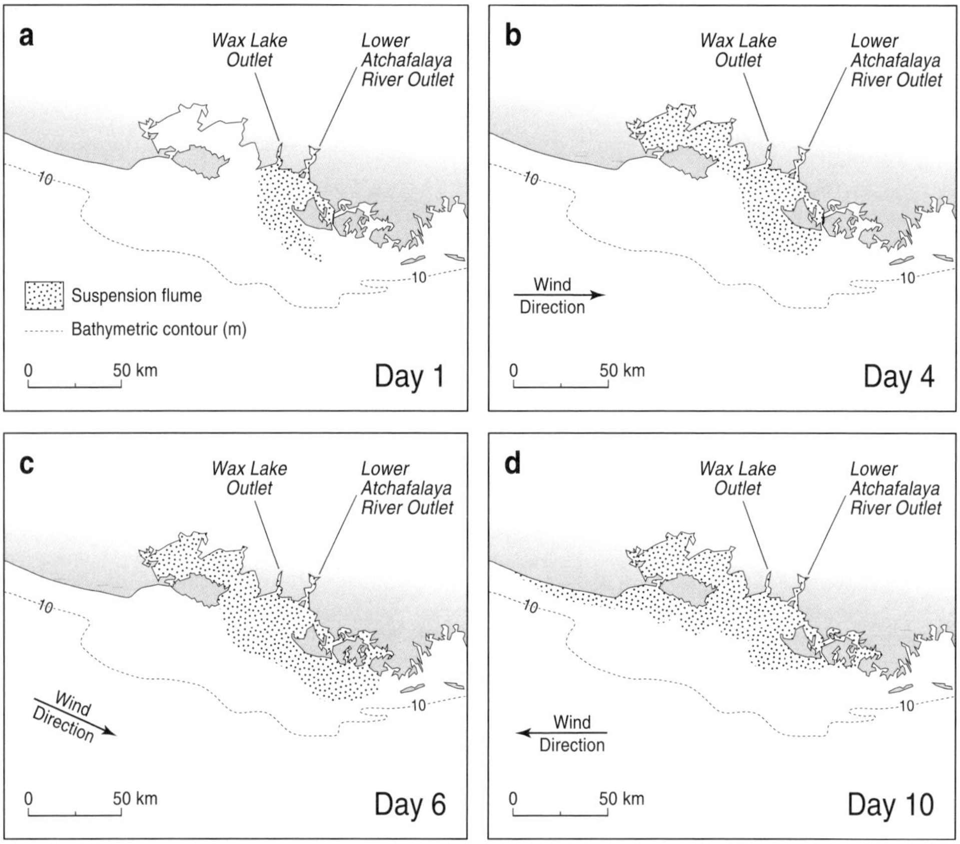

Fig. 11. Patterns of suspended sediment motion within and outside the Atchafalaya Bay area in response to changing wind direction (after Walker *et al.* 1999). (**a**) Plume in the absence of wind; (**b**) plume with westerly wind; (**c**) plume extending eastwards along the inner shelf and; (**d**) plume redistribution under a southeast wind.

in 1996 it was estimated that 200 000 t of suspended sediment was flushed onto the shelf. As indicated by Roberts *et al.* (1997), with 20–25 such cold front passages during the winter months of each year the effluent suspension plumes could account for the loss of as much as 15% of the annual load of sediment entering the Bay from the river systems.

The future of the Atchafalaya Bay deltas

The US Army Corps of Engineers has responsibility to maintain navigational access to Morgan City and the Atchafalaya. The dredging programmes with associated spoil dumping programmes have been referred to above. However, in order to anticipate future problems the Corps continues to attempt to predict the growth patterns for the delta, to assess potential dredging problems to come, and to prepare plans for any issues which may arise when the Atchafalaya discharges its waters and sediments directly into the Gulf. Numerical modelling techniques are used in this investigation. These models require detailed knowledge of the water circulation, the river discharges, the circulation and salinity variations within the Bay area, in addition to determining the bathymetric changes as they occur. Comparison with smaller crevasse-centred deltas of the present Mississippi Delta (Wells *et al.* 1984) suggests that any small delta lobe may grow to its greatest volumetric extension within a period of 60–100 years before being abandoned when an alternative outlet is followed. It is not easy to see that this analogy is entirely relevant for the principal outlet of a river, but it would be relevant to the Wax Lake Outlet Delta if structures were to be reinstalled to divert water and sediments away from this outlet. The impact achieved by re-engineering the Lower Atchafalaya River Outlet Delta may increase the life expectancy of the structure, which is expected to reach beyond the line of the Point au Fer shell reef by the year 2030 (Letter 1982; Wang 1985; McAnally *et al.* 1991; Donnell & Letter 1992).

The author wishes to gratefully acknowledge the courtesy and assistance received from many researchers, notably Drs W. H. McAnally, J. V. Letter Jr and W. McAdory during a period of study leave working with the U.S. Army Corps of Engineers at Vicksburg, and later at Louisiana State University, Baton Rouge, with Professors H. H. Roberts and N. D. Walker. Financial assistance from the Carnegie Trust for the Universities of Scotland is also gratefully acknowledged. The diagrams for the paper were prepared by G.Sandeman, Senior Cartographer, who is also gratefully acknowledged.

References

BARRY, J. M. 1998. *Rising Tide: The Great Mississippi Flood of 1927 and How it Changed America.* Touchstone, Simon & Schuster, New York.

COLEMAN, J. M. & GAGLIANO, S. M. 1965. Sedimentary structures: Mississippi River deltaic plain *In*: MIDDLETON, G. V. *Primary Sedimentary Structures and their Hydrodynamic Interpretation.* SEPM Special Publications, **12**, 133–148.

CRATSLEY, D. W. 1975. *Recent Deltaic Sedimentation, Atchafalaya Bay, Louisiana.* M.S. thesis, Department of Marine Sciences, Louisiana State University, Baton Rouge.

CUNNINGHAM, R. H. W. & GRYMES, J. 1997. *Atchafalaya Plume Response and Bathymetric Change Related to Long-term Meteorological Forcing.* Contract Report US Army Corps of Engineers Waterways Experiment Station, Vicksburg, Mississippi.

DONNELL, B. P. & LETTER, J. V. JR. 1992. *The Atchafalaya Delta; Report 13 Summary Report of Delta Growth Predictions.* Technical Report HL-82-15. US Army Corps of Engineers Waterways Experiment Research Station, Vicksburg, Mississippi.

FISK, H. N. 1952. *Geological Investigation of the Atchafalaya Basin and the Problem of Mississippi River Diversion,* Vol. 1. US Army Corps of Engineers, Mississippi River Commission, Vicksburg, Mississippi.

FISK, H. N., MCFARLAN, E, JR, KOLB, C. R. & WILBERT, L. J. JR. 1954. Sedimentary framework of the modern Mississippi delta. *Journal of Sedimentary Petrology,* **24**, 76–99.

KEMP, G. P., SUHAYDA, J. N., MASHRIQUI, H. S., VAN HEERDEN, I. L. & MARLBOROUGH, O. 1995. *Development of a Long-term Water and Sediment Distribution Plan for the Lower Atchafalaya Basin. Phase 1, Task 2: Assessment of the Impact of the Wax Lake Outlet Weir.* Report to Mayors of the Cities of Morgan City and Berwick, Louisiana.

KEOWN, M. P., DARDEAU, E. A. & CAUSEY, E. M. 1981. *Characterization of the Suspended-sediment Regime and Bed Gradation of the Mississippi River Basin.* Potomology Investigation Report 22-1. US Army Corps of Engineers, Waterways Experiment Station, Vicksburg, Mississippi.

KESEL, R. H. 1988. The decline in the suspended load of the Lower Mississippi River and its influence on adjacent wetlands. *Environmental Geology and Water Sciences,* **11**, 271–281.

KESEL, R. H. YODIS, E. G. & MCGRAW, D. J. 1992. An approxinmation of the sediment budget of the Lower Mississippi River prior to major human modification. *Earth Surface Processes and Landforms,* **17**, 711–722.

KOLB, C. R. 1962. *Distribution of Soils Bordering the Mississippi River from Donaldsonville to Head of Passes.* Technical Report 3-601, US Army Corps of Engineers, Waterways Experiment Station, Vicksburg, Mississippi.

KOLB, C. R. & VAN LOPIK, J. R. 1958. *Geology of the Mississippi River Deltaic Plain, Southeastern Louisiana.* Technical Report 3-483, US Army Corps of Engineers, Waterways Experiment Station, Vicksburg, Mississippi.

LETTER, J. V. JR. 1982. *The Atchafalaya River Delta; Report 3, Extrapolation of Delta Growth.* Technical Report HL-82-15. US Army Corps of Engineers, Waterways Experiment Station, Vicksburg, Mississippi.

MARTINEZ, J. D. & HAAG, W. G. 1987. *The Atchafalaya River and its Basin: A Field Trip.* Guidebook Series No. 4. Louisiana Geological Survey, Baton Rouge, Louisiana.

MAJERSKY, S., ROBERTS, H. H., CUNNINGHAM, R., KEMP, G. P. & JOHN, C. 1997. *Facies Development in the Wax Lake Outlet Delta: Present and Future Trends.* Basin Research Institute Bulletin, Louisiana State University, Baton Rouge.

MCANALLY, W. H. JR, HELTZEL, S. B. & DONNELL, B. P. 1991. *The Atchafalaya River Delta; Report 1, A Plan for Predicting the Evolution of Atchafalaya Bay, Louisiana.* Technical Report HL-82-15. US Army Corps of Engineers, Waterways Experiment Station, Vicksburg, Mississippi.

MCPHEE, J. 1989. *The Control of Nature.* The Noonday Press, Farrar, Straus & Giroux, New York.

MORGAN, J. P., VAN LOPIK, J. R. & NICHOLAS, J. G. 1953. *Occurrences and Development of Mudflats Along the Western Louisiana Coast.* Technical Report 2. Coastal Studies Institute, Louisiana State University, Baton Rouge.

MOSSA, J. & ROBERTS, H. H. 1990. Synergism of riverine and winter storm-related sediment transport processes in Louisiana's coastal wetlands. *Gulf Coast Association of Geological Societies Transactions,* **40**, 635–642.

PENLAND, S., WESTPHAL, K. A., ZGANJAR, C., CONNOR, P. & SEAL, R. W. 1996. *Beneficial Use of Dredged Material.* Monitoring Program 1995 Annual Report. US Army Corps of Engineers, New Orleans District.

ROBERTS, H. H. 1986. *A study of Sedimentation and Subsidence in the South-west Coastal Plain of Louisiana.* Final Report to US Army Corps of Engineers, Waterways Experiment Station, Vicksburg, Mississippi.

ROBERTS, H. H. 1999. *Atchafalaya Basin and Atchafalaya-Wax Lake Deltas. The AAPG Modern Deltas Field Seminar.* Coastal Studies Institute,

Louisiana State University, Baton Rouge, Louisiana.

ROBERTS, H. H. & COLEMAN, J. M. 1996. Holocene evolution of the deltaic plain: a perspective – from Fisk to present. *Engineering Geology*, **45**, 113–138.

ROBERTS, H. H., ADAMS, R. D. & CUNNINGHAM, R. H. W. 1980. Evolution of sand-dominant sub- aerial phase, Atchafalaya Delta, Louisiana. *AAPG Bulletin*, **64**, 264–279.

ROBERTS, H. H., WALKER, N. D., CUNNINGHAM, R., KEMP, G. P. & MAJERSKY, S. 1997. Evolution of sedimentary architecture and surface morphology: Atchafalaya and Wax Lake Deltas, Louisiana (1973–1994). *Gulf Coast Association of Geological Societies Transactions*, **47**, 477–484.

SHLEMON, R. J. 1972. *Development of the Atchafalaya Delta, Louisiana – Hydrologic and Geologic Studies of Coastal Louisiana*. Report 8. Center for Wetland Resources, Louisiana State University, Baton Rouge, Louisiana.

SHLEMON, R. J. 1975. Subaqueous delta formation – Atchafalaya Bay, Louisiana. *In*: BROUSSARD, M. L. (ed.) *Deltas*. Houston Geological Society, 209–221.

THOMPSON, W. C. 1951. *Oceanographic Analysis of Atchafalaya Bay, Louisiana and Adjacent Continental Shelf Area Marine Pipeline Problems*. Texas A & M Research Foundation, Department of Oceanography, Section 2, Project 25.

TYE, R. S. & COLEMAN, J. M. 1989. Depositional processes and stratigraphy of fluvially dominated lacustrine deltas:Mississippi delta plain. *Journal of Sedimentary Petrology*, **59**, 973–996.

VAN HEERDEN, I. L. 1983. *Deltaic Sedimentation in Eastern Atchafalaya Bay, Louisiana*. PhD Thesis, Department of Marine Sciences, Louisiana State University, Baton Rouge, Louisiana.

VAN HEERDEN, I. L. & ROBERTS, H. H. 1988. Facies development of Atchafalaya Delta, Louisiana: a modern bayhead delta. *AAPG Bulletin*, **72**, 439–453.

VAN HEERDEN, I. L., WELLS, J. T. & ROBERTS, H. H. 1981. Evolution and morphology of sedimentary environments, Atchafalaya Delta, Louisiana. *Gulf Coast Association of Geological Societies Transactions*, **31**, 399–408.

WALKER, N. D. 1996. Satellite assessment of Mississippi River plume variability: causes and predictability. *Remote Sensing of Environment*, **58**, 21–35.

WALKER, N. D. & HAMMACK, A. B. 1999. *Impacts of River Discharge and Wind Forcing on Circulation, Sediment Distribution, Sediment Flux and Salinity Changes: Vermillion/Cote Blanche System, Louisiana*. Report to US Army Corps of Engineers, Waterways Experiment Station, Vicksburg, Mississippi.

WALKER, N. D., DAY, J., RABELAIS, N., HUH, O., HAMMACK, A. B. & LANE, R. 1997. Circulation, river discharge plume, Louisiana. Extended Abstract, *Proceedings of the Fourth International Conference on Remote Sensing for Marine Coastal Environments, Orlando, Florida*, I-277–I-286.

WANG, F. 1985. *The Atchafalaya River Delta; Report N7, Analytical Analysis of the Development of the Atchafalaya River Delta*. Technical Report HL-82-15. US Army Corps of Engineers, Waterways Experiment Station, Vicksburg, Mississippi.

WELLS, J. T., CHINBURG, S. J. & COLEMAN, J. M. 1984. *The Atchafalaya River Delta; Report 4 Generic Analysis of Delta Development*. Technical Report 4, HL-82-15. US Army Corps of Engineers, Waterways Experimental Station, Vicksburg, Mississippi.

Fluviatile sediment fluxes to the Mediterranean Sea: a quantitative approach and the influence of dams

S. E. POULOS[1] & M. B. COLLINS[2]

[1]*Department of Geography and Climatology, Faculty of Geology, University of Athens, Panepistimioupoli, Zografou, Athens 15784, Greece*

[2]*School of Ocean and Earth Science, Southampton Oceanography Centre, University of Southampton, European Way, Southampton SO14 3ZH, U.K*

Abstract: The Mediterranean drainage basin incorporates more than 160 rivers with a catchment >200 km^2, of which only a few are larger than 50×10^3 km^2: this observation emphasizes the role of the smaller rivers. The present investigation, incorporating the analysis of data sets from 69 rivers, has estimated a total sediment flux of some 1×10^9 tonnes (t) year^{-1}; of this, suspended sediment contributes some two-thirds of the load, with the remaining third supplied by the combined dissolved and bed-load components. The magnitude of the sediment supply is best demonstrated by various observations: (i) some 46% of the total length of the Mediterranean coastline (46 133 km) has been formed by sediment deposition; (ii) many Mediterranean deltas have prograded in recent times by, at least, several metres per year; and (iii) Holocene coastal (inner shelf) deposits are some tens of metres in thickness. The construction of hundreds of dams around the Mediterranean Sea, especially over the last 50 years, has led to a dramatic reduction in the sediment supply- to approximately 50% of the potential (natural) sediment supply. Such a reduction is considered to be the primary factor responsible for the loss of coastal (mainly deltaic) land, with annual rates of erosion ranging from tens (Ebro, Po) to hundreds of metres (Nile).

The Mediterranean Sea (Fig. 1a) covers an area of 2501.5×10^3 km^2, with an average depth of 1536 m and a maximum depth of 5121 m (in the Hellenic Trench), comprising a water body of 3842×10^3 km^3 (Emelyanov & Shimkus 1986). The Mediterranean consists of an E–W-trending enclosed depression, extending almost 4000 km from the Straits of Gibraltar (in the west) to the coast of Lebanon (in the east). The total length of the coastline is 46 000 km; of this, some 33 750 km belong to the European margin, with the remaining 6550 km and 5700 km constituting the Asian and African margins, respectively (Barić & Cašparović 1992).

This sea can be divided geographically into three main areas (Carter *et al.* 1972): (i) the western Mediterranean, which includes the Alboran Sea, the Balearic Sea and the Algerian Basin; (ii) the central Mediterranean, consisting of the Tyrrhenian Sea, the Adriatic Sea and the Ionian Sea; and (iii) the eastern Mediterranean, incorporating the Aegean Sea (and the adjacent Sea of Marmara) and the Levantine Sea (Fig.1a). Water depths over the region are shown in Fig. 1(b), with those <200 m, between 200 and 2000 m, and >2000 m highlighted.

The present-day configuration of the surrounding Mediterranean hinterland is the result of three related factors: (a) crustal mobility; (b) climatic variability; and (c) sea-level change (Macklin *et al.* 1995). The Mediterranean is at the boundary between the Eurasian, African and Arabian plates (Fig. 2), the interaction of which resulted in the formation of the Alpine fold belt: this high relief (often >3000 m) feature extends from Gibraltar to the Middle East. The structure of the basin is extremely complex, incorporating a number of smaller secondary or microplates that have, often, very different geological histories to those of the major plates (Dewey *et al.* 1973). Two principal tectonic phases have been identified within this context: a N–S compression, culminating in the Late Miocene; this was followed by a tensional phase when large areas of the crust foundered during the Oligocene. As far as the landforms are concerned, such tectonic activity has resulted in the formation of three rather distinctive environments around the Mediterranean (Macklin *et al.* 1995): (i) the Precambrian African plate, underlying North Africa, nowadays forming a low desert environment (distinguished, in the eastern Mediterranean, by rifting in Sinai and in the Jordan Valley); (ii) the

From: JONES, S. J. & FROSTICK, L. E. (eds) 2002. *Sediment Flux to Basins: Causes, Controls and Consequences.* Geological Society, London, Special Publications, **191**, 227–245. 0305-8719/02/$15.00

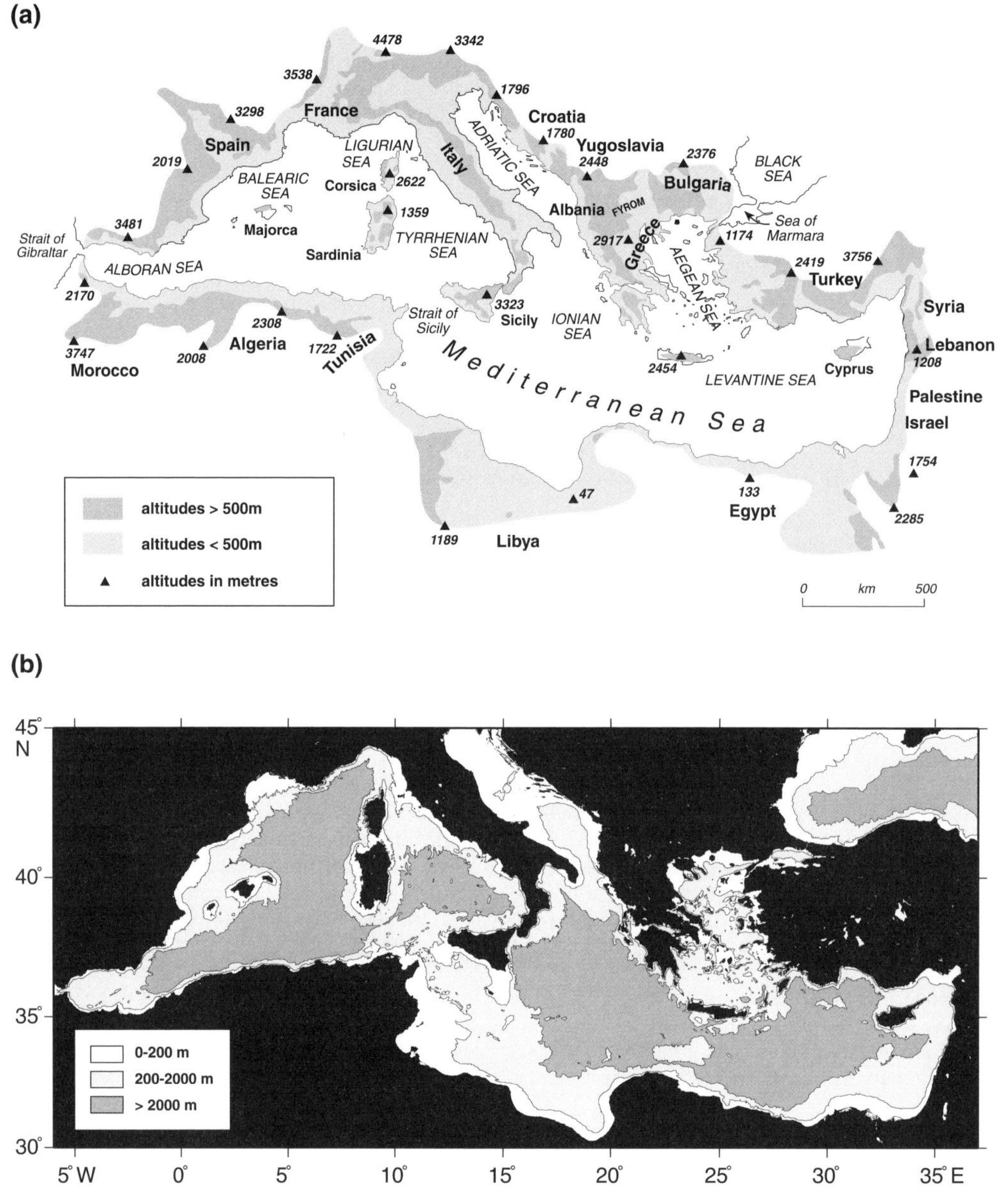

Fig. 1. (**a**) Principal geomorphological characteristics and geographical locations in the Mediterranean Basin. (**b**) Generalized bathymetry of the basin.

folded and partly metamorphosed Variscan Massifs of the Iberian Peninsula, Corsica and Sardinia – covered in eastern Spain, by flat-lying or gently folded Mesozoic and Cenozoic sediments; and (iii) the Alpine fold belt including the mountain chains (Pyrenees, Alps, Hellenides, Taurides) that form the northern morphological boundary of the Mediterranean drainage basin. Furthermore, the development and evolution of the fluvial systems in the circum-Mediterranean region have been dominated by the tectonics and the structural controls (Audley-Charles *et al.* 1977; Macklin *et al.* 1995). For example, the Ebro and Po rivers drain along a strike zone between mountain belts that

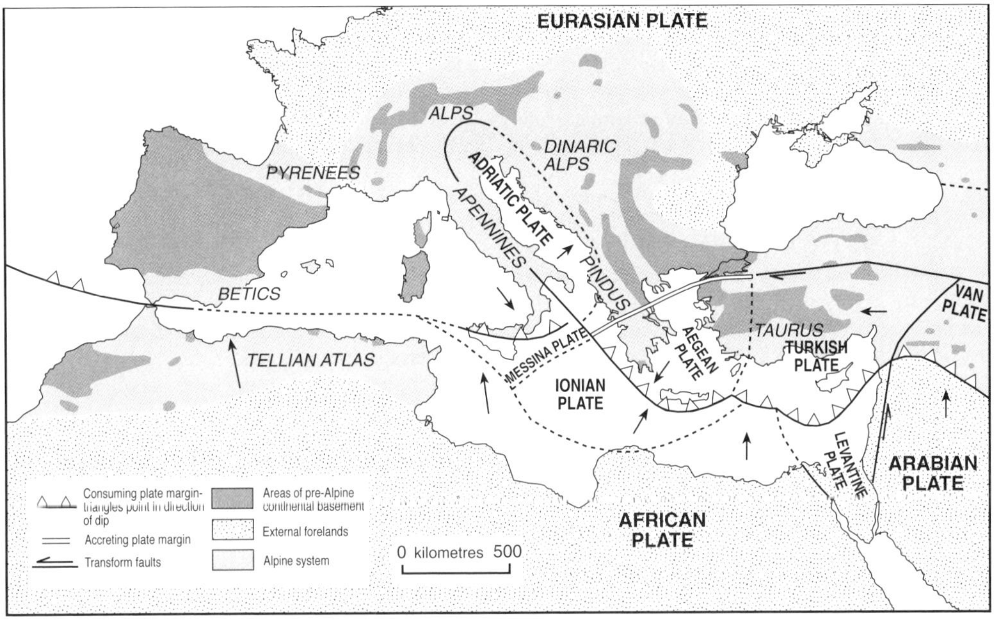

Fig. 2. Basic geological structure of the Mediterranean Basin (after Dewey *et al.* 1973; Windley 1984).

have been described as *intra-orogen* in type. In comparison, rivers in areas undergoing active normal faulting or compressional folding (e.g. the Aegean Sea and Betics) have been described as *trans-orogen* in type (Summerfield 1991).

The Mediterranean is characterized generally by relatively narrow continental shelves bounded by steep slopes (exceptions are the northern Adriatic and the eastern Tunisian coast): the deepest parts are associated with submarine trenches, with the deepest (5121 m) being the presently-active Hellenic Trench. Fluvial sediment inputs contribute to the development of the present bathymetry, particularly within the coastal zone and on the adjacent continental shelf. Furthermore, climatic changes, sea-level rise and/or alterations in the water and sediment volumes supplied by the rivers, have impacted upon the coastal zone; in particular in the lowland coastal areas, for example the coastal alluvial plains, deltas and wetlands. Vita-Finzi (1972) has claimed that between 20 000 and 10 000 BP, there was only limited fluvial input followed by two periods of high sediment delivery: (i) 10 000–2000 BP; and (ii) 300–0 BP. The first period occurred during the Flandrian transgression, the second occurred in historical times when increased soil erosion was associated with human activities (i.e. deforestation, agriculture, development, river regulation, etc.).

The present-day coastline of the Mediterranean is very rugged and indented in character. According to Barić and Cašparović (1992), some 54% of the coast is rocky with the remaining 46% consisting of various sedimentary accumulations; these are in the form of sand and rocky beaches, dunes, deltas, estuaries, wetlands and lagoons. Fluviatile sediment supplies have been extremely important, in relation to the evolution of the coastal zone and sedimentation on the adjacent sea floor. Furthermore, any modifications are likely to cause a series of socio-economic pressures on the system, as around 37% (some 133 million people) of the overall population of the Mediterranean countries live within the coastal zone (Grenon & Batisse 1989).

It has been estimated that some 350×10^6 t year^{-1} of suspended sediment are delivered by the Alpine rivers discharging along the European coastline (excluding the Greek and most of the Turkish rivers) of the Mediterranean (Milliman & Syvitski 1992), and 100×10^6 and 75×10^6 t year^{-1} of suspended sediments are provided by the Maghreb area (Morocco, Algeria) (Probst & Amiotte-Suchet 1992) and from the Greek mainland (Poulos & Chronis 1997), respectively.

Undoubtedly, the most pronounced anthropogenic influence on natural processes over the whole of the region has been the construction of dams for hydroelectric and irrigation purposes within recent decades; these have caused dramatic reductions in the overall fluviatile water-sediment supply to the Mediterranean Basin. The changes induced have led to the deterioration (e.g. retreat, seawater intrusion, etc.) of many of the coastal environments; in particular, deltaic sections of the coastline. Such changes have been enhanced in response to sea-level rise, due to the 'greenhouse effect'. Substantial coastal retreat has been reported from the major deltas following dam construction, for example the Nile (Fanos 1995), the Po (Simeoni & Bondesan 1997) and the Ebro (Jimenez *et al.* 1997). In a recent global appraisal concerning fluviatile sediment fluxes and the significance of the human interference in changing the delivery of sediment loads to the coastal zone, Walling & Webb (1998) have identified sediment fluxes as one of the key parameters in understanding the function and evolution of the 'Earth system' and its sensitivity to changes, that is climatic.

The present contribution is a synthesis based on extensive previously published and unpublished data sets of sediment fluxes from a large number of rivers, draining the countries bordering the Mediterranean Basin. The review incorporates a reassessment of existing sedimentological/oceanographic information for the coastal zone, offshore zone and the floor of the Mediterranean Sea. The aims of the investigation are: (a) the provision of a detailed description of the drainage basins of the Mediterranean Sea, focusing on aspects influencing fluvial sediment supply (such as geomorphological and climatological characteristics, rates of land erosion); (b) a quantitative evaluation of the total *fluvial sediment supply* from the different physiographical regions of the Mediterranean drainage basin, including an appraisal of the significance of the smaller river systems; (c) an assessment of the contribution of fluviatile sediment supply to the evolution of the coastal zone and, in particular, to coastal depositionary environments (i.e. delta systems); and (d) an evaluation of the reduction in sediment supply following the construction of large numbers of dams, together with a quantitative appraisal of present-day coastal retreat.

The Mediterranean drainage basin

Geomorphological setting

The Mediterranean drainage basin covers an area of some 1 335 000 km^2 excluding the catchment area of the River Nile, which accounts for a further 2 800 000 km^2 (i.e. similar, in size, to the continent of Australia), located to the southeast of the system (30–47°N and 5°W–37°E).

Apart from the presence of the Nile, the catchment area of the Mediterranean is somewhat narrow, with the enclosing watershed lying generally at only a short distance from the coastline; it consists of stretches of narrow coastal plains, with the general backdrop of extensive tracts of mountainous terrain of high relief. The only exception is the low-lying desert between Tunisia and Sinai. Thus, the 500 m topographic contour and selected elevations of the mountains surrounding the Mediterranean illustrate the mountainous character of the various catchment areas (Fig. 1a). This particular topographic contour has also been used elsewhere (Macklin *et al.* 1995) for the geomorphological description of the Mediterranean catchment area; likewise, as the division between *uplands* and *lowlands* in the elevation-based classification of world rivers (Milliman & Syvitski 1992). Within this tectonic/geomorphological setting, the drainage (with the exception of some of the larger river systems, e.g. Ebro, Po, Rhone and Nile) is dominated by rivers that are relatively short in length and have steep gradients (Macklin *et al.* 1995).

Climatic conditions and weathering

The Mediterranean Basin lies within the boundary between sub-tropical and mid-latitude atmospheric circulation patterns, characterized by climatic types that vary from 'temperate maritime' to 'arid subtropical'. On a seasonal basis, hot dry summers and mild damp winters distinguish the climatic conditions. Further, this type of climate influences the Mediterranean bordering lands at considerable distances (>1000 km) from the coast, moderating the summer temperatures. Accordingly, during winter, the juxtaposition of cold and dry continental air (with relatively warm waters) causes strong evaporation and atmospheric instability; these lead to local cyclone genesis (Gat & Magaritz 1980).

There are marked regional contrasts in the spatial distribution of the annual rainfall (Fig. 3a). Both the quantity of rainfall and the duration of the rainy season decrease from west to east, and from north to south, across the basin. A similar tendency characterizes the length and the intensity of the drought (Fig. 3b). Temporally, rainfall is distinctly irregular, both during the year and from one year to the next, especially in the drainage basin area to the south. There is a substantial seasonal difference in the pattern of the rainfall, with the wet period in the north during autumn; this occurs during winter in the south (Grenon & Batisse 1989). In general, the annual rainfall levels over the drainage basin vary from less than 200 mm in the south to in excess of 1500 m to the north and at high altitudes. Overall rainfall amounts are also augmented by altitude, most notably along the northern part of the basin, in particular along the western flank of the mountain chains, for example the Apennines, Dinaric Alps, and Pindus (Fig. 2).

Land degradation within the Mediterranean drainage basins is high, due to changing environmental conditions or to human occupation. Nowadays, large areas of the land bordering the Mediterranean are associated only with truncated (spatial) remnants of the original soil cover, with

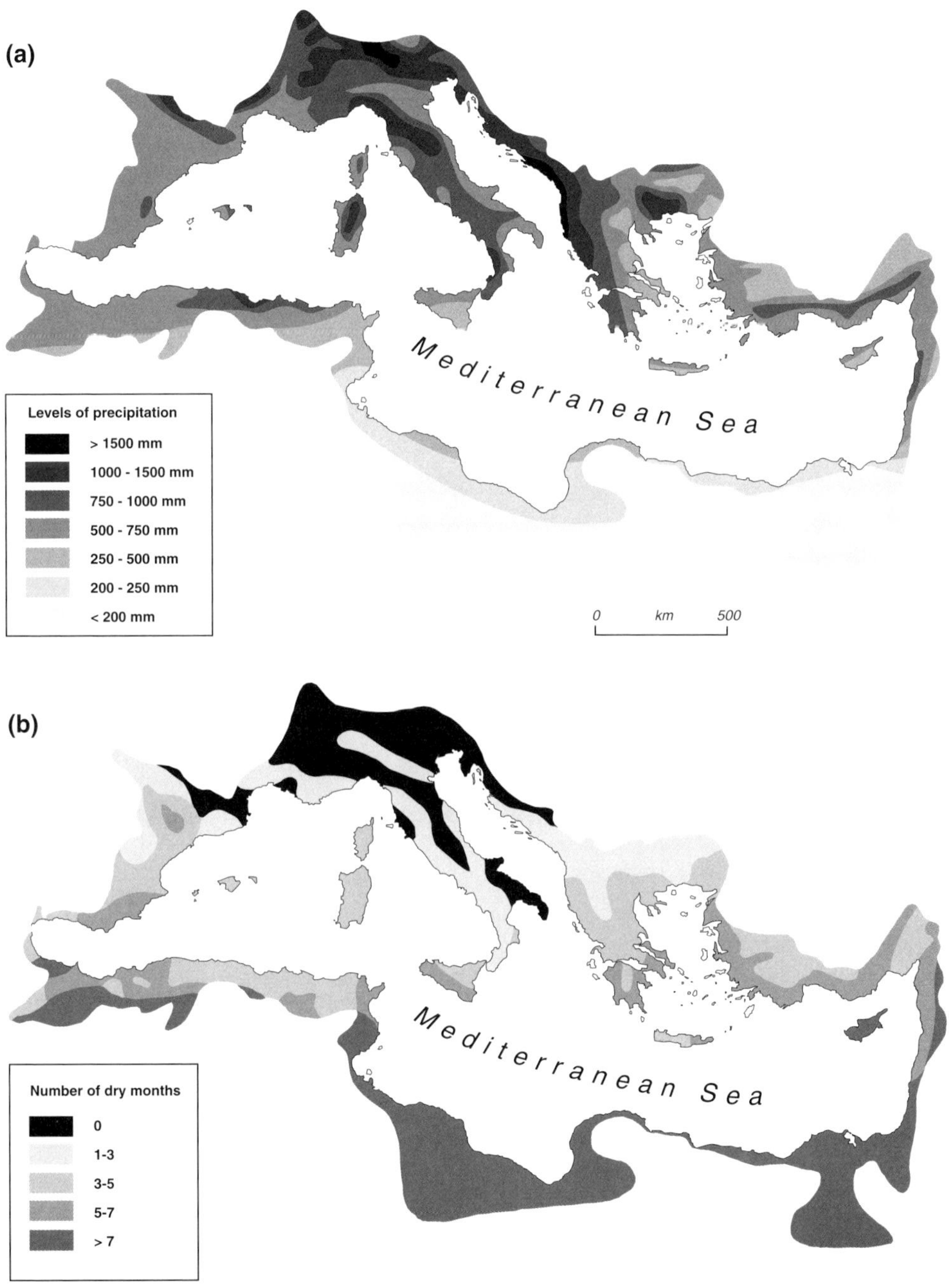

Fig. 3. (**a**) Rainfall over the Mediterranean drainage basin (after UNESCO 1978); (**b**) seasonal distribution of the drought period (after Grenon & Batisse 1989).

little potential for further degradation (Imeson & Emmer 1992). For example, within the former Yugoslavia there are 5661×10^4 km^2 of almost completely denuded karst, of which the 38% is totally barren (Zachar 1982). The annual erosion potentials (expressed as soil degradation, in t km^{-2}) within the drainage basins of the Mediterranean vary from around zero (slight), to as much as 5000 t km^{-2} (high) (UNEP 1984). The lowest values are in the semi-arid region of Tunisia and Libya, with the highest in the northeastern part of the watershed (the Balkans and Turkey). Furthermore, present-day degradation is expected to be greatest on colluvial and valley-fill deposits, related also to hydraulically-controlled erosion (runoff). This process is enhanced by the fact that more than half of the land area is unprotected by natural (vegetation) or man-made cover. Further, with respect to water erosion, two principal zones have been identified (De Walle *et al.* 1993): (a) the northern coast (Spain–Turkey, including Cyprus), where it varies from less than 62.5 to more than 500 t km^{-2} $year^{-1}$; and (b) the southern coast (Morocco–Syria), where the values are higher by an order of magnitude (<250 and >5000 t km^{-2} $year^{-1}$). The high rates of erosion occur within river basins underlain by unconsolidated sedimentary rocks, such as marls, siltstones and flysch. For example, in Turkey, where 70% of the land is vulnerable to erosion, it has been estimated that some 1000 million tonnes of topsoil is removed every year (Grenon & Batisse 1989).

The region encompasses a wider range of fluvial denudation rates than have been recorded elsewhere on a global scale (Walling & Webb 1983). Sediment yields over the different regions of the Mediterranean watershed area vary, on average, from <250 to >1000 t km^{-2} $year^{-1}$ (Woodward 1995). These yields are much higher on the scale of a particular river, occasionally exceeding 4000 t km^{-2} $year^{-1}$ (e.g. Albanian rivers): a maximum value of 5070 t km^{-2} $year^{-1}$ has been reported in the case of the Kasseb River, in Tunisia (Walling 1986). Similarly, extremely high rates of erosion (7 900–28 500 t km^{-2} $year^{-1}$) have been reported for the River Nekor, in the Rift Mountains of Morocco (McNeill 1992). Finally, in examining the importance of major factors controlling the magnitude of fluvial suspended sediment yields, precipitation levels and the maximum catchment elevation have been identified by Milliman & Syvitski (1992).

Mediterranean rivers

The Mediterranean drainage basin consists of the integrated catchment areas of more than 160 rivers, discharging along the coastline: this number may approach >300 when all the ephemeral rivers and torrential streams are considered. In general, Mediterranean rivers, with the exception of some of the large rivers, such as the Nile, Rhone and Po, tend to be short to intermediate in length; many are ephemeral (Stanley 1977).

The present investigation presents the analysis of data sets from 169 rivers, with catchment areas larger than 200 km^2 (Fig. 4). Most of these rivers (115) drain from the European continent, with the reminder belonging to Asia (23) and Africa (31). Many other rivers, smaller in size (watersheds <200 km^2) and with ephemeral flows and torrents, are related to mountainous coastal areas. The largest river draining into the Mediterranean catchment is the Nile. Overall, there are only a few rivers with drainage basins larger than 50×10^3 km^2: the Ebro (85×10^3 km^2) in Spain; the Rhone (96×10^3 km^2) in France; the Po (54×10^3 km^2), in Italy and the Moulouya (51×10^3 km^2) in Algeria.

The Mediterranean rivers and their catchment areas have been divided into five physiographical regions (Fig. 4) in relation to their geographical locations, their climatological conditions (related strongly to the prevailing precipitation levels), and the data available on their water and sediment fluxes. The areas of each of those physiographical regions, together with their basic geomorphological characteristics (e.g. relief) and the corresponding precipitation levels, are summarized in Table 1.

Interestingly, the frequency of occurrence of rivers within each of the physiographical regions is somewhat similar for Regions 1–4, where 0.1–0.3 rivers correspond to every 1000 km^2 of catchment area (Table 1). The highest density is associated with the west coast of the Balkan Peninsula, in response to the high levels of precipitation (>1500 mm $year^{-1}$) and associated high relief (>3000 m). In contrast, the lowest density (0.001 rivers per 10^3 km^2) is observed to be associated with Region 5 (eastern North Africa coast); this is explained by the exceptionally large (compared to the other Mediterranean rivers) area of the River Nile catchment, in relation to the absence of many other river systems within the prevailing dry climatological conditions.

The principal rivers (catchment area >10 000 km^2) are listed for each of the five physiographical regions in Table 2. The data sets reveal that the large river systems drain only around 50% of the associated physiographical region, with the exception of Region 5 (where the River Nile represents >95% of the area). This observation emphasizes the importance of the smaller (catchment <10 000 km^2) rivers, in terms of their water and sediment contribution to the Mediterranean Basin, as they drain some half of the total drainage area (apart from its southeastern part).

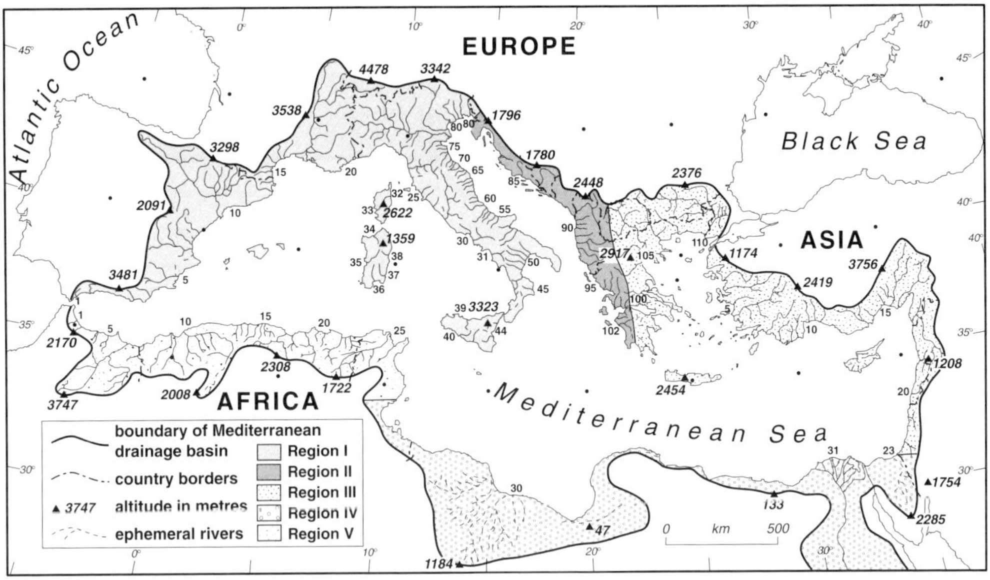

Fig. 4. Principal rivers and their drainage basins along the Mediterranean coast (after Macklin 1995; *The Times Atlas of the World* 1997). Rivers are ordered separately for each continent (Europe, 1–115; Asia, 1–23; Africa, 1–31) and are listed in the Appendix.

Fluvial sediment fluxes

The estimation of the total sediment load transported by the river networks draining the catchment of the Mediterranean Basin includes material that is transported in suspension (SSL, suspended sediment load), in solution (DL, dissolved load) and along the river bed (BL, bed load). Most of the available data refer to the suspended load, as only limited field measurements exist for the dissolved and bed-load components (even on a global basis). This limitation occurs due to the difficulties encountered in measuring, with any accuracy, the dissolved and (especially) the bed-load components of the total load.

Estimation of suspended sediment fluxes

Data from published information on field measurements of the suspended sediment fluxes of 69 rivers discharging along the coast of all the

Table 1. *Geomorphological characteristics and precipitation levels of the (five) physiographical regions of the Mediterranean drainage basin (for details, see text)*

Physiographical regions	Area (10^3 km^2)	Max. elevation (m)	Area >500 m of altitude (10^3 km^2)	Ratio to overall area (5)	No. of rivers	No of rivers per 10^3 km^2	Precipitation (based upon the data presented in Fig. 3) (mm) Min.	Max.	Mean
Region 1	490	4478	175.0	32	83	0.17	250	>1500	850
Region 2	60	2448	31.0	48	19	0.32	750	>1500	1100
Region 3	340	2917	12.0	36	36	0.10	250	>1000	650
Region 4	210	3747	67.5	32	21	0.10	250	>1000	600
Region 5 (incl. Nile)	235	1189	39.5	17	4	0.02	<200	500	~200
Region 5 (excl. Nile)	3035		2350.0	84	5	0.002			
Total (excl. Nile)	1335								
Total (incl. Nile)	4135		2635.5		169				

Table 2. *Principal rivers (catchment area >10^4 km^2), their catchment area and ratio relative to the total area of the drainage basin for each of the (five) physiographical regions*

		Principal river (>10^4 km^2) (associated country)	Catchment area (km^2)	Area drained, compared to the total area of the region (%)
Region 1		Segura (Spain)	15 000	3
		Ebro (Spain)	85 835	17
		Rhone (France)	96 000	19
		Tevere (Italy)	17 000	3
		Po (Italy)	54 290	11
	Total:		**268 125**	**54**
Region 2		Neretva (Croatia)	13 000	22
		Drin (Albania)	19 582	33
	Total:		**32 582**	**55**
Region 3		Pinios (Greece)	10 750	3
		Axios (Greece)	23 747	7
		Strimonas(Greece)	16 550	5
		Evros (Greece–Turkey)	27 465	8
		Cediz-Nehri (Turkey)	18 000	5
		Büyükmenderes (Turkey)	25 000	7
		Seyhan (Turkey)	14 015	4
		Ceyhan (Turkey)	19 095	6
	Total:		**154 622**	**45**
Region 4		Moulouya (Algeria)	51 000	24
		Chellif (Algeria)	44 000	21
		Mejerda (Tunisia)	22 000	11
	Total:		**117 000**	**56**
Region 5		Nile (Egypt)	2 800 000	92
	Total:		**2 800 000**	**92**
Overall total (excl. Nile)			**572 329**	**42**
Overall total (incl. Nile)			**3 372 329**	**81**

Mediterranean countries have been abstracted, in order to calculate the overall SSL and the suspended sediment yields for each of the five physiographical regions.

In general, estimation of the annual suspended sediment yields of the regions is based on calculation of the weighed average of the corresponding yields, from field measurements obtained at an appropriate number of rivers; these represent more than the two-thirds of the total catchment area for each of the five regions. This generalization does not apply to Region 3, where the available data incorporates only around 21% of the watershed; this is despite the fact that the data set relates to 11, out of 36, rivers. These values were then scaled by the area of each of the regions, assuming the runoff weathering processes to be similar throughout each region. On this basis, the mean potential annual sediment supply, to the adjacent coast, was calculated.

The west coast of the Balkan Peninsula (Region 2) is characterized by the highest sediment yield (1400 t km^{-2}), due to the association of high precipitation levels with topographic relief; in contrast, the lowest sediment yield (40 t km^{-2}) occurs in Region 5, despite the contribution of the River Nile (Table 3). Region 1 supplies the largest annual amount of suspended sediment (257×10^6 t), whilst the lowest fluxes are derived from Regions 2 and 3 ($<90 \times 10^6$ t). The overall (potential) annual suspended sediment load is around 650×10^6 t. Such an amount of suspended sediment load identifies the Mediterranean Basin as an area of high fluvial influx, when it is compared with an estimate of 20×10^9 t year^{-1} (SSL) on the basis of an analysis undertaken on a global scale by Milliman & Syvitski (1992).

Estimation of dissolved load

The annual dissolved load of the rivers of southern Europe (Regions 1 and 3) is, on average, >40% of the corresponding suspended sediment load (Table

Table 3. *Estimation of the annual suspended sediment loads and yields of the five physiographical regions of the Mediterranean drainage basin, based on field measurements*

	Suspended sediment load (10^6 t)	Suspended sediment (t km^{-2})	Area* (10^3 km^2)	Number of Rivers†
Region 1	257.2	525	299.1 (75%)	30 (83)
Region 2	84.00	1400	48.4 (81%)	11 (19)
Region 3	107.1	315	71.8 (21%)	11 (36)
Region 4	88.2	325 (420)‡	150.6 (71%)	16 (21)
Region 5 (excl. Nile)			235 (99%)	
Region 5 (incl. Nile)	121.4	40	3035 (99%)	1 (2)
Total (excl. Nile)	**536.5**	**402**	**804.9 (60.3)**	
Total (incl. Nile)	**657.9**	**150**	**3839.9 (92.7)**	**69 (169)**

*The area covered by the field measurements, together with the percentage relative to the total area of each of the physiographical regions.
†The number of rivers associated with the field measurements (in parenthesis, the total number of rivers is given (see Table 1)).
‡After Probst & Amiotte-Suchet (1992).

4). This contribution is much lower (<10%) in Region 5, where the soluble load of the dominant River Nile averages around 7%. Generally, within arid and semi-arid regions, where the vegetation is sparse and the sediment production is high, the dissolved load forms only a relatively small component of the total sediment load (e.g. Nile, 13.5%; Ganges, 12.3%; Yellow, 3.5%). When precipitation and runoff are higher and there is abundant vegetation, the contribution of the dissolved load increases (e.g. Amazon, 24.5%; Congo, 47%; and Changjiang, 34% (Meybeck 1988).

Taking into consideration the climatic differences over the catchment area of the Mediterranean Basin and the relatively high topographic relief (with 84% of the area with altitudes >500 m), it is estimated that the dissolved load is 30±5% of the mass of the suspended sediment load.

Estimation of bed load

In general, there is a world-wide dearth of comprehensive field measurements relating to bed-load transport by rivers; with only a few exceptions, this applies also to the Mediterranean Basin. Thus, between 1950 and 1970 the River Ebro transported 0.7×10^6 t as bed load (BL) (Guillén *et al.* 1992); over the same period, the corresponding suspended sediment load (SSL) was 2.8×10^6 t (Maldonado 1985). These data provide a ratio (SSL:BL) of 4 : 1 (i.e. bed load around 25% of the suspended load). Such transport for the Mediterranean rivers is

Table 4. *Published annual yields of dissolved load, based upon field measurements, together with their relationship to suspended sediment yields for various Mediterranean river systems*

River	Area (10^3 km^2)	SS (t km^{-2})	DL (t km^{-2})	Ratio (SS/DL)	Percentage (SS : DL,%)
Nile (Egypt)	2800	40	2.1*	19:1	5.2
Rhone (France)	96	583	177*	3.3:1	30.3
Arno (Italy)	8.2	99	99*	1:1	100
Tevere (Italy)	17	441	347*	1.3:1	77
Arachthos (Greece)	1.9	3941	401†	9.8:1	10
Acheloos (Greece)	5.4	614	280†	2.2:1	45
Aliakmon (Greece)	9.5	460	128†	3.6:1	28
Nestos (Greece)	6.2	160	128†	1.2:1	83
Average					47.25
Weighted average (excl. Nile)					42.20
Weighted average(incl. Nile)					7.00

Key: SS, suspended sediment; DL, dissolved load.
*After Milliman *et al.* (1995)
†After Skoulikidis (1993).

associated principally with extreme and ephemeral flood events. For example, detailed measurements undertaken in the Virginio, a tributary of the River Arno (Italy), has shown that during flood events the SSL flux varied between 1 and 16 kg s^{-1}; at the same time, the bed load was 0.1–2 kg s^{-1} (Tazioli & Billi 1987). On this basis, the percentage of bed load, relative to the suspended sediment load, ranged from 'minimal' to <15%. Similarly, an average value of the bed-load component of 15–20% of the total load was found for Albanian rivers (Pano 1992); interestingly, the latter are also characterized by the highest suspended sediment yields (>3000 t km^{-2}) over the region. Elsewhere, Dal Cin (1983) has estimated the bed load to be 25% of the total (solid) load in the case of the River Po. In general, other investigators (e.g. Blatt *et al.* 1972; UNESCO 1985) have accepted that some 10% of the total (solid) load is a reasonable estimate for the bed-load contribution. However, within particular areas this proportion may be much higher, as in the case of the River Shangyou (China), where the bed load reaches 30.5% of the total sediment load (Qian & Dai 1980).

Increased bed-load transport rates might be expected to occur during flood events in small ephemeral rivers, characterized by high relief and easily erodible lithologies; for example, rivers draining areas of desert. For example, within the Nahar Yatir (Israel) bed-load transport has been observed to be of the order of 1–10 kg s^{-1} m^{-1}. Desert rivers, when compared with other streams located within humid zones, appear to be some 400 times more efficient at transporting coarse-grained material, as bed load (Reid *et al.* 1985; Reid & Frostick 1987; Laronne & Reid 1993). On the basis of these observations, it seems appropriate to assume that bed load contributes some 10–15% of the total sediment load within Mediterranean river systems.

Estimation of 'total' sediment load

According to the above analysis and discussion, the catchment area of the the Mediterranean Basin represents a potential sediment flux that accounts for some 650×10^6 t of suspended sediment load, together with 195×10^6–260×10^6 t (i.e. 30±5% of the SSL) carried in solution and another 100×10^6–160×10^6 t transferred as bed load (i.e. 10–15%, relative to the total sediment load). Hence, the 'total' load may be estimated to be 1×10^9 t; of this, some two-thirds are carried in suspension, and a third as (combined) dissolved and bed-load contributions. This overall sediment budget means that the Mediterranean hinterland area (some 4.1×10^6 km^2 in area), has a sediment yield of about 250 t km^{-2}; this compares with a global estimate of 500 t km^{-2}. The latter figure is based on an estimate of 75×10^9 t $year^{-1}$ of potential sediment transportation (Pimental *et al.* 1995) from the continents (with a land area of 149×10^6 km^2) to the ocean. However, the mean value of 250 t km^{-2} for the Mediterranean becomes 400 t km^{-2} when Region 5 is excluded (the River Nile catchment), which is characterized by low sediment fluxes (with yields of approximately 50 t km^{-2}).

The sediment fluxes are not distributed evenly throughout the year, but have the same seasonal variability as the climate and, especially, the river runoff. Hence, the greater proportion of the sediment load is transferred during the wet season (September–May). Furthermore, the highest sediment discharges might be expected to occur at the beginning of this period (Fig. 5), when an increase

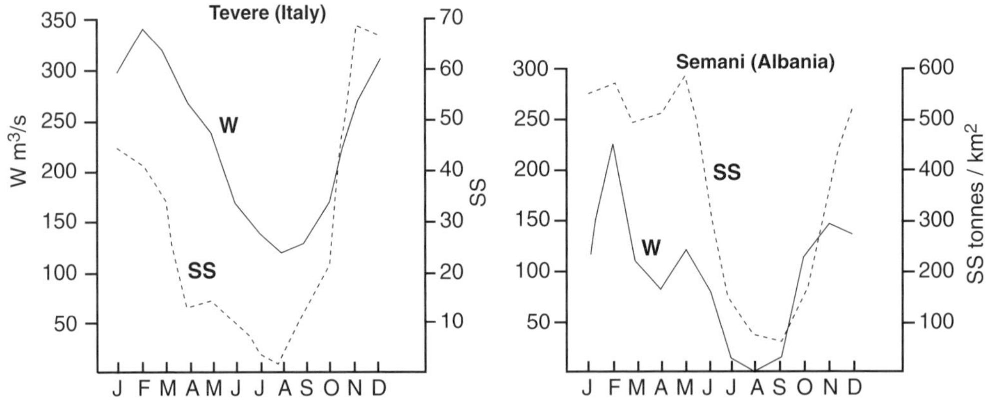

Fig. 5. Monthly distribution of water (W) and sediment (SS) fluxes of two typical Mediterranean rivers (data abstracted from UNESCO 1978 and Poulos *et al.* 1996).

in the water discharge is accompanied by a dramatic increase in the sediment concentrations (November–December). Towards the end of the (wet) season, when runoff begins to decrease gradually (March–May), the sediment discharges reduce rapidly and they are at a minimum during the dry (summer) season. This variation in the contributions to the overall sediment flux may be explained in terms of the sparse vegetation cover and desiccated soil conditions during summer; at such a time, the land surface becomes vulnerable to erosion. Subsequently, detritus is removed readily and is transported by the river network during the beginning of the wet season. Towards the end of the season, the sediment discharges decrease substantially in relation to river runoff as there is less weathered material available for transportation. This difference is related to dense plant cover, with a corresponding increase in soil cohesion and a reduction in erodibility.

Fate of fluviatile sediments

The greater parts of the high sediment loads transported by the Mediterranean rivers are delivered to the coastal zone (in relation to the proximity of their catchment areas, to the river mouth and their high hypsometric gradients: Milliman & Syvitski 1992) to form extensive river deltas and contribute to the development of alluvial coastal plains. Furthermore, the fine-grained sediments (silts and clays) are dispersed over the recent shelf/slope and deep basin, by various processes (Collins 1986).

Coastal and nearshore environment

Some 46% of the total length of the Mediterranean coastline (46 133 km) has been estimated to have been formed by sediment deposition (Barić and Cašparović 1992). There are more than 100 (moderate–large) deltas that have developed in their present form over the past 5000 years (Stanley 1997). This distribution relates to many factors such as sediment fluxes, deforestation vegetation of the hinterland and the bathymetry of the receiving basin. Such processes are exemplified by the large number of small deltas along the western coast of the Adriatic Sea and within shallow embayments in Greece and Turkey. The largest modern deltas are those of the Nile in Egypt, the Rhone in southern France, the Po in northern Italy, the Ebro in Spain, the deltaic complexes of the Ceyhan/Seyhan in southeastern Turkey, the Axios/Aliakmon in northwestern Aegean Sea and the Mejerda in northern Tunisia (Fig. 1a). On the other hand, the absence of any well-developed deltas along the Moroccan and Algerian coasts can be attributed to the steep and narrow adjacent continental shelf (Fig. 1b).

The influence of the Mediterranean rivers upon the evolution of the lower part of the coastal zone and, in particular, the coastline (especially during the Holocene, the last 5×10^3–6×10^3 years) is best demonstrated by: (i) the largest of the river systems, where coastal progradation is, at least, several metres per year; and (ii) the thickness of Holocene deposit, of up to tens of metres (Table 5). Such high rates of deposition and seaward progradation have led to the development of large deltaic plains along the Greek peninsula, where the ratio between the areas of the subaerial deltaic plain (D) and the catchment (C), D/C, reaches 14×10^{-2}, with a mean of 5×10^{-2} (Poulos & Chronis 1997). These values are comparable to those associated with other large river systems of the world – Amazon (7.4×10^{-2}), Mississippi (0.9×10^{-2}), Indus (3.1×10^{-2}), Niger (1.7×10^{-2}), and Hwanghe (4.9×10^{-2}) (Poulos *et al.* 1996). Further along the Aegean coast of Turkey, the annual deltaic growth during historical times (from 1000 BC and later) has been of the order of 1–10 m year^{-1}, in the case of the Kujuk Menderes (in antiquity, *Kaystros*), Buyuk Menderes (*Maiandros*), Daylan Cayi (*Kalbis*) and Karamenderes (*Scamandros*) (Brukner 1997). Similarly, along the Albanian coast, annual progradation since the beginning of the nineteenth

Table 5. *Rates of Holocene deposition and progradation of the major river–delta systems*

Rive-delta system	Holocene deposition rate (m)	Accretional rates of coastline progradation	Ratio ($\times 10^{-2}$): delta area/catchment area
Ebro	4–26 m[1]	0.14 km^2 year^{-1} (1750–1915)[6]	0.69
Rhone	35 m[2]	3.5 m year^{-1} (Holocene)[7]	0.03
Po	30 m[2]	8.5 m year^{-1} (Holocene)[7]	18.68
Axios/Aliakmon	20 m[3]	1.3 km^2 year^{-1} (1850–1987)[8]	6.25
Seyhan/Ceyhan	10 m[4]	2 m year^{-1} (Holocene)[7]	
Nile(Rosetta Promotory)	40 m[5]	20–40 m year^{-1} (since 1850)[9]	0.46

Data sources as follows: [1]Nelson (1990); [2]Chronis *et al.* (1991); [3]Poulos *et al.* (1996); [4]Bodur & Ergin (1992); [5]Stanley (1988); [6]Jimenez *et al.* (1997); [7]Evans (1971); [8]Poulos *et al.* (1994); [9]Sestini (1992*a*, *b*).

century has been estimated to be of the order of tens of metres – Vjose (15–47 m year^{-1}) and Semani (40–50 m year^{-1}) (Simeoni *et al.* 1997).

Increased sediment fluxes, within Classical times, have been related generally to increasing population, the conversion of woodlands to grazing ground or arable land and improvements in technology. Such use of land, since early Classical times, has led to soil erosion on an increasingly intensive scale in the eastern Mediterranean (Brice 1978). Furthermore, river diversions have been reported in historical sources (Semple 1932) and these have contributed to the ability of the rivers to transport and discharge sediments at the coastline.

The presence of dams

During the second half of the nineteenth century, progradation of the deltaic coastlines has been followed by retreat; in most areas, this is due to a man-induced reduction in the water and sediment supply of the rivers. The most pronounced human interference, in relation to the reduction of the natural supply of fluviatile sediments to the coast, has been the construction of dams. Overall, the Mediterranean countries have more than 3500 small (height ≥30 m) and high (height ≥60 m) dams along their river networks (Table 6). Over 98% of these dams have been constructed since the beginning of the nineteenth century, 84% within only the past 50 years. The dams are used for hydroelectric power generation, irrigation purposes or protection against flood events. Whatever their function, all of the bed load, most of the suspended load and some of the dissolved load is trapped behind the dams. Such retention has caused a dramatic reduction in the downstream supply of river sediments.

Nowadays, 94 of the (169) larger river systems (Fig. 4) are dammed with more than a single (dam) construction. As such, the original natural drainage basin area of the Mediterranean Sea has been reduced by some 40% (Fig. 6); this almost doubles to 78%, when the River Nile is included in the analysis. An estimate of the overall sediment reduction within each of the physiographical regions, in response to dam construction, shows that only 35% of the sediment provided by the Mediterranean catchment may enter the lower reaches of the drainage networks and, eventually, the coastal zone and offshore areas (Table 7). Indeed, this percentage may be even lower on the basis of the fact that the area downstream of the dam usually coincides with the lower part of the catchment area; this, in turn, is characterized by low relief and moderate to low slope gradients, contributing only small amounts of sediment as bed load and in suspension. Likewise, this area is not affected usually by floods as the river flow is regulated by the dam operation. On a global scale, it has been estimated that >40% of river discharges are intercepted today by large impoundments,

Table 6. *Dams of the countries of the Mediterranean Sea (abstracted from ICOLD 1998)*

	Prior 1900	1900–1949	Construction Periods 1950–1999				
			1950–1959	1960–1969	1970–1979	1980–1989	1990–1999
Albania			1	73	125	98	9
Algeria	3	15	3	9	6	40	31
Bosnia	1	0	3	9	4	7	1
Croatia	0	0	5	4	4	14	2
Cyprus	0	0	5	16	4	21	6
Egypt	0	4			1	1	
France	35	125	75	90	93	96	55
Greece	0	1	3	4	3	7	26
F.Y.R.O.M.*	0	1	2	8	3	2	
Italy	6	190	116	77	43	33	31
Lebanon	0	0	2	3	0	0	0
Libya					8	4	
Morocco	0	7	6	5	8	37	29
Spain	52	215	159	211	195	186	139
Syria				13	6	7	6
Tunisia						10	61
Turkey		3	6	29	72	151	339
Yugoslavia		5	10	10	11	27	2
		subtotal:	396	561	586	720	728
Total	**38**	**566**			**2991**		

*F.Y.R.O.M., Former Yugoslavia Region of Macedonia.

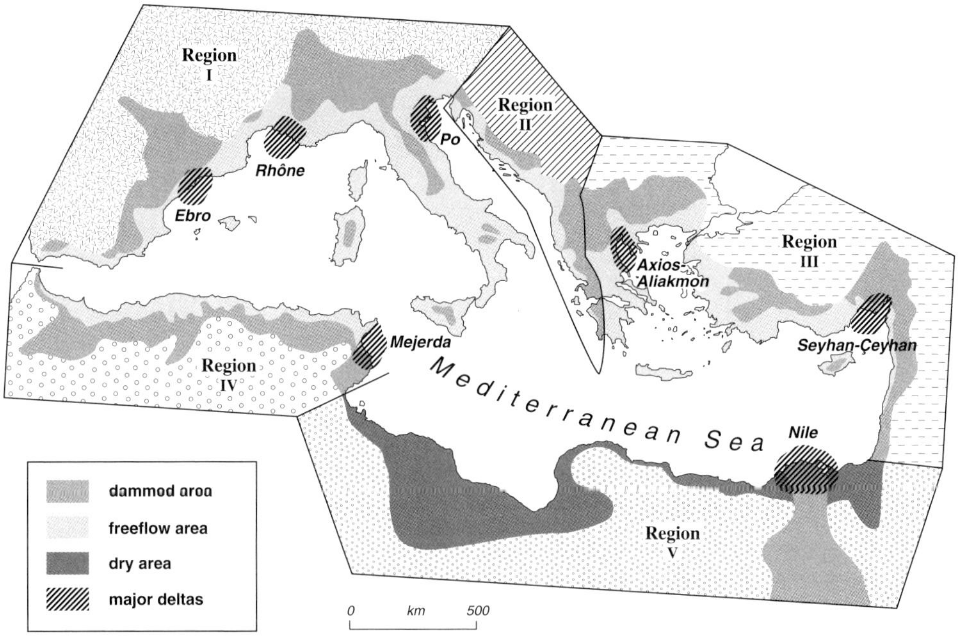

Fig. 6. Generalized map showing the dammed area of the Mediterranean drainage basin, together with the area subjected to dry climatic conditions.

causing a retention of the global sediment flux from the land to the ocean of >25% (Vorosmarty *et al.* 1997). This figure may be much higher within specific regions, as in the case of the west and central part of the African continent, where 'natural' sediment supply has been reduced by approximately 70% (Collins & Evans 1986).

The effects of the construction of hundreds of dams around the Mediterranean Sea is exacerbated by other anthropogenic modifications: the extraction of aggregates from the lower reaches of the river valleys; overpumping of coastal aquifers; and the natural subsidence of deltaic deposits. In response to all these modifications, the persistent prevailing nearshore hydrodynamic conditions (waves and currents) have often led to a dramatic loss of coastal land, particularly along deltaic areas. Several investigators have examined coastal retreat phenomena around the Mediterranean; for example the Italian coast (Colantoni *et al.* 1997; Bird 1988; Cocco *et al.* 1992), the Albanian coast (Ciavola *et al.* 1999), the coastal zone of the Greek mainland

Table 7. *Dammed area and sediment fluxes, before and after damming*

Physiographical region	Dammed area (10^3 km^2) (approx.)	Percentage of dammed to overall catchment area (%)	Before damming: SS flux (10^6 t)	Before damming: Total sediment flux (10^6 t)	After damming: SS flux (10^6 t)	After damming: Total sediment flux (10^6 t)
Region 1	*c.* 225	46	257.2	396	138.9	174
Region 2	*c.* 31	51	84.0	129–165	41.2	51
Region 3	*c.* 125	37	107.1	136	67.5	84
Region 4	*c.* 145	70	88.2	187	26.5	33
Region 5 (excl. Nile)						
Region 5 (incl. Nile)	*c.* 2800	92	121.4		9.7	12
Total	**3326**		**657.9**	**1012**	**283.8 (43%)**	**355 (35%)**

Table 8. *Rates of coastal retreat and related reduced sediment supply of specific rivers, over the last 50 years, following dam construction*

River/Delta	Subsidence	Coastal retreat and/or reduced progradation	Reduced sediment supply
Ebro	4–5 mm $year^{-1}$ (Holocene)[1]	From –10 to –60 m $year^{-1}$ (1957–1973)[4]	From 15×10^6–21×10^6 to 0.2×10^6 t $year^{-1}$ [7]
Rhone	3–7 mm $year^{-1}$ (Holocene)[1]	–423.5 m $year^{-1}$ (1954–1971)[9)]	40×10^6 (in 18th century) to 12 10^6 t $year^{-1}$ (1956–57) and 4×10^6–5×10^6 t $year^{-1}$ (in 1970)[10]
Po	1–3 mm $year^{-1}$ (Holocene) [(1)]	–10 m $year^{-1}$ (1954–1978)[5]	From 8.4×10^6 m^3 $year^{-1}$ (1945) to 6.4×10^6 m^3 $year^{-1}$ (1972)[5]
Axios	>2.5 m (1955–1985)[2]	From +4.3 (1941–1956) to +0.8 km^2 $yearr^{-1}$ (1956–1987)[11]	
Aliakmon		From +2.0 (1941–1956) to +0.6 km^2 $year^{-1}$ (1956–1987)11	
Seyhan		–25.3 km^2 $year^{-1}$ (1954–1995)[12]	From 5.3×10^6 t $year^{-1}$ (prior to 1954) to minimal (1998)12
Ceyhan		From +133 (1947–1954) to +54 km^2 $year^{-1}$ (1954–1995)[12]	
Nile (Rosetta Promotory)	1–5 m (Holocene)[1] and 0.6–1 m (since 1900)[3]	20 m $year^{-1}$ [6] (1900–1964) 120–240 m $year^{-1}$ [6] (1965–1991)	From 125×10^6 to 2×10^6 t $year^{-1}$ [8]

Data sources as follows: [1]Stanley (1997); [2]IGME (1989); [3]Stanley (1990); [4]Jimenez *et al.* (1997); [5]Simeoni & Bondesan (1997); [6]Fanos (1995); [7]Palanques & Drake (1990); [8]Degens *et al.* (1991); [9]Bird (1988); [10]Ookmens (1970); [11]Poulos *et al.* (1994); [12]Cetin *et al.* (1999).

(Poulos & Chronis 1997) and the coastlines of the major Mediterranean deltas, such as those of the Rhone (Bird 1988), Ebro (Jimenez *et al.* 1997), Nile (Stanley & Warne 1993; Fanos 1995) and Po (Simeoni & Bondesan 1997). Such studies have concluded that the decrease in sediment supply transported to the sea by rivers appears to be the most important factor controlling (coastal) retreat. Other factors act in a supplementary manner and only occasionally (and locally) to enhance substantially delta subsidence and coastal retreat. The annual rates of coastal retreat for some of the principal Mediterranean deltas (Table 8), in relation to the dramatic reduction in their sediment supply, vary (on an annual basis) from tens of metres (Ebro, Po) to hundreds of metres (Nile). Moreover, the influence of a reduction in supply is not restricted only to the deltaic coastline, but also affects the adjacent sections of the coastlines, as exemplified by the coast of Israel following the damming of the River Nile (Rohrlich & Goldsmith, 1984).

Conclusions

The Mediterranean Basin (4 150 000 km^2) receives water–sediment fluxes from >169 rivers, with catchment areas >200 km^2, together with those associated with hundreds of ephemeral rivers and torrential streams. There are 19 principal rivers (with catchment areas >10 000 km^2, including the River Nile) that drain an area of 3 370 000 km^2 (the River Nile accounts for some 2 800 000 km^2), comprising about 80% of the total drainage basin. This percentage is reduced to 42% if the River Nile is excluded, increasing the potential contribution of the smaller rivers to the overall system.

The Mediterranean Basin could (potentially) receive annually some 650×10^6 t of suspended sediment, i.e. 1×10^9 t annually, if the dissolved and bed-load components are taken into account. These high levels of sediment supply to the coastal zone are shown by: (i) the length of the (Mediterranean) coastline (45%) consisting of sedimentary deposits; and (ii) the formation of large deltas, associated with high rates of sediment accumulation over the adjacent shelf (10–40 m, within the Holocene).

However, since the second half of the nineteenth century, this progradation has either ceased or has been reversed and coastal erosion has become dominant as a response, mainly, to the construction of dams along the course of all of the major rivers and more than half of the smaller river systems. Such constructions have inhibited the transfer of sedimentary detritus to the ocean, as more than 40% of the total area of the Mediterranean catchment area (excluding the Nile) is located upstream of the dams; this becomes 80% when the River Nile is included in the analysis, influencing accordingly the water and sediment (total) supply. This dramatic reduction in sediment supply has caused extensive coastal retreat, which ranges locally from 10 (Ebro, Po) to 100 m year^{-1} (Nile). Such coastal erosion is not restricted to the deltaic systems themselves, but extends to adjacent coastal areas; it is likely to be enhanced further, in response to climatic change and associated sea-level rise.

Hence, sediment flux to the Mediterranean coastal zone and basin plays an important role not only in the evolution of the Mediterranean geosystem, but is also a key factor in determining its sensitivity to various environmental changes. For the Mediterranean coastal communities and countries, changes in sediment fluxes may have serious socio-economic implications.

The authors are grateful to Mrs K. Davis for the artistic presentation of the figures included, whilst thanks are extended to both reviewers (Prof. J. McManus and Prof. G. Evans) for their constructive comments.

Appendix

Names, size of catchment area and suspended sediment loads (SSL) of the major Mediterranean rivers (for locations see Fig. 4). Data were abstracted from UNESCO/IAHS (1974), Milliman *et al.* (1995), Milliman & Syvitski (1992) and Poulos & Chronis (1997).

	River	Area (km^2)	SS-flux (10^6 t)
AFRICA			
1	Martine (Mor)	1200	?
2	Lao		
3	Rhis		
4	Nekar (Neckor)	790	2.8
5	Kerte	3100	?
6	Moulouya	51 000	6.7
7	Tafna (Algeria)	6900	1.0
8	Sig (Sarno)		
9	Mebtouh		
10	Chellif	44 000	4.0
11	Mazafran	1800	3.0
12	El Harrach	390	0.6
13	Sebau	1500	1.2
14	Isser	3600	8.3
15	Bouguenm		
16	Soummam	8500	4.1
17	Agriun	660	4.8
18	Saf-Saf	300	0.4
19	Kebir O.	1100	0.2
20	Fessa		
21	Beni Zid		
22	Seybouse	5500	1.2
23	B. Namoussa	570	0.2

	River	Area (km^2)	SS-flux (10^6 t)
24	Garaet el Mekada		
25	Mejerda (Tunisia)	22 000	9.4
26	Meliane	2000	0.9
27	El Ghedahia (Libya)		
28	W. Zemzem		
29	W. Beiel el		
30	Chebir		
31	Nile (Egypt)	3 000 000	120.0
ASIA			
1	Karamenderes (Turkey)	1569.4	
2	Bakircay		
3	Cediz-Nehri	18 000	
4	Kujukmenderes	6900	
5	Buyukmenderes	25 000	
6	Dalaman	4481	0.7
7	Esencay	2458	2.3
8	Sarni_		
9	Alakir D.		
10	Aksu	1579	0.2
11	Kuprucay	1974	0.5
12	Manavgat	609.4	0.3
13	Goksu	10 561	2.5
14	Lamas		
15	Tarsus	1400	0.1
16	Seyhan	20 000	3.1
17	Ceyhan	20 900	7.4
18	Asi (Orontes–Syria)		
19	Kebir (Lebanon)		
20	Nahr al-Litani		
21	Qishon (Israel)		
22	Nahal-Besor		
23	W. el Arish (Egypt)		
EUROPE			
1	Guadares (Sp)		
2	Guadalteba		
3	Grande de Adru		
4	Aguas Blancas		
5	Segura	15 000	
6	Jucar (Xiquer)	2100	
7	Turia		
8	Mijares		
9	Ebro	85 835	22.0
10	Cardiner	4900	
11	Ter		
12	Fluvia	1124	
13	Tet (France)		
14	Agly		
15	Aude		
16	Orb	1800	
17	Herault		
18	Rhone	96 000	56.0
19	Argens		
20	Var		
21	Gole (COR)		
22	Tavaro (COR)		
23	Magra (Italy)	939	0.5
24	Edron		
25	Arno	8183	2.2
26	Omborne	2657	1.9
27	Albegna		
28	Fiora		
29	Tevere	17 000	7.5
30	Liri		
31	Volturno	5500	4.2
32	R. Sele		
33	Coghinas (Sardinia)		
34	Tirso (Sardinia)		
35	Mannu (Sardinia)		
36	Flumendosa (Sardinia)		
37	Cedrino (Sardinia)		
38	S. Leonando (Sicily)		
39	Platani (Sicily)		
40	Salso (Sicily)		
41	Gela (Sicily)		
42	Gornalugna		
43	Simeto(Sicily)		
44	Neto		
45	Crati	1332	1.2
46	Sinni	1142	2.5
47	Agri	278	0.1
48	Bassento	1400	
49	Bradano	2743	2.8
50	Late		
51	Ofando	2716	1.8
52	Cerrano		
53	Candearo		
54	Fortone	1126	1.5
55	Biferno	1290	2.2
56	Trigno	544	0.4
57	Sangro		
58	Pescara	3125	0.9
59	Tavo	213	0.0
60	Tronto	1200	1.1
61	Tesino	110	0.1
62	Aso	280	0.2
63	Ete Vivo	180	0.3
64	Tenna	490	0.5
65	Chienti	1300	1.3
66	Potenza	770	0.5
67	Misa	380	0.5
68	Esino	1200	0.3
69	Musona	1200	0.3
70	Metauro	1045	2.0
71	Foglia	603	1.4
72	Marrechia	357	1.6
73	Savo	597	1.4

	River	Area (km^2)	SS-flux (10^6 t)
74	Lamone	522	1.3
75	Reno	3410	2.7
76	Po	54 290	35.3
77	Adige	1200	0.5
78	Brenda	1563	0.2
79	Piave		
80	Silicia Cellina		
81	Tagliamento		
82	Isonzo		
83	Zrmanja (Croatia)		
84	Cetina		
85	Neretva	13 000	
86	Drin (Albania)	19 582	21.0
87	Mati	2450	2.0
88	Ishmi	670	2.0
89	Erseni	760	3.2
90	Shkumbi	2444	7.2
91	Semani	5649	24.0
92	Vijose	6706	23.2
93	Bajkaj		
94	Thiamis (Greece)	1826	2.4
95	Louros	785	
96	Arachthos	1895	7.5
97	Acheloos	5350	3.3
98	Evinos	1070	0.4
99	Mornos	1010	
100	Pinios (Peloponesos)	913	
101	Alfios	3610	
102	Sperchios	1780	
103	Pinios (Thessalia)	10 750	
104	Aliakmon	9455	4.4
105	Axios	23 747	
106	Gallikos	930	
107	Strimonas	16 550	
108	Nestos	6178	1.0
109	Evros (Greece–Turkey)	27 465	
110	Serrachis (Cyprus)		
111	Kouris		
112	Garillis		
113	Ag. Minas		
114	Yialias		
115	Pedieos		

References

AUDLEY-CHARLES, M., CURRAY J. R. & EVANS, G. 1977. Location of major deltas. *Geology*, **5**, 341–344.

BARIĆ, A. & CAŠPAROVIĆ, F. 1992. Implications of climatic change on the socio-economic activities in the Mediterranean coastal zones. *In*: JEFTIC, L., MILLIMAN, J. D. & SESTINI, G. (eds) *Climatic Change and the Mediterranean*. Edward Arnold, London, 129–173.

BIRD, E. C. 1988. *Coastline Changes: A Global Review*. Wiley, Chichester.

BLATT, H., MIDDLETON, G. V. & MURRAY, R. C. 1972. *Origin of Sedimentary Rocks*. Prentice Hall, Englewood Cliffs, NJ.

BODUR, M. N. & ERGIN, M. 1992. Holocene sedimentation patterns and bedforms in the wave-current-dominated nearshore waters of eastern Mersin Bay (eastern Mediterranean). *Marine Geology*, **108**, 73–93.

BRICE, W. C. 1978. *The Environmental history of the Near and Middle East Since the Last Ice Age*. Academic Press, London.

BRUKNER, H. 1997. Coastal changes in western Turkey. Rapid delta progradation in historical times. *In*: BRIAND, F. & MALDONADO, A. (eds) *Transformations and Evolution of the Mediterranean coastline*. Bulletin de l' Institut Oceanographique Special Publication, **18**, CIESM Science Series 3, 63–72.

CARTER, T. G., FLANAGAN, J. P., JONES, C. R., MARCHANT, F. L., MURCHISON, R. R., REBMAN, J. A., SYLVESTER, J. C. & WHITNEY, J. C. 1972. A new bathymetric chart and physiography of the Mediterranean Sea. *In*: STANLEY, D. J. (ed.) *The Mediterranean Sea: A Natural Sedimentation Laboratory*. Dowden, Hutchinson & Ross, Stroudsburg, PA, 1–24.

CETIN, H., BAL, Y. & DEMIRKOL, C. 1999. Engineering and environmental effects of coastline changes in Turkey, northeastern Mediterranean. *Environmental & Engineering Geoscience*, **3**, 315–320.

CHRONIS, G., LYKOUSIS, V. & BALOPOULOS, E. 1991. Comparative sedimentological studies in deltaic platforms of eastern and western Mediterranean: Deltaic platforms of Thermaikos (Greece), Rhone (France), Po (Italy). *Bulletin Geological Society of Greece*, **XXV/4**, 95–109 [in Greek].

CIAVOLA, P., MANTOVANI, F., SIMEONI, U. & TESSARI, U. 1999. Relation between river dynamics and coastal changes in Albania: an assessment integrating satellite imagery with historical data. *International Journal of Remote Sensing*, **20**, 561–584.

COCCO, E., MAGISTRIS, M. A., EFAICCHIO, M. T. & BOSCAINO, F. 1992. Geoenvironmental features of the Sele river plain littoral (Gulf of Salerno, southern Italy). *Bollettino di oceanologia Teorica ed Applicata*, **X**, 235–246.

COLANTONI, P., GABBIANELLI, G., MANCINI, F. & BERTONI, W. 1997. Coastal defence bt breakwaters and sea-level rise: the case of the Irtalian Northern Adriatic coast. *In*: BRIAND, F. & MALDONADO, A. (eds) *Transformations and Evolution of the Mediterranean Coastline*. Bulletin de l' Institut Oceanographique Special Publication, **18**, CIESM Science Series 3, 133–150.

COLLINS, M. B. 1986. Processes and controls involved in the transfer of fluviatile sediments to the deep ocean. *Journal of the Geological Society , London*, **143,** 915–920.

COLLINS, M. B. & EVANS, G. 1986. The influence of fuvial sediment supply on coastal erosion in West and Central Africa. *Journal of Shoreline Management*, **2**, 2–12.

DAL CIN, R. 1983. I litorali del delta del Po e alle foci dell' Adige e del Brenta:caratteri tesituarali e dispersione dei sedimenti, cause dell'arrtramento e previsioni sull'evoluzione futura. *Bulletin of the Geological Society of Italy*, **102**, 9–56.

DEGENS, E. T., KEMPE, S. & RICHEY, J. E. 1991. *Biogeochemistry of Major World Rivers*, SCOPE-42. Wiley, Chichester, 334–347.

DE WALLE, F. B., NIKOLOPOULOU-TAMVAKLI, M. & HEINEN, W. J. (EDS). 1993. *Environmental Condition of the Mediterranean Sea: European Community Countries*. Kluwer Academic, Dordeecht, 524.

DEWEY, J. F., PITMAN, W. C., RYAN, W. B. F. & BONNIN, J. 1973. Plate tectonics and the evolution of the Alpine system. *Geological Society of America Bulletin*, **84**, 3137–3180.

EMELYANOV, E. M. & SHIMKUS, K. M. 1986. *Geochemistry and Sedimentology of the Mediterranean Sea*. Dordrecht, D. Reidel.

EVANS, G. 1971. *Recent coastal sedimentation: a review. Vol. XXIII of the Colston Papers. Proceedings of the 23rd Symposium of the Colston Research Society*. Butterworth Scientific, London, 89–112.

FANOS, A. M. 1995. The impact of human activities on the erosion and accretion of the Nile Delta coast. *Journal of Coastal Research*, **11**, 821–833.

GAT, J. R. & MAGARITZ, M. 1980. Climatic variations in the eastern Mediterranean area. *Naturuissenschaften*, **67**, 80–87.

GRENON, M. & BATISSE, M. 1989. *Futures for the Mediterranean Basin: The Blue Plan*. Oxford University Press, Oxford.

GUILLÉN, J., DIAZ, J. I. & PALANQUES, A. 1992. Quantification and evolution of the Ebro River bedload sediment supplies during the XX century. *Review Society Geology España*, **5**, 27–37 (in Spanish).

ICOLD. 1998. *World Register of Dams*. International Commission on Large Dams, Paris.

IGME. 1989. *Geotechnical Study of Soil Subsidence in the village of Kalochori (Thessaloniki)*. Institute for Geology and Mineral Exploration Report, **86**, Athens (in Greek).

IMESON, A. C. & EMMER, I. M. 1992. Implications of climatic change on land degradation in the Mediterranean. *In*: JEFTIC, L., MILLIMAN, J. D. & SESTINI, G. (eds) *Climatic Change and the Mediterranean*. Edward Arnold, London, 111–128.

JIMENEZ, J. A., SANCHEZ-ARCILLA, A. & MALDONADO, A. 1997. Long to short term coastal changes and sediment transport in the Ebro delta; a multi-scale approach. *In*: BRIAND, F. & MALDONADO A. (eds) *Transformations and Evolution of the Mediterranean Coastline*. Bulletin de l' Institut Oceanographique Special Publication, **18**, CIESM Science Series 3, 169–185.

LARONNE, J. B. & REID, I. 1993. Very high rates of bedload sediment transport by ephemeral desert rivers. *Nature*, **366**, 148–150.

MACKLIN, M. G., LEWIN, J. & WOODWARD, J. C. 1995. Quaternary fluvial systems in the Mediterranean Basin. *In*: LEWIN, J., MACKLIN, M. G. & WOODWARD, J. C. (eds) *Mediterranean Quaternary River Environments*. A. A. Balkema, Rotterdam, 1–25.

MALDONADO, A. 1985. Evolution of the Mediterranean basins and a detailed reconstruction of the Cenozoic paleoceanography. *In*: MARGALEF, R. (ed.) *Western Mediterranean*. Pergamon Press, Oxford, 17–59.

MCNEILL, J. R. 1992. *The Mountains of the Mediterranean World: An Environmental History*. Cambridge University Press, Cambridge.

MEYBECK, M. 1988. How to establish and use world budgets of riverine materials. *In*: LERMAN, A. & MEYBECK, M. (eds) *Physical and Chemical Weathering in Geochemical Cycles*. Kluwer Academic, Dordrecht, 247–272.

MILLIMAN, J. D. & SYVITSKI, P. M. 1992. Geomorphic/tectonic control of sediment discharge to the ocean: the importance of small mountainous rivers. *Journal of Geology*, **100**, 525–544.

MILLIMAN, J. D., RUTKOWSKI, C. & MEYBECK, M. (eds). 1995. *River Discharge to the Sea: A Global River Index (GLORI)*. LOICZ Reports and Studies, Texel, The Netherlands.

NELSON, C. H. 1990. Estimated post-Messinian sediment supply and sedimentation rates on the Ebro continental margin, Spain. *Marine Geology*, **95**, 395–418.

OOKMENS, E. 1970. Depositional sequences and sand distribution in the postglacial Rhone delta complex. *In*: MORGAN, J. P. (ed.) *Deltaic Sedimentation: Modern and Ancient*. SEPM Special Publications, **15**, 198–212.

PALANQUES, A. & DRAKE, D. E. 1990. Distribution and dispersal of suspended particulate matter on the Ebro continental shelf, northwestern Mediterranean Sea. *Marine Geology*, **95**, 193–206.

PANO, N. 1992. Dinamica del litorale albanese (sintesi delle conoscenze). *Proceedings of the 19th A.I.G.I. Meeting, Genova, Italy*. G. Lang, Rome, 3–18.

PIMENTAL, D., HARVEY, C., RESOSUDARMO, P, SINCLAIR., K., KURZ, D., MCNAIR, M., CRIST, S., SHPRITZ, L., FITTON, L., SAFFOURI, R. & BLAIR, R. 1995. Environmental and economic costs of soil erosion and conservation benefits. *Science*, **267**, 1117–1122.

POULOS, S. & CHRONIS, G. 1997. The importance of the Greek river systems in the evolution of the Greek coastline. *In*: BRIAND, F. & MALDONADO, A. (eds) *Transformations and Evolution of the Mediterranean Coastline*. Bulletin de l' Institut Oceanographique Special Publication, **18**, CIESM Science Series 3, 75–96.

POULOS, S. E., COLLINS, M. & EVANS, G. 1996. Water sediment fluxes of Greek rivers, southeastern Alpine Europe: annual yields, seasonal variability, delta formation and human impact. *Zeitschrift für Geomorphologie*, **40**, 243–261.

POULOS, S. E., PAPADOPOULOS, A. & COLLINS, M. B. 1994. Deltaic progradation in Thermaikos Bay, northern Greece and its socio-economical implications, *Journal of Ocean and Shoreline Management*, **22**, 229–247.

PROBST, J. L. & AMIOTTE-SUCHET, P. 1992. Fluvial suspended sediment transport and mechanical erosion in Maghreb (North Africa). *Hydrological Sciences Journal*, **37**, 621–637.

QIAN, N. & DAI, D. Z. 1980. The problems of river sedimentation and the recent status of its research in China. *In*: Proceedings of the International

Symposium on River Sedimentation. Beijing, Guanghue Press, 3–39.

REID, I. & FROSTICK, L. E. 1987. Toward a better understanding of bedload transport. *In*: Ethridge, G., Flores, R. M. & Harvey, M. D. (eds) *Recent Developments in Fluvial Sedimentology*. SEPM Special Publications, **39**, 13–21.

REID, I., FROSTICK, L. E. & LAYMAN, J. T. 1985. The incidence and nature of bedload transport during flood flows in coarse-grained alluvial channels. *Earth Surface Processes and Landforms*, **10**, 33–34.

ROHRLICH, V. & GOLDSMITH, V. 1984. Sediment transport along the southeast Mediterranean: a geological perspective. *Geo-Marine Letters*, **9**, 99–103.

SEMPLE, E. C. 1932. *The Geography of the Mediterranean Region its Relation to Ancient History*. Constable, London.

SESTINI, G. 1992*a*. Implications of climatic changes for the Nile delta. *In*: JEFTIC, L., MILLIMAN, J. D. & SESTINI, G. (eds) *Climatic Change and the Mediterranean*. Edward Arnold, London, 533–601.

SESTINI, G. 1992*b*. Implications of climatic changes for the Po Delta and Venice Lagoon. *In*: JEFTIC, L., MILLIMAN, J. D. & SESTINI, G. (eds) *Climatic Change and the Mediterranean*. Edward Arnold, London, 428–494.

SIMEONI, U. & BONDESAN, M. 1997. The role and responsibility of man in the evolution of the italian Adriatic coast. *In*: BRIAND, F. & MALDOLADO, A. (eds) *Transformations and Evolution of the Mediterranean Coastline*. Bulletin de l' Institut Oceanographique Special Publication, **18**, CIESM Science Series 3, 75–96.

SIMEONI, U., PANO, N. & CIAVOLA, P. 1997. The coastline of Albania: morphology, evolution and coastal management issues. *In*: BRIAND, F. & MALDONADO, A. (eds) *Transformations and Evolution of the Mediterranean Coastline*. Bulletin de l' Institut Oceanographique Special Publication, **18**, CIESM Science Series 3, 151–168.

SKOULIKIDIS, N. 1993. Significance evaluation of factors controlling river water composition. *Environmental Geology*, **22**, 178–185.

STANLEY, D. J. (ed.). 1972. *The Mediterranean Sea: A Nartural Sedimentation Laboratory*. Dowden, Hutchinson & Ross, Stroudsburg, PA.

STANLEY, D. J. 1997. Mediterranean Deltas: subsidence as a major control of relative sea-level rise. *In*: BRIAND, F. & MALDONADO, A. (eds) *Transformations and Evolution of the Mediterranean Coastline*. Bulletin de l' Institut Oceanographique Special Publication, **18**, CIESM Science Series 3, 35–62.

STANLEY, D. J. 1988. Low sediment accumulation rates and erosion on the middle and outer Nile delta shelf off Egypt. *Marine Geology*, **84**, 111–117.

STANLEY, D. J. 1990. Recent subsidence and northeast tilting of the Nile delta, Egypt. *Marine Geology*, **94**, 147–154.

STANLEY, D. J. & WARNE, A. G. 1993. Nile Delta: recent geological evolution and human impact. *Science*, **260**, 628–634.

SUMMERFIELD, M. A. 1991. *Global Geomorphology*. Longman, Harlow.

TAZIOLI, G. S. & BILLI, P. 1987. Bedload transport measurements by the vortex-tube trap on Virginio creek, Italy. *In*: THORNE, C. R., BATHURST, J. C. & HEY, R. D. (eds) *Sediment Transport in Gravel-bed Rivers*. Wiley, Chichester, 154–167.

UNEP. 1984. *Pollutants from Land-based Sources in the Mediterranean*. UNEP/Regional Seas Reports and Studies, **32**.

UNESCO. 1978. *World Water Balance and Water Resources of the Earth*. UNESCO Studies and Reports in Hydrology, **25**.

UNESCO 1985. *Recent developments in erosion and sediment yield studies*. Technical Documents in Hydrology, UNESCO, Paris.

UNESCO/IAHS. 1974. *Gross Sediment Transport into Oceans*, First Preliminary Edition. UNESCO/IAHS, Paris.

VITA-FINZI, C. 1972. Supply of fluvial sediment to the Mediterranean during the last 20,000 years. *In*: STANLEY, D. J. (ed.) *The Mediterranean Sea: A Natural Sedimentation Laboratory*. Dowden, Hutchinson & Ross, Stroudsburg, PA, 43–46.

VOROSMARTY, C. J., WASSON, R. & RICHEY, J. E. 1997. *Modelling the Transport and Transformations of Terrestrial Materials to Freshwater and Coastal Ecosystems*. IGBP Report, **39**, IGBP, Stockholm, Sweden.

WALLING, D. E. 1986. Sediment yields and sediment delivery dynamics in Arab countries: some problems and research needs. *Journal of Water Resources*, **5**, 775–799.

WALLING, D. E. & WEBB, B. W. 1983. Patterns of sediment yield. *In*: GREGORY, K. L. (ed.) *Background to Palaeohydrology*. Wiley, Chichester, 61–100.

WALLING, D. E. & WEBB, B. W. 1998. Erosion and sediment yield: a global overview. *In*: *Proceedings of the Exeter Symposium, July 1996*. IAHS Publication, **236**, 3–19.

WINDLEY, B. F. 1984. *The Evolving Continents*. Wiley, Chichester.

WOODWARD, J. C. 1995. Patterns of erosion and suspended sediment yield in Mediterranean river basins. *In*: FORSTER, I. D. L., GURNELL, A. M. & WEBB, B. W. (eds) *Sediment and Water Quality in River Catchment*. Wiley, Chichester, 365–389.

ZACHAR, D. 1982. *Soil Erosion*. Developments in Soil Science, **10**. Elsevier, Amsterdam.

Sediment fluxes and the evolution of a riverine-supplied tectonically-active coastal system: Kyparissiakos Gulf, Ionian Sea (eastern Mediterranean)

SERAFIM E. POULOS[1], GEORGE VOULGARIS[2], VASILIS KAPSIMALIS[3], MICHAEL COLLINS[4] & GRAHAM EVANS[4]

[1]*Department of Geography and Climatology, Faculty of Geology, University of Athens, Panepistimiupoli, Zografou, Athens 15784, Greece*

[2]*Marine Science Program, Department of Geological Sciences, University of South Carolina, Columbia, SC 29208, USA*

[3]*Departement de Geologie et Oceanographie, UMR CNRS 5805, Universite Bordeaux 1, Avenue des Facultes, 33405 Talence Cédex, France*

[4]*School of Ocean and Earth Science, Southampton Oceanography Centre, University of Southampton, European Way, Southampton SO14 3ZH, UK*

Abstract: Kyparissiakos Gulf, located in the southwestern part of Greece (northeastern Ionian Sea), is a riverine–coastal system that has developed over the southern flank of the Alpine orogenic belt (Hellenides). Some 4 km in the vertical separates the heights of the mountain peaks to the depths of the adjacent offshore deep-water basin. This system extends horizontally over approximately 100 km. The area experiences intensive tectonism (e.g. seismicity), a Mediterranean type of climate and microtidal and moderate wave-energy oceanographic settings.

Large quantities of sediments (>2.5×10^6 t year^{-1}), transferred principally by the River Alfios, are the product of denudation of the high relief (in excess of 2000 m), developed on erodible lithology (with siliciclastics and carbonates >90%) under moderate climatological conditions. The large amounts of sediments produced in the hinterland, in association with land–ocean process interaction, have led to the formation of a coastal zone that includes deltaic plains and coastal barriers with dune fields, which enclose lagoons. The shape and morphological characteristics of the shore zone indicate, clearly: (a) the dominance of the wave activity; (b) an overall northward longshore sediment movement; and (c) a major depocentre at the northern, naturally sheltered, end of the Gulf.

Seawards, the coastal zone includes a narrow continental shelf covered with a blanket of recent sediments, which are terrigenous in origin; these extend down a steep slope, where materials is transferred to the deep (approximately 1800 m) offshore basin (the northward component of the Hellenic Trench) primarily by gravitational mass movements; these are triggered often by earthquake activity. To a first approximation, some 50% of the riverine sediment fluxes accumulate over the shelf, whilst another 25% is transported over the slope to the deeper ocean waters.

The construction of (two) dams has led to a dramatic reduction in the sediment supply to the coast, having already caused the retreat of the River Alfios deltaic coastline, by >100 m. Similarly, artificial drainage of the lagoons (for the development of agricultural land) has affected the overall ecosystem (altering the the fauna and flora) of the coastal zone

Thus, the Upper Quaternary (mostly Holocene) evolution of this particular coastal system is attributed, primarily, to processes and balances between sediment fluxes (e.g. terrestrial transportation, seaward dispersion) and, recently, to human interference.

Continental margins are the principal areas of sediment accumulation on Earth. As part of the sedimentation cycle, sedimentary particles are transported from a source to a sink, through a number of processes that include: hill slope erosion; transport by rivers; biological production; temporal storage in an alluvial plain; and, finally, seabed burial. These processes are of major societal importance, as they control natural hazards, pollutant transport, coastal erosion and preservation of

From: JONES, S. J. & FROSTICK, L. E. (eds) 2002. *Sediment Flux to Basins: Causes, Controls and Consequences.* Geological Society, London, Special Publications, **191**, 247–266. 0305-8719/02/$15.00

natural resources (Nittrouer & Driscoll 1999). All of the above processes are influenced by changes in the Earth's ecosystem, which are subsequently fingerprinted on the stratigraphic record. Besides, within the concept of global change, IGBP (the International Geosphere Biosphere Programme) has recently focused its interest to present and future fluvial sediment flux to the coastal ocean (Syvitski 2001).

The intensity of the processes, described above, depend upon the geological and geographical setting of the coastal area under consideration. For example, recent studies have revealed the importance of small mountainous rivers that discharge into active tectonic continental margins; these have large sediment yields and deliver significantly higher sediment loads to the sea. The high sediment yield is attributed to the overall tectonic milieu (i.e. high relief, fractured and brecciated rocks, oversteepened slopes, seismic and volcanic activity). The effectiveness in delivering the sediment to the ocean is related to the absence of estuaries, steeper river gradients, susceptibility to periodic floods and their proximity to the source material (Milliman & Syvitski 1992). Moreover, the sediment is more likely to escape narrow shelves, usually found in active margins, to be deposited into the deeper basins (Von Hueve & Scholl 1991).

The present study investigates the contribution of sediment fluxes to the evolution of Kyparissiakos Gulf, which belongs to the outer part of the Hellenic subduction zone and receives the sediment fluxes from a number of small rivers (catchment < 6000 km^2); these drain the proximal mountainous region of the southern part of the Greek orogenic belt (Fig. 1). The latter is also characterized by large sediment yields (Poulos & Chronis 1997; Poulos & Collins 2002). Furthermore, the present investigation focuses on sedimentary fluxes from source to sink and takes place within the concept of the total coastal system. Processes include denudation of the hinterland, accumulation and/or transport through the coast to the shelf (nearshore and offshore waters) and to deep waters (over the slope and adjacent trench).

The total coastal system consists of the terrestrial and the marine sub-systems. These sub-systems are interrelated, as they are the products of interaction between land, sea and air processes; any change in one influences the other. With respect to sediment fluxes, the terrestrial sub-system acts as the principal supplier of sediments, whilst the marine sub-system plays primarily the role of the receiver. Within this context, the coastal zone is a strip of land and sea, with its width depending on the localized nature of these environments. Nowadays, it is accepted widely that the coastal zone incorporates the coastal plains and the adjacent continental shelf (Cadee *et al.* 1994); it assumes the role of a 'buffer' of the prevailing hydrodynamic and sediment dynamic processes. These processes are also responsible for the formation of the shore (beach) zone, that incorporates the most rapidly changing morphological characteristic in time and in space, the coastline.

This work presents an integrated synthesis, spatial and temporal, of the sediment fluxes within the river-influenced and tectonically-active coastal system of the Kyparissiakos Gulf in terms of: (i) processes controlling the production and transfer of sediments from the terrestrial to marine environment; (ii) the formation and evolution of the coastal zone, in relation to land–air–ocean interaction; (iii) the configuration of the shore zone and its recent evolution, as a result of the processes of littoral sedimentation; (iv) the offshore transfer, dispersion and accumulation of sediments, in relation to the evolution of the continental margin (slope and basin); and (v) the consequences of human impact on the riverine sediment fluxes. The study of the above-mentioned processes is presented below, within the concept of a system's approach and with respect to the scale of the major components of the system, which in descending order are: total coastal system, coastal zone and shore (beach) zone.

The Kyparissiakos Gulf total coastal system

The Kyparissiakos Gulf coastal system (Fig. 2) covers an area of 6170 km^2 (Table 1). The system is divided into: (a) the *terrestrial sub-system*; and (b) the *marine sub-system*.

The *terrestrial sub-system* includes the land area that extends landwards from the coastline to the watershed of the drainage basins of the incoming rivers; it is the most extensive part (68%) of the total system, covering an area of some 4210 km^2. The sub-system includes basically the mountainous hinterland area, which is bordered towards the sea by the sub-aerial part of the coastal zone. The *hinterland* includes mainly the catchment area of the River Alfios (some 3600 km^2) and some other ephemeral rivers; the most important of these being the River Nedas (300 km^2) and the River Peristera (140 km^2) (Fig. 2). The sub-aerial part of the coastal zone includes low-lying marine terraces, the deltaic plains of the rivers, the intervening low-relief upper Quaternary coastal plains and beach–dune barriers, which enclose lagoons.

The *marine sub-system* accounts for some 1960 km^2, representing almost 32% of the total area of the Kyparissiakos Gulf coastal system (Table 1). The sub-system includes the inner/outer continental

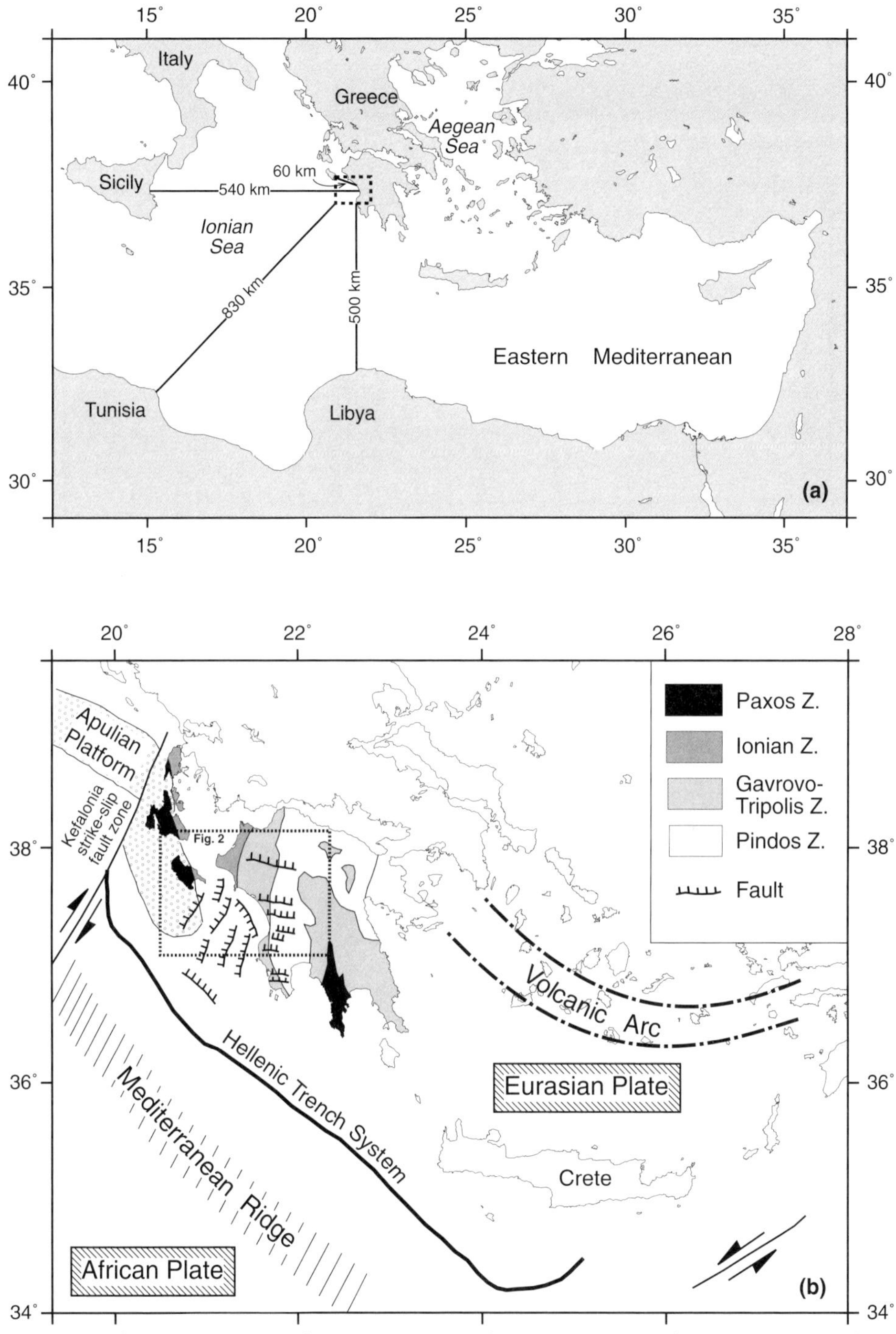

Fig. 1. Generalized map showing the geographical location (**a**) and the basic geotectonic environment (**b**) of the Kyparissiakos Gulf riverine coastal system.

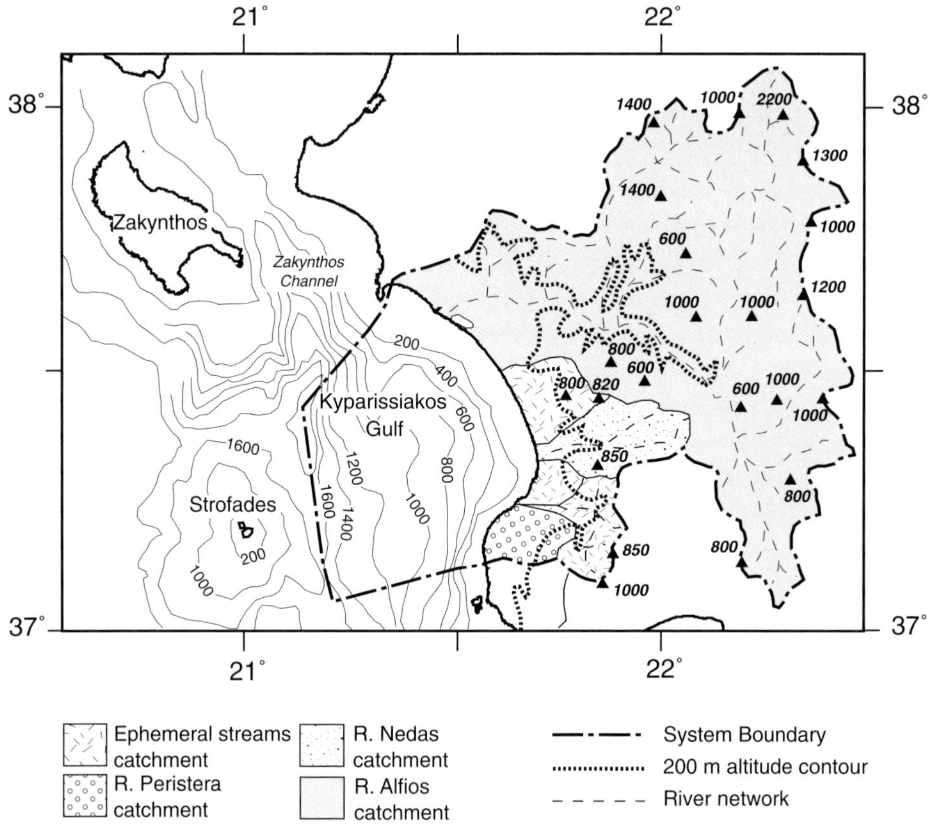

Fig. 2. Map showing the morphological characteristics of the terrestrial and marine environment of the Kyparissiakos Gulf coastal system.

shelf, with the slope being extended up to the deepest part (about 1850 m) of the marine environment; the latter is basically the longitudinal axis of a canyon-type valley, trending 140°N and sloping southwards. It is connected to the north with the Zakynthos Channel, whilst towards the south it is abutted on the deep Matapan Trench, where water depths are of about 5000 m.

Geological setting

The Kyparissiakos Gulf coastal system belongs to the western Peloponnese; the latter is the most highly seismically active zone of Greece, as it is formed by the convergence and subduction of the African plate (oceanic lithosphere) beneath the Eurasian plate (continental crust) (Fig. 1b). This

Table 1. *Morphometric characteristics of the different components of the Kyparissiakos Gulf coastal system (Ionian Sea, eastern Mediterranean)*

Coastal system	6170 km^2	(100.0%)
Terrestrial part (maximum elevation: 2224 m)	421 km^2	(68.2%)
Oceanic part (largest water depth: 1850 m)	1960 km^2	(31.8%)
Coastal zone	410 km^2	(6.6%)
Sub-aerial part (coastal plain with slopes <5%)	145 km^2	(2.3%)
Sub-aqueous part (from 0 to shelf-break)	275 km^2	(4.3%)
Coastline length	70 km	
Shelf break water depth	70–170 m	

large-scale mechanism causes a significant deformation in the overridden continental crust, with a complexity of kinematics (Sachpazi *et al.* 2000); this is expressed by a large number (>40, since 1900) of earthquake shocks with intensities >4 on the Richter scale (IGME 1989).

The Alpine basement of the Kyparissiakos Gulf coastal system consists of four geotectonic zones (Fig. 1b). The most seaward part is the Paxos Zone (mainly underwater); this is followed by the Ionian Zone, which consists of outcropping Eocene limestones in the southern part of the hinterland area. Eastwards and landwards, the Eocene flysch of the Gavrovou–Tripolis Zone covers much of the southwestern Peloponnese, whilst neritic Cretaceous limestones of the same zone crop out near Kaiafa Lagoon. Farther inland, the Olonos–Pindos Zone expands over the central Peloponnese thrust, on to the Gavrovos–Tripolis Zone. This zone is characterized by pelagic calcareous, pyritic and, locally, siliciclastic sediments, as well as a Tertiary siliciclastic flysch composed of rhythmic alternations of conglomerates, sandstones, mudstones and marls (Kaberis 1987).

Within the study area, as over most of the western Peloponnese, Neogene and Quaternary sediments cover wide areas; their thickness, in some areas (Amalias Basin, Katakolo Peninsula, southern area of the Alfios River catchment) exceeds 400 m. This post-Alpine siliciclastic sequence overlies unconformably the older rocks of the Ionian, Gavrovo–Tripolis and Olonos–Pindos Zones (Fig. 1b). The Early Miocene sediments are composed of coarse-grained siliciclastic material and lignite; these crop out near the Peristera River mouth (Fig. 2). During the Upper Miocene and Early Pliocene, sands and muds were deposited into the newly created shallow-water marine basins. In the Middle Pliocene, sedimentation was interrupted by intensive tectonism, which fragmented the Neogene relief. The depositional process started again in the Upper Pliocene, infilling shallow-marine basins and lagoons throughout the north-western Peloponnese. The upper part of this sequence is composed of alluvial and lacustrine sediments that have accumulated mainly in the central part of the Kyparissiakos Gulf. The transition from Pliocene to Pleistocene occurred without any interruption of sedimentation; thus, it is very difficult (and, in some cases, impossible) to differentiate between these deposits. Since the Early Pleistocene, diapiric movement of evaporites and eustatic sea-level changes have made the sequence complex and unstable. During this period, the Alfios River terraces were formed, together with the development of barriers, lagoons and widespread aeolian deposits, recording coastal movements (Kaberis 1987).

The marine sub-system forms part of an active outer-arc compression region (Underhill 1989) of the Hellenic Arc–Trench system (Fig. 1); it is bounded by the western Peloponnese slope and the eastern side of the Strophades Ridge (Fig. 2). An elongated graben follows the major structural direction of the area (NNW–SSE). The graben is divided into several segments by E–W ridges, creating a considerable morphological complexity consisting of almost isolated basins (e.g. Zakynthos Perched Basin), submarine terraces (e.g. the Upper and Lower Terrace of the middle Hellenic Slope) and local ridges (Vittori 1978; Le Quellec *et al.* 1980).

Weathering processes and sediment fluxes

The type and the intensity of the weathering processes are strongly related to the prevailing climatological conditions. The climate of the hinterland area may be characterized as 'terrestrial Mediterranean', with a mean annual rainfall ranging between 800 and 1200 mm, and air temperatures of between 18.5 and 19.0°C (Table 2). Over the coastal area the climate becomes more 'temperate', with the most frequent winds blowing from the west. On a seasonal basis, southwesterly winds are dominant during the summer period; whilst in winter easterlies dominate.

Under these climatic conditions, according to Leopold *et al.* (1964), prevailing type and the degree of intensity of the denudational processes are: (i) low to intermediate mechanical weathering; (ii) high to intermediate chemical weathering; (iii) intermediate to high denudation, due to running water; and (iv) high denudation in relation to mass movements caused by high relief. These denudation processes are associated with intensive soil degradation over the region (Walling & Webb 1998). Thus, the hinterland is subjected to intense weathering processes that produce significant amounts of sedimentary material, available for transport by the river network.

No direct measurements exist for the sediment fluxes of the rivers discharging along the coastline of the Kyparissiakos Gulf. However, monthly suspended sediment flux data are available for other Greek rivers discharging into the Ionian Sea, to the north of the study area (i.e. River Acheloos, 2.5×10^6 t; River Arachthos, 7.3×10^6 t; River Kalamas, 1.9×10^6 t), corresponding to annual yields of 600–4000 t km^{-2} (Poulos & Chronis 1997). Thus, in the absence of field measurements, an estimation has been made, based upon: an established relationship between the measured water and suspended sediment loads of Greek rivers (Poulos & Chronis 1997); and between the suspended sediment load and catchment area of

Table 2. *General characteristics of the rivers discharging along the coastline of the Kyparissiakos Gulf*

	River Alfios	Smaller rivers
Geomorphology		
Catchment area	3665 km^2	735 km^2
Maximum relief	2224 m	1421 m
Relief ratio	3.1×10^{-2}	5.6×10^{-2}
Deltaic plain	*c.* 110 km^2	*c.* 10 km^2
Lithology		
Quaternary deposits	41.3%	40.3%
Carbonates	47.2%	52.7%
Flysch	11.1%	7.0%
Metamorphic	0.4%	
Climate		
Mean annual temperature	18.5–19.0°C	
Mean annual rainfall	1000 mm	
Winds (dominant direction)	SW, W, NW, N	
Type	Mediterranean	
Hydrology*		
Mean annual discharge	67 $m^3 s^{-1}$	
Maximum annual discharge	145 $m^3 s^{-1}$	
Minimum annual discharge	21 $m^3 s^{-1}$	
High-water period (months)	Nov–Apr	
Low-water period (months)	May–Oct	
Sediment fluxes†		
Suspended sediment load	*c.* 2.5×10^6t $year^{-1}$	*c.* 0.7×10^6t $year^{-1}$
Total sediment load	$>3.0 \times 10^6$t $year^{-1}$	$>0.8 \times 10^6$t $year^{-1}$

*Measured values, after Therianos (1974).
†Estimated values, after Poulos & Chronis (1997)

smaller river systems draining Italy, Albania, Greece and Turkey (Poulos *et al.* 1996*a*). The estimate, so derived, suggests some 2.5×10^6 t $year^{-1}$ of suspended sediment load (i.e. SSL) for the River Alfios. Elsewhere, field measurements from (10) Greek rivers (Skoulikidis 1993) have shown the ratio of the dissolved load (DL) to the suspended sediment load (SSL) to be SSL : DL = 1 : 0.2. Likewise, published information on Albanian (Pano 1992) and the larger Mediterranean rivers (Po, Dal Cin 1983; and Ebro, Guillen *et al.* 1992) showed an average value of 10–25% of the total sediment load to be attributed to bed load (BL). Thus, the total sediment transported by the River Alfios may exceed 3×10^6 t $year^{-1}$. At the same time, another 0.7×10^6–0.8×10^6 t $year^{-1}$ can be attributed to other smaller ephemeral rivers discharging over the region (Fig. 2 and Table 2). Such high sediment loads can be explained by the easily erodible lithology (siliciclastics and carbonates >90%), the 'Mediterranean type' of climate, with precipitation levels in excess of 1000 mm $year^{-1}$ (Table 2), and the relatively sparse vegetation cover, especially during the (dry) summer period.

The sediment and water discharges in the region experience episodic, seasonal and longer-term variability. The highest sediment discharges occur at the beginning of the wet season (October–December). Towards the end of the wet season, even though there is still substantial runoff, the sediment flux decreases drastically to become minimal during the dry (summer) period (Poulos & Chronis 1997). The variation in sediment flux may be explained in terms of sparse vegetation and desiccated soil conditions that are associated with the dry season (June–September); during which the land surface becomes susceptible to erosion. Thus, at the beginning of the succeeding wet season, detritus is removed readily by heavy rains and is transported then by the river network. However, as the plant cover becomes dense, there is a corresponding increase in soil cohesion and a considerable reduction in its erodibility.

Episodic production of sediment is related to landslides, that occur commonly as the result of rapidly fluctuating rainfall (which might entail a seasonal variability) and regional tectonic (mainly seismic) activity. For example, a series of 47 earthquake-triggered landslides and seven sites of lateral flows of soil, caused by the earthquake shocks (magnitude of 5.2 on the Richter scale) over the Pyrgos area on 26 March 1993, have been mapped by Koukouvelas *et al.* (1996).

Long-term changes in sediment fluxes have also been attributed to human interference. Archaeological evidence indicates that there was deterioration in the vegetation over the hinterland and deforestation; these changes began in Classical times and continued into Roman times, leading to an increase in the supply of sediments during Hellenistic and Roman times. Alluvial deposition occurred in the valleys, with the ancient city of Olympus and Byzantine settlements all being buried. Since Medieval times (AD 326–1354) sediment supply has been reduced and previous alluvial deposits have been entrenched (Raphael 1973). During the last few decades, following the construction of the dams in both the Ladona (in 1955) and the Alfiousas (only a few tens of km from its mouth), there has been a dramatic reduction in sediment discharge and, consequently, rapid erosion of the fronts of the river delta (see below).

Oceanographic setting

The Kyparissiakos Gulf, as part of the NE Ionian Sea, is a micro-tidal marine environment with tides rarely exceeding 10 cm (Tsimplis *et al.* 1995). In terms of wind-induced surface-wave activity the Gulf is exposed to southerly, southwesterly, westerly and northwesterly winds. Winds from the south, southwest and west are associated with very long (hundreds of kilometres) wave fetches (Fig. 1a); they comprise 50% of the annual occurrence, with the strong winds (≥6 Beaufort) representing >90% of the annual frequency of occurrence (Ginis 1973).

Beyond the outer shelf area of the Kyparissiakos Gulf the water masses consist of, from the surface to the bottom (Malanotte-Rizzoli *et al.* 1997): (i) the Ionian Surface Waters (ISW), which are more easily defined during the summer period, when there is a maximum salinity as a result of the excess summer evaporation (T = 22–26°C; S = 38.6–39.1 psu); (ii) the Modified Atlantic Water (MAW), a sub-surface water mass (T =1 5–18°C; S = 38.0–38.6 psu) usually identified by a salinity minimum in water depths between 30 and 200 m; (iii) the Levantine Intermediate Water (LIW) which lies below the MAW, from which it is distinguished by its salinity maximum between 200 and 600 m (T = 14–15°C; S = 38.8–39.0 psu); (iv) between 600 and 1600 m is a transitional water mass, with physical properties between those of the overlying LIW and the underlying EMDW; and (v) the Eastern Mediterranean Deep Water (EMDW), which occupies the abyssal layers below 1600 m (T = 13.0–13.3°C; S = 38.6–38.7 psu) and which originates from the Atlantic Deep Water and is colder and less saline (compared to the LIW).

The offshore area is characterized by a slow northward water movement of the upper few hundred metres (Nittis *et al.* 1993; Malanotte-Rizzoli *et al.* 1997). In contrast, near-bed current meter measurements (obtained in January 1981) have revealed a southerly near-bed flow (Ferentinos *et al.* 1985; Cramp *et al.* 1987).

Sub-aqueous sedimentary deposits

The sub-bottom stratigraphy of the continental margin consists of two stratigraphic sequences (Vittori *et al.* 1981): a lower seismic sequence that is overlain by an upper sequence, separated by an unconformity of Upper Miocene or Early Pliocene age.

The lower seismic sequence consists of many hundred metres of stratified muds and marls of the post-Alpine siliciclastic sequence; these were formed within the Upper Miocene and unconformably overlie the rocks of the Alpine geotectonic zones of Gavrovo–Tripoli and Pindos (Fig. 1b). This complex sequence, on the upper and middle continental slope, is due to the intensive influence of coastal processes and gravitational sedimentation.

Overall, the upper seismic sequence consists of two distinct units, with different acoustic features: (a) a lower (transparent) unit, characterized by generally weak internal reflections and poor lateral continuity (although a more internally reflective response is recorded near areas of sedimentary input) – normal faults linked possibly to the faulted underlying basement, are observed locally; (b) an upper unit with parallel, high-amplitude reflectors and numerous other sedimentary structures (lenticular layers, slump scars, slides and locally irregular cross-bedded strata), indicating that these sediments have been reworked and deposited by gravitational processes. This upper unit infills the terraces and basins of the middle and lower slope, with its thickness usually exceeding 500 m.

On the upper sub-marine slope, a variety of mass failures can be recognized. Translational slides, showing negligible internal deformation, are associated with fault escarpments and show slip along unconformities or bedding planes; these lie parallel to the slope (Ferentinos 1991). Debris flow deposits have accumulated on the overthrown sides of the faults, forming rollover structures, or drape the upper slope with a sheet-like layer. Liquified/fluidized flows and turbidity currents have also been identified on the basis of the seismic profiles and gravity cores (Got *et al.* 1981). Furthermore, Vittori *et al.* (1981) have proposed the *cascade feeding* sedimentation model for the sediments, moving down-slope in response to gravitational processes along the Hellenic Trench (i.e. sediment transport by spill-over on the upper terrace, sediment accumulation in a peripheral depression and subsequent transport towards the lower terrace). In addition, the continued failure and short residence time, of sediments on the steep and seismically-affected slope, produce textural and compositional homogenization of those deposits (Stanley 1981).

The Holocene sediments that cover the continental shelf and upper slope (although their base is usually masked by gases) lie unconformably on a Plio-Pleistocene acoustic basement. This boundary (Reflector A, in Fig. 3) is characterized by a high acoustic impedance contrast and forms an irregular surface; this is probably as a result of sub-aerial erosion during the pre-Holocene and the post-glacial transgression (Swift 1968; Vail *et al.* 1991). This surface is diachronic, created as the shoreline moved landward from its maximum seaward position (Swift *et al.* 1991), about 18 000 BP, since the low levels of the last glaciation.

The River Alfios and, to a lesser extent, the rivers Nedas and Peristera, deposits have formed several wedge-like bodies, with a seaward sigmoid-oblique configuration. These bodies consist of three distinct units: (I) an uppermost dark-colour horizon; (II) a well-stratified unit; and (III) a semi-transparent sequence (Fig. 3).

The upper highly backscattered horizon (Unit I; Fig. 3) consists of very fine–fine sand and contains a low percentage of carbonates (Kapsimalis *et al.* 1997); it shows very strong reflectors in the nearshore areas at the mouths of the rivers. The thickness of Unit I in front of the Alfios River delta is at least 3 m; in front of the mouth of the Nedas and Peristera River delta it is 2 and <1 m, respectively. This unit thins seaward and dies out at a water depth of 20–30 m. It has a discordant and, sometimes, wavy acoustic character that can be attributed to the relatively intense hydrodynamic regime (mainly waves and longshore currents) near the coast and/or to the variable sediment supply from the rivers (Coleman *et al.* 1983).

The stratified series (Unit II; Fig. 3) shows continuous alternations of high- and low-amplitude reflectors that correspond to sand and mud layers, respectively (Kapsimalis 1993); it is characterized by parallel or sub-parallel layers, wedging out seaward. The maximum thickness of the deposits at the delta-front and inner shelf is difficult to ascertain, due to the presence of gas smearing. The identifiable thickness varies from 25 m at the delta-front of the River Alfios, to zero at the edges of the prodelta zone. Studies on analogous deltas in the Gulfs of Patras (Chronis *et al.* 1991) and Corinth (Perissoratis *et al.* 2000) attribute this facies to progradational deposits of the present sea-level high-stand. The date of the base of the unit is approximately 6000 BP, when the sea approximately reached its present-day level (Lambeck 1996).

The acoustically semi-transparent zone (Unit III; Fig.3) is characterized by only a few moderate sub-parallel reflections. This faintly stratified unit represents disturbed bottom-set deltaic clays and silts (Kapsimalis 1993) that rest unconformably on Reflector A, the Plio-Pleistocene acoustic basement of the Holocene.

The presence of gases of biogenic origin (probably methane) has modified the geotechnical properties of the sediments, causing a variety of seafloor instabilities. Rotational slumps and contemporaneous growth faults occur; these are related to commonly localized scarps with amplitudes ranging from 20 to 250 m off the River Alfios. Slump fault planes often display a concave-upward geometry, affecting only the well-stratified Unit I to a maximum depth of approximately 25 m below the sea floor (Fig. 3). Similar phenomena occur in many present-day deltaic Greek environments, formed in the Gulfs of Patras, Corinth, Themaikos and Amvrakikos (Ferentinos 1991; Lykousis 1991; Papatheodorou & Ferentinos 1993).

On the basis of the interpretation of seismic profiles obtained from the shelf area (Fig. 3), the accumulation rates of Holocene sediments are at their maximum just to seawards of the various river mouths (i.e. 3.8 m ka^{-1} for the River Alfios, 2.1 m ka^{-1} for the River Nedas and 1.2 m ka^{-1} for the River Peristera). In comparison, the river-influenced continental shelf of Thermaikos Gulf (northwestern Aegean Sea) (Poulos *et al.* 2000) is characterized by accumulation rates between 3 m ka^{-1}, close to the river mouths, and 0.2–0.5 m ka^{-1}, over the prodelta area (Lykousis & Chronis 1989). Further offshore and over the lower part of the slope in the Kyparissiakos Gulf, along the Hellenic Trench, accumulation rates are of the order of 0.07–0.14 m ka^{-1} (Cramp *et al.* 1987). Once again, this latter value is similar to that reported for the deep (> 1000 m) Sporades Basin (adjacent to the Thermaikos Gulf continental slope), of 0.05–0.15 m ka^{-1} (Lykousis & Chronis 1989).

Coastal zone

The coastal zone represents that part of the coastal system where the interaction between land–ocean–air processes is most evident. Moreover, its sedimentological evolution is related primarily to the influx of the terrigenous sediments. The coastal zone covers an area of some 610 km^2, representing only 6.6% of the total area of the coastal system,

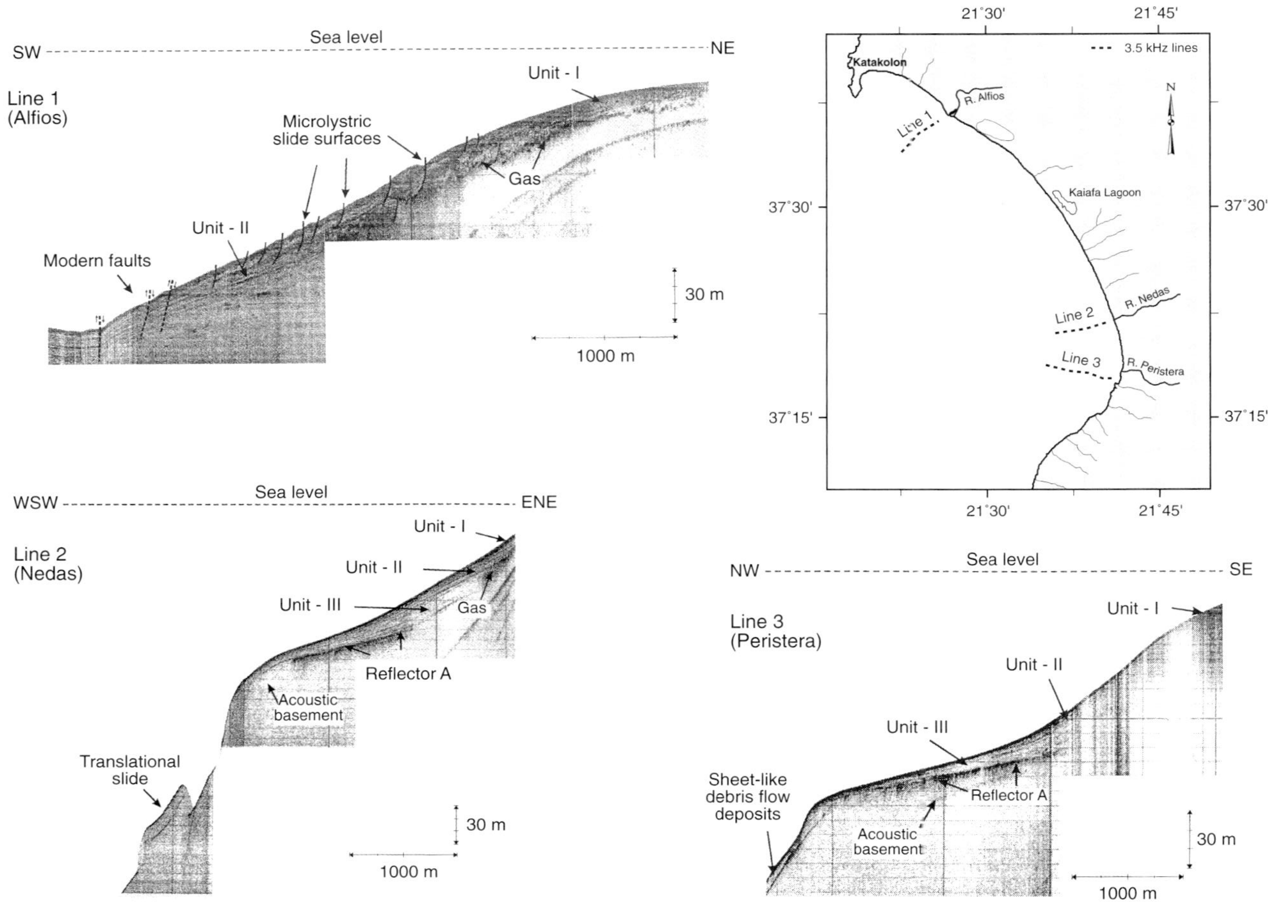

Fig. 3. Shallow seismic (3.5 kHz) profiles across the continental shelf and offshore at the mouths of the principal rivers discharging into the Kyparissiakos Gulf.

and is divided, by the coastline, into: (i) a sub-aerial part and (ii) a sub-aqueous part.

Sub-aerial part

The sub-aerial part of the coastal zone consists of a lowland area (with a gradient of <10% and at elevations mostly <20 m), formed of Upper Quaternary (mostly Holocene) fluvio-deltaic sediments; it covers an area of only some 145 km^2, representing less than a third of the coastal zone and only 2.3% of the total coastal system (Table 1).

The terrestrial part of the coastal zone incorporates a variety of sedimentary environments, with the deltaic coastal plains being dominant. These coastal plains have developed within the Holocene, during and after the rise in sea level. Fluvio-deltaic environments, flanked by beach–dune barriers and lagoons, were established; these have buried the Pleistocene deposits and earlier rocks, with aggradational and progradational deposits. Furthermore, these deltaic deposits have well-developed levees and point bars, and, sometimes, show evidence of avulsion and dramatic changes of the river course, being bordered by fluvial marshes (Raphael 1973). The coast is fringed by beach–dune low-relief barrier–ridge complexes (10–300 m in width), that enclose discontinuous lagoons (e.g. the Kaiafa Lagoon, and those of Agoulinitsa and Mouria that were drained in the 1970s); these, in turn, pass landwards into drained lacustrine plains (Fig. 4).

Fluvio-marine Holocene terraces occur along the River Alfios, which lie only a few metres above sea level; these are the product of tectonic activity and/or salt diapirism. Similar terraces occur on the Kyllili Peninsula, to the north of Kyparissiakos Gulf (Maroukian *et al.* 2000).

The uplands, against which various elements of the coastal plain abut, are fringed by Pleistocene alluvial fans. These are composed of red gravels that developed where the rivers debouch on to the coastal plain and plunge below the Holocene deposits. In the southern part of the Gulf, where the coastal beach–dune barrier is absent, the fluvial deposits and the deposits of these fans reach the coast, to form low cliffs. Elsewhere, there are rocky cliffs of Neogene rock that enclose some pocket beaches at the mouths of ephemeral streams.

The coastline appears to have stabilized around 2400 BP, then to have prograded seawards during Hellenistic–Roman times. However, during the Mediaeval (AD 326–1354) and later Turkish (1354-1821 AD) and Modern Greek (post-1821) times, the fluvio-deltaic deposits were entrenched and erosion became dominant (Raphael 1973). This erosion of the beach–dune barriers often reveals Roman remains enclosed in beach rock. Sand is being driven landward by aeolian action. Where the fluvio-deltaic deposit and alluvial fan deposits reach the coast, they have been cliffed and are capped by dunes that are advancing inland.

Landward of the present-day coastline and to the north of the mouth of the River Nedas, Ferentinos *et al.* (1987) have claimed that a palaeo-coastline formed at the beginning of the Holocene. However, the junction between the pre-Holocene rocks and the Holocene sediments is likely to have been produced by alluvial–lagoonal drowning of the former as the Flandrian transgression progressed, and does not mark the former position of a former open coastline. The width of the modern deposits (from the shore zone to the commencement of the hinterland) varies between a few hundred metres to more than 2 km. The largest widths occur at the northern half of the Gulf due to progradation during the Holocene of the deltaic plain of the River Alfios, and the development of the beach–dune barrier and their associated lagoons (Fig. 4).

Sub-aqueous part

The sub-aqueous part of the coastal zone extends up to the shelf break, including the whole of the continental shelf (Cadee *et al.* 1994; IGBP 1995). It is rather narrow, with widths ranging from 2.5 to 9 km, and quite steep, with a gradient of between 2% and 5%, covering an area of some 275 km^2; this represents more than two-thirds of the total area of the coastal zone, but <5% of the total coastal system.

The shelf break is tectonically controlled by a fault system lying almost parallel to the coastline (Fig. 1b); it varies in depth from 95 to 175 m in the north, whilst to the south it is more uniform with water depths between 125 and 130 m. Beyond the shelf break is a slope that has steep gradients, in some areas in excess of 15%. In terms of the prevailing hydrodynamics, the shelf can be further divided into the inner and outer shelf area, with respect to the existing wave regime (as tides are insignificant). The inner shelf covers an area with water depths less than the half of the maximum wavelength ($d < L/2$).

Deep-water wave characteristics (i.e. wave height and period) were used to calculate the wavelengths to define the seaward extent of the inner shelf. A 10-year period wind record (Ginis 1973), measured at Strofades Island (for the location see Fig. 2), was used for hindcasting (CERC 1984) the wave climate. A 24-h wind duration, which represents the maximum duration for this area, was used during the hindcasting process (Ferentinos *et al.* 1987).

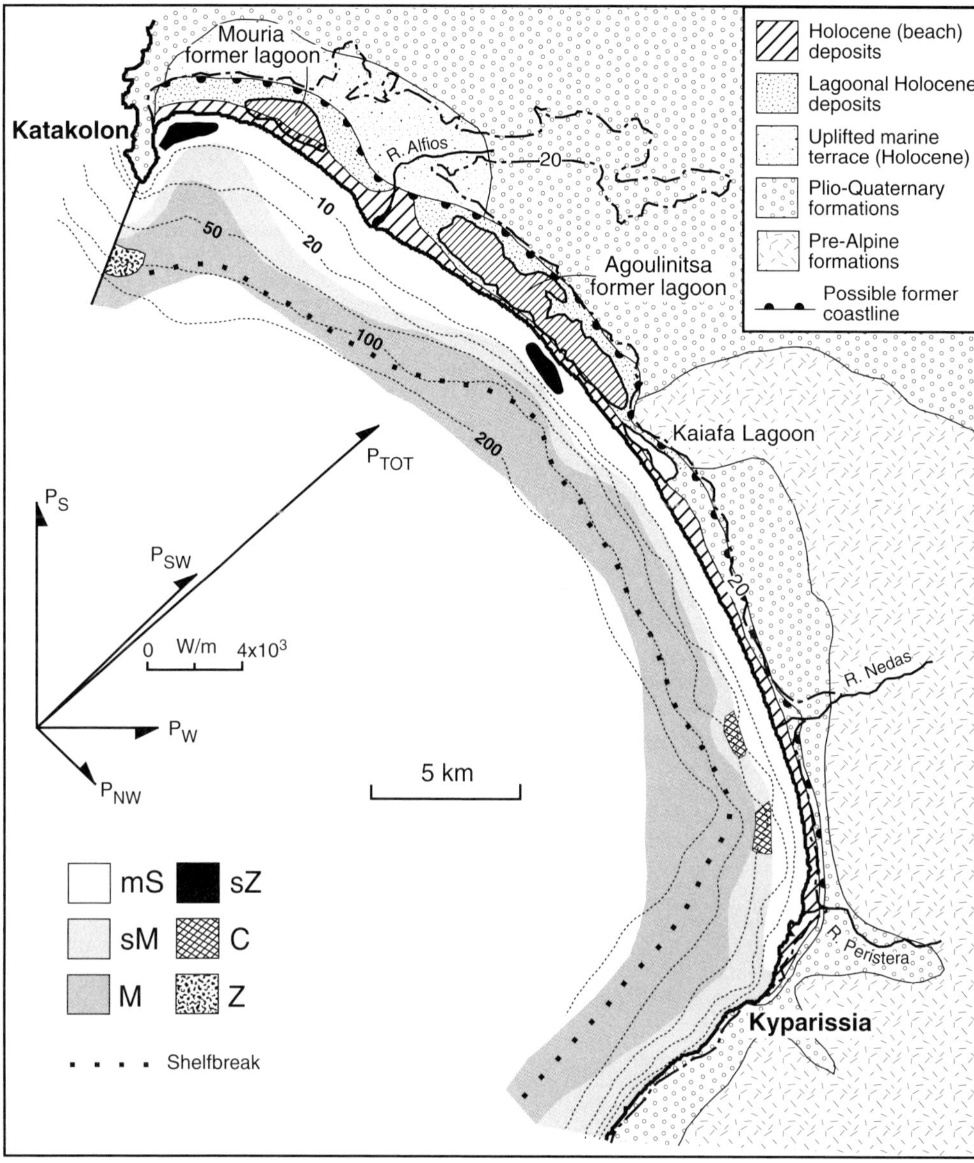

Fig. 4. Map showing the physical characteristics, sedimentary provinces and processes operating within the sub-aerial and sub-aqeous parts of the Kyparissiakos Gulf coastal zone sub-system. Key: mS, muddy sand; sM, sandy mud; M, mud (silt+clay); sZ, sandy clay; C, clay; Z, silt.

Estimated wavelengths varied for the moderate to strong winds from 16 (NW) to 80 m (SW), and for the very strong winds between 40 (NW) and 120 m (SW); these correspond to water depths at the seaward limit of the inshore zone (equal to *L*/2) of 8–20 m and 20–60 m, respectively. Thus, the inner shelf area adjacent to the beach is controlled by the incoming wave activity and the associated longshore currents; these form specific wave-related rythmic beach (sub-aqueous) morphology.

The surface water circulation in the outer shelf/slope area is generally weak (generally <10 cm s^{-1}), being wind-driven and usually towards the north. Under certain wind conditions, from westerly and northwesterly directions, an anticyclonic type of circulation may be responsible for the transfer of only the very fine-grained material towards the south, over the outer shelf-edge region. However, under weak to moderate wind conditions over the inner shelf, the buoyant plumes are diverted northwards, in response to the geostrophic flow.

In general, the coarse-grained siliciclastic gravels and sands of the shoreface zone (as outlined below) pass seawards into siliciclastic muds (silt and clay). On the basis of grain-size analysis, Kapsimalis (1993) and Kapsimalis *et al.* (1997) classified the shelf/upper slope surficial bottom sediments into three sedimentary zones (Fig. 4b). (1) An elongated zone, extending from the coastline to the mean fair-weather wave-base, which coincides with the 20-m bathymetric contour. This zone is characterized by intense and continuous disturbance, by oscillatory and shoaling waves, as well as rip and longshore currents. Only fine sand and muddy sand are deposited here, due to the intense hydrodynamic regime. (2) A zone that extends from the 20 m bathymetric contour to the mean storm wave-base, where alternations of low- and high-energy conditions occur. Frequent changes between fair-weather and storm conditions result in a cyclic and/or mixed deposition of sand and fine-grained material. In this narrow zone, restricted to the middle shelf, sandy mud and sandy silt are the dominant facies. (3) A zone covering the outer shelf and upper slope. Limited bottom disturbance occurs here and gravitational settling of suspended particles, originating from the seaward dispersion of the riverine fine-grained sediments, are characteristic of this zone. It is covered by muddy deposits, incorporating clay (35–70%), silt (30–60%) and sand fractions (<8%). The absence of any relict sediments on the outer shelf and at the shelf break indicates that, during the Holocene, the deposition of siliciclastic sediment has been sufficient to completely blanket the shelf. This situation is very typical of narrow Mediterranean shelves, bordering mountainous areas.

Shore (beach) zone – coastline

The shore zone also includes a sub-aerial part (the backshore zone) and a sub-aqeous part (the inshore zone). This zone and the associated coastline are the most dynamic components of any coastal system, as they change almost constantly in space and time in response to the prevailing hydroynamic and sediment dynamic conditions. In terms of spatial extent, the shore zone represents only a very small part of the coastal zone area, which becomes almost negligible in comparison to the scale of the total area of the coastal system.

Sub-aerial part (backshore zone)

The sub-aerial part of the shore zone is formed of a strip of unconsolidated sandy beach–dune deposits, whose width varies between 30 and 60 m in the central part of the Gulf to >70 m at its northern part, where the River Alfios delta has prograded. At the southernmost part of the Gulf, the beach zone is either absent or only a few metres wide; it is usually associated with pocket beaches, developed in coastal cliffs, of indurated cohesive conglomerates (Neogene in age). In addition, in some places between the River Peristera and the city of Kyparissia, beach rocks composed of breccio-conglomerate are exposed (Fig. 5a).

The steep beach face is often succeeded to landward by a backshore, which often has at least three berms; these reach to 1–2 m above sea level. The sedimentary material of the shoreface varies from place to place, it is a mixture of sand and fine gravel (present in percentages ranging from <1%, up to >50%). In general, the relatively coarse-grained material is found in the northern part of the Gulf and near the active river mouths (Fig. 5b). Thus, the shoreface between the Ak. Kamariki and to the south of the mouth of the River Nedas consists of sediments of gravelly sand to sandy gravel, with an $Mz = 0\phi$. Further to the north (between the River Nedas and Kaiafa Lagoon) the sediments become finer with an $Mz = 2.5\phi$. At the mouth areas of the rivers Nedas and Alfios, the grain sizes are highly variable, with a mean $Mz = -0.5\phi$. The finest material is observed within the northernmost part of the Gulf, where the beach consists of fine–very fine sand. The sediments around the river mouths are characterized by poor sorting ($\sigma_I = 1\phi$), whilst the areas remote from these are characterized by the fine sand with good sorting ($\sigma_I = 0\phi$) (Fig. 5b).

The curved shoreline is interrupted over its central and southern part by some small-scale protrusions, associated with the presence of giant cusp-like forms. Such features occur every 350–400 m and their width is less than 10 m. The narrow shelf and steep gradients of the inner shelf permit most of the incoming offshore wave power to reach the inshore zone, where it develops strong longshore currents. The simple and almost straight configuration of the coastline can be attributed to the intensive wave activity, capable of distributing all the riverine sediment. However, around the mouth of the River Alfios, despite the large sediment flux ($>2.5\ 10^6$ t year^{-1}), the shape of the delta is cuspate; this is similar to those of the River San Francisco and the Rhone where, in the absence of a tide, wave activity overcomes the riverine processes (Galloway & Hobday 1983; Briggs *et al.* 1997).

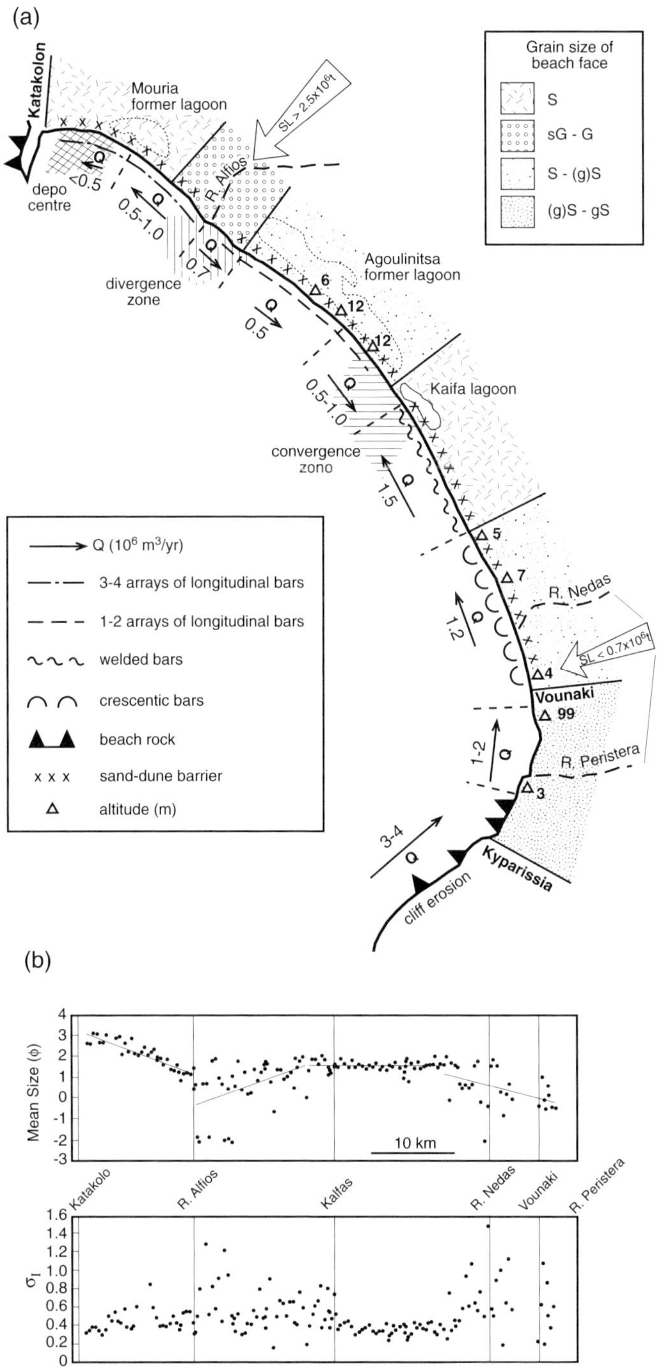

Fig. 5. (**a**) Map showing the physical characteristics, sedimentary provinces and hydrological processes operating within the Kyparissiakos Gulf shore (beach) zone sub-system. (S, sand; G, gravel; sG, sandy gravel; (g)S, slightly gravelly sand; gS, gravelly sand (after Folk 1980)). (**b**) The sedimentary texture, in terms of the distribution of mean size (Mz in ϕ units) and of the standard deviation (σ_I), of the sediments along the shoreface of the Kyparissiakos Gulf.

Sub-aqueous part (inshore zone)

This particular part of the beach zone is controlled hydrodynamically by the incoming wave activity and the associated longshore currents; its width is regarded to extend to water depths (d) of 4–10 m ($d < L/20$) of the highest incoming waves (under storm conditions).

Intensive wave activity (wave heights >5 m) and associated strong longshore sediment transport, due to the long fetch distances (500–800 km) (Fig. 1a) in relation to the prevailing westerly winds (4–8 Beaufort), are responsible for the distribution of sediments within the nearshore zone where most of the sand is trapped. Here, the incoming intensive wave has formed sub-aqueous rhythmic sedimentary forms, consisting of coarse-grained material (mostly coarse-grained sand, with some fine gravel). Thus, in the southern part (near the mouth of the River Nedas) there are crescentic bars with lengths of 360–400 m and widths of 200 m (Fig. 5a). In the middle part of the Gulf (from the River Nedas to the Kaiafa Lagoon) welded bars exist, with a length of 120–140 m and a separation of 180–200 m. Finally, a system of two–four parallel longitudinal bars is present along the northern part of the Gulf, extending seawards by up to 200 m and between 30 and 50 cm in height (Fig 5a).

The longshore sediment transport has been investigated on the basis of the calculation of the potential annual transport rate (Q_t), induced by the prevailing wind directions. Transport is related to the simulated wave longshore (P_l) power distribution along the coast, using the equation given by Komar (1976): for wave propagation and refraction in shallow waters a numerical model, based upon the Griswold (1963) equations, was used. The longshore pattern of potential sediment transport was calculated for each of the wave conditions hindcasted using the wind directions and speeds from the Strofades Ilsand records (Ginis 1973). The net annual longshore pattern was calculated by weighing each result with the frequency of occurrence of the generating winds and vector summation in the alongshore direction. Although the accuracy of the values predicted by this method might be questionable due to the errors included in the assumed wind duration, this is not considered important here. The important issues are the gradients and directions that are indicative of areas of convergence and divergence. These relative trends are considered to be relatively robust to errors in wind duration.

The longshore potential sediment transport indicates the existence of a zone of divergence in the river mouth area of the Alfios delta. Likewise, a convergence zone exists in the middle part of the Gulf, near the Kaiafa Lagoon (Fig. 5a). Furthermore, the highest values of potential sediment transport (1×10^6–3.5×10^6 m^3 year^{-1}) occur in the central and southern parts of the Gulf; in contrast, the lowest values (0.1×10^6–1.2×10^6 m^3 year^{-1}) occur at its northernmost part. The overall potential transport is towards the north, being of the order of some 2×10^6 m^3 year^{-1}. These observations, in relation to the presence of the major sediment sources along the coast, explain the nearshore bathymetry (Fig. 4) and the distribution of sediments on the beachface. The lack of fine-grained sediment over the southern part of the Gulf is related to the intensive northward longshore transport, in combination with the absence of sediment influx to the area. Where the convergence zone occurs, there is a sub-aqueous concentration of sediments (Fig. 5a), but there has not been any substantial coastline progradation. Such an absence is probably because of the relatively small amounts of sediments delivered by the small and ephemeral River Nedas, discharging to the south, and the relatively reduced sediment transport rates in the northern part of the Gulf. Together, these factors combine with the overall northward transport to inhibit the transport of large amounts of sediment southwards, from the mouth of River Alfios. In contrast, the beach of the northernmost part of the Gulf is characterized by very fine sandy material. This area is sheltered from northerly winds and has long been recognized as a major depocentre of the fine sediments delivered by the River Alfios, then transported northwards. Continuous extraction of sediments from the shallow waters in the latter area is necessary, to retain the functioning of the harbour at Katakolo.

Human impact

Human interference to the sediment fluxes is related mainly to the denudation processes and the transport of sediments within the terrestrial part of the system. Within historical (Hellenistic and Roman) times, periods can be identified associated with high river sediment loads. These loads were the product of deforestation and the development of new agricultural land. In contrast, over the last century, the construction of (two) dams along the course of the River Alfios (the principal river) has led to a dramatic reduction in sediment supply to the coastal zone (primarily, the shore zone).

The first dam was constructed for hydroelectric purposes on the River Ladon, one of the main tributaries of the River Alfios, in 1955. This structure is likely to have reduced the original sediment flux by about 30%, according to the catchment area enclosed behind the dam (some 1000 km^2). The second dam was constructed in 1967, some tens of kilometres from the mouth of the River Alfios; this was for irrigation purposes, allowing only a water discharge of *c*. 40 $m^3 h^{-1}$, throughout the year. Despite this constant outflow, all of the bed load and the majority of the suspended load had been deposited behind the dams. Examination of aerial photographs collected around the mouth of the River Alfios, in 1945 and 1972 (prior and after the dam construction, respectively), indicate that erosion has taken place along the deltaic coastline; at the same time, even the dune barrier had started to be affected. Recent field measurements (in April 2001) have shown that the mouth and the northern part of the deltaic coastline has retreated by as much as 200 m, with active erosion of the sand dunes of the coastal barrier.

Human interference extends to the coastal zone ecosystem through the artificial drainage (in the 1960s) of the lagooons of Mouria (2–6 km^2) and Agoulinitsa (2.5–3.7 km^2), with maximum depths of <2 m; these were converted into agricultural land. Finally, within the northern part of the Gulf the offshore transport of marine sediments is enhanced by the artificial dredging of sediments from the entrance of the harbour at Katakolo (Fig. 5a), prior to the subsequent dumping of these sediments on the outer part of the adjacent continental shelf.

Synthesis

The riverine coastal system of the Kyparissiakos Gulf lies on the outer-arc area of the Hellenic Arc–Trench system; this is formed by the convergence and subduction of the African plate beneath the Eurasian plate (Fig. 1a). The system is characterized by high relief (with about 4 km, as the vertical distance between the highest elevation on land and the deepest water depth) and intensive tectonic (seismic) activity; as such, it resembles the typical characteristics of an active margin.

The present configuration of this coastal system is controlled by the geological structure of the region, which has evolved over hundreds of millions of years. The major evolution of the area was due to processes that took place during the Alpine tectogenesis, which had its paroxysmic phase during the Oligocene (38–24.6 Ma) and terminated at the Miocene (24.6–5.1 Ma). Since then, the geological evolution of the system has been the result of a balance between *endogenetic* and *exogenetic* processes (Haslett 2000) operating on the Earth's surface (Fig. 6). During the Pliocene and throughout the Quaternary, deposition of unconsolidated sediments has occurred in both the terrestrial and marine environment. The initial stage of formation of the present-day coastal zone took place within the Upper Quaternary. The present morphological configuration was acquired mostly within the Holocene, when sea level reached its present level. Within this period (i.e. Upper Quaternary–Holocene) there have been three major morphological phases, influenced by natural processes and human impact (Fig. 6): (i) an aggradational phase, associated with the misuse of

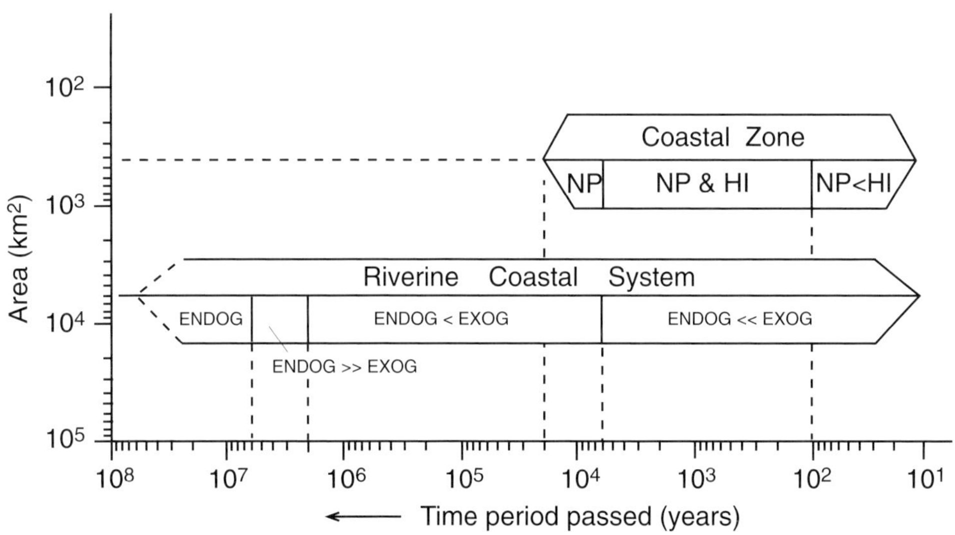

Fig. 6. Schematic presentation of the evolution of the major components of the Kyparissiakos Gulf riverine coastal system, relative to their spatial area and the period passed. Key: NP, natural processes; HI, human impact; ENDOG, endogenetic processes, i.e. tectonism, diapirism; and EXOG, exogenetic processes, i.e. climate, sea-level changes.

land by man; (ii) a more recent erosional period, related to human interferance to the natural riverine flows (i.e. river regulation, dam construction); and (iii) the most recent phase related to sea-level rise, associated with climate change. The beach zone, which consists of the part of the system most sensitive to human impact and vulnerable to changes, has been developed and modified throughout the Holocene by the nearshore wave-dominated forcing. The present-day configuration of the system has developed within the last 2000–3000 years, although significant local changes may have occurred relatively rapidly at the meso-scale (i.e. within a few months to a few years only).

Sediment fluxes play a significant role in the evolution of the river-influenced coastal system of the Kyparissiakos Gulf (Fig. 7a). The hinterland area is characterized by high sediment yields (>500 t km^{-2} $year^{-1}$), produced by intensive denudation processes; these are associated with its high topographic relief, erodible lithology, temperate climatic conditions and seismic activity. Most of this large amount of sediment (>3×10^6 t $year^{-1}$) reaches the coastal zone, due to the high gradients of the mountainous small river networks and its proximity to the source area. Here, part of the sediment load participates in the formation and evolution of the sub-aerial part of the coastal zone; the rest is dispersed seawards to cover the continental shelf and slope and eventually to reach the deepest parts of the coastal system (i.e. the (Hellenic) Trench; Fig. 1b).

The terrestrial part of the coastal zone, developed by interaction of the land (sediment fluxes) and marine (sea-level rise and nearshore hydrodynamics) processes, includes fluvio-deltaic sedimentary environments; these are flanked along the shoreline by beach–dune low-relief barriers and lagoons. The northern tectonic uplift has also led to the presence of a marine terrace a few metres above the sea level. In the southern part of the Gulf, fluvial deposits approach the sea and form low coastal cliffs.

The nearshore zone, the most changeable part of the coastal zone (sub-system), consists of unconsolidated sandy deposits and presents a simple and almost straight configuration of the coastline; this is attributed to the strong longshore sediment transport. The latter feature is induced by the intensive wave and current activity within the nearshore zone; here, they also form a specific wave-related rythmic beach (subaqueous) morphology.

The seaward dispersion and deposition of sediments of terrigenous origin is extended seawards over the shelf (the marine part of the coastal zone); this can be attributed to the offshore spreading of the sediment-laden river plumes (usually associated with flood events) and the intensive wave- and wind-driven current activity, especially over the inner shelf (in water depths <30 m; Fig. 7a). The absence of relict sediments over the shelf and the zonal distribution of the surficial bottom sediments (fining seawards) provide evidence that adequate amounts of terrigenous sediments are being transported offshore by the marine processes (e.g. waves, wind-induced circulation), operating over a rather narrow and steep-sloping continental shelf (Fig. 7b). Furthermore, the interpretation of seismic profiles has revealed the high-energy mode of the recent deposits and the existence of reworking processes over the shelf area. Likewise, the prograding-seawards deltaic prisms consist of relatively coarse-grained sediments (siliciclastic muds); these were deposited over interfluve regions during the high stand of sea level. Here, Holocene accumulation rates are of the order of 1–4 m ka^{-1}; these are in comparison with those observed in other tectonically-active embayments in Greece, for example: Thermaikos Gulf (>3 m ka^{-1}; Lykousis & Chronis 1989); Gulf of Corinth (>10 m ka^{-1}; Poulos *et al.* 1996*b*). Such restricted growth of the deltaic lobes may be attributed to: (a) the narrowness of the adjacent shelf, together with its relative steep gradients; (b) the regional tectonism (seismicity), which prevents the build-up of sediments (especially over the shelf edge area); and (c) the redistribution of sediments, due to the highly energetic hydrological conditions.

Subsequently, from the shelf-edge region, sediments are transported over the slope to the deepest part of the system by gravitational mass movements (e.g. slides, debris flows, turbidites); these are often triggered by earthquake shocks (Ferentinos 1992), associated with the subduction zone. Such gravitational mass movements are common along the flanks of the Hellenic Trench system and have also been reported for the region of the dextral transformation, the Kefallonia Fault (to the north of Kyparissiakos Gulf) (Ferentinos 1992; Poulos *et al.* 1999) and to the south of the study area (Got *et al.* 1981; Got 1984).

Hence, in terms of sediment advection into the marine sub-system of the Kyparissiakos Gulf; this can be regarded as a semi-enclosed system, where most of the sediments originate from its terrestrial part. Such a source is supplemented in the deeper waters (e.g. beyond the outer shelf) by biologically-produced carbonate within the water column. The predominance of siliciclastic sediment of terrigenous origin, in combination with the oligotrophic status of the Mediterranean Sea (in general), causes the biogenic contribution to be limited. Some material is expected also to be advected from the north, by the deep-water southward-moving near-bed currents observed at the Zakynthos Channel

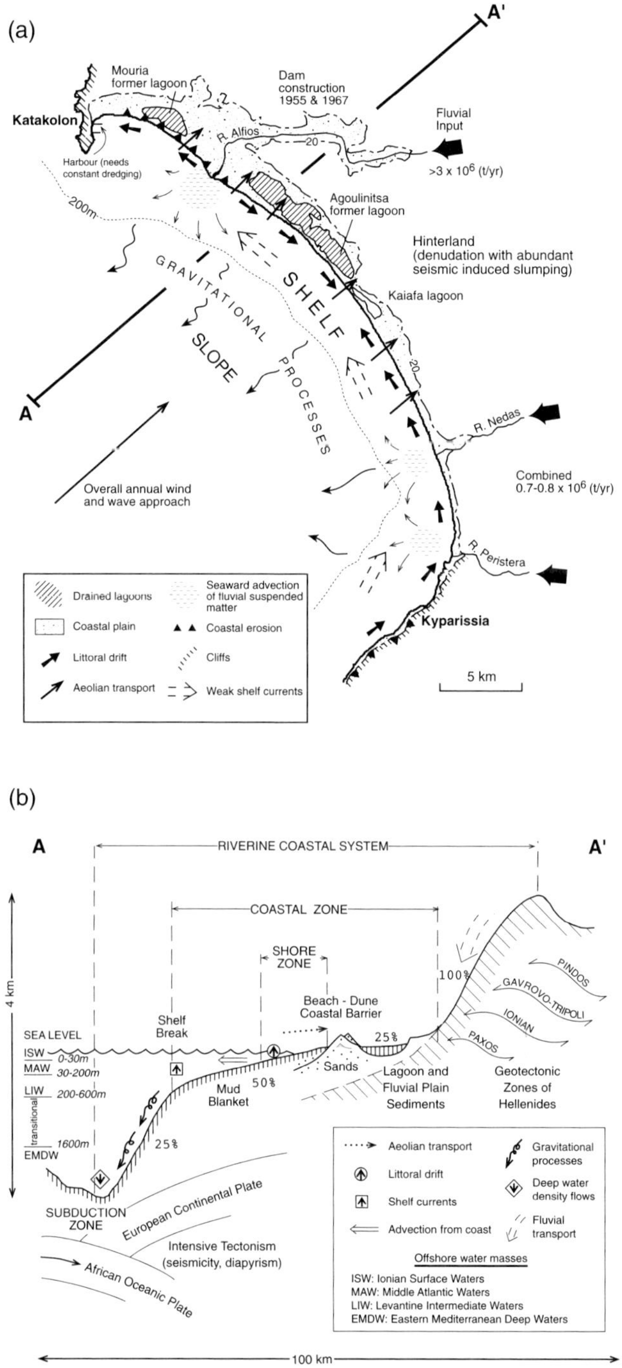

Fig. 7. Conceptual model showing, in a plan view (**a**) and along the cross section A–A′ (**b**), the geological, sedimentological, oceanographic characteristics, and associated processes governing the dispersion and accumulation of the fluvial sediment fluxes within the various components of the tectonically-active Kyparissiakos Gulf coastal system.

(Ferentinos *et al.* 1985). Conversely, the northward-moving near-surface general circulation of the water masses produces negligible transfer of terrigenous material, as there are no major sedimentary sources to the south of the Kyparissiakos Gulf area.

In terms of transport and deposition of the river-derived sediments ($>3 \times 10^6$ t year^{-1}; this corresponds to $>1 \times 10^9$ m^3 ka^{-1}, assuming a sediment density of 2650 kg m^{-3}) within the various components of the coastal system, it would appear that since the sea-level high-stand and, to a first approximation, 50% has been deposited over the shelf, whilst another 25% has been transported over the slope to the deepest part of the system (the Hellenic Trench). The remaining amount is incorporated into the formation of the sub-aerial part of the coastal zone (e.g. the fluvio-deltaic plains) and to the shore zone. These estimates are based on assumed accumulation rates for the shelf area (where a mean value of 1 m ka^{-1} has been adopted (see Fig. 3)) and for the deep-slope region (<0.2 m ka^{-1}; Cramp *et al.* 1987), with respect to their spatial location (see Table 1). Such percentages can be compared with those presented elsewhere by Sternberg (1986) for the Washington continental margin, which receives the sediment input from the River Columbia. Here, 67% of the annual sediment discharge is deposited over the mid-shelf, with a further 17% transported over the open slope and within the various canyon systems. Differences in the budgets may well be explained in terms of the width of the Washington continental shelf (50–60 km) in comparison to the narrow shelf (<10 km) of the Kyparissiakos Gulf (<10 km), which also experiences intensive tectonic (seismic) activity.

On the basis of the above summary, the Upper Quaternary (mostly Holocene) evolution of this coastal system is attributed, primarily, to processes related to balances between sediment fluxes (e.g. terrestrial transportation, seaward dispersion); these occur within the context of a mountainous terrain, related to a tectonically-active margin and a narrow continental shelf with restricted delta formation. Recently, this particular coastal system has also been affected by human interference, in the form of hydroelectric/irrigation dam construction and the drainage of coastal lagoons; these have modified the supply and flux of sediment along the hydrological pathways of both the terrestrial and marine parts of the Kyparissiakos Gulf coastal system. Finally, on a global scale (LOICZ 2001) the present investigation incorporates consideration of: (a) establishment of the recent sediment flux to the coast; (b) the effect of fluxes on system evolution; and (c) the influence of man and/or climatic change.

The authors are grateful to Mrs K. Davis (SOES, SOC) for the artistic preparation of the figures. Likewise, Dr A. Velegrakis (SOES) is aknowledged for his constructive remarks and recommendations, on an earlier version of the manuscript.

References

BRIGGS, D., SMITHSON, P., ADDISON, K. & ATKINSON, K. 1997. *Fundamentals of the Physical Environment*, Second Edition. Routledge, London.

CADEE, N., DRONKERS, J., HEIP, C., MARTIN, J.-M. & NOLAN, C. 1994. *European Land–Ocean Interaction Studies*. Science Plan, EUR 15608 EN.

CERC (COASTAL ENGINEERING RESEARCH CENTRE). 1984. *Shore Protection Manual*. US Army Corps of Engineers, Washington.

CHRONIS, G., PIPER, D. J. W. & ANAGNOSTOU, C. 1991. Late Quaternary evolution of the Gulf of Patras, Greece: tectonism, deltaic sedimentation and sea level change. *Marine Geology*, **97**, 191–209.

COLEMAN, J., PRIOR, D. & LINDSAY, J. F. 1983. Deltaic influences on continental instability processes. *In*: STANLEY, D. J. & MOORE, G. T. (eds) *The Shelfbreak: Critical Interface on Continental Margins*. SEPM Special Publications, **33**, 121–137.

CRAMP, A., COLLINS, M. B. & WAKEFIELD, S. J. 1987. Sedimentation in the Zakynthos Channel – a conduit link to the Hellenic Trench, eastern Mediterranean. *Marine Geology*, **76**, 71–87.

DAL CIN, R. 1983. I litorali del delta del Po e alle foci dell' Adige e del Brenta:caratteri tesituarali e dispersione dei sedimenti, cause dell'arrtramento e previsioni sull'evoluzione futura. *Bulletin of the Geological Society of Italy*, **102**, 9–56.

FERENTINOS, G. 1991. Offshore geological hazards in the Hellenic Arc. *Marine Geotechnology*, **9**, 261–277.

FERENTINOS, G. 1992. Recent gravitative mass movements in a highly tectonically active arc system: the Hellenic Arc. *Marine Geology*, **104**, 93–107.

FERENTINOS, G., COLLINS, M. B., PATTIARATCHI, C. B. & TAYLOR, P. G. 1985. Mechanisms of sediment transport and deposition in a tectonically active submarine valley/canyon system: Zakynthos Straits, NW Hellenic Trench. *Marine Geology*, **65**, 243–269.

FERENTINOS, G., KONTOPULOS, N., SAMBO, V. & VOULGARIS, G. 1987. Wave conditions and littoral drift along the Kyparissiakos Gulf coastline. *Hydrotechnica*, **3**, 1–10 [in Greek].

FOLK, R. L. 1980. *Petrology of Sedimentary Rocks*. Hemphill Publishing Company, Austin, Texas.

GALLOWAY, W. E. & HOBDAY, D. K. 1983. *Terrigenous Clastic Depositional Systems*. Springer, New York.

GINIS, S. 1973. *The Wind Regime in the Ionian Sea*. PhD Thesis, University of Athens (in Greek).

GOT, H. 1984. Sedimentary processes on the west Hellenic Arc Margin. *In*: STOW, D. V. A. & PIPER, D. J. W. (eds) *Fine-grained Sediments*. Geological Society, London, Special Publications, **15**, 169–183.

GOT, H., MONACO, A., VITTON, J., BRAMBATI, A., CATANI, G., MASOLI, M., PUGLIESE, N., ZUCCHI-STOLFA, M.,

BELFIORE, A., GALLO, F., MEZZADRI, G., VERNIA, L., VINCI, A. & BONADUCE, G. 1981. Sedimentation on the Ionian active margin (Hellenic Arc) – Provenance of sediments and mechanisms of deposition. *Sedimentary Geology*, **28**, 243–272.

GRISWOLD, G. M. 1963. Numerical calculation of wave refraction. *Journal of Geophysical Research*, **68**, 1715–1723.

GUILLEN, J., DIAZ, J. I. & PALANQUES, A. 1992. Quantification and evolution of the Ebro River bedload sediment supplies during the XX century. *Review Society Geology España*, **5**, 27–37 (in Spanish).

HASLETT, S. K. 2000. *Coastal Systems*. Routledge, London.

IGBP. 1995. *Land–Ocean Interactions in the Coastal Zone: Implementation Plan*. PERNETTA, J. C. & MILLIMAN, J. D. (eds) International Geosphere Biosphere Programme Report, **33**. IGBP, Stockholm.

IGME. 1989. *Seismotectonic Map of Greece, 1:500 000*. Institute for Geological and Mineralogical Exploration, Athens. IGME, Athens.

KABERIS, E. 1987. *Geological and Petrologic Studies in the Northwestern Peloponnese*. PhD Thesis, Technical University of Athens.

KAPSIMALIS, V. 1993. *Sedimentological and Geochemical of the Recent Sediments on Continental Shelf and Upper Continental Slope of Kyparissikos Gulf, Greece*. PhD Thesis, Technical University of Athens.

KAPSIMALIS, V., KONISPOLIATIS, N., CHRONIS, G., LYKOUSIS, V., FILIPPAS, D. & PANAGOS, A. 1997. Distribution of surface sediments of Kyparissiakos Gulf. *In*: *Proceedings of the 5th Hellenic Symposium Oceanography and Fisheries., Kavala*, National Centre for Marine Research, Athens, 379–382.

KOMAR, P. D. 1976. *Beach Processes and Sedimentation*. Prentice-Hall, Englewood Cliffs, NJ.

KOUKOUVELAS, I., MPRESIAKAS, A., SOKOS, E. & DOUTSOS, T. 1996. The tectonic setting and earthquake ground hazards of the 1993 Pyrgos earthquake, peloponnese, Greece. *Journal of the Geological Society, London,* **153**, 39–49.

LAMBECK, K. 1996. Sea-level change and shore-line evolution in Aegean Greece since Upper Palaeolithic time. *Antiquity,* **70**, 588–611.

LEOPOLD, L. B., WOLMAN, G. M. & MILLER, J. P. 1964. *Fluvial Processes in Geomorphology*. W. H. Freeman, San Francisco.

LE QUELLEC, P., MASCLE, J., VITTORI, J. & GOT, H. 1980. Seismic structure of the southern Peloponnesus continental margin. *AAPG Bulletin*, **64**, 242–263.

LYKOUSIS, V. 1991. Submarine slope instabilities in the Hellenic Arc region, northeastern Mediterranean Sea. *Marine Geotechnology*, **10**, 83–96.

LYKOUSIS, V. & CHRONIS, G. 1989. Mechanisms of sediment transport and deposition: sediment sequences and accumulation during the Holocene on the Thermaikos Plateau, the continental slope, and basin (Sporades Basin), northwestern Aegean Sea, Greece. *Marine Geology*, **87**, 15–26.

MALANOTTE-RIZZOLI, P., MANCA, B. B., D'ALCALA, M. R., THEOCHARIS, A., BERGAMASCO, A., BREGANT, D., BUDILLON, G., CIVITARESE, G., GEORGOPOULOS, D., MICHELATO, A., SANSONE, E., SCARAZZATO, P. & SOUVERMEZOGLOU, E. 1997. A synthesis of the Ionian Sea hydrography, circulation and water mass pathways during POEM-Phase I. *Progress in Oceanography*, **39**, 153–204.

MAROUKIAN, H., GAKI-PAPANASTASIOU, K., PAPANASTASIOU, D. & PALYVOS, N. 2000. Geomorphological observations in the coastal zone of Kyllini Peninsula, NW Peloponnesus–Greece, and their relation to the seismotectonic regime of the area. *Journal of Coastal Research*, **16**, 853–863.

MILLIMAN, J. D. & SYVITSKI, P. M. 1992. Geomorphic/tectonic control of sediment discharge to the ocean: the importance of small mountainous rivers. *Journal of Geology,* **100**, 525–544.

NITTIS, K., PINARDI, N. & LASCARATOS, A. 1993. Characteristics of the summer 1987 flow field in the Ionian Sea. *Journal of Geophysical Research*, **98**, 10 171–10 184.

NITTROUER, C. A. & DRISCOLL, N. 1999. Source to sink. *MARGINS Newsletter*, **3**, 2–3.

PANO, N. 1992. Dinamica del litorale albanese (sintesi delle conoscenze). *Proceedings of the 19th A.I.G.I. Meeting, Genova, Italy*. G. Lang, Rome, 3–18.

PAPATHEODOROU, G. & FERENTINOS, G. 1993. Sedimentation processes and basin-filling depositional architecture in an active asymmetric graben: Strava graben, Gulf of Corinth, Greece. *Basin Research*, **5**, 235–253.

PERISSORATIS, C., PIPER, D. J. W. & LYKOUSIS, V. 2000. Alternating marine and lacustrine sedimentation during late Quaternary in the Gulf of Corinth rift basin, central Greece. *Marine Geology,* **167**, 391–411.

POULOS, S. E. & COLLINS, M. B. 2002. Fluviatile sediment fluxes to the Mediterranean Sea: a quantitative approach and the influence of dams. *In*: JONES, S. J. & FROSTICK, L. E. (eds) *Sediment Flux to Basins: causes, controls and consequences*. The Geological Society, London, Special Publications, **191**, 227–245.

POULOS, S. & CHRONIS, G. 1997. The importance of the Greek river systems in the evolution of the Greek coastline. *In*: BRIAND, F. & MALDOLADO, A. (eds) *Transformations and Evolution of the Mediterranean Coastline*. Bulletin de l' Institut Oceanographique Special Publication, **18**, CIESM Science Series 3, 75–96.

POULOS, S. E., COLLINS, M. & EVANS, G. 1996*a*. Water sediment fluxes of Greek rivers, southeastern Alpine Europe: annual yields, seasonal variability, delta formation and human impact. *Zeitschrift für Geomorphologie*, **40**, 243–261.

POULOS, S. E., COLLINS, M. B., PATTIARATCHI, C., CRAMP, A., GULL, W., TSIMPLIS, M. & PAPATHEODOROU, G. 1996*b*. Oceanography and sedimentation in the semi-enclosed, deep-water Gulf of Corinth (Greece). *Marine Geology,* **134**, 213–235.

POULOS, S. E., CHRONIS, G., COLLINS, M. & LYKOUSIS, V. 2000. Thermaikos Gulf Coastal-System, NW Aegean Sea: An overview of water/sediment fluxes in relation to air-land-ocean interactions and human activities. *Marine Systems*, **25**, 47–76.

POULOS, S. E., LYKOUSIS, V., ROHLING, I., COLLINS, M. B. & PATTIARATCHI, C. 1999. Sedimentation processes in a tectonically active environment: the Kerkyra–Kefalonia submarine valley system (NE Ionian Sea*). Marine Geology*, **160**, 25–44.

RAPHAEL, N. 1973. Late Quaternary changes in coastal Elis. *Geographical Review*, **63**, 73–89.

SACHPAZI, M., HIRN, A., CLEMENT, C., HASLINGER, F., LAIGLE, M., KISSLING, E., CHARVIS, P., HELLO, Y., LEPINE, J.-C., SAPIN, M. & ANSORGE, J. 2000. Western Hellenic subduction and Cephalonia Transform: local earthquakes and plate transport and strain. *Tectonophysics*, **319**, 301–319.

SKOULIKIDIS, N. 1993. Significance evaluation of factors controlling river water composition. *Environmental Geology*, **22**, 178–185.

STANLEY, D. J. 1981. Unifites: structureless muds of gravity flow origin in Mediterranean basins. *GeoMarine Letters,* **1,** 77–83.

STERNBERG, R. W. 1986. Transport and accumulation of river-derived sediment on the Washington continental shelf, USA. *Journal of the Geological Society, London*, **143,** 945–956.

SWIFT, D. J. P. 1968. Coastal erosion and transgressive stratigrapgy. *Geology,* **76**, 444–456.

SWIFT, D. J. P., PHILLIPS, S. & THORNE, J. A. 1991. Sedimentation on continental margins. VI: Lithofacies and depositional systems. *In*: SWIFT, D. J. P., OERTEL, G. F., TILLMAN, R. W. & THORNE, J. A. (eds) *Shelf Sand and Sandstone Bodies: Geometry, Facies and Sequence Stratigraphy.* International Association of Sedimentologists, Special Publications, **14**, 89–152.

SYVITSKI, J. P. M. 2001. Supply and flux of sediment along hydrological pathways: Anthropogenic influences at the global scale. *LOICZ Newsletter*, **20**, 4–7.

TSIMPLIS, M. N., PROCTOR, R. & FLATHER, R. A. 1995. A two-dimensional tidal model for the Mediterranean Sea. *Journal of Geophysical Research*, **C8**, 223–216.

THERIANOS, A. D. 1974. The geographical distribution of the river water supply in Greece. *Bulletin Geological Society, Greece*, **11**, 28–58 (in Greek).

UNDERHILL, J. R. 1989. Late Cenozoic deformation of the Hellenide foreland, western Greece. *Geological Society of America Bulletin*, **101**, 613–634.

VAIL, P. R., AUDERMAND, J., BOWMAN, S. A., EISNER, P. N. & PEREZ-CRUZ, C. 1991. The stratigraphic signatures of tectonics, eustasy and sedimentology: an overview. *In*: EINSELE, G., RICKEN, W. & SEILACHER, A. (eds) *Cycles and Events in Stratigraphy*. Springer, Berlin, 617–659.

VITTORI, J. 1978. *Caracteres Structuro-sedimentaires de la Couverture Plio-quaternaire au Niveau des Pentes et des Fosses Helleniques du Peloponnesus (Grece)*. PhD Thesis, Univesite Paul Sabatier, Toulouse.

VITTORI, J., GOT, H., LE QUELLEC, P., MASCLE, J. & MIRABILE, L. 1981. Emplacement of the recent sedimentary cover and processes of deposition on the Matapan Trench margin (Hellenic Arc). *Marine Geology,* **41**, 113–135.

VON HUEVE, R. & SCHOLL, D. W. 1991. Observations at convergent margins concerning sediment subduction, subduction erosion, and the growth of continental crust. *Review Geophysics*, **29,** 279–316.

WALLING, D. E. & WEBB, B. W. 1998. Erosion and sediment yield: a global overview. *In*: *Proceedings of the Exeter Symposium, July 1996*. IAHS Publication, **236**, 3–19.

Late Quaternary turbidite input into the east Mediterranean basin: new radiocarbon constraints on climate and sea-level control

MICHAEL S. REEDER[1], DORRIK A. V. STOW[1] & R. GUY ROTHWELL[2]

[1]*School of Ocean and Earth Sciences, University of Southampton, Southampton Oceanography Centre, European Way, Southampton SO14 3ZH, UK (e-mail: davs@soc.soton.ac.uk)*

[2]*Challenger Division for Seafloor Processes, Southampton Oceanography Centre, European Way, Southampton SO14 3ZH, UK*

Abstract: The Late Pleistocene–Holocene (0–30 ka BP) allochthonous sedimentation in the Herodotus Basin of the eastern Mediterranean has been controlled, in part, by a combination of regional climatic change and eustatic sea-level fluctuation. A new series of radiocarbon dates, made on planktonic foraminifers and pteropod shells taken from the pelagic and hemipelagic intervals between individual turbidite units, has given bracketing dates for each major turbidity current event that deposited sand and mud on the Herodotus Basin plain. Two partly independent cycles are evident. Climate-induced cycles have lead to an alternation of periods of turbidites sourced from the Nile delta–fan system with those from the North African shelf and Anatolian rise. These correlate with pluvial and inter-pluvial climatic periods recognized in the Nile hinterland. Sea-level cycles have tended to focus turbidite emplacement, from whatever source, at periods of sea-level fall within the latest Wisconsin and sea-level rise from the Wisconsin–Holocene period. In addition to the Herodotus Basin Megaturbidite (HBM) described previously, six other beds with volumes in excess of 25 km^3 and wide lateral extent across the basin can be termed megaturbidites. There is no simple sea-level or climate control on the timing of these events, so we must conclude that triggering and emplacement of megaturbidites is independent and variable.

The Herodotus Basin is a SW–NE-trending elongate depression that lies to the NW of the Nile Cone and has a maximum water depth in excess of 3000 m. It is the largest basin in the eastern Mediterranean, with the 3000-m isobath delineating an area of approximately 27 000 km^2. The detailed physiography of the basin is presented by Reeder *et al.* (1998, 2000), who have calculated the areal extent of Late Pleistocene ponded sediment to be in the order of 40 000 km^2 (Fig. 1). This is significantly greater than the area based on the 3000-m isobath, as sediment has overspilled this level by up to at least 2700 m in places.

Herodotus Basin sedimentation

The Late Pleistocene sedimentary characteristics of the Herodotus Basin are reported in detail by Cita *et al.* (1984), Lucchi & Camerlenghi (1993) and Reeder *et al.* (1998, 2000). The most recent work by Reeder *et al.* (1998, 2000) was based on a longitudinal basin transect of five giant piston cores, collected on the *Marion Dufresne* Cruise 81 (Table 1). These cores penetrated to a maximum depth of 24.2 m and emplacement of the oldest sediments was estimated at less than 30 ka BP. Hence, all the sediments are from the Late Pleistocene and mainly comprise turbidites from four discrete sediment sources (Fig. 1). Individual turbidites are readily correlated between cores across the Herodotus Basin on the basis of their particular compositional characteristics. Sources have been determined both by mineralogical content (Cita *et al.* 1984; Lucchi & Camerlenghi 1993) and by the measured decrease of turbidite and sand-fraction thicknesses with increased transportation (Reeder *et al.* 1998). Individual turbidites have been named ‘a’–‘p’, where ‘a’ represents the youngest and ‘p’ the oldest of the events.

Some 50% of the named turbidites were derived from an organic-rich Nile Cone source. These have

From: JONES, S. J. & FROSTICK, L. E. (eds) 2002. *Sediment Flux to Basins: Causes, Controls and Consequences.* Geological Society, London, Special Publications, **191**, 267–278. 0305-8719/02/$15.00

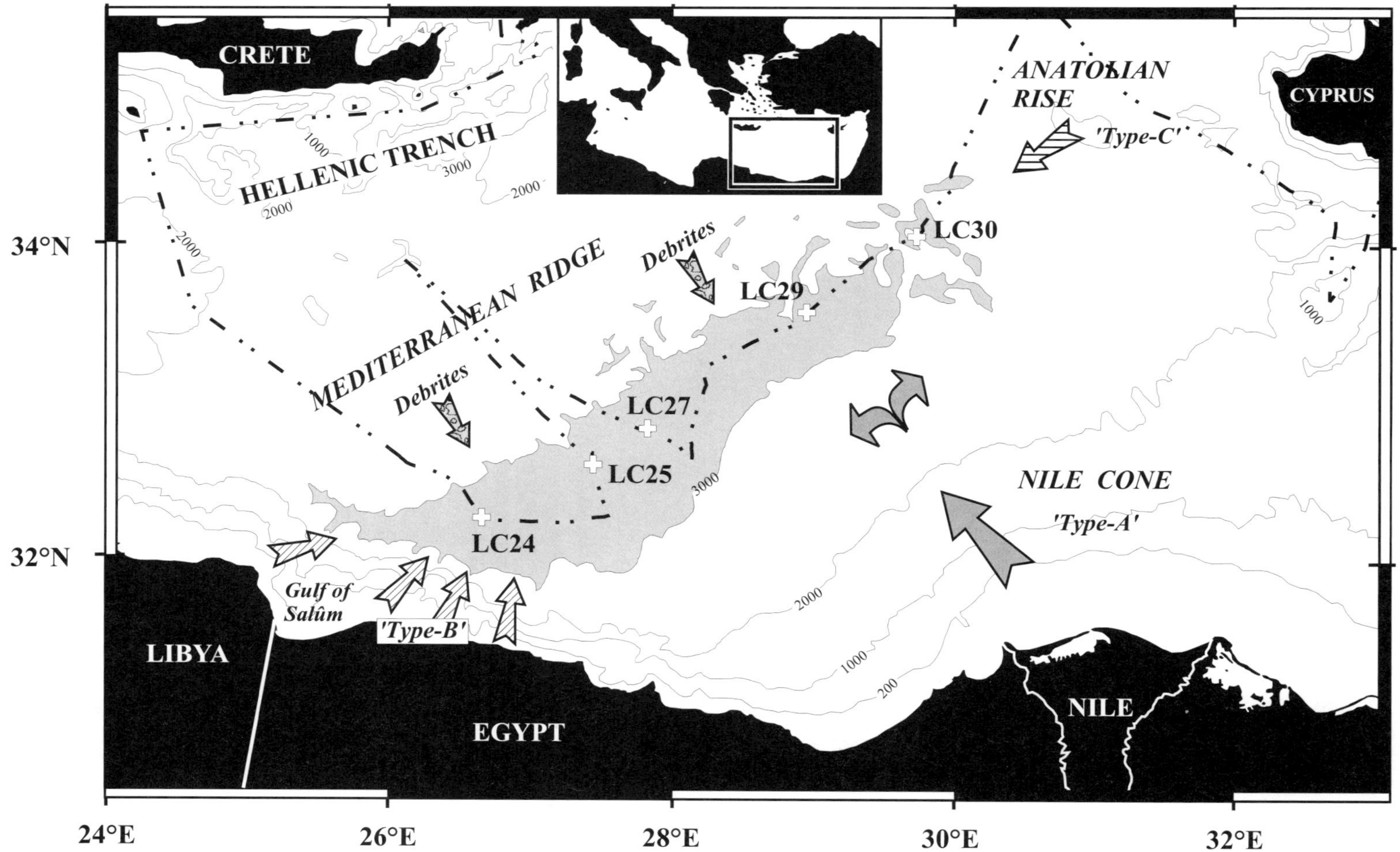

Fig. 1. Location of the Herodotus Basin, showing the main sediment sources, core positions and the navigational track of the *Marion Dufresne* Cruise 81 (MD81).

Table 1. *Water depth, positions and lengths for the five long-piston cores collected during* Marion Dufresne *Cruise 81 (MD81)*

Core No.	Water depth (m)	Latitude	Longitude	Core length (m)
LC24	3191	32°17.71′N	26°37.95′E	18.31
LC25	3129	32°36.01′N	27°23.25′E	13.71
LC27	3131	32°48.91′N	27°40.45′E	14.95
LC29	3138	33°35.63′N	28°56.32′E	24.66
LC30	3144	34°04.70′N	29°42.72′E	25.82

been termed 'Type A' turbidites, following the nomenclature of the first description of allochthonous sediments in the basin by Cita *et al.* (1984). They range from 15 to 750 cm in thickness and are fine-grained turbidites with thin (maximum of 30 cm thick) silty and sandy bases. Estimates of their total volume range from 0.1 to 126 km^3, with a possible maximum for one of the deepest turbidites 'o' of around 190 km^3 (Reeder *et al.* 1998). The second most abundant source is the NE African shelf ('Type B' turbidites) with thicknesses ranging from 6 cm to over 16 m (Fig. 2), and estimated volumes from 0.1 to 400 km^3. The largest of these, turbidite 'n' is referred to as the Herodotus Basin Megaturbidite (HBM). These more carbonate-rich (typically 40–50 % $CaCO_3$) sediments are easily recognizable from their Nile Cone counterparts by their lighter coloration. The light olive–grey 'Type B' turbidites reflect a relative abundance of carbonate shelf organisms, whilst the 'Type A' Nile Cone-derived turbidites are dark brown–grey and reflect an abundance of plant and other terrigenous material. Two further discrete sources have been described by Reeder *et al.* (1998), being 'Type C' turbidites derived from the Anatolian Rise to the NE of the Herodotus Basin, and the debris flow deposits (debrites) from the Mediterranean Ridge on the northern flank of the basin (Fig. 1).

The petrology, geochemistry, correlation and stratigraphy of Late Pleistocene sedimentation in the Herodotus Basin are presented in detail elsewhere (Cita *et al.* 1984; Lucchi & Camerlenghi 1993; Reeder *et al.* 1998, 2000) and summarized partially here in Fig. 2. Although the Late Pleistocene stratigraphy of the Herodotus Basin is known in general terms, precise dating of individual turbidite events, and hence elucidation of their relationship with climatic change and eustatic sea-level fluctuation, has not previously been attempted. The nannofossil zonation reported previously indicates that all turbidites were deposited within the last 60 ka. The thickness of hemipelagic accumulation between turbidites was then used to estimate tentative emplacement dates for all event deposits, thereby concluding that the oldest sediments recovered were around 30 ka (Reeder *et al.* 1998).

In this paper we present absolute dates for turbidite emplacement on the Herodotus Basin plain, based on radiocarbon dating of the pelagic and hemipelagic intervals between turbidites. These are then discussed in relation to climatic changes and fluctuation in eustatic sea level during the latest Pleistocene.

AMS radiocarbon (^{14}C) dating

By sampling the planktonic foraminifera from pelagic horizons overlying and underlying turbidites, bracketing emplacement dates were obtained for each event. The AMS radiocarbon dating method of Thomson & Weaver (1994) was used, which entailed sampling the whole of the pelagic interval, as well as the top 10–15 cm of the underlying turbidite, in order to include any forams reworked by burrowing fauna. A continuous trench sampling technique was employed.

In total, 10 samples were collected from the Herodotus Basin cores, at the core depths shown in Fig. 2. Larger planktonic foraminifera are expected to be the least susceptible component of the pelagic interval to be redistributed by currents and thus the most suitable for dating. Samples were therefore sieved to obtain the >150-μm size fraction, and the clean planktonic foraminifera and pteropods were selected from this. All other bioclastic and lithic material, together with forams and pteropods contaminated with sediment, were removed from the >150-μm sieved fraction by hand picking.

Corrections to the raw ^{14}C dates were made to account for the relationship between the radiocarbon and calendar timescales by applying the correction factor of:

$$\text{Calendar years} = 1.24 \times (^{14}\text{C years}) - 440$$

(Bard *et al.* (1993) (Table 2). The values include a correction of 400 years to account for the offset ^{14}C age of the atmospheric and oceanic mixed-layer reservoir (Stuiver 1990).

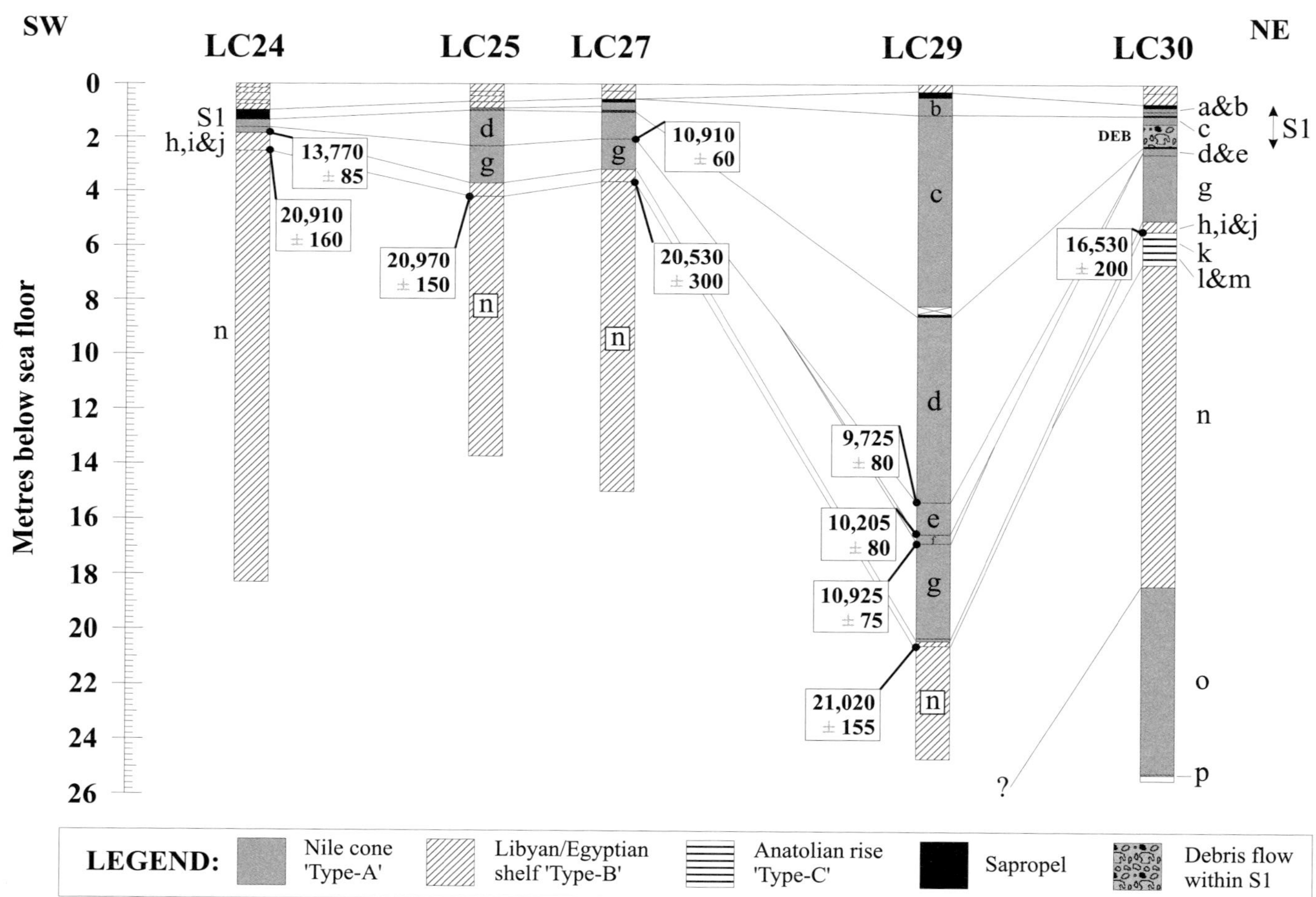

Fig. 2. Stratigraphy, correlation and conventional radiocarbon dates of sediments in the five long-piston cores from the Herodotus Basin. Note: The transect from LC24 to LC30 is approximately 325 km in length.

Table 2. *Details from the AMS ^{14}C dating of the 10 samples from the Herodotus Basin cores, with calculated calendar ages*

Publication code	Sample identifier	^{14}C enrichment (% modern ± 1σ)	Carbon content (5 dry wt)	$\delta^{13}C_{PDP}$ ±0.1 0/00	$\delta^{18}O_{PDP}$ ±0.1 0/0C	Convetional radiocarbon age (years BP ± 1σ)	Calculated calendar age* (years BP)
AA-24162	LC29.7.83-99	29.80 ± 0.30	10.2	0.1	1.3	9725 ± 80	11 620
AA-24163	LC29.6.64-77	28.07 ± 0.28	n/a	0.4	n/a	10 205 ± 80	12 215
AA-24164	LC29.6.873-100	25.66 ± 0.25	10.7	–0.3	3.1	10 925 ± 75	13 105
AA-24165	LC29.3.18-31	7.30 ± 0.15	n/a	–1.5	n/a	21 020 ± 155	25 625
AA-24166	LC25.7.62-75	7.35 ± 0.15	10.2	1.4	3.1	20 970 ± 150	25 565
AA-24167	LC27.9.49-64	7.76 ± 0.29	10.6	1.5	3.4	20 530 ± 300	25 015
AA-24168	LC30.16.105-116	12.77 ± 0.32	10.7	1.2	3.7	16 530 ± 200	20 055
AA-24169	LC24.13.10-25	18.02 ± 0.20	9.7	1.3	2.7	13 770 ± 85	16 635
AA-24170	LC24.13.49-63	7.40 ± 0.15	10.5	1.5	3.7	20 910 ± 160	25 490
CAMS-388650	LC27.10.22-53	25.71 ± 0.17	10.6	0.7	1.3	10 910 ± 60	13 090

*Following the correction of Bard *et al.* (1993), where: Calendar age = 1.24 × (^{14}C age) –440.

Turbidite emplacement vs sea level

The results of the radiocarbon dating are presented in Fig. 3. By bracketing the emplacement of the turbidite an average date has been obtained for each event (Table 3). We can identify four distinct periods of time characterized by cyclic differences in turbidite source and then compare these with the eustatic sea-level curve (Fig. 3) (Haq *et al.* 1987).

Sea-level low-stand (pre-28 ka BP)

Only two turbidites from the present core suite (core LC30) fall within this time period, and both of these are Type-A turbidites from the Nile source. Stanley & Maldonado (1979) also reported Nile-derived turbidites between 55 and 28 ka BP, relating to a transgressive period of slightly higher sea level within the main Wisconsin low stand system tract. Based on the limited data available in these earlier studies, we estimate a relatively low average accumulation rate of between 0.02 and 0.05 m ka^{-1} during this interval.

Lowering sea level (28–17 ka BP)

By contrast, the Late Wisconsin regression, when sea level dropped to its lowest value of around 125 m below present depths, is characterized by an absence of Type A turbidites. This period begins with deposition of the Herodotus Basin Megaturbidite (Type B) at just over 27 ka, followed by three smaller volume Type C turbidites from a northern source (Anatolian Rise), and then three small Type B turbidites from the North African margin. Although the numbers are somewhat limited, this represents turbidite emplacement on to the basin plain of one event every 1.4 ka approximately. The thickness of the megaturbidite can be estimated at just over 20 m proximally, decreasing to some 12.5 m where the base was cored at site LC30. This clearly makes up the major part of an average accumulation rate of between 1.3 and 2 m ka^{-1}.

Rising sea level (17–6 ka BP)

Following the lowest sea-level stand of the last glacial (around 18–17 Ma BP), the pattern of sedimentation changed notably as sea level rose rapidly at a rate of approximately 12.5 m ka^{-1} (1.25 cm a^{-1}). During this period, and in contrast to the period preceding the glacial, the Herodotus Basin stratigraphy was dominated by thicker Nile Cone-derived turbidites (Type A), rather than the thin North African shelf and Anatolian Rise-derived turbidites.

The Nile Cone received very large quantities of sediment and entered a period of major accumulation (Mart 1993). Several of the larger-scale turbidity currents bypassed the fan to reach the basin plain, entering the relatively flat basin adjacent to the location of core station LC29. This is reflected in the thickest accumulation of Type A turbidites centred around this area (Fig. 2) and also indicated by 3.5 kHz seismic evidence of distributary channels in this vicinity (Reeder *et al.* 1998). Towards the end of the sea-level rise, anoxic bottom water conditions occurred across the eastern Mediterrranean basin leading to deposition of the organic-rich S1 sapropel. During this period only small-volume Nile Cone turbidites ('a', 'b' and 'c') were deposited, continuing the pattern of mono-source input on the Herodotus Basin plain. Turbidites reached the basin plain on average once every 1.8 ka during this phase of rising sea level, giving an accumulation rate of 1.82 m ka^{-1} in the main depocentre (core LC29) and only some 0.1 m ka^{-1} at the SW end of the basin (core LC24).

Sea-level high-stand (6–0 ka BP)

The record of sedimentation after the deposition of the S1 sapropel is poorly preserved in the MD81 cores. There is an inherently large amount of deformation caused by the piston coring technique, which may have removed a small quantity of the most recent sediments. However, the sequence observed is dominated by the presence of thin Type B turbidites (un-named) and thin hemipelagic intervals (Fig. 2). Turbidite emplacement varied from one event in 6 ka to one event every 1.5 ka, and the average rate of accumulation from 0.07 to 0.15 m ka^{-1}.

Summary

The most apparent relationship between turbidite emplacement and sea-level is an increase in the thickness of turbidites reaching the Herodotus Basin plain, and hence in the average rate of accumulation, during both the Late Wisconsin sea-level fall to its lowest stand and the Late Wisconsin–Holocene sea-level rise. In the former period the turbidites were exclusively Types B and C, whereas in the latter period they were Nile Type A.

Turbidite emplacement vs climate

Considering the same four periods of time, we can consider instead the climatic variation that may also have had some effect on both the rate of turbidite supply and the source of sediment (turbidite type). The principal variation likely to affect sedimentation is between pluvial and inter-pluvial episodes

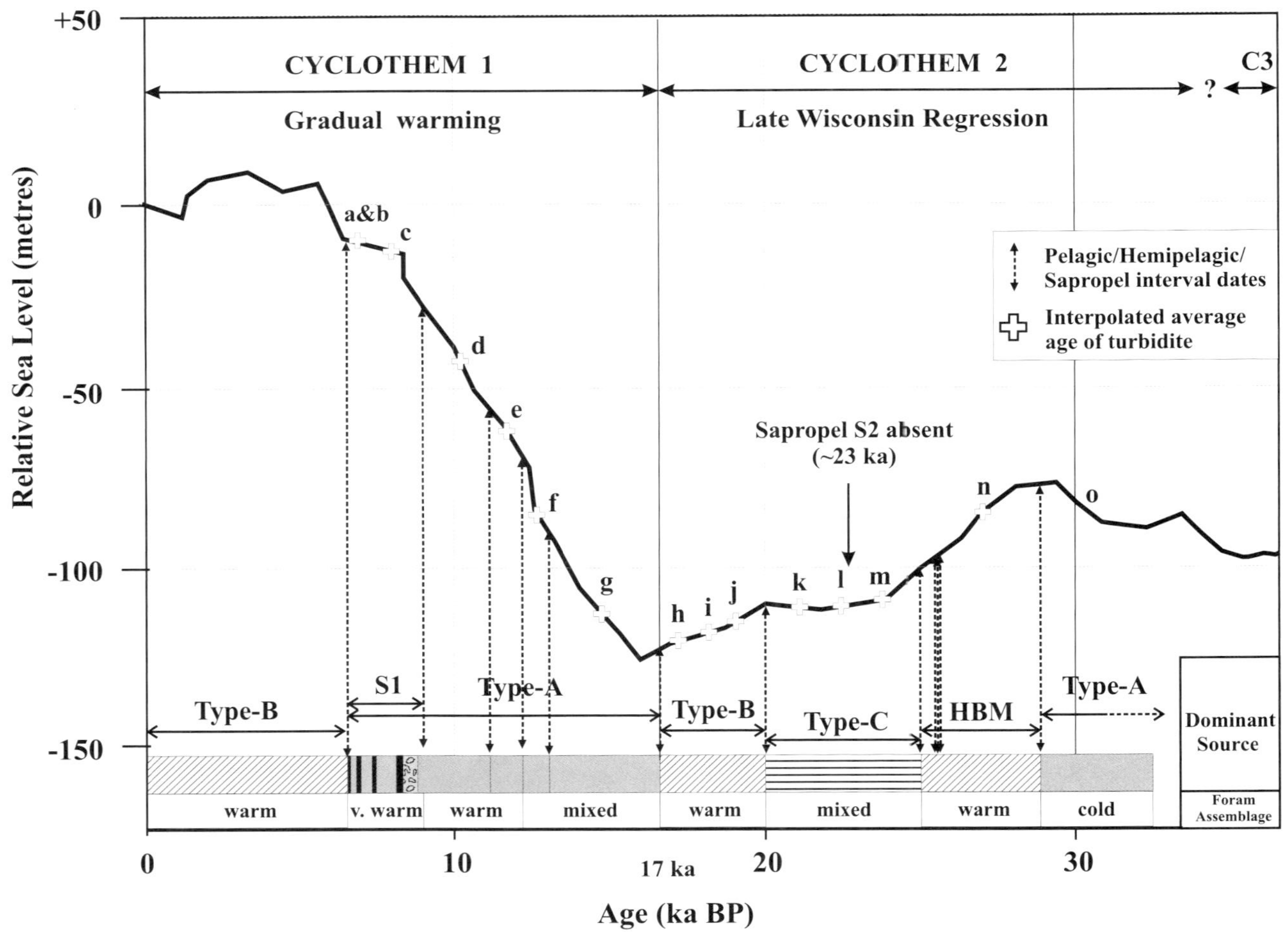

Fig. 3. Eustatic sea-level curve for the past 30 ka (after Shackleton 1987) showing individual turbidites as well as turbidite types based on source area. Two possible cyclothems are indicated, but see text for discussion. Foram assemblages after Kahler & Dossi (1996).

Table 3. *Interpolated depositional ages for the Late Pleistocene Herodotus Basin turbidites*

Turbidite	Type	Approximate volume*	Age of overlying dated pelagic interval (years BP)	Age of underlying dated pelagic interval (years BP)	Interpolated depositional age† (years BP)
Turbidite 'a'	A	0.1			7000
Turbidite 'b'	A	6.0	6500 (S1)	9500 (S1)	7500
Turbidite 'c'	A	80.0			8000
Debrite		2.4			8500
Turbidite 'd'	A	126.0	9500 (S1)	11 620	10 560
Turbidite 'e'	A	25.2	11 620	12 215	11 920
Turbidite 'f'	A	0.1	12 215	13 105	12 660
Turbidite 'g'	A	72.0	13 105	16 635	14 870
Turbidite 'h'	B	2.2			17 490
Turbidite 'i'	B	0.8	16 635	20 055	18 345
Turbidite 'j'	B	0.8			19 200
Turbidite 'k'	C	8.0			21 400
Turbidite 'l'	C	0.04	20 055	25 425	22 740
Turbidite 'm'	C	4.0			24 740
Turbidite 'n'	B	400	25 425[HBM]	28 825[HBM]	27 125[HBM]
Turbidite 'o'	A	190?	28 825	?	>28 825
Turbidite 'p'	B	0.1?	?	?	?

*Volumes from Reeder *et al.* (1998).
†Mean date of the overlying and underlying pelagic intervals.
[HBM]Dates for the Herodotus Basin Megaturbidite.

that brought alternately wetter and drier conditions, respectively, to the upper reaches of the Nile River (or Neonile as the Late Pleistocene river is generally known) (Said 1990). These were not exactly time equivalent to interglacial and glacial climatic intervals.

Kubbanivan pluvial (c. *50–28 ka*)

The earliest period represented by our sediment records from the Herodotus Basin fell during the Kubbanivan pluvial episode – a relatively wet but still quite cold period. The Neonile would have carried much sediment to the delta region, thus providing an adequate source for downslope resedimentation by turbidity currents. The Herodotus Basin turbidites recovered from this period are all Type A (Nile-source).

Interpluvial (c. *28–17 ka*)

During this period, the climate in the Nile hinterland was cold and dry, and much less sediment would have been supplied to the Nile Delta. The river was subject to rare storm-related flooding, but otherwise lower discharge. No Nile-derived turbidites reached the Herodotus Basin during this period, which is characterized by Type B (North African shelf) and Type C (Anatolian) turbidites only.

Nabtian pluvial (c. *17–6 ka*)

Following the Late Wisconsin glacial maximum, the climate of the region began to ameliorate and the vegetation pattern changed from a cold, dry ecosystem to a humid, lake and marsh environment dominated by a herbaceous flora (Cheddadi *et al.* 1991; Cheddadi & Rossignol-Strick 1995). This pluvial period also led to an increase in sedimentary output onto the Nile Cone as a direct result of increased precipitation in the Nubian Massif (Ethiopian highlands) to the south. There is still uncertainty about the exact timing of climate change in the region, with some authors supporting more arid conditions in parts of the Nile drainage basin from 13 to 10 ka (Cheddadi & Rossignol-Strick 1995). The Nabtian pluvial *sensu stricto* is fairly well documented and generally considered to have lasted for a rather shorter time period, from about 10 to 5 ka, and to include three much shorter dry episodes (Said 1990). The generally more humid conditions and increased vegetation cover are confirmed by the high organic detritus content recorded in sediments collected from the Nile Cone and Herodotus Basin dating from this period (Ross *et al.* 1978; Cita *et al.* 1984; Lucchi & Camerlenghi 1993). The turbidites are all of Type A (Nile source).

Also, at approximately 10 ka BP reforestation occurred in the hinterlands of the Mediterranean, leading to a rejuvenation in supply of nutrients to the eastern Mediterranean basin (Cheddadi *et al.*

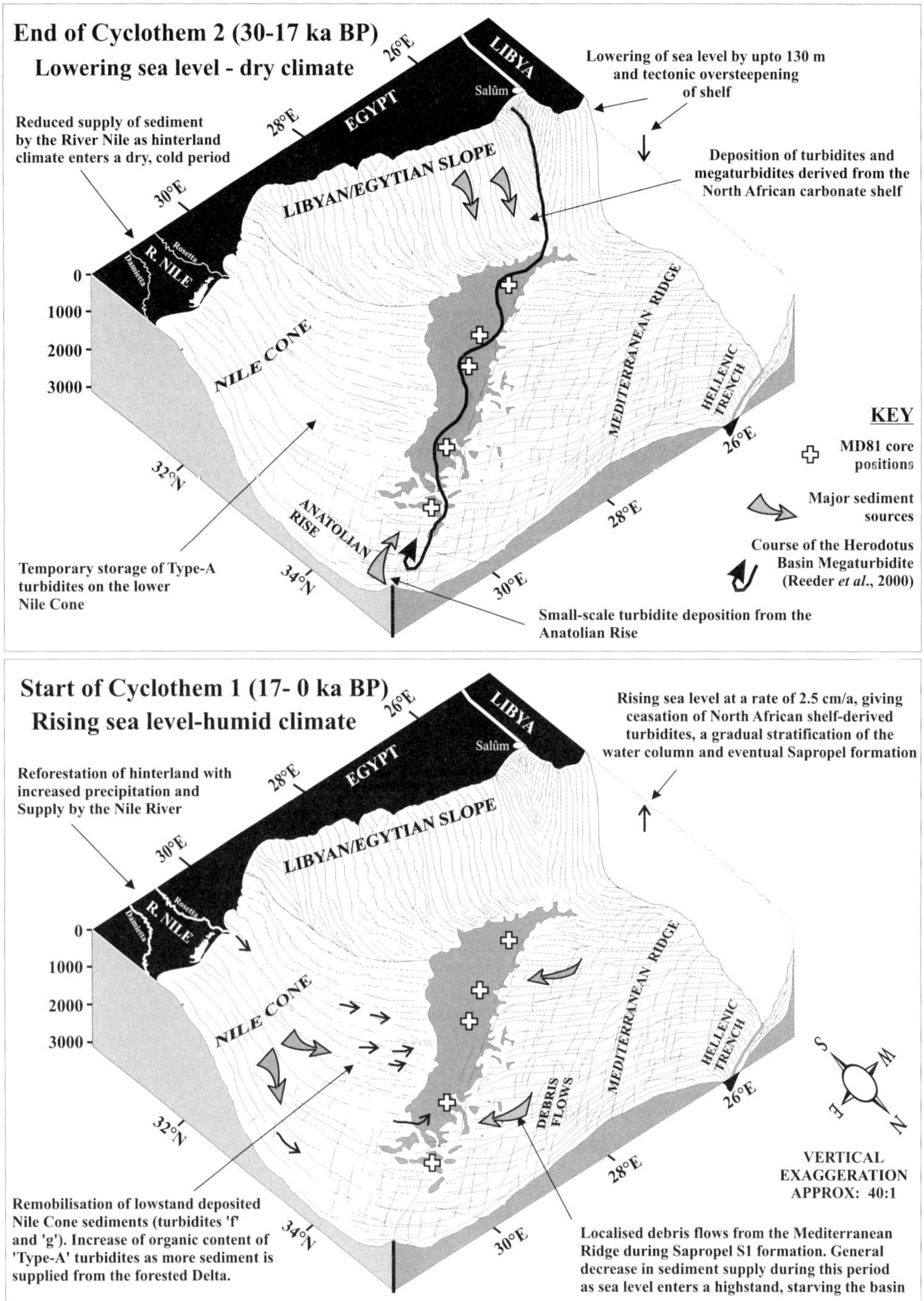

Fig. 4. Three-dimensional schematic box models for the climate-controlled variation in source for Herodotus Basin turbidites over the past 30 ka. (**a**) During the period from 30 to 17 ka, the stratigraphy is dominated by 'Type B' and 'Type C' turbidites, derived from the North African shelf and slope and the Anatolian Rise, respectively. This correlates with a dry, inter-pluvial climate. (**b**) During the period from 17 to 0 ka, the sequence is dominated first by rejuvenated sedimentary discharge from the River Nile, with 'Type A' turbidites abundant in the Herodotus Basin cores, and then later by reduced sedimentation during the Late Holocene.

1991). An increase of fresh water from influxes of snowmelt and river discharge is thought to have caused a disruption to the thermohaline currents of the Levantine and Ionian Seas, and initiated an increase in the primary productivity (Maldonado & Stanley 1978; Stanley & Maldonado 1979; Bethoux 1993; Rohling 1994; Higgs *et al.* 1994; Aksu *et al.* 1995). Stratification of the water column occurred, leading to stagnation of the bottom waters and deposition of the organic-rich sapropel S1 at the very end of this period.

Holocene inter-pluvial (6 – 0 ka)

The climate of the North African margin and Nile hinterland has remained fairly constant since about 6 ka BP, with very warm and arid conditions prevailing. Less material was supplied to the Nile Delta leading to lower rates of accumulation on the Nile Cone and an absence of Type A turbidites from the Herodotus Basin. Only thin Type B turbidites have been recovered.

Summary

There seems to be a clear relationship between turbidite type, indicative of sediment source, and climatic conditions. Pluvial episodes provided greater influx from the Nile River system to the Nile Delta. This in turn lead to greater downslope resedimentation to the Herodotus Basin. By contrast, the inter-pluvial episodes saw an absence of Nile-derived turbidites and dominance of the carbonate-rich North African shelf and Anatolian sources.

It is interesting to note, however, that foraminiferal assemblages from all types of turbidite show generally warm-water conditions post-28 ka BP, except for turbidites 'f' and 'g', which show mixed cold–warm assemblages (Kahler & Dossi 1996). Turbidite 'o', dating from around 30 ka BP, shows a complete cold assemblage of planktonic foraminifers, demonstrating deposition during the height of the Wisconsin glacial. The two mixed assemblage turbidites may result from the mixing of warm- and cold-water forms, at least one of which would have been reworked older turbidites or hemipelagic sediments deposited during different climatic conditions.

Discussion

Sediment cyclicity and controls

Based on an analysis of 65 piston and gravity cores, which collectively penetrated up to 55 ka into the Nile cone record, Maldonado & Stanley (1976, 1978) and Stanley & Maldonado (1979) noted a cyclic repetition of sedimentary facies, which they correlated with variations in the palaeo-climate and sea level. The timing of their three cycles or cyclothems more or less correlates with our broad temporal divisions based on turbidite types, although they recognize only Nile-derived turbidites. For Cycle 3 (55–28 ka) they record from base-up relatively fine-grained turbidites, sapropel 3, fine-grained turbidites and hemipelagites, and a calcareous ooze. Cycle 2 has coarser-grained turbidites, sapropel 2, turbidites/hemipelagites and calcareous ooze. Cycle 1 shows, once more, fine-grained turbidites, sapropel 1, hemipelagites and calcareous ooze. They do note that these cycles vary in detail between cores and regions, some being more dominated by hemipelagic and pelagic sediments ('suspensites') and some by turbidites and associated downslope facies ('gravitites').

In comparing the Herodotus Basin data with those described above mainly from the Nile Cone, we could consider the post-17 ka sequence of the Herodotus Basin as a single cyclothem (Fig. 3) consisting of three main constituents: (1) a lower dark brown–grey and turbidite-dominated sequence (Type A, Nile-derived) with thin hemipelagic muds, overlain by (2) a dark grey–black sapropelic layer (S1), and (3) an upper pale olive–grey mud section with thicker hemipelagic muds and thin turbidites (Type B, African shelf-derived). This sequence is, at first sight, similar to Cycle 1 of Stanley & Maldonado (1979), although the thin upper turbidites are from a different source. Possible earlier cyclothems (Fig. 3) are still less comparable, in that the lowermost Type A turbidites are not fully penetrated, the overlying sapropel (S2) is missing and the pale-coloured turbidites are from two different sources.

Therefore, although we do recognize some degree of cyclicity in cores from the Herodotus Basin, the cycles differ significantly from those described previously from the region. We suggest that both sea level and climate have influenced sedimentation, but that the cyclic response to each control has been partly independent. The climate-induced cycles are principally an alternation of periods dominated by Nile-sourced, organic-carbon-rich, muddy turbidites (Type A) and periods during which turbidites were derived from elsewhere, notably from North African and Anatolian biogenic, carbonate-rich, sources (Type B and Type C turbidites). These periods appear to correspond closely with pluvial and inter-pluvial episodes, respectively, that affected the African–Nile hinterland (Fig. 4).

Sea-level induced cycles are less evident, although from the data we have it would appear that turbidity current input into the Herodotus Basin, from whatever source, was especially common during both falling and rising sea level. In particular, sea-level fall to the latest Wisconsin low-stand is characterized by turbidites derived from the North African shelf and Anatolian sources, including the Herodotus Basin Megaturbidite, whereas sea-level rise through the Late Wisconsin–Holocene is characterized by Nile-derived turbidites. Only the most recent Holocene period (post-6 ka) has fewer and thinner turbidites and more pelagic–hemipelagic sedimentation.

We cannot comment on sapropel cyclicity, but note that whereas sapropel S1 is well developed throughout the basin, sapropel S2 is absent from all basin cores. This seems more likely to be because it was never deposited than subsequently eroded, as the only turbidity current capable of eroding such a layer over a distance of at least 400 km is the HBM. Current dating puts the HBM event several thousands of years prior to S2.

Megaturbidites

Much has been written in the past 30 years on very large-scale turbidites – variously called megaturbidites or megabeds – and yet still no generally accepted rigorous definition exists. In our recent paper on the Herodotus Basin Megaturbidite (Reeder *et al.* 2000) we do comment on this problem but still prefer to steer away from too quantitative a definition at this stage. The early definitions of >1 m thick or >1 km^3 (Ricchi Lucchi & Valmori 1980; Mutti *et al.* 1984) are both too small in light of recent work on modern examples. The HBM, for example, is up to 20 m thick and some 400^3 km in volume. In fact, 10 of the 16 turbidites in this study might also be considered as megaturbidites as they display total sediment volumes in excess of 1 km^3, and six of these have volumes in excess of 25 km^3. This is perhaps typical of basin plain turbidite systems. We propose the following as a working definition, without placing strict limits on bed dimensions.

A *megaturbidite* is a generic term that describes a very thick, large-volume sedimentary unit (or bed), with a large areal extent, that has been deposited by a single turbidity current event. The principal characteristics of such beds are: (1) they are very thick in comparison with associated turbidites, at least in some part of the basin; (2) they have a very large volume, again relative to the associated turbidites; and (3) they are laterally extensive and, hence, tend to make good stratigraphic and seismic marker beds. This definition incorporates most of the qualitative characteristics for megaturbidites proposed by Bouma (1987), although we do not agree that they must be of 'different composition from the host rock', nor that they must necessarily 'lack submarine fan geometries'.

The HBM was deposited at about 27.1 ka BP during a period of lowering sea level from an already glacial low-stand. Of the other megaturbidites revealed in this study, turbidite 'o' was also deposited during the Wisconsin low-stand, whereas turbidites 'g', 'e', 'd' and 'c' occurred during rising sea level, the last two at the onset of the Holocene. In Reeder *et al.* (2000) we concluded that a combination of factors caused the catastrophic slope collapse that led to emplacement of the HBM, including lowered sea level leading to sediment destabilization, a weak glide plane horizon within the sedimentary succession on an over-steepened slope and a probable seismic trigger. Clearly not all these factors could have contributed to deposition of the other megaturbidites, and others may have been involved instead. It is reasonable to conclude, therefore, that the triggering and emplacement of megaturbidites is very variable, and that individual events occur independently of a particular sea-level stand or climatic condition.

The authors would like to thank the NERC Scientific Services, Radiocarbon Laboratory, East Kilbride, for their support in obtaining the radiocarbon dates (^{14}C Dating Allocation No. 664/0896). Thanks also to NERC's British Ocean Sediment Core Repository (BOSCOR) based at the SOC, where description and sampling was completed. M. S. Reeder's PhD is NERC funded: grant GT4/95/288/E. This work was supported by the European Union Marine Science and Technology Programme contract No. MAS2-CT93-0051.

References

AKSU, A. E., YASAR, D. & MUDIE, P. J. 1995. Paleoclimatic and paleoceanographic conditions leading to development of sapropel layer S1 in the Aegean Sea. *Palaeogeography, Palaeoclimatology, Palaeoecology*, **116**, 71–101.

BARD, E., ARNOLD, M., FAIRBANKS, R. G. & HAMELIN, B. 1993. ^{230}Th-, ^{234}U and ^{14}C obtained by mass spectrometry on corals. *Radiocarbon*, **35**, 191–199.

BETHOUX, J.-P. 1993. Mediterranean sapropel formation, dynamic and climatic viewpoints. *Oceanologica Acta*, **16**, 127–133.

BOUMA, A. H. 1987. Megaturbidite: an acceptable term? *Geo-Marine Letters*, **7**, 63–67.

CHEDDADI, R. & ROSSIGNOL-STRICK, M. 1995. Eastern Mediterranean quaternary paleoclimates from pollen and isotopic records of marine cores in the Nile cone area. *Paleoceanography*, **10**, 291–300.

CHEDDADI, R., ROSSIGNOL-STRICK, M. & FONTUGNE, M. 1991. Eastern Mediterranean palaeoclimates from 26 to 5 ka B.P. documented by pollen and isotopic analysis of a core in the anoxic Bannock Basin. *Marine Geology*, **100**, 53–66.

CITA, M. B., BEGHI, C., CAMERLENGHI, A., KASTENS, K.

A., McCoy, F. W., Nosetto, A., Parisi, E., Scolari, F. & Tomadin, L. 1984. Turbidites and megaturbidites from the Herodotus Abyssal Plain (Eastern Mediterranean) unrelated to seismic events. *Marine Geology,* **55**, 79–101.

Haq, B. U., Hardenbol, J. & Vail, P. P. 1987. Chronology of fluctuating sea levels since the Triassic. *Science*, **235**, 1156–1166.

Higgs, N. C., Thomson, J., Wilson, T. R. S. & Croudace, I. W. 1994. Modification and complete removal of Eastern Mediterranean sapropels by postdepositional oxidation. *Geology*, **22**, 423–426.

Kahler, G. & Dossi, M. 1996. Micropalaeontology. *In*: Rothwell, R. G. (ed.) R/V Marion Dufresne *Cruise 81 – Mediterranean Giant Piston Coring Transect*. Cruise Report, **40–63**.

Lucchi, R. & Camerlenghi, A. 1993. Upslope turbiditic sedimentation on the southeastern flank of the Mediterranean ridge. *Bollettino di Oceanologia Teorica ed Applicata*, **11**, 3–25.

Maldonado, A. & Stanley, D. J. 1976. The Nile Cone: Submarine fan development by cyclic sedimentation. *Marine Geology*, **20**, 27–40.

Maldonado, A. & Stanley, D. J. 1978. Nile Cone depositional processes and patterns in the Late Quaternary. *In*: Stanley, D. J. & Kelling, G. (eds) *Sedimentation in Submarine Canyons, Fans, and Trenches.* Dowden, Hutchinson & Ross, Stroudsburg, PA, 239–257.

Mart, Y. 1993. The sedimentologic and geomorphic provinces of the Nile Fan. *In*: Rhodes, E. G. & Moslow, T. S. (eds) *Marine Clastic Reservoirs.* Springer, New York, 101–112.

Mutti, E., Ricci Lucchi, F., Seguret, M. & Zanzucchi, G. 1984. Seismoturbidites: a new group of resedimented deposits. *Marine Geology,* **55**, 103–116.

Reeder, M. S., Rothwell, R. G. & Stow, D. A. V. 2000. Influence of sea level and basin physiography on emplacement of the late Pleistocene Herodotus Basin Megaturbidite, SE Mediterranean Sea. *Marine and Petroleum Geology,* 17, 199–218.

Reeder, M. S., Rothwell, R. G., Stow, D. A. V., Kahler, G. & Kenyon, N. H. 1998. Turbidite flux, architecture and chemostratigraphy of the Herodotus Basin, Levantine Sea, south-eastern Mediterranean. *In*: Stoker, M. S., Evans, D. & Cramp, D. (eds) *Geological Processes on Continental Margins: Sedimentation, Mass Wasting and Sability*, Geological Society, London, Special Publications, **129**, 19–41.

Ricci Lucchi, F. & Valmori, E. 1980. Basin-wide turbidites in a Miocene, over-supplied, deep-sea plain: a geometrical analysis. *Sedimentology,* **27**, 241–270.

Rohling, E. J. 1994. Review and new aspects concerning the formation of Eastern Mediterranean Sapropels. *Marine Geology*, **122**, 1–28.

Ross, D. A., Uchupi, E., Summerhayes, C. P., Koelsch, D. E. & El Shazly, E. M. 1978. Sedimentation and structure of the Nile Cone and Levant Platform area. *In*: Stanley, D. J. & Kelling, G. (eds) *Sedimentation in Submarine Canyons, Fans and Trenches*, Dowden, Hutchinson and Ross, Stroudsburg, PA, 261–275.

Said, R. 1990. Geomorphology. *In*: Said, R. (ed.) *The Geology of Egypt*. A. A. Balkema, Rotterdam, 9–25.

Shackleton, N. J. 1987. Oxygen isotopes, ice volume and sea level, *Quaternary Science Reviews*, **6**, 183–190.

Stanley, D. J. & Maldonado, A. 1979. Levantine Sea – Nile Cone lithostratigraphic evolution: Quantitative analysis and correlation with paleoclimatic and eustatic oscillations in the Late Quaternary. *Sedimentary Geology*, **23**, 37–65.

Stuiver, M. 1990. Timescales and telltale corals. *Nature*, **345**, 387–388.

Thomson, J. & Weaver, P. P. E. 1993. An AMS radiocarbon method to determine the emplacement of recent deep-sea turbidites. *Sedimentary Geology*, **89**, 1-7.

Index